MW00714980

Borates

Borates

Handbook of Deposits, Processing, Properties, and Use

Donald E. Garrett
SALINE PROCESSORS, INC.
OJAI, CALIFORNIA

ACADEMIC PRESS
San Diego London Boston New York Sydney Tokyo Toronto

Front cover photograph: The 20-mule team used to haul borax from Death Valley, to Mojave, 1883–1890. (For more details, see Chapter 4, Figure 7.)

This book is printed on acid-free paper. ♾

Transferred to digital printing 2005

Academic Press
a division of Harcourt Brace & Company
525 B Street, Suite 1900, San Diego, California 92101-4495, USA
http://www.apnet.com

Academic Press Limited
24-28 Oval Road, London NW1 7DX, UK
http://www.hbuk.co.uk/ap/

Library of Congress Catalog Card Number: 98-86055

International Standard Book Number: 0-12-276060-3

PRINTED IN THE UNITED STATES OF AMERICA
98 99 00 01 02 03 EB 9 8 7 6 5 4 3 2 1

Contents

Chapter 3 Calcium, Magnesium, or Silicate Buried Deposits

Chapter 4 Calcium or Magnesium Surface (Playa or Mantle) Deposits

Chapter 5 Lake or Brine Deposits

Preface

The borates are among the most interesting of the world's industrial minerals, having been known and used since the earliest recorded history, first for precious metal working and later in ceramics. They are an unusually large grouping of minerals, but the number of commercially important borates is limited, and their chemistry and crystal structure are both unusual and complex. There are only a few large deposits, with many smaller formations, and the number of noncommercial occurrences in other rocks or brines is very large. The accounts of the early exploration, mining, and processing of borates are fascinating, because their remote locations often led to unusual difficulties, hardships, and resourcefulness in recovering the desired products. This varied from workers wading into Himalayan lakes to harvest the "floor" and then transport the borax in saddlebags on sheep across the Himalaya's to the markets, to the "Dante's Inferno" of the Larderello boric acid fumaroles, to the colorful 20-mule teams of the western United States. Such operations helped in the development of these remote areas and slowly transformed borates from minor high-cost minerals into the large-volume industrial commodities they are today.

Boron's chemistry and reactivity are also fascinating because they form a wide variety of oxygen compounds (the borates) that occur in an essentially unending variety of simple to exceedingly complex molecules. Determining their crystal structures has given rise to a separate subfield of crystallography. Boron has only two isomers, ^{10}B and ^{11}B, and is somewhat unique in that the ratio of these isomers varies widely in nature. The isotopes' very different reactivity during both physical and chemical changes has allowed their determination to be a tool in predicting many geologic and other events, again forming a specialized field in geology. The ^{10}B isomer has an unusually large neutron-capture cross section, which gives it a unique nuclear application and a potent medical use (another subdiscipline). Finally, boron's ability to form organic compounds gives it many specialized uses (a much underdeveloped field) and another subdiscipline. There have been many symposia and publications in each of these fields.

It is the purpose of this book to bring all of these subjects together as they relate to borates as industrial commodities. This means that the very scattered literature of the early-operated deposits is compiled and summarized, and the voluminous recent literature on the deposits in Argentina, Turkey, China, and Russia is reviewed and condensed. The larger of the borate occurrences (nondeposits) and minor borate types are also reviewed. The crystal structures of the borates are considered, and the isotopic data are compiled to allow their incorporation into the geologic data. From this information, comprehensive theories are developed on how the borate deposits formed and what determined their crystal form (i.e., massive beds, nodules, and disseminated crystals).

Attention is next turned to the past and present mining practices for borate ores, followed by similar reviews of their processing methods. This includes the development of mining and processing techniques, from the simplest playa operations to the most complex of the very large deposits, along with details of the current technology and the basic technical factors involved. The major uses of borates, followed by some of the industry's production, consumption, ore reserves, and product specification statistics, are then reviewed. Finally, a compilation of the major phase data for the sodium borates is given. The following brief history of the borate industry may provide a useful guide to these details.

Since the early days of civilization, borax has been a valuable flux and bonding agent for silver and goldsmithing. Later, borax also became an important frit and glaze component for ceramics, and then it developed consumer uses as a medicinal agent (eye wash, mild germicide) and as a cleaning compound. More recently, its use as a cleaning agent (including perborates) has increased, but its largest use came to be its inclusion as a critical ingredient in high-performance glass.

It is quite possible that borax was first used in Babylon more than 4000 years ago. Literature references from that period use words similar to "borax" (boorak, bayrach, borar, etc.) or "tincal" (a former synonym, e.g., tincar) from the "Far East." This would appear to mean only Tibet because China, India (except Kashmir), and Persia (Iran) had no borax sources. Unfortunately, the word "borax" has also been loosely used to mean several other things [i.e., natron (trona), niter (sodium nitrate), and many fluxes such as "chrysocolla" ("gold glue," a crude copper ammonium salt)]. Sometimes the use of the word eliminates borax, but at other times (as in the Babylonian example), borax appears to be what was actually meant. Considering the vigor of the early traders and the ease of recovering borax from alkali lakes, it quite obviously has been used for a very long time. However, the first fairly positive written description of its use came from the Arabic centers around Mecca, Medina, and Baghdad in the year AD 762. Borax was introduced to China soon after that date, but not in Europe until the Arabic writings on "chemistry" were translated in the 12th and 13th centuries.

By the 15th century, the Venetians had a monopoly on the borax trade in Europe, which lasted more than 200 years. Their source and refining methods were closely guarded secrets. Each batch of crude borax was different, with both initial and added impurities (such as animal fat to reduce water damage on the trip from Tibet across the Himalayas), so that producing consistent, colorless, large crystals (the purest and most desired product) was difficult. By the 17th century, the Dutch also had acquired a source and processing know-how, and because of border disputes in the Venetian's source area, the Dutch became dominant for about 100 years. By the 1770s they had competition from the French (with their supply from India), and from England in 1798. Since that time, there has been more normal competition. From about 1840 when large-scale Italian boric acid became available, some of the Tibetan borax refining has been done in India in an effort to be more competitive.

During the mid to late 1800s, many new sources of borates were discovered, and the use of borates was greatly expanded. The traditional supply from Tibet continued, as did that from Italy, and from 1864 to 1868 borax was also recovered from Borax Lake, California, which initiated the first major borate production in the United States. Some borax was later recovered from the nearby Little Borax Lake, and in 1870 the much more significant "playa borate" period began when cotton-ball ulexite ($NaCaB_5O_9 \cdot 8H_2O$) was discovered and began to be harvested from several small Nevada playas. In 1871 much larger amounts were mined and processed at Columbus Marsh, and in 1872 surface borax crusts were discovered and processed at Teels Marsh, then at Searles Lake in 1873, and at Fish Lake Marsh in 1874. Other playas were quickly put into production for ulexite or borax at Rhodes Marsh, Alvord Valley, Saline Valley, Death Valley, and others. Many ulexite playas also were later found in the high Andes of Argentina, Bolivia, Chile, and Peru, and by the 1880s fairly large amounts of ulexite also were being produced from them. Very likely, the production of ulexite and magnesium borates from the playas of China began at about the same time.

In 1862 small amounts of priceite ($Ca_4B_{10}O_{19} \cdot 7H_2O$) were mined from a deposit at Chetco, Oregon, and in 1865 some previously discarded nodules of priceite began to be recovered from a gypsum mine in Susurluk, Turkey. In 1887 small-scale underground mining for priceite commenced at the nearby Sultancayiri mine. Colemanite ($Ca_2B_6O_{11} \cdot 5H_2O$) was discovered in Death Valley in 1882 and at Calico in 1883. Large-scale mining began at Calico in 1888, which initiated the "colemanite" period. This mine's (and soon others') low-cost colemanite forced the closure of many playa operations, and when the Lila C mine in Death Valley opened in 1907, the borax price was further cut, putting additional stress on the other playa producers. After the Lila C closed, the Death Valley Ryan complex was opened, with some competition from the Lang and Frazier mines in California and the Muddy Mountain, Nevada mines.

In 1925 the large Kramer (or Boron) borax deposit was discovered and immediately developed, closing the colemanite mines and beginning the "present," or large-tonnage borate production period. In 1957 Kramer was converted to an open pit mine, but as production costs increased, borax prices also rose, allowing many other deposits to be developed. This started with Searles Lake in 1919, on the basis of their shared-product evaporation and later carbonation or liquid extraction processes. The Inder deposit in Russia opened in the 1930s, and in the 1950s the very large Turkish colemanite deposits were developed. Finally, in the mid-1970s the shallow, extensive Kirka open pit borax deposit was opened in Turkey. There has also been much smaller production from other Death Valley colemanite deposits, the Argentinean Tincalayu and Loma Blanca open pit borax mines, and the Sijes colemanite–hydroboracite–inyoite deposit. In the 1990s borate prices became high enough to reopen the Puna region's ulexite playas, and the Chinese and Russian skarn deposits were developed. During this period, borates became large-volume industrial minerals.

Donald E. Garrett

Acknowledgments

The author expresses his gratitude to the many institutions and individuals who provided assistance in preparing this book. Professor Cahit Helvaci was very generous with his time, comments, and assistance in obtaining articles on the very important Turkish borate deposits, as was Professor Ricardo N. Alonso with articles on the Argentine deposits. Mr. Robert B. Kistler of U.S. Borax supplied very informative comments on their deposits. The author is particularly appreciative of the North American Chemical Company and their Trona Divisions's Senior Vice President of Technology and Development, Mr. Jake Haung, for allowing a visit, discussions, and pictures of their borate operations and for supplying sales brochures. Mr. Manuel G. Mata provided very interesting details on their solvent extraction and West End plants, as did Mr. Angelo DeNuz on their borate processing operations. It was very gratifying to have the most technically advanced and sophisticated of the three major borate producers be so open and encouraging. Etibank's sales department very generously supplied product brochures and pictures, and Mr. Ali Arat provided recent Etibank annual reports. The author also appreciates the several publications and pictures supplied by the public relations department of U.S. Borax.

Donald E. Garrett

Chapter 1 Borate Minerals and the Origin of Borate Deposits

About 230 naturally occurring borate minerals were identified as of 1996 (Table 1.1), and the increasing sophistication of analytical instruments, computer assistance, and crystallographic identification ensures that many new ones will be found in the future. Most of the newly reported minerals contain multiple cations or anions, are very large molecules, or have changed cation or anion proportions in large families of borates such as borosilicates, rare earths, boracites, and others. The number of nonmineral borates produced in the laboratory is also very large. The reason for this abundance is boron's ability to form boron–oxygen compounds (the borates) in many molecular and polymer forms. Its tri- and tetra (negatively charged)-bonded groups with oxygen can combine in a very large number of geometric combinations and polymer types. Also, the borates can combine with any cation, as well as form double or multiple salts with many other compounds. Boron readily crystallizes with silicates, and can replace aluminum or silicon in varying proportions in some minerals. Because of this, there are more than 8 Na-borate minerals ($>$27 in the laboratory) and more than 23 Ca-borates with no other cations or anions.

However, despite these long lists, only a comparative few of the borates are important in commercial deposits. This includes the hydrogen borate sassolite (H_3BO_3); the two sodium borates, borax ($Na_2B_4O_7 \cdot 10H_2O$) and kernite ($Na_2B_4O_7 \cdot 4H_2$); the calcium borates, colemanite ($Ca_2B_6O_{11} \cdot 5H_2O$), inyoite ($Ca_2B_6O_{11} \cdot 13H_2O$), and priceite ($Ca_4B_{10}O_{19} \cdot 7H_2O$); the sodium–calcium borates, ulexite ($NaCaB_5O_9 \cdot 8H_2O$) and probertite ($NaCaB_5O_9 \cdot 5H_2O$); the magnesium borates, szaibelyite (ascharite; $Mg_2B_2O_5 \cdot H_2O$), inderite or kurnakovite ($Mg_2B_6O_{11} \cdot 15H_2O$; monoclinic or triclinic), and pinnoite ($MgB_2O_4 \cdot 3H_2O$); the magnesium–calcium borate, hydroboracite ($CaMgB_6O_{11} \cdot 6H_2O$); the borosilicates datolite ($Ca_2B_2Si_2O_9 \cdot H_2O$) and ludwigite ($Mg_2FeBO_5$); and the magnesium chloride double salt, boracite ($Mg_3B_7O_{13}Cl$). Each of these minerals is being, or has been, mined and processed on a commercial scale. A number of other borates that are present in limited quantity (such as danburite, $CaB_2Si_2O_8$), or that result from weathering may be mined with the major minerals. Also, other borates such as howlite ($Ca_4Si_2B_{10}O_{23} \cdot 5H_2O$) are fairly common, but have not yet become commercial minerals.

Many of the less common borate minerals were formed by the high-temperature contact and reaction of boron with different elements in rocks during

Table 1.1

Known Boron Minerals as of 1996[a]

Name (Group; form)	Formula	Molecular weight[b]; $\%B_2O_3$[c]; hardness; density; crystal system
1. Adamontite (d., mcallisterite)	$Mg(H_2O)_6\{Mg[B_6O_7(OH)_6]_2\}\cdot 2H_2O$ $MgB_6O_{10}\cdot 7H_2O$	375.27; 55.66%; —; 1.82; monoclinic
2. Aksaite	$Mg[B_6O_7(OH)_6]\cdot 2H_2O$ $MgO\cdot 3B_2O_3\cdot 5H_2O$ $MgB_6O_{10}\cdot 5H_2O$	339.24; 61.57%; 2.5; 1.97–2.07; orthorhombic
3. Albite, boron (=Reedmergnerite) (Feldspar)	$NaAlSi_3O_8$-$NaBSi_3O_8$ $Na_2O\cdot B_2O_3\cdot 6SiO_2$	246.05; 14.15%; 6–7; 2.70–2.78; triclinic
4. Aldzhanite	$(Ca, Mg)_2BO_3Cl\cdot nH_2O$	(n = 7) 284.75; 12.22%; —; 2.21; orthorhombic
5. Ameghinite	$Na[B_3O_3(OH)_4]$ $Na_2O\cdot 3B_2O_3\cdot 4H_2O$ $NaB_3O_5\cdot 2H_2O$	171.45; 60.91%; 2.5; 2.03–2.04; monoclinic
6. Ammonioborite	$(NH_4)_3[B_5O_6(OH)_3\cdot OB_5O_6(OH)_2\cdot OB_5O_6(OH)_3]\cdot 4H_2O$ $3(NH_4)_2O\cdot 15B_2O_3\cdot 16H_2O$ $(NH_4)_3B_{15}O_{24}\cdot 8H_2O$	744.39; 70.15%; —; 1.76–1.77; monoclinic
7. Apophyllite, boron bearing	$KCa_4(B, Si)_8O_{20}(F, OH)\cdot 8H_2O$	837.11; 16.63%
8. Aristarainite	$Na_2Mg[B_6O_8(OH)_4]_2\cdot 4H_2O$ $Na_2O\cdot MgO\cdot 6B_2O_3\cdot 8H_2O$ $Na_2MgB_{12}O_{20}\cdot 8H_2O$	664.13; 62.90%; 3.5; 2.01–2.10; monoclinic
9. Ascharite (=Szaibelyite; α Ascharite)	$Mg_2OH[B_2O_4(OH)]$ $2MgO\cdot B_2O_3\cdot H_2O$; $(MgHBO_3)$ $Mg_2B_2O_5\cdot H_2O$	168.24; 41.38%; 3–3.5; 2.60–2.76; monoclinic
10. β Ascharite (=Camsellite)		

11. Avogadrite (i., barite)	(K, Cs) BF_4 End Member, KBF_4	172.81; 20.14%; —; 2.50–2.51 125.90; 27.65%; orthorhombic
12. Axinite	(Ca, Mn, $Fe^{+2})_3Al_2Si_2O_6 \cdot [Si_2O_6BO_3(OH)]$ 6(Ca, Mn, $Fe^{+2})O \cdot 2Al_2O_3 \cdot B_2O_3 \cdot 8SiO_2 \cdot H_2O$ (Ca, Mn, $Fe^{+2})_6Al_4Si_8B_2O_{31} \cdot H_2O$	1169.95; 5.95%; 6.5–7.3; 3.19–3.25; triclinic
13. Azoproite (Ludwigite)	(Mg, $Fe^{+2})_2(Fe^{+3}$, Ti, Mg)$[O_2BO_3]$ 4(Mg, Fe)O·(Fe, Ti, Mg)$_2O_3 \cdot B_2O_3$ (Mg, $Fe^{+2})_2(Fe^{+3}$, Ti, Mg)BO_5	213.64; 16.29%; 6.5; 3.63–4.15; orthorhombic
14. Bakerite (Gadolinite)	$Ca_4[B_4(SiO_4)_3BO_3OH(OH)_4]$ $8CaO \cdot 5B_2O_3 \cdot 6SiO_2 \cdot 5H_2O$ $Ca_8B_{10}Si_6O_{35} \cdot 5H_2O$	1247.30; 27.91%; 4.5; 2.89–2.94; monoclinic
15. Bandylite	$Cu^{+2}Cl[B(OH)_4]$ $CuClBO_2 \cdot 2H_2O$	177.84; 19.57%; 2.5; 2.81–2.82; tetragonal
16. Barberiite[d]	NH_4BF_4	104.84; 33.20%; —; 1.86; orthorhombic
17. Bechilite	$Ca[B_4O_6(OH)_2] \cdot 3H_2O$ $CaB_4O_7 \cdot 4H_2O$	267.38; 52.08%
18. Behierite	(Nb, Ta)$[BO_4]$ End Member, $TaBO_4$	211.74; 16.44%; 7–7.5; 6.67–8.0 255.76; 13.61%; tetragonal
19, 20, 21 Berborite 1T,2T,2H[d]	$Be_2(OH, F)[BO_3] \cdot H_2O$[d] $4BeO \cdot B_2O_3 \cdot 3H_2O$ End Member, $Be_4B_2O_7 \cdot 3H_2O$	EM, 223.71; 31.12%; 3; 2.04–2.20; 2H hexagonal, 1T, 2T rhombohedral
22. Biringuccite (=Hoeferite)	$Na_4[(OH)B_5O_8 \cdot OB_5O_7(OH)] \cdot 2H_2O$ $2Na_2O \cdot 5B_2O_3 \cdot 3H_2O$ $Na_4B_{10}O_{17} \cdot 3H_2O$	526.10; 66.17%; —; 2.30–2.38; monoclinic
23. Blatterite[e] (Pinakiolite)	$(Mn^{+2}, Mg)_2(Mn^{+3}, Sb^{+3}, Fe^{+3})O_2[BO_3]$ End Member, $Mn_2^{+2}Mn^{+3}BO_5$	247.56; 14.06%; —; 2.28 255.62; 13.62%; orthorhombic
24. Bonaccordite[f] (=Ludwigite)	$Ni_2Fe^{+3}O_2[BO_3]$ $Ni_2Fe^{+3}BO_5$	264.04; 13.18%; 7; 5.19; orthorhombic

continues

Table 1.1 (*continued*)

Name (Group; form)	Formula	Molecular weight[b]; %B_2O_3[c]; hardness; density; crystal system
25. Boracite (=Stassfurite)	$Mg_3[B_3O_5]_2[BO_3]Cl$	392.04; 62.16%; 7–7.5; 2.89–2.97; β orthorhombic,
a. Low temp., β[e]	$5MgO \cdot MgCl_2 \cdot 7B_2O_3$	α cubic (>265°C)
b. High temp., α[d]	$Mg_3B_7O_{13}Cl$	
26. Borax (=Tincal)	$Na_2[B_4O_5(OH)_4] \cdot 8H_2O$	381.372; 36.510%; 2–2.5; 1.711–1.715; monoclinic
	$Na_2O \cdot 2B_2O_3 \cdot 10H_2O$	
	$Na_2B_4O_7 \cdot 10H_2O$	
27. Borcarite	$Ca_4Mg(CO_3)_2[B_4O_6(OH)_6]$	545.92; 25.51%; 5; 2.77–2.79; monoclinic
	$4CaO \cdot MgO \cdot 2CO_2 \cdot 2B_2O_3 \cdot 3H_2O$	
	$Ca_4Mg(CO_3)_2B_4O_9 \cdot 3H_2O$	
28. Boromuscovite	$KAl_2Si_3BO_{10}(OH, F)_2$	384.13; 9.06%; —; 2.13–2.90
29. 1M, 2M[4] (Mica)	End Member, $KAl_2Si_3BO_{10}(OH)_2$	282.14; 9.11%; monoclinic
30. Braitschite	$12B_2O_3 \cdot 6(Na_2O, CaO) \cdot RE_2O_3 \cdot 6H_2O$	—; typical 48.2%; —; 2.84-2.90; hexagonal
	$(Ca, Na_2)_6RE_2B_{24}O_{45} \cdot 6H_2O$	
31. Braitschite-Ce[d]	$(Ca, Na_2)_7(Ce, La)_2B_{22}O_{43} \cdot 7H_2O$	1632.15; 46.92%; rhombohedral
32. Braunite, boron	$Mn_7O_9SiO_3—Mn_7O_9BO_3$	587.37; 5.93%
33. Buergerite[f] (Tourmaline)	$NaFe_3^{+3}Al_6O_3(OH, F)_4[BO_3]_3Si_6O_{18}$	1052.35; 9.92%; 7–8; 3.29–3.34; rhombohedral
	End Member, $NaFe_3^{+3}Al_6B_3Si_6O_{30}F$[g]	
34. Cahnite	$Ca_2[B(OH)_4]AsO_4$	297.92; 11.68%; 3; 3.06–3.18; tetragonal
	$4CaO \cdot B_2O_3 \cdot As_2O_5 \cdot 4H_2O$	
	$Ca_2BAsO_6 \cdot 2H_2O$	
35. Calciborite	$Ca[BO_3 \cdot BO]$	125.70; 55.39%; 3.5; 2.88; orthorhombic
	$CaO \cdot B_2O_3$	
	CaB_2O_4	
36. Canavesite	$Mg_4[B_2O_5](CO_3)_2 \cdot H_2O$	258.90; 13.47%; —; 2.00; monoclinic
	$Mg_2(CO_3)(HBO_3) \cdot 5H_2O$[d]	

37. Cappelenite[f]-(Y)[g]	$(Ba, Ca, Ce)_3(Y, Ce, La)_3[(BO_3)_6$ $Si_3O_9]$; $Ba(Ce, Y)_6B_6Si_3O_{24}F_2$[g] End Member, $BaY_6B_6Si_3O_{24}F_2$	1395.49; 14.97% (typical 16.9–17.2%); 6–6.5; 4.41; hexagonal (rhombohedral ?)[g] 1241.87; 16.82%
38. Carboborite	$MgCa_2(CO_3)_2[B(OH)_4]_2 \cdot 4H_2O$ $2CaCO_3 \cdot MgO \cdot B_2O_3 \cdot 8H_2O$ $Ca_2Mg(CO_3)_2B_2O_4 \cdot 8H_2O$	454.22; 15.33%; 2; 2.09–2.12; monoclinic
39. Caryocerite[f]	$(Ca, Na)_4(RE, Th, Ce)_6[(Si, B)O_4]_6F_2 \cdot 5H_2O$	—; (typical 1.50–4.70%); 4.5–6; 4.13–4.45; hexagonal
40. Chambersite (Boracite)	$Mn_3[B_3O_5]_2[BO_3]Cl$ $5MnO \cdot MnCl_2 \cdot 7B_2O_3$ $Mn_3B_7O_{13}Cl$	483.94; 50.35%; 7; 3.47–3.49; orthorhombic
41. Charlesite[d] (Ettringite)	$Ca_6(Al, Si)_2(SO_4)_2B(OH)_4(O, OH)_{12} \cdot 26H_2O$; End Member, $Ca_6Al_2S_2BO_9(OH)_{15} \cdot 26H_2O$	1164.91; 2.99%; —; 1.69 1236.28; 2.81%; rhombohedral
42. Chelkarite	$CaMg[B_2O_4]Cl_2 \cdot 7H_2O$	347.01; 20.06%; —; 2.94; orthorhombic
43. Chestermanite[e] (Ludwigite)	$Mg_2(Fe^{+3}, Mg, Al, Sb^{+5})O_2[BO_3]$ End Member, $Mg_2Fe^{+3}BO_5$	196.64; 17.70%; —; 3.76–3.80 195.27; 17.83%; orthorhombic
44. Chlorite, boron bearing (cf., Manandonite $2H_2$)[d]	$Li_2Al_4[AlBSi_2O_{10}(OH)_8]$ $Li_{1.25}Al_{4.66}B_{1.35}Si_{2.13}O_{10}(OH)_8$[g] End Member, $Li_2Al_5BSi_2O_{14} \cdot 4H_2O$	511.82; 6.80%; —; 2.53–2.89; monoclinic (triclinic)[g] 504.88; 9.31%
45. Chromdravite[a] (Tourmaline)	$NaMg_3(Cr, Fe^{+3})_6[BO_3]_3(Si_6O_{18})(OH)_4$ End Member, $NaMg_3Cr_6B_3Si_6O_{27}(OH)_4$	1120.39; 9.32%; —; 3.42 1108.84; 9.42%; rhombohedral
46. Clinokurchatovite[d] (d., kurchatovite)	$Ca(Fe^{+2}, Mg, Mn)B_2O_5$ End Member, $CaMgB_2O_5$	186.73; 37.28%; —; 3.02–3.40 166.00; 41.94%; monoclinic
47. Colemanite (=Borocalcite)	$Ca[B_3O_4(OH)_3] \cdot H_2O$ $2CaO \cdot 3B_2O_3 \cdot 5H_2O$ $Ca_2B_6O_{11} \cdot 5H_2O$-I	411.09; 50.81%; 4.5; 2.42–2.43; monoclinic
48. Congolite (Boracite) (d., ericaite)	$(Fe^{+2}, Mg, Mn)_3[B_3O_5]_2[BO_3]Cl$ $(Fe, Mg, Mn)_3B_7O_{13}Cl$ End Member, $Fe_3B_7O_{13}Cl$	454.21; 53.65%; 7.5; 3.57–3.58; rhombohedral 486.66; 50.07%

continues

Table 1.1 (*continued*)

Name (Group; form)	Formula	Molecular weight[b]; $\%B_2O_3$[c]; hardness; density; crystal system
49. Danburite (r., paracelsian)	$Ca[BSiO_4]_2$	245.87; 28.32%; 7–7.5; 2.95–3.09; orthorhombic
	$CaO \cdot B_2O_3 \cdot 2SiO_2$	
	$CaB_2Si_2O_8$	
50. Datolite (Gadolinite)	$Ca_4[B_4(SiO_4)_4(OH)_4]$	319.96; 21.76%; 5–6; 2.97–3.02; monoclinic
	$4CaO \cdot 2B_2O_3 \cdot 4SiO_2 \cdot 2H_2O$	
	$Ca_2B_2Si_2O_9 \cdot H_2O$	
51. Diomignite[e]	$Li_2[B_4O_6]$ $(O_7?)$[d]	153.12; 90.93%; —; 2.21; tetrahedral
52. Dravite[f] (Tourmaline)	$NaMg_3Al_6(OH, F)_4[BO_3]_3(Si_6O_{18})$	962.74; 10.85%; 7–8; 2.98–3.15
	End Member, with $(OH)_4$[d]	958.75; 10.89%; rhombohedral
53. Dumortierite	$(Al, Mg, Ti, Fe)Al_6BSi_3O_{16}(O, OH)_2$[e]	572.94; 6.08%; 7–8.5; 3.21–3.48; orthorhombic
	$(Al, Fe)_7O_3[(BO_3)(SiO_4)_3]$	
	End Member, $Al_{6.75}Fe_{0.25}BSi_3O_{17.25}(OH)_{0.75}$[g]	579.90; 6.00%
54. Ekaterinite	$Ca_2[B_4O_7](Cl, OH)_2$ $(2H_2O?)$[d]	287.86; 48.37%; —; 1.97; hexagonal
55. Elbaite[f] (Tourmaline)	$Na(Li, Al)_3Al_6(OH, F)_4[BO_3]_3(Si_6O_{18})$;	940.70; 11.10%; 7–7.5; 3.00–3.11; rhombohedral
	End Member, $NaLi_{1.5}Al_{7.5}B_3Si_6O_{27}(OH)_4$[g]	936.72; 11.15%
56. Ericaite (Boracite) (d., congolite)	$Fe_3[B_3O_5]_2[BO_3]Cl$; $(Fe^{+2}, Mg, Mn)_3$[d]	486.66; 50.07%; 7; 3.17–3.57; orthorhombic
	$5FeO \cdot FeCl_2 \cdot 7B_2O_3$	
	End Member, $Fe_3B_7O_{13}Cl$	
57. Estroncioginorite	$(Sr, Ca)_2B_{10}O_{17} \cdot 7H_2O$	633.91; 54.91%
	End Member, $Sr_2B_{10}O_{17} \cdot 7H_2O$	681.45; 51.08%
58. Ezcurrite (Nasinite)	$Na_2[B_5O_7(OH)_3] \cdot 2H_2O$	598.17; 58.19%; 3–3.5; 2.05; triclinic
	$2Na_2O \cdot 5B_2O_3 \cdot 7H_2O$	
	$Na_4B_{10}O_{17} \cdot 7H_2O$	
59. Fabianite	$Ca[B_3O_5(OH)]$	339.03; 61.61%; 6; 2.77–2.79; monoclinic
	$2CaO \cdot 3B_2O_3 \cdot H_2O$	
	$Ca_2B_6O_{11} \cdot H_2O$-II	

60. Fedorovskite[h] (Roweite)	$Ca_2Mg_2(OH)_4[B_4O_7(OH)_2]$; $(Mg, Mn)_2$[d] End Member, $Ca_2Mg_2B_4O_{10}\cdot 3H_2O$	386.04; 36.07%; —; 2.67; orthorhombic
61. Feldspar, potassium, boron	$KAlSi_3O_8$-$KBSi_3O_8$	262.16; 13.28%; 6; 2.9–3.0; monoclinic
62. Ferroaxinite[f] (Axinite)	$Ca_2FeAl_2(Si_2O_6)[Si_2O_6BO_3(OH)]$ $Ca_2FeAl_2Si_4BO_{15}(OH)$	570.12; 6.11%; 6.5–7.5; 3.19–3.35; triclinic
63. Ferruccite (i. anhydride)	$NaBF_4$	109.79; 31.70%; 3; 2.50–2.51; orthorhombic
64. Ferruvite[d] (Tourmaline)	$Ca(Fe^{+2}, Mg)_3(Al, Mg)_6[BO_3]_3(Si_6O_{18})(OH)_4$ End Member, $CaMgFe_3^{+2}Al_5\ B_3Si_6O_{27}(OH)_4$	1015.12; 10.29%; —; 3.15–3.31; rhombohedral 1067.79; 9.78%
65. Fluoborite	$Mg_3(OH, F)_3[BO_3]$ $3MgO\cdot 3Mg(OH, F)_2\cdot B_2O_3$ End Member, $Mg_3BO_3F_3$	185.73; 18.74%; 3.5; 2.79–3.00; hexagonal (rhombohedral)[g] 188.72; 18.45%
66. Foitite[d] (Tourmaline)	$Fe^{+2}(Al, Fe^{+3})_3Al_6[BO_3]_3(Si_6O_{18})(OH)_4$; End Member, $Fe^{+2}Fe^{+3}Al_7B_3Si_6O_{27}(OH)_4$	1042.94; 10.01%; —; 3.17–3.30 1001.52; 10.43%; rhombohedral
67. Fredrikssonite[g] (Ludwigite)	$Mg_2(Fe^{+3}, Mn^{+3})O_2[BO_3]$ End Member, $Mg_2Mn^{+3}BO_5$	194.81; 17.87%; —; 3.80 194.36; 17.91; orthorhombic
68. Frolovite	$Ca[B(OH)_4]_2$ $CaO\cdot B_2O_3\cdot 4H_2O$ $\beta CaB_2O_4\cdot 4H_2O$-III	197.76; 35.20%; 3.5; 2.14–2.18; triclinic
69. Gadolinite[f] (Datolite)	$Y_4Fe_2[(Be, B)_4(SiO_4)_4O_4]$	939.29; 7.41%
70. Garrelsite	$NaBa_3Si_2B_7O_{16}(OH)_4$ $Na_2O\cdot 6BaO\cdot 4SiO_2\cdot 7B_2O_3\cdot 4H_2O$ $NaBa_3Si_2B_7O_{18}\cdot 2H_2O$	890.84; 27.35%; 6; 3.68–3.88; monoclinic
71. Gaudefroyite	$Ca_4Mn_3^{+3}[O_3(BO_3)_3CO_3)]$; $(O, OH?)_3$[d] End Member, $Ca_4Mn_3^{+3}CO_3B_3O_{12}$	561.56; 18.60%; 6–6.5; 3.44–3.50; hexagonal
72. Ginorite	$Ca_2[B_6O_9(OH)_2\cdot OB_6O_8(OH)\cdot OB(OH)\cdot OB(OH)_2]\cdot 5H_2O$ $2CaO\cdot 7B_2O_3\cdot 8H_2O$ $Ca_2B_{14}O_{23}\cdot 8H_2O$	743.62; 65.54%; 3.5; 2.07–2.14; monoclinic

continues

Table 1.1 (*continued*)

Name (Group; form)	Formula	Molecular weight[b]; %B_2O_3[c]; hardness; density; crystal system
73. Gowerite	$Ca[B_5O_8(OH)\cdot B(OH)_3]\cdot 3H_2O$ $CaO\cdot 3B_2O_3\cdot 5H_2O$ $CaB_6O_{10}\cdot 5H_2O$	355.01; 58.83%; 3; 1.98–2.00; monoclinic
74. Grandidierite (cf., andalusite)	$(Mg, Fe^{+2})\ Al_3[BO_3(SiO_4)O_2]$ End Member, $MgAl_3BSiO_9$	303.91; 11.45%; 7–7.5; 2.9–3.09 288.14; 12.08%; orthorhombic
75. Halurgite	$Mg_2[B_4O_5(OH)_4]_2\cdot H_2O$ $2MgO\cdot 4B_2O_3\cdot 5H_2O$ $Mg_2B_8O_{14}\cdot 5H_2O$[4]	449.17; 62.00%; 2.5–3; 2.14–2.25; monoclinic
76. Hambergite	$Be_2[BO_3(OH, F)]$ $4BeO\cdot B_2O_3\cdot H_2O$ End Member, $Be_4B_2O_7\cdot H_2O$	187.68; 37.09%; 7–7.5; 2.28–2.37; orthorhombic
77. Harkerite (cf., sakhaite)	$Ca_{24}Mg_8(O, OH_{16})_2(Cl, H_2O)(CO_3)_8(BO_3)_8$ Al_2Si_8[g] $96CaO\cdot 30MgO\cdot 2MgCl_2\cdot 3Al_2O_3\cdot 36CO_2\cdot$ $15B_2O_3\cdot 24SiO_2\cdot 12H_2O$ End Member, $Ca_{24}Mg_8Al_2(CO_3)_8B_8Si_8O_{55}\cdot 2H_2O$	2932.94; 9.49%; —; 2.94-2.96; cubic (rhombohedral)[g] 2917.52; 9.55%
78. Heidornite	$Ca_3Na_2Cl(SO_4)_2[B_5O_8(OH)_2]$ $2Na_2O\cdot 2NaCl\cdot 6CaO\cdot 4SO_3\cdot B_2O_3\cdot 2H_2O$ $Ca_3Na_2Cl(SO_4)_2B_5O_9\cdot H_2O$	609.86; 28.54%; 4–5; 2.71–2.75; monoclinic
79. Hellandite[i]-(Y) (i., tadzhikite)	$(Ca, Y)_6(Al, Fe^{+3})Si_4[B_4O_{20}(OH)_4]$ End Member, $Ca_{3.5}Y_{2.5}AlSi_4B_4O_{20}(OH)_4$[g]	917.97; 14.33%; 4.5–6.5; 2.95–3.55 933.12; 14.92%; monoclinic
80. Hellandite (RE, Th)[g]	$(Ca, RE, Th, U)_6(Al, Fe^{+3})Si_4[B_4O_{20}(OH)_4]$	—
81. Henmilite[e]	$Ca_2Cu(OH)_4[B(OH)_4]_2$ $Ca_2CuB_2O_6\cdot 6H_2O$	369.41; 18.85%; —; 2.53; triclinic
82. Hexahydroborite	$Ca[B(OH)_4]_2\cdot 2H_2O$ $CaO\cdot B_2O_3\cdot 6H_2O$ $\alpha CaB_2O_4\cdot 6H_2O$-I	233.79; 29.78%; —; 1.88; monoclinic

83. Hilgardite,	$Ca_2Cl[B_5O_9]\cdot H_2O$	331.67; 52.48%; 5; 2.69–2.71; 4M monoclinic; 1A,
84. 3A, 1A (=[illegible]d-	$3CaO\cdot CaCl_2\cdot 5B_2O_3\cdot 2H_2O$	3A triclinic
85. tretskite), and 4M (or 1Tc)	$Ca_2B_5O_9Cl\cdot H_2O$-I	
86. Holtite (Dumortierite)	$Al_6(Al, Ta)[BO_3][(Si, Sb)O_4]_3(O, OH)_3$[e]; (Al, Sb, Ta) $(O, OH)_{2.25}$[d]	1058.90; 3.29%; 8.5; 3.01–3.90; orthorhombic
	End Member, $Al_{6.75}Ta_{0.25}Sb_{0.75}BSi_{2.25}O_{17.25}$[g]	668.67; 5.21%
87. Homilite[f] (Datolite)	$Ca_4Fe_2[B_4(SiO_4)_4O_4]$; $(Fe^{+2}, Mg)_2$[d]	747.58; 18.63%; 5; 3.34–3.58
	End Member, $Ca_2FeB_2Si_2O_{10}$	373.79; 18.63%; monoclinic
88. Howlite	$Ca_2[B_3O_4(OH)_2][SiB_2O_5(OH)_3]$	391.33; 44.48%; 3.5–4; 2.56–2.62; monoclinic
	$4CaO\cdot 5B_2O_3\cdot 2SiO_2\cdot 5H_2O$	
	$Ca_4Si_2B_{10}O_{23}\cdot 5H_2O(Ca_2SiHB_5O_{12}\cdot 2H_2O)$	
89. Hulsite (i., magnesiohulsite)	$Mg_{0.7}Fe^{+2}_{1.5}Fe^{+3}_{0.6}Sn^{+4}_{0.2}[O_2(BO_3)]$	248.84; 13.99%; 3; 4.5–4.6; monoclinic
	End Member, $Fe^{+2}_{2.5}Sn_{0.5}BO_5$[g]	289.78; 12.13%
90. Hungchaoite	$Mg[B_4O_5(OH)_4]\cdot 7H_2O$	341.68; 40.75%; 1–2; 1.70–1.71; triclinic
	$MgO\cdot 2B_2O_3\cdot 9H_2O$	
	$MgB_4O_7\cdot 9H_2O$-II	
91. Hyalotekite	$(Pb, Ca, Ba)_4(F, OH)[(BO_3)Si_6O_{14}]$	1250.29; 11.14%; 5–5.5; 3.81–3.97; triclinic
	$(Pb, K, Ba)_4Ca_2(Si, B, Be)_{12}O_{28}F$[d]	
	End Member, $Pb_2Ba_2Ca_2Be_{0.5}Si_{9.5}B_2O_{28}F$[g]	1529.15; 4.55%
92. Hydroboracite	$CaMg[B_3O_4(OH)_3]_2\cdot 3H_2O$	413.33; 50.53%; 2–3; 2.167–2.173; monoclinic
	$CaO\cdot MgO\cdot 3B_2O_3\cdot 6H_2O$	
	$CaMgB_6O_{11}\cdot 6H_2O$	
93. Hydrochlorborite	$Ca_2[B_3O_3(OH)_4\cdot OB(OH)_3]Cl\cdot 7H_2O$	936.02; 29.75%; 2.5; 1.84–1.88; monoclinic
	$3CaO\cdot CaCl_2\cdot 4B_2O_3\cdot 21H_2O$	
	$Ca_4B_8O_{15}Cl_2\cdot 21H_2O$	
94. Hydroxyascharite	$Mg_2\{B_{2-x}H_{3x}O_5]\cdot H_2O$; $x\sim 0.18$	166.84; 37.97%
95. Idocrase, boron-containing (=Vesuvianite) (Amesite)	$Ca_{10}Mg_2(Al, B)_4(SiO_4)_5(Si_2O_7)_2(OH)_4$	1389.75; 5.01%

continues

Table 1.1 (*continued*)

Name (Group; form)	Formula	Molecular weight[b]; $\%B_2O_3$[c]; hardness; density; crystal system
96. Inderborite	$CaMg[B_3O_3(OH)_5]_2 \cdot 6H_2O$	503.41; 41.49%; 3.5; 1.928–1.934; monoclinic
	$CaO \cdot MgO \cdot 3B_2O_3 \cdot 11H_2O$	
	$CaMgB_6O_{11} \cdot 11H_2O$	
97. Inderite (=Lesserite) (d., kurnakovite)	$MgB_3O_3(OH)_5 \cdot 5H_2O$	559.70; 37.32%; 2.5–3; 1.78–1.79; monoclinic
	$2MgO \cdot 3B_2O_3 \cdot 15H_2O$	
	$Mg_2B_6O_{11} \cdot 15H_2O$	
98. Inyoite	$Ca[B_3O_3(OH)_5] \cdot 4H_2O$	555.21; 37.62%; 2; 1.87–1.88; monoclinic
	$2CaO \cdot 3B_2O_3 \cdot 13H_2O$	
	$Ca_2B_6O_{11} \cdot 13H_2O$	
99. Iquiqueite	$Na_8K_6Mg_2Cr_2^{+6}B_{48}O_{87} \cdot 25H_2O$	2932.37; 56.98%; —; 2.06; hexagonal
100. Ivanovite	$Ca_xB_2O_3Cl_y$	(x, y = 1)145.15; 47.96% monoclinic
101. Jeremejevite	$Al_6[(BO_3)_5(OH, F)_3]$	1013.91; 34.33%; 6.5; 3.21–3.28; hexagonal
	$6Al_2O_3 \cdot 5B_2O_3 \cdot 3H_2O$; $Al_{12}B_{10}O_{33} \cdot 3H_2O$	
	End Member, $Al_6B_5O_{15}F_3$[g]	512.93; 33.93%
102. Jimboite (Kotoite)	$Mn_3^{+2}[BO_3]_2$	282.43; 24.65%; 5.5; 3.98–4.12; orthorhombic
	$3MnO \cdot B_2O_3$	
	$Mn_3B_2O_6$	
103. Johachidolite	$CaAl[B_3O_7]$	211.49; 49.38%; 7.5; 3.37–3.44; orthorhombic
104. Kalborsite[e]	$K_6Al_4Si_6O_{20}[B(OH)_4]Cl$	945.31; 3.68%; —; 2.48; tetragonal
	$K_6Al_4Si_6BO_{22}Cl \cdot 2H_2O$	
105. Kaliborite (=Paternoite)	$HKMg_2[B_5O_7(OH)_3 \cdot OB(OH)_2]_2 \cdot 4H_2O$	1433.15; 58.29%; 4–4.5; 2.11–2.12; monoclinic
	$K_2O \cdot 4MgO \cdot 12B_2O_3 \cdot 19H_2O$	
	$K_2Mg_4B_{24}O_{41} \cdot 19H_2O$	
106. Karlite	$Mg_7[(Cl, OH)_5(BO_3)_3]$	477.71; 21.86%; —; 2.91–3.24; orthorhombic
	$Mg_7(Cl, OH)_5B_3O_9$; $(Mg, Al)_7$[d]	
	End Member, $Mg_{14}B_6O_{23} \cdot 5H_2O$[g]	863.20; 24.20%

107. Kernite (=Rasorite)	$Na_2[B_4O_6(OH)_2]\cdot 3H_2O$ $Na_2O\cdot 2B_2O_3\cdot 4H_2O$ $Na_2B_4O_7\cdot 4H_2O$	273.28; 50.95%; 2.5–3; 1.906; monoclinic
108. Korerupine (Prismatine)	$Mg_3Al_6O_4(OH)(Si_2O_7)[(Al, Si)_2(Si, B)O_{10}]$; (Mg, Fe^{+2})(Mg, Fe^{+3}, Al)$_9$(Si, Al, B)$_5$(O, OH, F)$_{22}$	852.24; 6.80%; 6–7; 3.24–3.45; orthorhombic
	End Member, $Mg_3Al_7Si_4BO_{21}(OH)$	727.93; 4.90%
109. Korzhinskite	$Ca[BO_3\cdot BO]\cdot H_2O$ $CaO\cdot B_2O_3\cdot H_2O$ $CaB_2O_4\cdot H_2O$	143.71; 48.44%
110. Kotoite (cf., jimboite)	$Mg_3[BO_3]_2$ $3MgO\cdot B_2O_3$;(Mg, Mn)$_3B_2O_6$ End Member $Mg_3B_2O_6$[g]	109.53; 36.54%; 6.5; 3.04–3.10; orthorhombic
111. Kurchatovite, α	(Mg, Fe)Ca$[B_2O_5]$; (Mg, Mn, Fe?)[d]	181.77; 38.30%; —; 3.02–3.27
112. Kurchatovite, β	$Mg_5Ca_6Mn[B_2O_5]_6$	1026.65; 40.69%; orthorhombic
(d., clinokurchatovite)	End Member, $MgCaB_2O_5$	166.00; 41.94%
113. Kurgantaite	(Sr, Ca)$_2[B_4O_7(OH)_2]$ 2(Sr, Ca)O$\cdot 2B_2O_3\cdot H_2O$ (Sr, Ca)$_2B_4O_8\cdot H_2O$	316.95; 43.93%; 5–7; 2.8–2.9; triclinic
114. Kurnakovite (d., inderite)	$Mg[B_3O_3(OH)_5]\cdot 5H_2O$ $2MgO\cdot 3B_2O_3\cdot 15H_2O$ $Mg_2B_6O_{11}\cdot 15H_2O$-II (triclinic)	559.70; 37.32%; 3; 1.85–1.86; triclinic
115. Lagonite	$Fe_2B_6O_{12}\cdot 3H_2O$	422.60; 49.42%
116. Larderellite	$NH_4[B_5O_7(OH)_2]\cdot H_2O$ $(NH_4)_2O\cdot 5B_2O_3\cdot 4H_2O$ $NH_4B_5O_8\cdot 2H_2O$	236.12; 73.71%; —; 1.88–1.91; monoclinic
117. Lesserite	$Mg[B_3O_3(OH)_5]\cdot 5H_2O$ $Mg_2B_6O_{11}\cdot 10H_2O$	469.62; 44.47%
118. Leucosphenite	$Na_4BaTi_2O_2[(B, Si)_4Si_8O_{28}]$ $Na_4BaTi_2O_{30}B_2Si_{10}$[d]	1107.51; 6.29%; 6–6.5; 3.05–3.09; monoclinic

continues

Table 1.1 (*continued*)

Name (Group; form)	Formula	Molecular weight[b]; $\%B_2O_3$[c]; hardness; density; crystal system
119. Liddicoatite[d] (Tourmaline)	$Ca(Li, Al)_3Al_6[BO_3]_3(Si_6O_{18})(O, OH, F)_4$ End Member, $CaLi_2Al_7B_3Si_6O_{27}(OH)_4$	955.12; 10.93%; —; 3.02–3.06; rhombohedral 943.79; 11.06%
120. Ludwigite	$(Mg, Fe^{+2})_2Fe^{+3}O_2[BO_3]$ $4(Mg, Fe)O \cdot Fe_2O_3 \cdot B_2O_3$ End Member, $Mg_2Fe^{+3}BO_5$[d]	226.81; 15.35%; 6–7.5; 3.62–4.41; orthorhombic 195.27; 17.83%
121. Luneburgite	$Mg_3(PO_4)_2[B_2O(OH)_4] \cdot 6H_2O$ $3MgO \cdot P_2O_5 \cdot B_2O_3 \cdot 8H_2O$; $(OH)_6$[g] $Mg_3(PO_4)_2B_2O_3 \cdot 8H_2O$ ($9H_2O$?)[e]	476.60; 14.61%; 1–2; 2.07–2.11; monoclinic (triclinic)[g]
122. Magnesioaxinite[e] (Axinite)	$Ca_2MgAl_2(Si_2O_6)[Si_2O_6BO_3(OH)]$	538.58; 6.46%; 6.5–7; 3.20–3.34; triclinic
123. Magnesiodumortierite[d] (Dumortierite)	$MgAl_6Si_3B(O, OH)_2$ $MgAl_6Si_3BO_{0.33}(OH)_{1.67}$[g]	314.27; 11.08%; —; 1.82 314.94; 11.05%; orthorhombic
124. Magnesiohulsite[d] (i., hulsite)	$(Mg, Fe^{+2})_2(Fe^{+3}, Sn^{+4}, Mg)B_2O_5$ End Member, $Mg_{2.5}Sn^{+4}_{0.5}BO_5$	248.06; 28.07%; —; 2.38 210.93; 16.50%; monoclinic
125. Manandonite-$2H_2$[g] (cf., B-rich chlorite) (Amesite)	$Li_2Al_4(Si_3, Al, B)O_{10}(OH)_8$ End Member, $Li_2Al_5Si_2BO_{10}(OH)_8$[g]	449.18; 2.58%; —; 2.53–2.89; monoclinic 511.82; 6.80
126. Manganaxinite[f] (Axinite)	$Ca_2MnAl_2(Si_2O_6)[Si_2O_6BO_3(OH)]$	569.21; 6.12%; —; 3.31; triclinic
127. Mcallisterite (=Trigonomagneborite) (d., admontite)	$Mg_2[B_6O_7(OH)_6]_2 \cdot 9H_2O$ $2MgO \cdot 6B_2O_3 \cdot 15H_2O$ $Mg_2B_{12}O_{20} \cdot 15H_2O$	768.56; 54.35%; 2.5; 1.866–1.867; triclinic (rhombohedral)[g]
128. Melanocerite[f] (Ce)[d] (Apatite?)	$CaRE_4BSi_3O_{14}(F, OH)$ $(Ce, Ca)_5(B, Si)_3O_{12}(F, OH) \cdot nH_2O$[g] End Member, $Ce_5SiB_2O_{12}(OH) \cdot nH_2O$[g]	(n = 1) 736.84; 7.09% (typical 0.05–4.85%); 5; 3.42–3.45 977.30; 7.12%; hexagonal
129. Metaborite (=Metaboric acid)α, β[d]	$[BO(OH)]$ $B_2O_3 \cdot H_2O$ HBO_2; $H_2B_2O_4$	87.64; 79.44%; 5; 2.47–2.49; cubic

130. Meyerhofferite	$Ca[B_3O_3(OH)_5]\cdot H_2O$	447.12; 46.71%; 2; 2.118–2.120; triclinic
	$2CaO\cdot 3B_2O_3\cdot 7H_2O$	
	$Ca_2B_6O_{11}\cdot 7H_2O$	
131. Moydite[3]-(Y)[d]	$(Y, RE)CO_3[B(OH)_4]$	227.76; 15.28%; —; 3.06; orthorhombic
	End Member, $YCO_3BO_2\cdot 2H_2O$	
132. Nagashimalite[e] (Taramellite)[d]	$Ba_4(V^{+4}, Ti)_4Si_8B_2O_{27}(O, OH)_2Cl$	(x = 1) 1474.03; 4.72%; —; 2.07; orthorhombic
	$Ba_4(V^{+3}, Ti, Fe^{+3}, Mg)_4Si_8B_2O_{27}O_2Cl_x$[d]	
	End Member, $Ba_4V_4^{+4}Si_8B_2O_{28}(OH)Cl$[g]	1499.82; 4.64%
133. Nasinite (Ezcurrite)	$Na_2[B_5O_8(OH)]\cdot 2H_2O$	562.14; 61.92%; —; 2.13; orthorhombic
	$2Na_2O\cdot 5B_2O_3\cdot 5H_2O$	
	$Na_4B_{10}O_{17}\cdot 5H_2O$	
Neocolemanite	(alatropic form of Colemanite)	411.09; 50.81%
134. Nifontovite	$Ca_3[B_3O_3(OH)_6]_2\cdot 2H_2O$	521.22; 40.07%; 3.5; 2.35–2.36; monoclinic
	$3CaO\cdot 3B_2O_3\cdot 8H_2O$	
	$Ca_3B_6O_{12}\cdot 8H_2O$	
135. Nobleite (i., tunellite)	$Ca[B_6O_9(OH)_2]\cdot 3H_2O$	337.00; 61.98%; 3; 2.09–2.10; monoclinic
	$CaO\cdot 3B_2O_3\cdot 4H_2O$	
	$CaB_6O_{10}\cdot 4H_2O$	
136. Nordenskioldine (i., dolomite, tusionite)	$CaSn^{+4}[BO_3]_2$	276.41; 25.19%; 5.5–6; 4.12–4.22; triclinic (rhombohedral)[g]
	$CaO\cdot SnO_2\cdot B_2O_3$	
	$CaSnB_2O_6$	
137. Okanoganite-(Y)[d]	$(Na, Ca)_3(Y, Ce, Nd, La)_{12}Si_6B_2O_{27}F_{14}$	2519.21; 2.76%; —; 4.66; rhombohedral
138. Olenite[d] (Tourmaline)	$NaAl_9Si_6O_{18}[BO_3]_3(O, OH)_4$	964.77; 10.82%; —; 3.07
	End Member, $NaAl_9Si_6B_3O_{30}(OH)$[g]	963.76; 10.84%; rhombohedral
139. Olshanskyite	$Ca_3[B(OH)_4]_4(OH)_2$	469.61; 29.65%; 4; 2.23; monoclinic[g] **(triclinic)**
	$3CaO\cdot 2B_2O_3\cdot 9H_2O$	
	$Ca_3B_4O_9\cdot 9H_2O$	
140. Orthopinakiolite (Fredrikssonite, Takeuchiite)	$(Mg, Mn^{+2})_2Mn^{+3}[O_2(BO_3)]$[e]	224.99; 15.47%; —; 3.92–4.03; orthorhombic
	$3MgO\cdot MnO\cdot Mn_2O_3\cdot B_2O_3$	
	End Member, $Mg_2Mn^{+3}BO_5$	194.36; 17.91%

continues

Table 1.1 *(continued)*

Name (Group; form)	Formula	Molecular weight[b]; $\%B_2O_3$[c]; hardness; density; crystal system
141. Oyelite[d] (Tobermorite)	$Ca_{10}Si_8(B, Al)_2O_{29} \cdot 12.5H_2O$	1352.43; 2.57%; —; 2.71; orthorhombic
Paigeite (=Vonsenite)	$(Fe^{+2}, Mg, Mn^{+2})_2Fe^{+3}BO_5$	236.72; 14.71%; 5; 4.39–4.81
	End Member, $Fe_2^{+2}Fe^{+3}BO_5$	306.96; 11.34%; orthorhombic
142. Painite[e]	$Zr_{1.03}Ca_{0.86}Al_{9.11}O_{18}$	673.03; 5.17%; 8; 3.54–4.01
	End Member, $CaZrAl_9BO_{18}$[e]	672.94; 5.17%; hexagonal
Pandermite	See Priceite	
143. Parahilgardite (=Hilgardite 3Tc)	$[Ca_2ClB_5O_9 \cdot H_2O]_3$	331.67; 52.48%; 5; 2.69–2.71; triclinic
	$3CaO \cdot CaCl_2 \cdot 5B_2O_3 \cdot 2H_2O$	
	$Ca_2B_5O_9Cl \cdot H_2O$-II	
Paraveatchite, p Veatchite	See Veatchite, p	
144. Paternoite	$MgB_8O_{13} \cdot 4H_2O$	390.85; 71.25%
145. Penobosquisite[e]	$Ca_2Fe[B_9O_{13}(OH)_6]Cl \cdot 4H_2O$	650.85; 48.14%; monoclinic
	$Ca_2FeB_9O_{16}Cl \cdot 7H_2O$	
146. Pentahydroborite	$Ca[B_2O(OH)_6] \cdot 2H_2O$	215.77; 32.27%; 2.5; 2.00–2.03; triclinic
	$CaO \cdot B_2O_3 \cdot 5H_2O$	
	$CaB_2O_4 \cdot 5H_2O$	
147. Peprossiite (Ce)[d]	$(Ce, La)\ Al_2B_3O_9$	369.90; 28.23%; —; 3.57
	End Member, $CeAl_2B_3O_9$	370.51; 28.19%; hexagonal
148. Pinakiolite (p, fredrikssonite; cf. ortho pinakiolite)	$(Mg, Mn^{+2})_2(Mn^{+3}, Sb^{+3})[O_2BO_3]$[e]	224.99; 15.47%; 6; 3.97–4.03; monoclinic
	$(Mg, Mn^{+2})_2Mn^{+3}BO_5$	
	End Member, $Mg_2Mn^{+3}BO_5$	194.36; 17.91%
149. Pinnoite	$Mg[B_2O(OH)_6]$	163.97; 42.46%; 3.5–4; 2.24–2.29; tetragonal
	$MgO \cdot B_2O_3 \cdot 3H_2O$	
	$MgB_2O_4 \cdot 3H_2O$	
150. Poudretteite[e] (Osumilite)	$KNa_2B_3Si_{12}O_{30}$	934.52; 11.17%; —; 2.53; hexagonal

151. Povondravite (=Ferridravite) (Tourmaline)	$(Na, K)(Fe^{+2}, Mg, Fe^{+3})_3Fe_6^{+3}[BO_3]_3$ (Si_6O_{18}); $(O, OH)_4$[d] End Member, $NaFe_3^{+2}Fe_6^{+3}B_3Si_6O_{30}(OH)$[g]	1201.07; 8.69%; —; 3.54; rhombohedral 1223.55; 8.54%
152. Preobrazhenskite (=Preobratschenskite)	$HMg_3[B_9O_{12}(OH)_4 \cdot [O_2B(OH)_2]_2$ $6MgO \cdot 11B_2O_3 \cdot 9H_2O$ $Mg_6B_{22}O_{39} \cdot 9H_2O$	1169.79; 65.47%; 4–5.5; 2.45–2.46; orthorhombic
153. Priceite (=Pandermite)	$Ca_2[B_5O_7(OH)_5] \cdot H_2O$ $4CaO \cdot 5B_2O_3 \cdot 7H_2O$ $Ca_4B_{10}O_{19} \cdot 7H_2O$	698.52; 49.83%; 3–3.5; 2.41–2.48; triclinic
154. Pringleite[e] (Boracite) (d., ruitenbergite)	$Ca_9[B_{20}(OH)_{18}][B_6O_6(OH)_6]Cl_4 \cdot 13H_2O$ $Ca_9B_{26}O_{46}Cl_4 \cdot 25H_2O$	1969.95; 45.94%; —; 2.13; triclinic
Prismatine[d] (=Kornerupine)	(Mg, Fe) $(Mg, Fe, Al)_9(Si, B, Al)_5(O, OH, F)_{22}$ End Member, $Mg_3Al_6Si_4BO_{21}(OH)$	852.64; 6.80%; 6–7; 3.24–3.45; orthorhombic 710.95; 4.90%
155. Probertite (=Kramerite) (Ulexite)	$CaNa[B_5O_7(OH)_4] \cdot 3H_2O$ $Na_2O \cdot 2CaO \cdot 5B_2O_3 \cdot 10H_2O$ $NaCaB_5O_9 \cdot 5H_2O$	351.19; 49.56%; 3–3.5; 2.13–2.14; monoclinic
156. Qilianshanite[a]	$Na_{1.07}Ca_{0.01}H_{8.86}C_{0.71}B_{1.06}O_8$	181.91; 20.28%; —; 1.63; monoclinic
Reedmergnerite	See Albite	
157. Rhodizite (=Rhodicite)	(Cs, K, Rb) $Be_4Al_4[B_{11}O_{26}(O, OH)_2]$ $(K, Cs)Be_4Al_4B_{12}O_{28}$[e]; $(B, Be)_{12}$[d] End Member, $CsBe_5Al_4B_{11}O_{28}$	807.69; 51.72%; 8.5; 3.44–3.59; cubic 852.80; 44.90%
158. Rivadavite[e]	$Na_6Mg[B_6O_7(OH)_6]_4 \cdot 10H_2O$ $3Na_2O \cdot MgO \cdot 12B_2O_3 \cdot 22H_2O$ $Na_6MgB_{24}O_{40} \cdot 22H_2O$	1458.02; 57.30%; 3.5; 1.91–1.92; monoclinic
159. Roweite (Frederovskite)	$Ca_2(Mn, Mg)_2(OH)_4[B_4O_7(OH)_2]$ $Ca_2(Mn, Mg)_2B_4O_{10} \cdot 3H_2O$; Mn_2[d] End Member, $Ca_2Mn_2B_4O_{10} \cdot 3H_2O$	416.68; 33.42%; 4.5; 2.94–2.96; orthorhombic 447.32; 31.13%
160. Ruitenbergite (Boracite) (d., pringleite)	$Ca_9[B_{20}(OH)_{18}][B_6O_6(OH)_6]Cl_4 \cdot 13H_2O$ $Ca_9B_{26}O_{46}Cl_4 \cdot 25H_2O$	1969.95; 45.94%; —; 2.13; monoclinic

continues

Table 1.1 (*continued*)

Name (Group; form)	Formula	Molecular weight[b]; %B_2O_3[c]; hardness; density; crystal system
161. Sakhaite (cf., harkerite)	$Ca_3Mg[(BO_3)_2CO_3]\cdot O.36H_2O$	1266.87; 19.23%; 5; 2.78–2.83; cubic
	$Ca_{12}Mg_4(BO_3)_7(CO_3)_4(OH)_2Cl\cdot nH_2O^g$;(n<1)	
	$Ca_{12}Mg_4B_7O_{21}(CO_3)_4Cl\cdot H_2O$; $(2H_2O?)^e$	
162. Santite	$K[B_5O_6(OH)_4]\cdot 2H_2O$	293.21; 59.36%; 2.5; 1.77; orthorhombic
	$K_2O\cdot 5B_2O_3\cdot 8H_2O$	
	$KB_5O_8\cdot 4H_2O$	
163. Sassolite	$B(OH)_3$	61.833; 56.297%; 1; 1.48–1.50; triclinic
	$B_2O_3\cdot 3H_2O$	
	H_3BO_3	
164. Satimolite	$KNa_2(Al_4[B_2O(OH)_6]_3(OH)_6Cl_3)\cdot H_2O$	838.42; 24.91%; —; 1.70; orthorhombic
	$K_2O\cdot 2Na_2O\cdot 3Al_2O_3\cdot 2AlCl_3\cdot 6B_2O_3\cdot 26HO$	
	$KNa_2Al_4B_6O_{15}Cl_3\cdot 13H_2O$	
165. Sborgite	$Na[B_5O_6(OH)_4]\cdot 3H_2O$	295.12; 58.98%; 3.5; 1.711–1.713; monoclinic
	$Na_2O\cdot 5B_2O_3\cdot 10H_2O$	
	$NaB_5O_8\cdot 5H_2O$	
166. Schorl[g] (Tourmaline)	$Na(Fe^{+3}, Mn^{+3})_3Al_6(OH, F)_4[BO_3]_3$	1056.00; 9.89%; 7–8; 2.92–3.43; rhombohedral
	(Si_6O_{18}); Fe_3^{+3}, $(OH)_4{}^d$	
	End Member, $NaFe_3^{+3}Al_6B_3Si_6O_{27}(OH)_4$	1053.38; 9.19%
167. Seamanite	$Mn_3^{+2}(OH)_2[B(OH)_4]PO_4$	372.64; 9.34%; 4; 3.08–3.13; orthorhombic
	$6MnO\cdot P_2O_5\cdot B_2O_3\cdot 6H_2O$	
	$Mn_3PO_4BO_3\cdot 3H_2O$	
168. Searlesite	$NaBSi_2O_5(OH)_2$	203.98; 17.07%; 3.5; 2.44–2.46; monoclinic
	$Na_2O\cdot B_2O_3\cdot 4SiO_2\cdot 2H_2O$	
	$NaBSi_2O_6\cdot H_2O$	
169. Serendibite (Aenigmatite)	$(Ca, Mg)_5(AlO)_5[BO_3(SiO_4)_3]$	729.08; 9.55%; 6.5; 3.42–3.52; triclinic
	$(Ca, Na)_2(Mg, Al, Fe)_6(Si, Al, B)_6O_{20}$	
	End Member, $Ca_2Mg_3Al_{4.5}B_{1.5}Si_3O_{20}$	694.95; 7.51%

170. Shabynite	$Mg_5[(Cl, OH)_2(OH)_5(BO_3)]\cdot 5H_2O$ $Mg_{10}(Cl, OH)_4B_2O_{11}\cdot 10H_2O$; $(9H_2O?)$[d]	725.74; 9.59%; —; 2.24; monoclinic
171. Sibirskite	$2CaO\cdot B_2O_3\cdot H_2O$ $Ca_2B_2O_5\cdot H_2O$-II	199.79; 34.85%; orthorhombic
172. Sinhalite (i. olivine)	$MgAlBO_4$	126.10; 27.61%; 6.5; 3.45–3.50; orthorhombic
173. Solongoite (Boracite)	$Ca_2Cl[B_3O_4(OH)_4]$ $Ca_2ClB_3O_6\cdot 2H_2O$	280.07; 37.29%; 3.5; 2.49–2.51; monoclinic
174. Spencite[g] (=Tritomite) (=RE tadzhikite)	$(RE, Ca)_6(B, Al)_4[(Si, B)O_4]_6(F, OH)_2\cdot 8H_2O$ $(Y, Ca, La, Fe^{+2})_5(Si, B, Al)_3(O, OH, F)_{13}$[d]	695.90; 5.00% (typical 10.04–11.12%); 5.5–6.5; 3.05–3.40; hexagonal
175. Stillwellite-Ce	$(Ce, La)BO(SiO_4)$; $(Ce, La, Ca?)$[d] $(Ce, La)_2O_3\cdot B_2O_3\cdot 2SiO_2$ End Member, $CeSiBO_5$	258.40; 13.47%; 6.5; 4.57–4.70; hexagonal (rhombohedral)[g] 259.01; 13.44%
176. Strontioborite	$Sr[B_6O_9(OH)\cdot OB(OH)\cdot OB(OH)_2]$ $SrO\cdot 4B_2O_3\cdot 2H_2O$ $SrB_8O_{13}\cdot 2H_2O$	418.13; 66.60%; —; 2.38–2.81; monoclinic
177. Strontioginorite (Ginorite)	$Sr_2[B_6O_9(OH)_2\cdot OB_6O_8(OH)\cdot OB(OH)\cdot OB(OH)_2]\cdot 5H_2O$; $(Sr, Ca)?$[d] $2SrO\cdot 7B_2O_3\cdot 8H_2O$; End Member, $Sr_2B_{14}O_{23}\cdot 8H_2O$	838.70; 58.11%; 2–3; 2.25–2.26; monoclinic
178. Strontiohildgardite-1Tc (Hilgardite)	$(Ca, Sr)_2B_5O_9(OH, Cl)\cdot H_2O$ $3(Cr, Ca)O\cdot (Sr, Ca)Cl_2\cdot 5B_2O_3\cdot 2H_2O$ $CaSrB_5O_9Cl\cdot H_2O$-II	379.22; 45.90%; 5–7; 2.99–3.02; triclinic
179. Studenitsite (Ulexite)	$NaCa_2[B_9O_{14}(OH)_4]\cdot 4H_2O$ $Na_2O\cdot 4CaO\cdot 9B_2O_3\cdot 8H_2O$ $NaCa_2B_9O_{16}\cdot 4H_2O$	528.50; 59.28%; —; 2.33; monoclinic
180. Sturmanite[d] (Ettringite)	$Ca_6(Fe^{+3}, Al, Mn^{+2})_2(SO_4)_2[B(OH)_4](OH)_{12}\cdot 25H_2O$ End Member, $Ca_6Fe_2^{+3}(SO_4)_2BO_8\cdot 33H_2O$	1,257.75; 2.77%; —; 1.78; rhombohedral 1277.60; 2.72%

continues

Table 1.1 (*continued*)

Name (Group; form)	Formula	Molecular weight[b]; $\%B_2O_3$[c]; hardness; density; crystal system
181. Suanite	$Mg_2[B_2O_5]$; $2MgO \cdot B_2O_3$ $Mg_2B_2O_5$	150.23; 46.34%; 5.5; 2.91–2.92; monoclinic
182. Sulfoborite (=Sulphoborite)	$Mg_3(OH, F)_2SO_4[B(OH)_4]_2$[e] $Mg_3SO_4 \cdot B_2O_5 \cdot 5H_2O$[d]; $((OH)_9F)$[d]? End Member, $Mg_3SO_4B_2(OH)_9F$[g]	360.67; 19.30%; 4–4.5; 2.40–2.44; orthorhombic 362.67; 19.20%
183. Sussexite (Szaibelyite)	$Mn_2(OH)[B_2O_4(OH)]$ $2MnO \cdot B_2O_3 \cdot H_2O$ $Mn_2B_2O_5 \cdot H_2O$	229.51; 30.33%; 3.5–5; 3.0–3.43; monoclinic
Szaibelyite (=Ascharite)	$Mg_2(OH)[B_2O_4(OH)]$ $2MgO \cdot B_2O_3 \cdot H_2O$ $Mg_2B_2O_5 \cdot H_2O$	168.24; 41.38%; 3–3.5; 2.60–2.76; monoclinic
184. Tadzhikite[f]-(Y)[d] (=Y spencite) (i., hellandite)	$(RE, Ca)_6(B, Al)_4[(Si, B)O_4]_6$ $(F, OH)_2 \cdot 8H_2O$; $Ca_3(Y, Ce)_2$ $(Ti, Fe, Al)B_4Si_4O_{22}$[d] End Member, $Ca_3Y_2TiB_4Si_4O_{22}$[g]	900.40; 15.46% (typical 10.04–11.12%); 6–6.5; 3.73–3.84; hexagonal (monoclinic)[g] 853.50; 16.31%
185. Tadzhikite-(Ce)[d]	$Ca_3(Ce, Y)_2(Ti, Al, Fe)B_4Si_4O_{22}$ End Member, $Ca_3Ce_2TiB_4Si_4O_{22}$[g]	900.40; 15.46%; monoclinic[g] 955.92; 14.57%
186. Takedaite[d]	$Ca_3B_2O_6$	237.85; 29.27%; —; 3.09; rhombohedral
187. Takeuchite (p., fredrikssonite)	$Mg_{1.59}Mn^{+2}_{0.42}Mn^{+3}_{0.78}Fe^{+3}_{0.19}Ti^{+4}_{0.01}[O_2(BO_3)]$; $(Mg, Mn^{+2})_2(Mn^{+3}, Fe^{+3})BO_5$[e] End Member, Mg_2MnBO_5[g]	206.47; 16.86; —; 1.85–2.15 225.44; 15.44%; orthorhombic 194.36; 17.91%
188. Taramellite[e]	$Ba_4(Fe^{+3}, Ti)_4B_2Si_8O_{29}Cl_x$; $(Fe^{+3}, Ti, Fe^{+2}, Mg)_4$[d] End Member, $Ba_4Fe_4^{+3}B_2Si_8O_{28}Cl(OH)$[g]	(x = 1) 1502.51; 4.63%; —; 4.13–4.34; orthorhombic 1579.45; 4.58%
189. Teepleite	$Na_2Cl[B(OH)_4]$ $Na_2O \cdot 2NaCl \cdot B_2O_3 \cdot 4H_2O$ $Na_2ClBO_2 \cdot 2H_2O$	160.27; 21.72%; 3–3.5; 2.076–2.082; tetragonal

190. Tertschite	$Ca_4[B_5O_7(OH)_5]_2 \cdot 15H_2O$ $4CaO \cdot 5B_2O_3 \cdot 20H_2O$ $Ca_4B_{10}O_{19} \cdot 20H_2O$	932.72; 37.32%; monoclinic
191. Teruggite	$Ca_4Mg[B_6O_7(OH)_6 \cdot OAsO_3]_2 \cdot 14H_2O$ $4CaO \cdot MgO \cdot 6B_2O_3 \cdot As_2O_5 \cdot 20H_2O$ $Ca_4MgAs_2B_{12}O_{28} \cdot 20H_2O$; $(O_{30}?)$[d]	1272.48; 32.83%; 2.5; 2.15–2.20; monoclinic
192. Tienshanite	$Na_2BaMnTi(SiO_2)_6B_2O_8$ $KNa_9Ba_6Ca_2Mn_6^{+3}Ti_6Si_{36}B_{12}O_{123}(OH)_2$[d]	796.25; 8.74%; 6–6.5; 3.13–3.3 4909.80; 8.51%; hexagonal
193. Tincalconite (=Mohavite)	$Na_2[B_4O_5(OH)_4] \cdot 3H_2O$ $Na_2O \cdot 2B_2O_3 \cdot 5H_2O$ $Na_2B_4O_7 \cdot 5H_2O$	291.296; 47.800%; 2.5; 1.88–1.91; hexagonal[g] (rhombohedral)
194. Tinzenite[f] (Axinite)	$(Ca, Mn)_2MnAl_2(Si_2O_6)[Si_2O_6BO_3(OH)]$; $(Ca, Mn, Fe?)$[d] End Member, $CaMn_2Al_2Si_4BO_{15}(OH)$[g]	584.07; 5.96%; 6.5–7; 3.26–3.51; triclinic
195. Titantarmellite[g] (Tarmellite)	$Ba_4(Ti, Fe^{+3})_4B_2Si_8O_{29}Cl_x$; $(Fe^{+3}, Ti, Fe^{+2}, Mg)_4(OH, Cl)$[d] End Member, $Ba_4Ti_2Fe_2^{+3}B_2Si_8O_{28}Cl(OH)$[g]	(x = 1) 1502.51; 4.63%; —; 4.11; orthorhombic 1503.52; 4.63%
Tourmaline (group)	$Na_2O \cdot 6(Mg, Fe)O \cdot 6Al_2O_3 \cdot 3B_2O_3 \cdot 12SiO_2 \cdot 8(OH, F)$(Dravite type) $Na(Mg, Fe)_3B_3Al_6Si_6O_{27}(OH, F)_4$	1010.05; 10.34%; 7–7.5; 3.10–3.15; hexagonal (rhombohedral)[g]
196. Trembathite (Boracite)	$(Mg, Fe)_3[B_3O_5]_2[BO_3]$ Cl End Member, $Mg_3B_7O_{13}Cl$	439.35; 55.46%; —; 2.92–3.27 392.04; 62.16%; rhombohedral
197. Tretskite-1A[g] (cf., hilgardite)	$Ca_2B_5O_9(OH) \cdot H_2O$ $Ca_4B_{10}O_{19} \cdot 3H_2O$	626.46; 55.57%; triclinic
198. Tritomite[f] (Ce)[g]	$(Ca, Na)_4(RE, Th)_6[(Si, B)O_4]_6F_2 \cdot 6H_2O$ $(Ce, La, Y, Th)_5(Si, B)_3(O, OH, F)_{13}$[d]	1033.66; 5.05% (typical 7.31–8.37%); 5.5; 4.16–4.66; hexagonal
Tritomite (Y) (=Spencite)[d]	$(Y, Ca, La, Fe^{+2})_5(Si, B, Al)_3(O, OH, F)_{13}$	695.90; 5.00%; —; 4.44; hexagonal
199. Tsilaisite[f] (Tourmaline)	$NaMn_3Al_6(OH, F)_4[BO_3]_3(Si_6O_{18})$	1054.63; 9.90%; 7; 3.09–3.31; rhombohedral

continues

Table 1.1 (*continued*)

Name (Group; form)	Formula	Molecular weight[b]; $\%B_2O_3$[c]; hardness; density; crystal system
200. Tunellite (i., nobleite)	$Sr[B_6O_9(OH)_2]\cdot 3H_2O$ $SrO\cdot 3B_2O_3\cdot 4H_2O$ $SrB_6O_{10}\cdot 4H_2O$	384.54; 54.31%; 2.5; 2.38–2.40; monoclinic
201. Tusionite (i., dolomite, norden-skioldine)	$Mn^{+2}Sn^{+4}[BO_3]_2$ $MnSnB_2O_6$	291.27; 23.90%; —; 4.81; rhombohedral
202. Tuzlaite (Ulexite)	$NaCa[B_5O_8(OH)_2]\cdot 3H_2O$ $Na_2O\cdot 2CaO\cdot 5B_2O_3\cdot 8H_2O$ $NaCaB_5O_9\cdot 4H_2O$	333.18; 52.24%; —; 2.24; monoclinic
203. Tyretskite (−1A)[d] (cf., hilgardite)	$Ca_2(OH)[B_5O_8(OH)_2]$ $4CaO\cdot 5B_2O_3\cdot 3H_2O$ $Ca_4B_{10}O_{19}\cdot 3H_2O$	626.46; 55.57%; —; 2.55–2.56; triclinic
204. Ulexite (=Boronatrocalcite)	$NaCa[B_5O_6(OH)_6]\cdot 5H_2O$ $Na_2O\cdot 2CaO\cdot 5B_2O_3\cdot 16H_2O$ $NaCaB_5O_9\cdot 8H_2O$	405.24; 42.95%; 2.5; 1.955–1.961; triclinic
205. Uralborite (d., vimsite)	$Ca_2[B_3O_3(OH)_5\cdot OB(OH)_3]$ $CaO\cdot B_2O_3\cdot 2H_2O$ $CaB_2O_4\cdot 2H_2O$-I	161.73; 43.05%; 4; 2.58–2.60; monoclinic
206. Uvite[f] (Tourmaline)	$(Ca, Na)(Mg, Fe^{+2})_3Al_5Mg(OH, F)_4$ $[BO_3]_3(Si_6O_{18})$[d] End Member, $CaMg_4Al_5B_3Si_6O_{27}(OH)_4$[g]	1015.92; 10.28%; 7; 3.05–3.08; rhombohedral 973.16; 10.73%
207. Veatchite (t., veatchite-A and p-veatchite)	$Sr_2[B_5O_8(OH)]_2\cdot B(OH)_3\cdot H_2O$ $4SrO\cdot 11B_2O_3\cdot 7H_2O$ $Sr_4B_{22}O_{37}\cdot 7H_2O$-I	1306.41; 58.62%; 2; 2.66–2.67; monoclinic
208. Veatchite-A	$Sr_2[B_5O_8(OH)]_2\cdot B(OH)_3\cdot H_2O$ $4SrO\cdot 11B_2O_3\cdot 7H_2O$ $Sr_4B_{22}O_{37}\cdot 7H_2O$-I	As above

209. p-Veatchite	$Sr_2[B_5O_8(OH)]_2 \cdot B(OH)_3 \cdot H_2O$ $4SrO \cdot 11B_2O_3 \cdot 7H_2O$ $Sr_4B_{22}O_{37} \cdot 7H_2O$-II	1306.41; 58.62%; —; 2.60–2.72; monoclinic
210. Vicanite (Ce)[d]	$(Ce, RE, Th)_{15}As^{+5}As^{+3}_{0.5}Na_{0.5}Fe^{+3}B_4$ $Si_6O_{40}F_7$	rhombohedral
211. Vimsite (d. uralborite)	$Ca[B_2O_2(OH)_4]$ $CaO \cdot B_2O_3 \cdot 2H_2O$ $CaB_2O_4 \cdot 2H_2O$-II	161.73; 43.05%; 4; 2.49–2.54; monoclinic
212. Vistepite[d]	$Mn_5SnSi_5B_2O_{20}$	875.44; 7.95%; —; 4.23; monoclinic
213. Volkovskite (=Volkovite)	$(Sr, Ca)_2[B_6O_9(OH)_2 \cdot OB_6O_8(OH) \cdot OB(OH) \cdot$ $OB(OH)_2] \cdot 5H_2O$ $KCa_4[B_5O_8(OH)]_4[B(OH)_3]_2Cl \cdot 4H_2O$[d] End Member, $KCa_4B_{22}O_{37}Cl \cdot 9H_2O$[g]	791.16; 61.60%; —; 2.30–2.35; monoclinic (triclinic)[g] 1226.82; 62.42%
214. Vonsenite (=Paigeite) (Ludwigite)	$(Fe^{+2}, Mg)_2Fe^{+3}[O_2BO_3]$ $4(Fe, Mg)O \cdot Fe_2O_3 \cdot B_2O_3$ End Member, $Fe_2^{+2}Fe^{+3}BO_5$[d]	226.81; 15.35%; 5–6; 4.39–4.81; orthorhombic 258.35; 13.47%
215. Wardsmithite	$Ca_5Mg[B_4O_5(OH)_4]_6 \cdot 18H_2O$ $5CaO \cdot MgO \cdot 12B_2O_3 \cdot 30H_2O$ $Ca_5MgB_{24}O_{42} \cdot 30H_2O$	1696.59; 49.24%; 2.5; 1.88; hexagonal
216. Warwickite	$(Mg, Fe^{+2})_3TiO_2[BO_3]_2$ $(Mg, Fe^{+3}, Ti, Al)_2BO_4$[d] End Member, $Mg_{1.5}Ti_{0.5}BO_4$[g]	152.32; 22.85%; 4; 3.3–3.43; orthorhombic 135.21; 25.75%
217. Wawayandaite[d]	$Ca_{12}Mn_4^{+2}Be_{18}B_2Si_{12}O_{46}(OH, Cl)_{30}$	2744.43; 2.54%; —; 3.11; monoclinic
218. Werdingite[e]	$Al_{12}(Fe^{+2}, Mg)_2(Al, Fe^{+3})_2(Al, B)_2B_2Si_4O_{37}$[d]; $Al_{14}(Fe^{+2}, Mg)_2B_4Si_4O_{37}$ End Member, $Al_{14}Mg_2B_4Si_4O_{37}$	1205.46; 11.55%; —; 3.00–3.08; triclinic 1173.92; 11.86%
219. Wightmanite	$Mg_5[O(OH)_5BO_3] \cdot 2H_2O$ $10MgO \cdot B_2O_3 \cdot 9H_2O$ $Mg_{10}B_2O_{13} \cdot 9H_2O$	634.80; 10.97%; 5.5; 2.59–2.70; monoclinic (orthorhombic)[g]

continues

Table 1.1 *(continued)*

Name (Group; form)	Formula	Molecular weight[b]; %B_2O_3[c]; hardness; density; crystal system
220. Wiserite	$(Ca, Mg, Mn)_4[B_2O_5(OH, Cl)_4]$	365.63; 19.04%; —; 3.42–3.81
	$(Mg, Mn)_{14}(Si, Mg)B_8O_{22}(OH)_{10}Cl$[g]	1224.90; 22.74%; tetragonal
	End Member, $Mg_{0.5}Mn_{14}Si_{0.5}B_8O_{22}(OH)_{10}Cl$[g]	1439.33; 19.35%
221. Yuanfuliite[d] (Warwickite)	$Mg(Fe^{+3}, Al)BO_4$	140.53; 24.77%; —; 1.75–1.93
	End Member, $MgFe^{+3}BO_4$	154.96; 22.46%; orthorhombic
	Unnamed, not recorded[d,g]	
222. Unknown 1[d] (=Mineral X)	Mn-borate	—; 6; 2.78; hexagonal
223. Unknown 2[g] (Colemanite)	$Ca[B_3O_4(OH)_3]$	375.06; 55.69%; —; 2.54; triclinic
	$Ca_2B_6O_{11}\cdot 3H_2O$	
224. Unknown 3[d] (Metamict)	$Th_6B_8Si_7O_{38}$	2283.29; 12.20%; —; 4.12
225, 6, 7. UK53 (Kalsilite), and a, b	$NaBSiO_4$	125.88; 27.65%; —; 1.49; monoclinic; a, b hexagonal
228. UK48	$CaY_{2x}(Si, Be, B)_{44}(O, OH)_{10}\cdot 2H_2O$	(x = 1) 1121.61; 45.52%; —; 3.4
229. Unnamed (Hellandite)	$(Ca, RE, Th, U)(Al, Fe)(OH)_4 [Si_8B_8(OH)_4O_{40}]$	—
230. Unnamed (Gadolinite)	$(Ca, Y, RE)_2(Be, B)_2(Fe, Si)_{2.5}O_4(O, OH)_6$;	393.90; 8.84%; monoclinic
	End Member, $CaYBeFe_{0.5}Si_2BO_9(OH)$	

[a] Erd, 1980; Heller, 1986.

[b] When alternate elements occurred, equal amounts were assumed for molecular weight calculations. The lower formulas were used, with NBS 1989 atomic weights.

[c] %B may be obtained by multiplying by 0.3105708.

[d] Anovitz and Grew, 1996.

[e] Hawthorne, Burns, and Grice, 1996.

[f] Erd, 1980.

[g] Anovitz and Hemingway, 1996 (source of all End Member formulas).

[h] Malinko, Shashkin, and Yurkina, 1976.

[i] Mellini and Merlino, 1977.

[j] cf. = compare with, d. = dimorphous with, i. = isostructural with, p. = polymorphous with, r. = related to, t. = trimorphous with.

the intrusion of hydrothermal solutions (accompanying magma) into other formations, or the weathering or water entry and reaction at the outcrops, edges, or near faults of other borate deposits. Boron can form very different compounds by comparatively small changes in the solution's mineral content, concentration, pH, time to reach equilibrium, temperature, water availability or pressure. Thus, the very wide variety of borate minerals can be easily accepted, even if it is not known why one forms instead of another.

Boron is not a major element in the earth's crust and waters, averaging only 3–20 ppm in the upper continental crust as estimated by different authors (i.e., 15 ppm, 27th element in abundance; Anovitz and Grew, 1996). However, it is present as a minor component in many rocks and fluids, so the possibility of concentration in certain geologic situations is quite high. Being both mobile and volatile, boron tends to accumulate in the last phase of magmatic flows or eruptions. When such magma and its accompanying mineralized solution intrudes into carbonate formations (called a skarn), boron mineralization is common, and a few commercial deposits have resulted. When such fluids have intruded into other rocks, almost no commercial deposits have formed, but borates are a common minor component. When boron flowed to the surface into closed, arid basins, the world's large lacustrine borate deposits were produced.

Boron is also readily leached from other rocks by very hot water, creating by far the most common current boron source, surface water geothermal springs. They have formed most of the world's small borate deposits and perhaps two large ones, Inder and Searles Lake. It is assumed that magma fluid springs formed the larger deposits because they were accompanied by basalt flow and ash eruptions. The larger deposits are invariably interlayered, or in a matrix with tuff, and basalt is under (the most common), adjacent to (also very common), or over (occasionally) the borate beds. Among the smaller lacustrine borate deposits there is no basalt and usually no tuff, but the boron source is always inferred to have been geothermal springs.

1.1 BORATE CRYSTAL STRUCTURE

The crystal structure of borate minerals and synthetic borates has been rather thoroughly studied, providing some general "rules" about their structure (Christ and Clark, 1977). Crystallographers also have systematized the possible atomic arrangements of the borates and helped to explain the very large number of borate compounds (Hawthorne, Burns, and Grice, 1996). Unfortunately, these studies have done little to explain why the common borates are the preferred structures in nature. All borates contain combinations of the three- (triangular) or four-bond (tetragonal; negatively charged) B—O struc-

1. Monoborates

1: Δ, isolated or 1Δ[a] or Δ[a]
$[B(OH)_3]$

Boric acid

1: T, isolated or 1□[a] or ■[a]
$[B(OH)_4]^-$

Teepleite

2. Diborates

2: Δ, isolated
partially hydrated
$[B_2O_4(OH)]^{3-}$

or ▲—▲[a] or 1Δ1Δ[a] or 2Δ[a]

Szaibelyite
(Ascherite)

2: T, isolated
$[B_2O(OH)_6]^{2-}$

■--■[a] or 1□1□[a] or 2□[a]

Pinnoite

3. Triborates

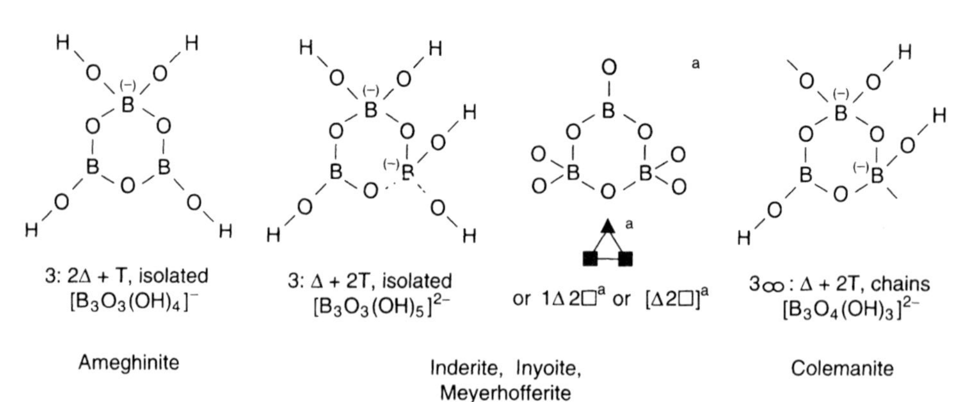

Figure 1.1 Examples of the first three groups of borates. (From [a]Hawthorne, Burns, and Grice, 1996; Heller, 1986; reproduced courtesy of Springer-Verlag, New York, Inc.)

tures (Fig. 1.1). Boron always maintains its valence state of +3, so the tetrahedral bonding has a negative charge and seeks attachment to a cation (which helps to make its structures more stable). Cations may occasionally replace hydrogen on some of boron's OH groups at very high temperatures (i.e., szaibelite). Hydrogen bonding is also especially strong with the borates, forming many hydrates, double salts and polymers. In all cases except for the comparatively few mono- and diborates, the B—O grouping tends to form one or more ring structures, allowing electrons to resonate around the ring, thus strengthening its bonds. A series of symbols have been assigned to help identify groupings: Δ for triangular (boron with 3 OH or O bonds) or T (or □) for tetragonal bonding (4 OH or O bonds); 1,2,3 for the number of boron atoms in a fundamental building block (FBB); isolated means that a unit is connected to another only by hydrogen bonds and not oxygen bridges; ⟨ and ⟩ means a ring structure; A and B mean the upper and lower rings, respectively; and the symbol ∞ is for chains, ∞_2 for sheets, and ∞_3 for solid network structures. A second number, α or β, or a Roman numeral at the end denotes isomers (Heller, 1986).

Sassolite (H_3BO_3) is thus noted 1:Δ isolated or one three-bonded boron, and written $[B(OH)_3]$. There is no single four-bonded, negatively charged tetrahedron mineral, but there are several double salts such as teepleite: 1:T isolated, written as $NaCl{\cdot}Na[B(OH)_4]$. NaCl is attached as if it were a hydrate, only much stronger. The diborate szaibelyite is a double salt with Mg_2OH^{3+}, and is 2:2Δ isolated, partially hydrated [meaning that three H's from OH groups have been ionized (removed)], $[B_2O_4(OH)]^{\equiv}$, $Mg_2OH[B_2O_4(OH)]$ or $Mg_2B_2O_5{\cdot}H_2O$. The 2:2T isolated $[B_2O(OH)_6]^{=}$ group appears to be particularly stabilized by Mg, such as pinnoite or $MgB_2O_4{\cdot}3H_2O$. The first of the ring structures occurs with the triborates, such as ameghinite, 3:2Δ + T, isolated $Na[B_3O_3(OH)_4]$, or $NaB_3O_5{\cdot}2H_2O$. Inderite is 3:Δ + 2T, $Mg[B_3O_3(OH)_5]{\cdot}5H_2O$ or $Mg_2B_6O_{11}{\cdot}15H_2O$. Colemanite is 3∞:Δ + 2T $Ca[B_3O_4(OH)_3]{\cdot}H_2O$, or $Ca_2B_6O_{11}{\cdot}5H_2O$. Borates through B_6 are common, and there are many higher borates, geometric combinations, polymers, isomers, and combined groupings (Fig. 1.2).

Simple variables such as temperature can significantly change the borate's structure, as with borax 4:2Δ + 2T, isolated, $Na_2[B_4O_5(OH)_4]{\cdot}8H_2O$ or $Na_2B_4O_7{\cdot}10H_2O$; and kernite 4∞:2Δ + 2T, chains, $Na_2[B_4O_6(OH)_2]{\cdot}3H_2O$ or $Na_2B_4O_7{\cdot}4H_2O$. Anhydrous borax is a $8\infty_3$ (5:3Δ + 2T, 3:2Δ + T) network. Tincalonite has the same structure as borax, with fewer molecules of H_2O (a higher temperature form). It is easier to crystallize from borax solutions and can form by a solid-phase dehydration of borax. The still higher temperature transformation to kernite takes more time to crystallize and is the only borax mineral formed in warm solutions with the longer residence times available in nature. Mayerhofferite changes from isolated to colemanite's chains to fabianite's sheets, whereas with Na-borates and

Triborate

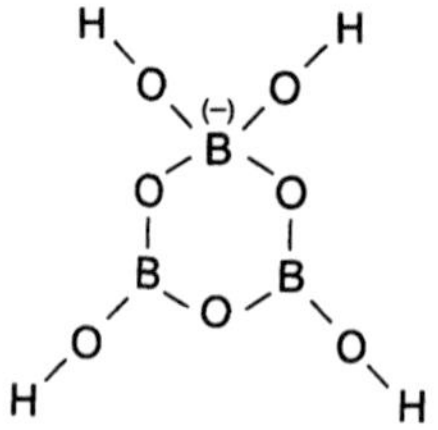

3: $2\Delta + T$, isolated
$[B_3O_3(OH)_4]^-$
Ameghinite
$Na_2B_6O_{10} \cdot 4H_2O$

Tetraborates

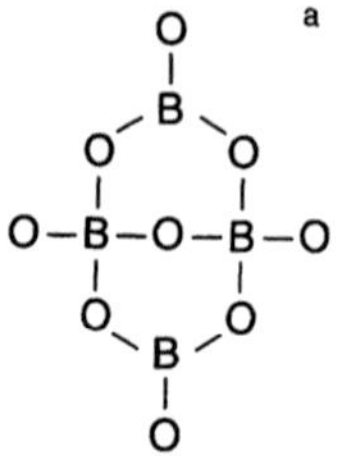

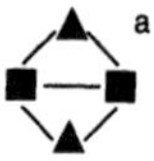

or $2\Delta 2\Box$[a]
or $<\Delta 2\Box>=<\Delta 2\Box>$[a]
Borax
$Na_2B_4O_7 \cdot 10H_2O$

and

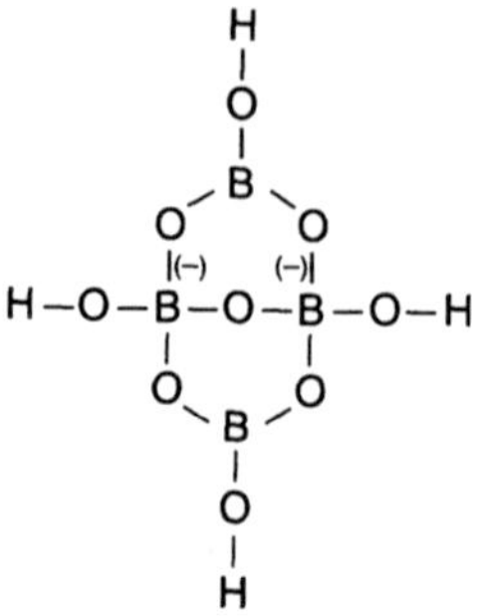

4 : $2\Delta + 2T$, isolated
$[B_4O_5(OH)_4]^{2-}$
Tincalonite
$Na_2B_4O_7 \cdot 5H_2O$

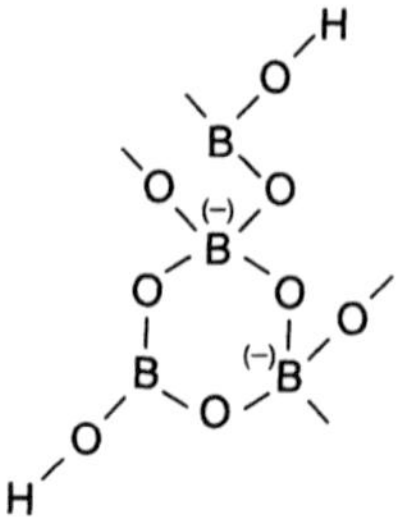

4∞: $2\Delta + 2T$, chains
$[B_4O_6(OH)_2]^{2-}$
Kernite
$Na_2B_4O_7 \cdot 4H_2O$

Pentaborates

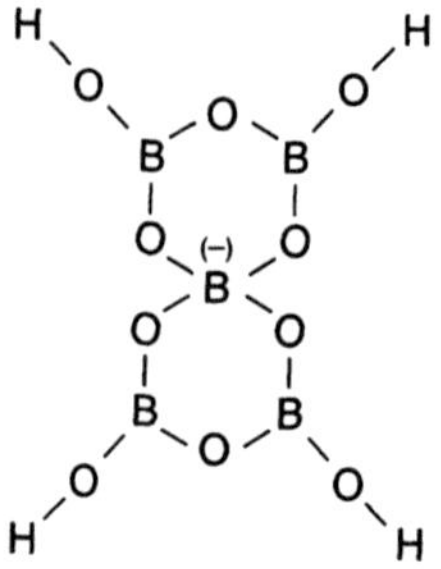

5: $4\Delta + T$, isolated
$[B_5O_6(OH)_4]^-$
Sporgite
$Na_2B_{10}O_{16} \cdot 10H_2O$

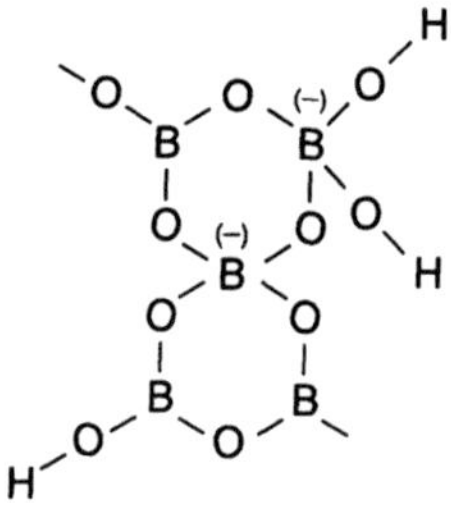

$5\infty 2$: $3\Delta + 2T$,
chains $A \neq B$
$[B_5O_7(OH)_3]^{2-}$
Ezcurrite
$Na_4B_{10}O_{17} \cdot 7H_2O$

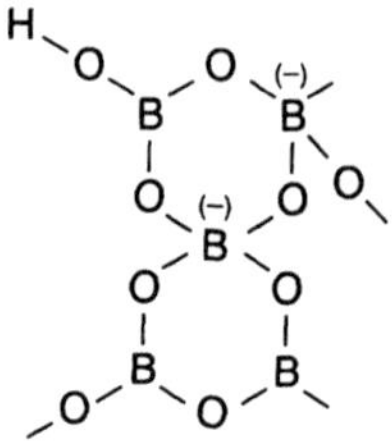

$5\infty 3$: $3\Delta + 2T$,
sheets $A \neq B$
$[B_5O_8(OH)]^{2-}$
Nasinite
$Na_4B_{10}O_{17} \cdot 5H_2O$

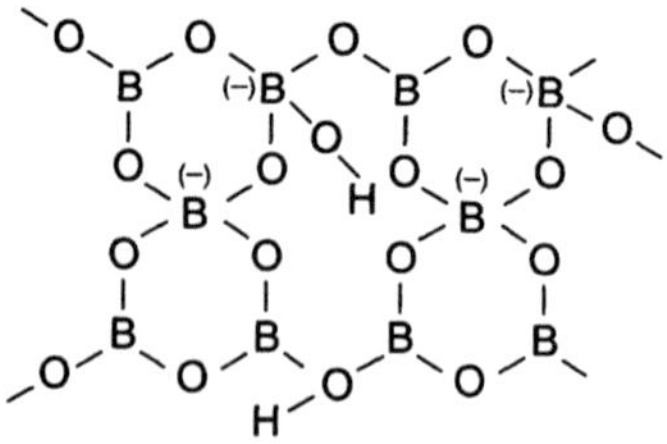

$5\infty 2$: $3\Delta + 2T$,
sheets, dimeric
$[B_{10}O_{16}(OH)_2]^{4-}$
Biringuccite
$Na_4B_{10}O_{17} \cdot 3H_2O$

different Δ and T's, sporgite changes to ezcurrite to nasinite to biringuccite, or from isolated to chains to sheets to dimere sheets. All of these changes are dehydration steps.

The borate crystal structures provide information to indicate why different forms are most stable at each solution composition and temperature (Fig. 1.3). The first mineral to crystallize in most of the large deposits has been colemanite, primarily from very dilute, warm calcium and boron waters. It is a divalent triborate with the simplest and smallest single-ring structure, giving it a strong resonant electron configuration. With the random motion of ions in a dilute solution, the positive divalent calcium would usually attract two negative monovalent borate tetrahedras, and then occasionally a neutral triangular borate ion would collide with this grouping to form the stable ring structure. Slightly elevated temperatures would result in a mild dehydration, causing the rings to combine as chains to form colemanite, which would then crystallize because of its very low solubility. As the solution became much more concentrated, the increased number of ions would cause commensurately more random grouping of ions to form increasingly complex rings. With medium-to-high sodium, and still some calcium, the next simplest structure would be ulexite's end-to-end rings. Sodium would aid calcium in stabilizing this trivalent grouping, and interfere with potential colemanite rings. Finally, when the boron was quite concentrated, with little calcium, the third simplest form, the back-to-back ring structure of borax would form.

The mechanism of polymerization for borates is by the elimination of water from the molecules, caused by the increased motion in warmer solutions (as the change from borax $Na_2[B_4O_5(OH)_4]{\cdot}8H_2O$ to kernite $Na_2[B_4O_6(OH)_2]{\cdot}3H_2O$). First, a bond in one of the rings is momentarily cleaved, allowing a hydroxyl group to combine with hydrogen from a nearby OH group to form water. The loose bonds from the cleaved ring and the OH group, and a similarly cleaved second borax molecule then instantaneously combine to form the polymeric chain structure. As this kernite crystallized four (instead of borax's ten) H_2O molecules would be loosely attached by hydrogen bonding. Sheets and networks are formed in a similar manner. With lesser temperature increases only the water of hydration is reduced, as with inorganic crystals. Even higher temperatures, and/or a change in the solution's concentration (often a dilution) may further cleave bonds in the molecule, reduce its water content, and form multiple ring structures. With extreme temperatures, hydrogen may cleave from some of the OH groups.

Figure 1.2 Crystal structure of several sodium borates. (From [a]Hawthorne, Burns, and Grice, 1996; Heller, 1986; reproduced courtesy of Springer-Verlag, New York, Inc.)

A. Low calcium, concentrated solutions (often high $CO_3^{=}$)

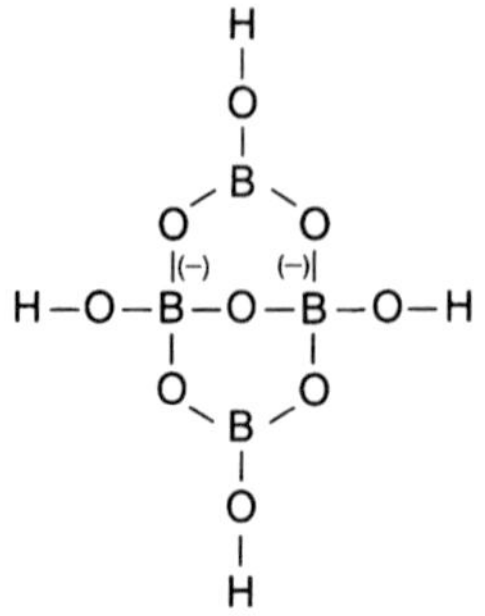

4 : 2Δ + 2T, isolated
$[B_4O_5(OH)_4]^{2-}$
1. Borax
$Na_2B_4O_7 \cdot 10H_2O$

(higher temperatures)

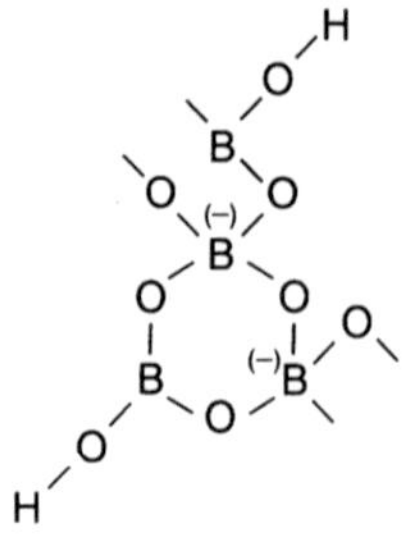

4∞: 2Δ + 2T, chains
$[B_4O_6(OH)_2]^{2-}$
2. Kernite
$Na_2B_4O_7 \cdot 4H_2O$

B. High sodium, medium calcium, more concentrated solutions

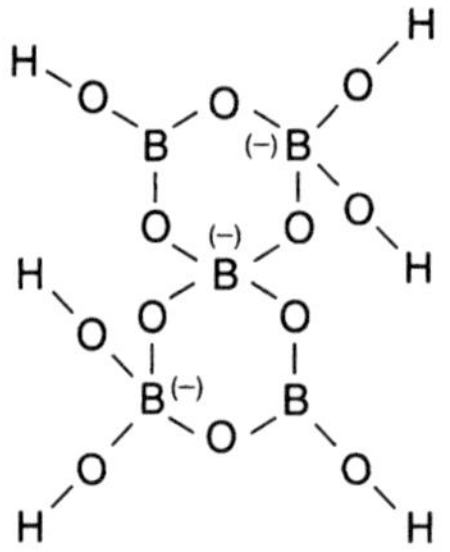

5: 2Δ + 3T, A=B, isolated
$[B_5O_6(OH)_6]^{3-}$
1. Ulexite
$NaCaB_5O_9 \cdot 8H_2O$

(higher temperature)

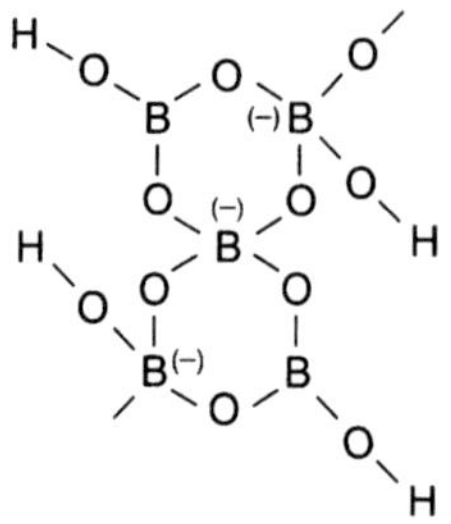

5∞ : 2Δ + 3T, A= B, chains
$[B_5O_7(OH)_4]^{3-}$
2. Probertite
$NaCaB_5O_9 \cdot 5H_2O$

C. Medium to high calcium, dilute solutions

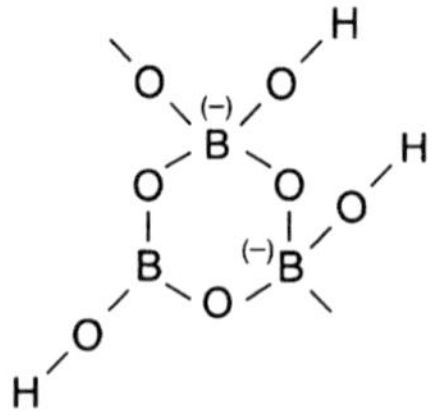

3∞: Δ + 2T, chains
$[B_3O_4(OH)_3]^{2-}$
1. Colemanite (higher temperature)
$Ca_2B_6O_{11} \cdot 5H_2O$

2. Inyoite (colder); $Ca_2B_6O_{11} \cdot 13H_2O$; structure as D2

D. Medium to high magnesium, dilute solutions

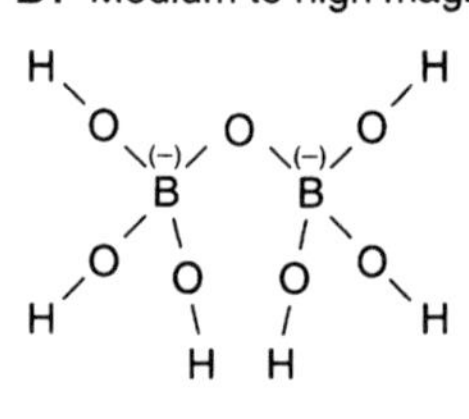

2: T, isolated
$[B_2O(OH)_6]^{2-}$
1. Pinnoite
$MgB_2O_4 \cdot 3H_2O$
(higher temperature)

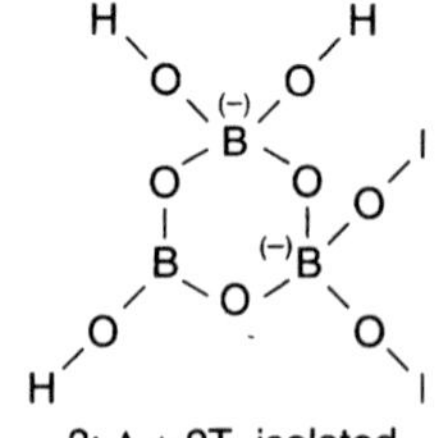

3: Δ + 2T, isolated
$[B_3O_3(OH)_5]^{2-}$
2. Inderite, Kurnakovit
$Mg_2B_6O_{11} \cdot 15H_2C$
(colder)

E. Medium calcium, magnesium, dilute solutions
1. Hydroboracite (structure as C1)

1.2 BORON PHASE CHEMISTRY

The phase chemistry of the borates is highly unusual because of these structural changes. Adding NaOH to a boric acid [($B[OH]_3$), H_3BO_3] solution starts with boric acid's solubility curve, but as the pH increases, sodium pentaborate ($5:4\Delta + T$, isolated [$B_5O_6(OH)_4^-$], $Na_2B_{10}O_{16}$) is formed with its comparatively high solubility. In the pH range of about 7–10, sodium tetraborate [$4:2\Delta + 2T$, isolated ($B_4O_5(OH)_4^=$), $Na_2B_4O_7$] is next formed with a more modest solubility, followed by the pH range greater than about 10 when sodium metaborate [1 : T, isolated ($B[OH]_4^-$)] is present with its much higher solubility. This means that over the pH range found in nature, there are four compounds (not including various hydrates and Suhr's and Auger's borates), and more than two maximums in the borate solubility curve. However, only the less soluble forms, boric acid and borax, are common minerals. The solubility of each species is fairly normally affected by temperature, and boric acid's solubility is increased by the presence of other salts in solution.

Many laboratory studies have been made to determine which ions are present in borate solutions, using primarily sodium hydroxide titration of boric acid, but also employing conductivity, freezing point lowering, pH determinations, ion exchange or chelation reactions, alcohol and diol distribution, and several other indicators. In addition to pH, the solution's concentration, the cations that are present, and many other factors influence which borate ions are present (Nies and Campbell, 1980). Figure 1.4 illustrates the borate polyions that might be present in a sodium perchlorate, 13.93g/lB_2O_3 solution, giving an excellent example of their complexity, but it does not predict the polyions present in industrial or natural deposit-forming waters.

Crystallization theory assumes that clusters of molecules in a solution randomly form and occasionally attach to each other as the solution approaches saturation. At saturation, the number of clusters that randomly collide reaches the point at which some are large enough to form a stable solid phase, or in other words, to nucleate and then grow on their own or other solid surfaces. If other nuclei, larger crystals, or even a compatible surface is present at saturation the random collisions with it result in some of them partially fitting into its lattice structure to obtain hydrogen-type bonding, and crystal growth begins. As the surface area of receptive sites increases, the supersaturation required for nucleation or growth decreases, and larger crystals are formed.

This model indicates that the clusters of molecules in the solution must be predominantly the same as found in the crystal. Of course, other groups also

Figure 1.3 The structure of the most common borates in sedimentary deposits, as a function of their brine's composition (From Heller, 1986; reproduced courtesy of Springer-Verlag, New York, Inc.)

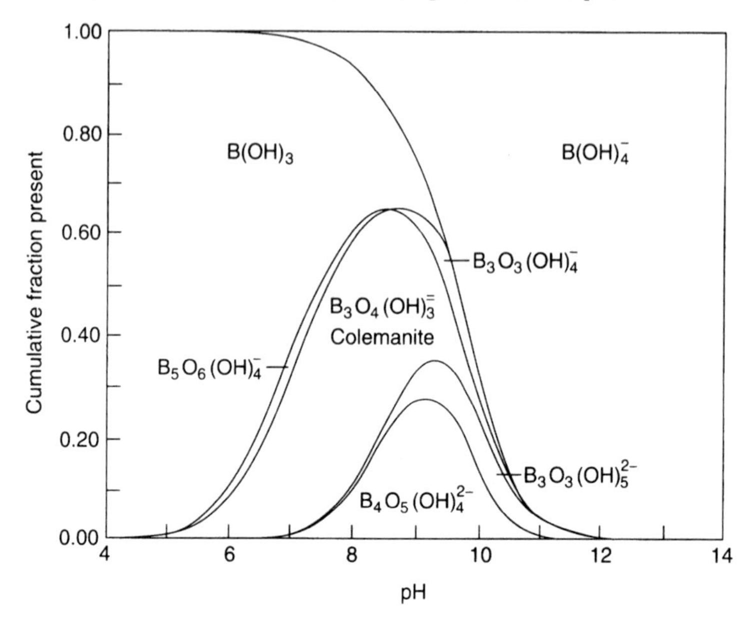

Figure 1.4 An estimate of the borate polyanions present in a sodium perchlorate–Borate solution (B_2O_3 = 13.93g/l). (From Ingri, 1963; reprinted by permission of Addison-Wesley Longman Ltd.)

	Crystalline Forms of Borate Ions:
$B(OH)_3$	Sassolite boric acid
$B(OH)_4^-$	Sodium metaborate tetrahydrate
$B_5O_6(OH)_4^-$	Sporgite, pentaborate
$B_3O_3(OH)_4^-$	Ameghinite
$B_4O_5(OH)_4^{-2}$	Borax, tincalonite
$B_3O_3(OH)_5^{-2}$	Inderite, inyoite, meyerhofferite, kurnakovite

could randomly hit a crystal surface, partially attach, and then rearrange their structure for a better fit with the available crystal sites, but this type of growth is much less probable. However, for the much later, extremely slow alteration of borates in deposits at outcrops, faults, or edges, the building of more complex molecules by changing the structure after their attachment to nuclei would be much more possible, and thus many different minerals could be formed. Their solutions also would contain far different companion ions, concentrations, and pH than the original depositing solutions. The ease of borate compounds to supersaturate in solutions is probably also caused by the increased concentration required for random collisions of smaller groups to form borates' more complex stable ring structures.

With all of the borates, hydrogen bonding (and to a lesser extent Van der Waals attraction) is particularly strong. It provides the force to hold the isolated, chain, and sheet borate structures together, as well as to attach the double salts and water of hydration. It also keeps the rings stable while they are cleaved and OH and H groups form water during dehydration, thus creating more complex structures. With some borates, the environment in which they were formed helps to explain their structures, such as teepleite $NaCl \cdot Na[B(OH)_4]$, which has been found in the NaCl-saturated, high-pH (10 to >11, and thus metaborate conditions), concentrated Searles Lake brines. Boracite $[MgCl \cdot Mg_2(B_7O_{13})]$ and its analogs have formed in concentrated $MgCl_2$ sea water end liquors, or brines, with the magnesium replaced by cations in the formation rocks (i.e., Ca from calcite: "dolomitization"). The high concentration of $MgCl_2$ (or $CaCl_2$, often >26%) and long residence times, even with a low boron concentration, has resulted in the multiring structure of the very insoluble boracites. Anhydrous or low-hydrate borates are formed under high temperature (perhaps from deep burial) and/or saturated brine conditions. In many cases the secondary borates formed from the conversion (weathering) of common borates during long residence times have an increasingly complex polymer structure, such as that of colemanite converting to priceite, gowerite, nobleite, and ginorite.

Ring structures can also form with borates and many organic compounds (in the laboratory, not in minerals), especially those with adjacent OH or groups that are similar (=O, $-C{=O \atop -OH}$, etc.). If the borate is first converted to boric acid in dilute solution, the polyol reaction is so strong that a hydrogen ion is liberated as the three-bonded boron structure is changed to a four-bonded ion. With the addition of a correctly configured polyol such as mannitol, a titration with dilute NaOH becomes the analytical method for determining borate concentrations. The equilibrium constants for two organic OH groups reacting is about 10,000, and 82,000 when four OH groups react.

1.3 SPECIFIC ORIGIN THEORIES

All of the large borate deposits were formed in one or more of three general solution concentration ranges. Because the colemanite and inderite groups of minerals are so insoluble (see Table 6.4), they may be assumed to have formed from very dilute solutions. This is further indicated by the freshwater organisms that are often present, and similar indicators in the deposits. The evaporation ratio for the geothermal brine borate source was probably only 0 to 10, or the boron concentration was 40–280 ppmB, and >15–110 ppmCa for colemanite to have crystallized (following, but similar to calcite deposition, and well before gypsum). For ulexite to crystallize, the boron concentration must be 460–630 ppmB, with >30 ppmCa, and the sodium concentration often 10% to saturated (based on present-day ulexite playas). For borax, the boron concentration must be much higher, >2100 ppmB, <20 ppmCa, and the NaCl and equivalent Na_2CO_3 concentrations (based on recent borax deposits) would need to be near their saturation points. These concentration differences greatly affect which borate anions are present in solution, with the more dilute solutions favoring the simpler structural forms.

1.3.1 Sodium Borates

1.3.1.1 Playa Surface Deposits

The mechanism that formed these deposits can be inferred quite reliably because somewhat similar deposits are currently forming, or recently have been formed. The deposits generally are, or have been, comparatively small, of recent or currently depositing age, and of only a low to medium purity. In the United States, the borax playas were Teels Marsh, Searles Lake, Fish Lake Marsh, Saline Valley, and portions of Rhodes Marsh, in that order of their borax content and importance. Many such playas also have been reported in Tibet. These deposits generally formed as efflorescent crusts from high-carbonate, low-boron springs that evaporated by capillary action, or from shallow short-lived sheets of brine. Material balances show that geothermal water formed the Searles Lake deposit, and in each of the other playas, the leaching of drainage basin rocks alone would not have formed a borax playa. In each basin except for Searles, most of the spring and runoff water entered the playa from aquifers, and most of the evaporation was by capillary action. Borax, being quite soluble, was one of the last salts to crystallize, but because all of the exposed water evaporated, there was always a mixture of salts in the crust. In the summer most of the borax crystallized from daytime evaporation, and in the evenings or winter, cooling was a factor in the shallow surface or interstitial brine.

The major salts crystallized with the borax were glauber salt ($Na_2SO_4 \cdot 10H_2O$) and natron ($Na_2CO_3 \cdot 10H_2O$) in the winter or evenings, and halite ($NaCl$), trona ($NaHCO_3 \cdot Na_2CO_3 \cdot 2H_2O$), and burkeite ($Na_2CO_3 \cdot 2Na_2SO_4$) in the summer. In most cases there was some segregation of the salts in various portions of the playa, forming areas with more concentrated, or even fairly pure individual salts. As the years went by, the surface crusts of borax-mixed salts accumulated, sometimes extending to a depth of 30–46 cm (12–18 in.). In all of the playas, there was some regrowth of the deposits from residual brines after the initial harvesting of the surface salts, but this amount was usually limited, because the borax input was small or cut off, and very little exists on the U.S. playas at present.

1.3.1.2 Playa Near-Surface Crystal Cluster Deposits

The formation of large clusters of borax crystals in the mud at a moderate depth has also been fairly common in modern carbonate–borax or Ca-depleted ulexite playas. Such deposits have been commercially worked at Borax Lake and Searles Lake, California, as well as Rhodes Marsh, Nevada, and have been observed in a number of playas in Argentina (Acazoque, Cauchari, Centenario, Lagunita, Lina Lari, Rincon, and Turi Lari). Again, the origin of these deposits is fairly easy to understand.

As low-calcium, boron-containing brines evaporate in shallow lakes during the summer, there is always some seepage of the lower, stronger brines into the underlying sediments. If the near-surface sediments are fairly porous, the strong brine will seep to a level of more impervious clay and be partially trapped (if it is fairly continuous over at least portions of the basin). The strong brine will slowly penetrate the more impervious clay, but at a very reduced rate. The strongest brine formed in the driest years displaces any weaker brine, always making this deep interstitial brine high in density and concentration. During the coldest winters, some borax would crystallize from this brine as the cold penetrated to its depth. Because borax is very sluggish to crystallize and form new nuclei, the preferred growth sites would be existing borax crystals, so they would grow to be quite large. For the winter cooling to be effective, these borax clusters generally were limited to a 1.2- to 2-m (4- to 6-ft) depth. If the brine was strong enough at intermediate depths, smaller crystals would also form there because the cooling would be greater. In the upper part of the lower impervious mud, there would be less brine and a reduced cooling effect, almost stopping nucleation and making its few crystals very large. In playas with fairly porous mud in much of its area (the normal case), the strong brine would more rapidly and continuously seep away, so that when the unusually cold winters occurred, only a few scattered borax crystals would form. Subsequent seepage, if not saturated with borax, would also be able to dissolve these crystals, limiting the occurrence of substantial borax crystal cluster in most borax playas.

1.3.1.3 Shallow Lake, Bedded Deposits

Among the world's currently depositing borax formations, only a few appear to be, or recently have been crystallizing massive bedded borax or kernite deposits. Several of the early literature reports on Tibetan borax lakes described the harvesting process as workers going into shallow lakes and digging "a kind of pavement under their feet," or "large masses which they later break into small pieces." Each lake reportedly had been harvested for a long time (perhaps centuries), and the beds of borax fairly quickly formed again after being removed. All were also fed by "foul-tasting" geothermal springs.

Kernite also has been depositing for a long time (>500 years, as estimated by current growth rates) over large <0.8-km^2 aprons from hot springs in the Puga Valley, Kashmir. It is ~44% pure and is forming at a rate of 550 tons/yr of equivalent pure kernite. Both the Tibet and Puga brines are of a strong carbonate type, and Puga's brines had a high temperature and boron to total dissolved salts (B/TDS) ratio. Most of its other salts, along with much of the boron, flowed off as a residual brine into a nearby river.

The purity of the large borax deposits implies that their geothermal brine source contained comparatively few salts other than boron, and/or that there was sufficient seepage for the other salts to escape before becoming saturated and crystallizing. Borax has a steep solubility curve (much less soluble at low temperatures), often allowing it to cool and crystallize in the winter before the other salts are saturated, and in the summer enabling it to stay in solution while other salts (such as NaCl) crystallize. Also, the NaCl and sulfate salts often segregate in separate areas of the lake (i.e., shallower, nearer their brine entry point, etc.). Because of this, a large amount of pure borax could be deposited in localized areas of a basin without the other salts crystallizing.

The near-surface borax clusters of Searles and Borax Lakes, Teels and Rhodes Marsh, and other playas are known to have formed from very concentrated multisalt brines, so a mechanism such as this is needed to explain their purity. Their geothermal brine sources had a reasonably high B/TDS ratio, but even so, it would appear that only brine seepage, separate depositional zones, periodic overflow, or other escape route for the nonborate salts can explain the borax's purity. It is known that this type of residual brine removal has happened in all of the world's potash and halite deposits (because of the absence of $MgCl_2$ and $MgSO_4$ salts; major ions in the original sea water), and it best explains the world's quite pure massive bedded trona deposits. Thus it apppears that this was a general phenomenon in the geology of most of the world's saline mineral deposits, and that it occurred with all of the large bedded borax deposits. Sanford and Wood (1991) described such seepage or flushing as a general phenomenon.

1.3.1.3.1 Tincalayu, Loma Blanca, Kirka, Boron

The mechanism for the origin of the four large buried borax deposits can only be theorized, and each one is slightly different. The *Tincalayu* deposit is the third largest, and also the simplest because it contains very few Ca- and Na,Ca-borate minerals, except for a small quantity of ulexite in the mudstone above the borax. It has a large underlying halite bed, but the NaCl appears to have had no bearing on the borax deposit because the basin could not possibly have been deep enough to accumulate borax as the halite was crystallizing. If present, the borax ultimately would have been coprecipitated, lost in the salt's interstitial brine, or seeped away. There also was no tuff with the salt, indicating that its input stopped just as the volcanic activity started and a high borate–carbonate spring opened (because the borax deposit has tuff interbeds and/or matrix, and some basalt). The presence of a high-carbonate spring is indicated by the absence of gypsum and Ca- or Na, Ca-borates with the borax, and by the presence of some searlesite (a final tuffaceous glass conversion product in an alkaline brine).

It appears that the massive borax beds were formed by the evaporation of the boron-containing geothermal brine in a shallow lake (Alonso and Robertson, 1992). As the brine cooled in the winters and evenings it crystallized borax beds, and occasionally the interbedded clay contained borax crystals of the same structure as that formed in the playas discussed earlier. Also, early in the depositional period at times of restricted runoff water inflow, pools of geothermal brine were hot enough to deposit small amounts of kernite in local areas. Conceivably, the kernite could have formed after deep burial when small zones of the deposit were penetrated by groundwater sufficiently hot to recrystallize some of the borax. However, there is a lack of deep burial indicators (doming, plastic flow, high pressure gas pockets or inclusions, anhydrite rather than gypsum over the deposit, etc.) or collapse structures above the kernite (because of its smaller volume than borax and the unsupported roof as the dissolving–recrystallizing took place). Also, the lack of bedding plane disruption and inlet or excess water (from the conversion) flow paths appear to make warm depositing conditions in geothermally fed brine pools a much more likely source of the kernite. Such hot, salts-crystallizing lakes are not uncommon at the present time. Furthermore, there almost certainly was a high $NaCl/Na_2CO_3$ concentration in the borax-crystallizing brine, and at $NaCl/Na_2CO_3$ saturation the kernite formation temperature is reduced to 37.5°C, a value very obtainable in modern, shallow brine lakes (even in the Puna region). If the kernite was formed by deep burial, a low-salts solution would then have been present, requiring >58°C for the conversion, or a $\sim$3000-m burial depth.

As the borax deposit reached its upper limit, the geothermal spring was cut off, and high-calcium runoff water (as now enters in one portion of the adjacent Salar de Hombre Muerto) mixed with the strong lake brine. This crystallized the ulexite nodules in the overlying mudstone, along with gypsum and occasional layers of inyoite, a combination that now occurs in the area's Salar de Surrie and Lagunita playas. It appears that the trace borate minerals were formed from postburial conversion by intruding groundwater. The other salts entering with the basin's water must have seeped away (or at the end were flushed out?), because the borax is so pure.

The *Kirka and Loma Blanca* deposits share the feature of the borax being layered between ulexite beds, with colemanite (in one case inyoite) beds usually before or after the ulexite. This would have required a quite different type of boron-containing geothermal brine–groundwater combination, with the carbonate content limited, and some calcium, but less than the stoichiometric amount of carbonate, bicarbonate, and borate. With this composition, calcite would crystallize first, followed by colemanite or inyoite. In periods of greater evaporation, the sodium concentration would increase, allowing ulexite to be crystallized. During periods of much greater evaporation, borax would next crystallize from the now Ca-depleted, more saturated brine until a larger

Ca-containing water supply again deposited ulexite, and later perhaps colemanite, causing the cycle to be repeated. This type of cycle is seen today in the small Lagunita playa and the ulexite–borax portion in at least seven Argentina playas.

At the *Loma Blanca* deposit, most of the minerals grew as nodules or disseminated crystals in the near surface mud of what must have been a small playa that only periodically became a lake. It now has older marine sediments on two of its sides, so very likely its basin's small drainage area had moderately high-Ca waters. This water would have at least partly seeped through the adjacent alluvial fans and entered the basin as an aquifer flowing upward through porous near-surface playa sediments. The ulexite, colemanite, and inyoite nodules would then have been formed as the descending or stationary boron-containing (with some carbonate) geothermal water met the rising Ca-water. The boron largely should have entered the playa at the surface as the springs rose through fault lines at the basin's edges (or elsewhere), and should have joined the smaller portion of the runoff water that also flowed onto the surface. Some calcite would precipitate, and some evaporation would occur, allowing a stronger boron-containing brine to seep downward through the porous surface sediments. Much of the surface water must have had a high enough geothermal content to remain hot (as now occurs at Lake Magadi with its 2 billion tons of crystallized trona and 50–86°C playa brine pools), so the borate that precipitated when the two brines met would have been colemanite. There is no colemanite phase data, but it is estimated that only >30–35°C were required for colemanite formation, values easily met in today's warm-weather solar ponds and playas. Also, colemanite is so insoluble and has such a favorable structure that it very likely crystallized metastably at much lower temperatures (as do several major potash salts). Because all borates are slow to nucleate, once the first crystals were deposited, they would become easier crystallizing sites, and thus preferentially grow.

The upward flow of Ca-rich water passing over the first crystals and meeting the descending (or stationary) boron-rich water on their surface, with clay also competing for surface sites, would cause the new growth to be layered, and thus form nodules instead of larger crystals. When the initial crystals formed as clusters, a geodal structure would tend to occur. A similar growth pattern at Loma Blanca would account for the inyoite (formed after the surface bine had cooled), and the ulexite, which formed in periods of moderately dry weather with less aquifer calcium and more concentrated seepage brine with a higher sodium and boron content. During periods of drier weather, there would be less aquifer brine and higher surface evaporation to allow winter cooling and subsurface borax crystal growth from the seepage brine. These weather cycles would then be repeated two more times for ulexite–colemanite, with one intermediate borax period. Several times during the inyoite and colemanite cycles there was enough runoff water entering the surface to allow bedded deposits to form on the then flooded playa. After the last inyoite cycle, the high-borate geothermal spring closed, and its place was taken by high-calcium springs that deposited travertine over part of the deposit.

An alternate formation mechanism could have the calcium coming from either the runoff or geothermal water, with the two waters mixing on the surface. Cooling could have caused the evaporated and, if supersaturated, seepage brine to crystallize and form the borate nodules. However, with this theory nothing explains the nodule form rather than crystal formation, and much more bedded Na-Ca or calcium borates on the lake's floor, or disseminated crystals in the mud would have been expected. Also, the nodule minerals have a relatively flat solubility curve, making significant supersaturation much less likely (note that when the crystals are disseminated, they are very small). These factors make it appear that a general requirement for nodule formation of any borate mineral is an upward rising flow of one reactant (usually the cation-containing solution) and a descending or stationary flow of the other (usually containing boron). Porous near-surface and/or aquifer strata and a semi-impermeable lower strata or similar structure also are required to allow passage and containment of the brines. As always, there must have been enough seepage laterally or down, or periodic overflow to allow the other minerals in basin's waters to escape.

The *Kirka* deposit is much larger than Loma Blanca, and was formed in a basin with broad, shallow margins and a smaller central zone where the borax crystallized. Also, in it much of the ulexite and colemanite formed as massive bedded deposits above, below, and at the sides of the borax zone. Apparently, as the borax and carbonate-containing geothermal springs first opened, they flowed into a shallow lake with a modest Ca content. They somewhat heated the lake, and as evaporation occurred, calcite first precipitated until colemanite became saturated, and then beds of colemanite were deposited. The climate later became more arid, the lake's size diminished, and as the brine's Ca content was reduced the sodium concentration increased. This caused ulexite to be deposited in massive beds over the colemanite. As the lake further evaporated, the Ca content became very low in the central zone (most of it precipitated in the margins where the runoff water entered the lake), the concentration of boron increased, and borax crystallized.

The lake periodically reflooded, with the stratified strong brine and the runoff water's mud layers protecting the borax from dissolving. Enough sodium (from the concentrated brine) mixed with the new runoff water so that only ulexite was deposited during this wetter period. Because of the much higher evaporation rate at the shallow shores and the higher Ca content from the entering groundwater near the shores, most of the ulexite was deposited in the outer areas, and only a small amount, or none, on the borax bed. This cycle was repeated one more time, and finally the borax spring was cut off. A much thicker layer of ulexite then formed over the borax and was finally replaced by a last layer of colemanite. Local small cooler zones in the original lake allowed some inyoite to form instead of colemanite, just as local areas in which the entering run-off water contained an excess of magnesium allowed some hydroboracite, and later inderite, to form. Postdepositional groundwater intrusion also resulted in limited zones of many other borates being formed by alteration of the primary borates. In this and all of the Turkish deposits, every study has indicated that the majority of the borax, ulexite, and colemanite are primary minerals.

The *Boron* (*Kramer*) deposit is similar to Kirka in that it is surrounded (on three sides; the Western Fault cuts off the south side) as well as over- and underlain by ulexite and colemanite, except that it does not have colemanite under it, nor does it have significant ulexite beds within it. It also has alternating zones of lower-grade borax consisting of crystals in a matrix of clay between the four beds of high-grade borax. It further has a significant amount of kernite in the Basal bed, and somewhat scattered amounts in the Lower and Middle borax beds. The lower ulexite-clay layer rests on a fairly thin barren clay zone deposited directly on a basalt flow which forms the basin for the deposit. The mechanism of formation would thus appear to be similar to that of the Kirka deposit, except that there was sufficient precipitation in the shallow shore areas so that colemanite did not extend to the borax zone, and the ulexite was also most concentrated above, below, and adjacent to the borax (and much less in the low-borate beds). The borax in the latter beds was in the form of individual crystals, indicating a mechanism similar to that for the crystal clusters in the mud of modern playas.

Because of kernite's somewhat random location and occasional fairly distinct bedding planes, it would appear to be most logically explained by random pools of fresh geothermal springs having kept those areas hot enough to crystallize kernite instead of borax, just as in the world's other large borax deposits. The great crystal size of some of the kernite is a natural consequence of kernite being slow to nucleate and grow, consequently requiring comparatively long crystallization times, which unavoidably produced many large crystals. In a similar manner, random hot zones would have allowed probertite to crystallize instead of ulexite, and with each mineral, some of the hot brine could occasionally seep through fractures and crystallize in them or in lower formations. This active geothermal effect is further evidenced by the fact that colemanite (a warm water borate) was the calcium borate formed.

It appears that the areas with fragmented and disrupted kernite bedding in the lower section of the deposit were produced by the partial collapse and disruption that must have occurred with the postburial leaching of the very lowest borate zone. Its unusual "sucrosic" appearance and very small crystal size probably indicates that it was dissolved long after burial, then later recrystallized rapidly (e.g., by tectonic movement shocking a supercooled solution) to form a quite different structure. While dissolved the roof over the brine pool would have been only partially supported, and some slumping and bed disruption of the overlying kernite would have occurred. Finally, some of the crack-filling kernite veins may be secondary, nucleated by the primary kernite. It appears that the many minor borates, as with most deposits, were formed as they became saturated (i.e., the As and Sb borates), in local isolated environments (i.e., searlesite), or from postdepositional alteration by intruding groundwater.

The prevailing theory for the kernite and probertite in the Boron deposit are that long after burial the deposit was deep and hot enough to convert borax to kernite. Because borax cannot be dehydrated to kernite by a solid phase transformation, it is implied that a water source leached the borax and recrystallized it as kernite. However, some of the kernite bedding planes are still intact, and some kernite is in discontinuous masses, so leaching and recrystallization would have needed to be accomplished, at least in part, on a crystal-by-crystal basis. Perhaps an occasional minute amount of occluded brine in the borax (none has been reported) or other formation water (again, none reported) could have been the water source to initiate this dissolving, slow recrystallization process. Presumably the excess water from the borax-to-kernite conversion remained under the lowest kernite as large pools of saturated borax (unsupported?), and when the deposit was again elevated and cooled, it somehow was transformed to sucrosic borax. Alternately, the excess water may have escaped, and some of the kernite later was leached to form the sucrosic borax.

There is no present-day unusual thermal gradient under the deposit (i.e., the mines were at 21°C, approximately the area's mean average temperature), so the normal 1.8°C/100 m (1°F/100 ft) of depth temperature increase would have required a burial depth of more than 2600 m [8600 ft; $(58 - 21) \times 100/1.8 =$ m] for kernite to be the stable phase in a borate solution at 58°C (it would be highly unlikely for much NaCl or other salts to be present in the hydrate or aquifer water to lower the transition temperature). Those favoring this theory claim a 5.4°C/100-m thermal gradient (870-m burial depth without any present-day basis for such a high heat flow). However, at either depth, plastic flow, doming, and other deep-burial indicators should be present, but there are none. Also, there should have been some slumping and caving of the now unsupported overlying low-grade bed during the long crystallization period (to form the large kernite crystals), as during solution mining, but again there is none.

There is also no surface evidence of any significant burial depth for the deposit. The basement rock, the Saddleback basalt outcrops nearby, and its weathering appear to have been modest (nothing like the claimed 900–2600 m), and evidence of the extensive disturbance that might have occurred with such deep burial and rising appears to be minimal or nonexistent. For the Saddleback Mountains to have stood still and the borax alone to have been lowered and raised is hard to imagine. There is now only an average of about 150 m (500 ft) of arkosic sediments above the deposit, and again no evidence of great surface weathering of thousands of meters of sediments. Also, the deposit itself is relatively undisturbed, with some vertical faults, but no sign of the folding, doming, mixing, modification, and so on that have accompanied all other deeply buried and raised saline deposits. Only modest subsidence, faulting, and displacement appear to have occurred.

As a second point, the kernite and probertite masses occur at different levels (the kernite's upper surface varies from 600–1000 m in depth) and locations in the deposit, requiring that if they formed during deep burial, the heat effect and leaching–recrystallizing were localized and spotty, which again seems unlikely. It is also highly improbable that severe bed tilting (and later partial leveling) could have put all of the kernite occurrences at the same lowest level during deep burial because they are so scattered and presently at such different elevations. Finally, there is a complete absence of any pathway for the entering conversion water to have reached the kernite (and the numerous isolated crystal masses), or for the excess dehydration water to have escaped from any of the kernite or probertite zones in this soft, impervious, clay-surrounded deposit. These problems appear to make the deep burial theory unlikely. Original warm depositing conditions (to form colemanite) and thermal pools to form the kernite and probertite, as in the current Rift Valley (depositing trona in high-temperature pools; Garrett, 1992) and Puga kernite deposit models, appear to be much more logical for this and for most of the borate deposits in which extensive amounts of higher-temperature minerals are present.

Boron isotopic data have shown that geothermal brines such as might have been the boron source of the world's large borate deposits have $\delta^{11}B$ values (a change in the $^{11}B/^{10}B$ isotopic ratio compared with that of kernite from the Kramer deposit) in the range of -10 to $+30$, with an average of about $+6$ (see Table 6.5). When much of the boron in the geothermal water crystallized (as can be assumed in the world's large borax deposits), there would have been little preferential ^{10}B crystallization, [normally there is about 2% greater reactivity for ^{10}B compared with ^{11}B (a $\Delta^{11}\delta B$ of 25; see Table 6.6)], and the average $\delta^{11}B$ value of the borax would have been nearly the same as that of the original geothermal brine: about -4 at Kirka, 0 at Boron, and $+5$ at Searles Lake. Now, if the colemanite was formed in these deposits as the first borate crystallized, because it would be only a small fraction of the total boron present, then considerable ^{10}B preferential crystallization would occur. It appears that this is what happened: Colemanite's $\delta^{11}B$ is about -13 at Kirka and -12 at Boron. The ulexite would next be crystallized, consuming the residual calcium, and again only part of the borax would be precipitated (but including the colemanite, the cumulative removal would be greater), so its

$\delta^{11}B$ should be somewhat less negative from the now slightly higher $\delta^{11}B$ geothermal waters. Again, that is what has happened: Ulexite's $\delta^{11}B$ is -7 at Kirka and -4 at Boron. If either of these minerals had been formed by replacing borax, they would have kept borax's $\delta^{11}B$ value, which they did not (unless much of the boron in the conversion brine escaped, which is very unlikely). Likewise, if colemanite replaced ulexite, it would have kept its $\delta^{11}B$ value, which again it did not. There is only a small difference between the kernite and borax $\delta^{11}B$ values, but the kernite's is smaller (Boron: borax $+0.1$, kernite 0.0; Kirka: borax -3.8, kernite -7.0), again indicating on the average a slightly earlier crystallization (with some preferential ^{10}B crystallization) and not a replacement reaction. This strongly supports the mechanism described earlier, and indicates that at least the bulk of the kernite, colemanite, and ulexite in the world's borax deposits is primary. The fact that the colemanite on the top of each deposit also has a low $\delta^{11}B$ indicates that only part of the borax in those waters crystallized. This suggests that at that stage much of the last remaining borax escaped, which is what had to have happened to discharge the other salts in the entering waters, since they did not crystallize.

1.3.2 Colemanite, Ulexite, and Probertite Deposits

The evolution of theories on the formation of the world's colemanite, ulexite, proberite, and kernite deposits represents an interesting example of how difficult it is to establish an origin theory for a family of minerals when the first deposits considered are complex, highly faulted, and altered (primarily those of Death Valley in this case). In an attempt to explain how these and similar deposits formed, Foshag (1921) suggested that all of the U.S. colemanite deposits originated by the initial deposition of ulexite (as was seen occurring in the U.S. and Puna region's playas), and that after deep burial it was transformed to colemanite or probertite by "intruding warm borax solutions." No explanation was given as to where these conversion liquors came from with such uniformity and regularity, how they totally and completely penetrated the deposits, what their (very large) calcium source was, or where their residual brines went. There was no consideration for several of the deposits' having fresh water features, and colemanite's low solubility (also indicating freshwater precipitation), while ulexite always crystallizes from NaCl-saturated brines. There also was no concern for the lack of bed disruption or slump structures above the deposit, nor questions why priceite, ginorite, nobleite, and gowerite (common colemanite alteration minerals) were not present.

Foshag's theory was widely accepted, primarily because all of the Death Valley deposits were highly fractured and appeared to be altered (somehow it was not considered that the borates in such highly faulted adjacent rocks should be fractured, with much opportunity for local alteration near the fractures). This theory was later supported by Christ and Garrels (1959), who added that kernite in the Kramer deposit was similarly transformed from borax by the heat of deep burial, and that probertite was produced there and in Death Valley from ulexite. Muessig (1959) also suggested that all of the world's lacustrian borate deposits were initially formed as the highest hydrates, and that they were later dehydrated by the heat of burial. These "conversion" theories were universally accepted, and many later authors added evidence to support them by describing obviously altered portions of the deposits or by quoting the ease of borate conversion suggested by hypothetical phase diagrams.

This thinking slowly began to change in the mid-1970s when students of the large, much less disturbed Turkish colemanite and colemanite–ulexite deposits found no evidence of conversion, and each major mineral appeared to be primary. In the mid-1980s the same thing occurred with the buried Argentine borate deposits, and by the mid-1990s even a few U.S. authors questioned the theory. Colemanite is by far the least soluble of the common commercial borates [as little as 42 ppmB (>52 ppmCa); see Table 6.4], so in warm temperature areas such as Turkey and the United States, with warm to very hot geothermal brine as the boron source, colemanite might be expected to be the primary borate deposited from dilute, low-sodium (<1000 ppmNa) waters. Its simple triborate (Δ2T), six-membered, divalent ring structure should also make it the easiest borate formed. The calcium could have come from either the geothermal or runoff water, but the boron had to be in the geothermal springs. Normally, most volcanic area runoff waters are quite dilute and low in sodium, whereas Ca-containing waters are formed from many rocks such as marine sediments. For many of the colemanite deposits it is likely that much of the calcium and boron came from different sources, as is required when the colemanite occurs as nodules or when the geothermal water is high in carbonates.

Many of the conversion proponents noted that colemanite has not been reported in any of the world's existing borate playas. However, two of the three extensive borate playa areas are in the very cold [3000- to 5000-m (13,000- to 15,000-ft)] elevations of the Andes or Himalayan mountains or plateaus, and the third playa area (the U.S.) has generally high-salt, and often high-carbonate brines. Also, there are almost no hot geothermal springs currently flowing into the borate playas. Only the high hydrates should form in the cold China and Puna regions, and ulexite should dominate in all of the high-salt (noncarbonate) playas. However, the partially dehydrated meyerhofferite has been found in Death Valley's southern playa. Some inyoite exists in the high-Ca Puna region, and its magnesium counterparts, kurnakovite and inderite (both $Mg_2B_6O_{11}\cdot 15H_2O$), are common in the high-$MgCl_2$ Chinese playas. A higher-temperature magnesium borate, pinnoite ($MgB_2O_4\cdot 3H_2O$), is found in some of these playas in warmer, concentrated lower-brine zones.

If any dilute-B and Ca, low-Na warm geothermal waters existed in warm-weather shallow lake playas in the United States or Turkey, it would be highly probable that colemanite would be forming in them today. (The southern Death Valley playa and Teels Marsh have areas with waters of the correct type, and probably contain some colemanite, but searches for it have never been made.) However, even though such warm geothermal lakes and playas may not exist now, the combination appears to have been common during some periods of high-boron springs and active volcanoes in very limited areas of Turkey, the United States, and Argentina. The mechanism of formation for the large buried colemanite deposits was thus basically similar to that for ulexite and other borates in modern playas, except for their nearly freshwater boron source and higher temperatures. Colemanite in its deposits has crystallized in three different massive forms: nodules, beds, and disseminated crystals, in that order of frequency. Nodules are quite common in many of the deposits, such as at Emet, Turkey. Massive beds are also common, such as in Death Valley. Disseminated crystals are more rare, but were dominant in the Calico "muds." The formation of beds and nodules was discussed previously, but perhaps when there was only a limited boron supply and the playas' permeability was very low, only small crystals could form, which were quickly buried in the mud.

All of the factors discussed on ulexite or colemanite nodule formation, except for different brine compositions, would apply equally to the other borates. Unfortunately, there is little data on groundwater and brine analyses, as well as sediment porosity for most of the world's borate playas in order to test this nodule formation theory. However, for two playas, Teels Marsh and Hombre Muerto, the mechanism is fairly apparent, and for Lagunita it can be inferred. Both Teels Marsh and Hombre Muerto have one limited area of high-Ca groundwater entering the playa, and in both the water sinks into the alluvial fan and enters the playa from below. This has led to the formation of extensive ulexite nodules in areas where the high-Ca groundwater enters or can easily flow. The local springs contain some boron that has been concentrated in the playa brine, which is also saturated with salt.

In the Lagunita playa, the small area with high-Ca water also enters through the alluvial fan, and is directly opposite the former geothermal springs. In this playa, a sequence of inyoite, bedded and nodular ulexite, and borax all have formed, apparently following the depositional model suggested earlier. The deposit also has areas in which salt and glauber salt have crystallized, following the common playa pattern of some salts crystallizing in segregated areas.

As several authors have mentioned, another indication of high-Ca and boron-containing water entering separately above and below the nodules is the presence of veins containing a brine feeding into the nodules from below. Most authors also have noted a compression of the clay surrounding the nodules, indicating that the nodules grew in place and progressively in size, with thin layers of clay between layers of crystals, or clay entrapped as the nodules grew. Thus something induced the nodule to form, and then grow in the near-surface of borate lake-playas for a long time. A descending or stationary high-borate water in the underlying muds, and an ascending calcium water that flowed over the nodule surface would meet this requirement. If the water from below (or above) was already saturated with colemanite or ulexite, it is difficult to visualize what was unique about the near-surface mud zone for the crystallization to occur only as nodules, and in that location.

The few authors who noted the playas' permeability also observed that the nodules always occurred in the porous zone just above the first less permeable layer through which groundwater would slowly be rising (assuming artesian aquifers for much of the playas' water input), or that there was a more porous zone (for ground water entry) just below or with the nodules. The low-permeability mud would help create a reservoir for the boron-containing brine seeping down from the surface and delay its seeping further and escaping from the playa. In at least two priceite nodular deposits, Ca-water intruding (rising) into an existing boron-containing rock formation (instead of a playa) has been indicated. The small Chectco, Oregon deposit lies under and in a decomposed serpentine bed. There were brine feeder veins under all of the priceite nodules, indicating a Ca-depositing brine flow from below. It would have met a boron solution formed by the leaching of serpentine, which permeated (and was descending through) the upper muds, causing priceite nodules to form (the reverse direction of these flows is also possible, but the two oppositely flowing reactants appear to be certain). In a similar manner, it appears that priceite and other borate nodules were formed from the boron liberated from anhydrite as it was hydrated to form gypsum in various marine deposits. In this case the ground water would have a high calcium content (from the gypsum), and the nodules would grow as the liberated boron flowed to their area. Nodule feeder veins also have been reported.

1.3.3 Marine Deposits

Several marine potash deposits, such as the Zeichstein (each of the Z1–Z4 periods), and some of the Russian deposits have abnormally high boron contents, with numerous borate minerals. When the potash ore was processed by being completely dissolved, the borates remained behind in the mud residues, and at one time some of this borate (in Germany, primarily boracite, $Mg_3B_7O_{13}Cl$) was recovered as a small-tonnage by-product. The average boron content in all of the marine deposits is 20 ppm in gypsum/anhydrite, 10 ppm in halite, and 33 ppm in potash salts (see Table 6.1; Kitano, Okumura, and Idogaki, 1978). These very low values have been confirmed by experimental data, thus indicating an outside boron source for those marine deposits in which larger amounts of borates have been found. High-boron geothermal waters must have later penetrated or flowed into the evaporating seawater in these deposits. Borate concentration by leaching and reprecipitation of deposited formations, or very concentrated brine crystallization, however, could explain localized high-borate zones.

The only large marine borate deposit is the Inder formation in Russia, which was fairly obviously formed by intruding high-borate geothermal springs. Its borates are in localized areas, primarily in the overlying gypsum of a very large salt dome, occurring as 2–50% B_2O_3, and often ~20%B_2O_3 ore. The total boron in the deposit is 10^6–10^{14} more than could have entered in seawater, based on the mass of underlying halite. The localized high-B_2O_3 ore and the large number of borate minerals also help to indicate a later-intruding geothermal spring. An example of the geothermal water being added to an evaporating potash-forming brine is the high boracite content (1.5–8.5% B_2O_3) in the sylvinite and carnallite of the Thailand potash deposit (Garrett, 1995). The borate content again is far greater than from seawater alone, and the presence of massive quantities of tachyhydrate ($2MgCl_2 \cdot CaCl_2 \cdot 12H_2O$) required the presence of hot geothermal brine pools during its evaporative crystallization.

Other examples of geothermal water in marine deposits can be cited. In the sylvinite of the Carlsbad potash deposit, lueneburgite, $Mg_3(PO_4)_2 B_2O_3 \cdot 8H_2O$, is present. Both the phosphate and boron are out of proportion to seawater's supply, and there also is a large excess of boron in the deposit's underlying end liquor. In the Cane Creek (Moab, Utah) potash deposit, braitschite ($[Ca, Na_2]_6RE_2B_{24}O_{45} \cdot 6H_2O$) occurs in the halite, sylvinite, dolomite, and anhydrite of the upper potash and its overlying rock. The rare earth content far exceeds that in seawater, and the rare earth distribution is similar to that found in adjacent rocks. Again, it appears that a geothermal source would have been necessary. In the German potash deposits an amazingly large variety of borate minerals (>30) have been found in the calcite, dolomite, gypsum, anhydrite, hartsalz, sylvinite, carnallite, and all of the other nearby formations. The borates occur primarily along fault, shear, or other openings in the formation, and much of the potash has been transformed, some to high-temperature forms (e.g., langbeinite). It almost inescapably appears that geothermal waters entered the formations, depositing borates and altering the potash. Despite these examples, very few potash deposits have any borate minerals.

The numerous borate minerals (>21) found in the gypsum/anhydrite, halite, or sylvinite of various other marine formations (e.g., those in Newfoundland, Nova Scotia, U.S. Gulf Coast, Oklahoma, etc.) offer a somewhat similar but much less certain case for geothermal brines. In some cases the borate in the halite or gypsum is not necessarily out of proportion to that entering in seawater, but such deposits are rare, and the boron concentration in the evaporated seawater should have only been <32 ppm when halite started crystallizing, and <195 ppm when potash began to deposit (see Table 6.2). The $MgCl_2$ concentration also would have been quite low, with both values much less than required for boracite precipitation, at least in the first 70% or so of the halite formation. However, boracite has been found in the lower cores, cuttings, and salt-dissolved muds of a few halite deposits, along with some calcium and silicate borates. These latter salts would have needed to be alteration products or from nonseawater brines because there is very little Ca or SiO_2 in highly concentrated seawater. Each of these factors tends to indicate some geothermal brine addition or later penetration into the halite deposits containing borates.

The origin of the few marine gypsum–anhydrite deposits with borates is equally uncertain. Here again, the amount of boron is far out of proportion to the boron in the early-stage gypsum deposits, but not necessarily disproportionate to the gypsum deposited after the halite. Also, in each case the borates were found in gypsum that recently had been converted from anhydrite. If the anhydrite were leached of its boron content (the newly formed gypsum is very low in B), and the boron reprecipitated in restricted areas (it is generally only found near fault lines), then it would be concentrated. Its high $\delta^{11}B$ value, like that of seawater, would tend to support this. However, the amount of borates is often larger than would be expected, and borates are found only in a few of the anhydrite to gypsum formations. These factors again would tend to indicate the addition of high-B fluid in at least some of these occurrences.

1.3.4 Location of Borate Deposits

There are only a few areas of the world with large borate reserves, and the reason for these locations being favorable sites is totally unknown. It has been suggested that they are all in the uplifted side of rift subduction zones, as shown in Fig. 1.5. This criteria fits the Himalayan Mountain reserves quite well because there are borax hot springs and playas for much of its length (primarily along the eastern edges of its plateau). The Andes in South America also are a rift-zone uplift, but here the borate area is very limited, covering only ~10% of the Ande's length. The east side of United States' rift-uplifted Sierra Nevada mountains (or the Coastal Range) also has formed deposits, but again in a very limited area [Searles Lake, Lang, Frazier, Boron, Alvord Valley, and Chetco (west side)]. However, the main borate deposit line runs 340 km perpendicular to the Sierras (Lang-Frazier, Boron, Four Corners, Calico–Death Valley). A second line is also about the same length perpendicular to the Sierras [Borax Lake (west side), Owens Lake, Saline Valley, Nevada playas–Muddy Mountains]. Some of the Turkish and Yugoslav deposits are near the Red Sea–Aegean rift, but again most of them are on a 300-km line perpendicular to the rift (Bigadic, Emet, and Kirka). There is no "uplifted side" or large mountains near the deposits. The Chinese and Russian skarn

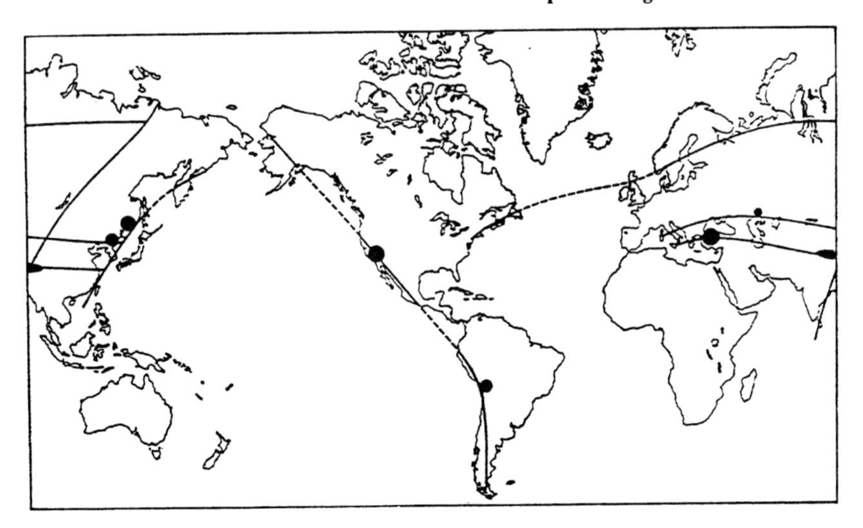

Figure 1.5 Suggested uplifted rift-zone correlation of borate deposits. (Adapted from Ozol, 1977; Morgan, 1980; reprinted by permission of *Int. Geol. Rev.* **20**(6), 692–698. ©V.H. Winston & Sons, Inc., 360 S. Ocean Blvd., Palm Beach, FL 33480. All rights reserved. Also reprinted by permission of Addison-Wesley Longman Ltd.)

deposits are apparently related to ancient rift basins, but not necessarily the uplifted portion. The other deposits are widely scattered.

The uplifted rift mountain theory for borate deposits thus appears to be far too general, and does little to explain their presence. Very few rift areas have borate deposits, and many deposits appear to be far from rifts. What is common to the deposits, however, is that all appear to have been formed by geothermal springs, and the large bedded, buried deposits in areas with active volcanoes (basalt and tuff). In the latter case, it appears that the high-boron geothermal fluid accompanied magma flow, either directly as a skarn or from the surface as springs flowing into closed basins with an arid climate. What differentiated the borate deposit-forming volcano, magma flow, and hot spring areas from the extremely large number of others in the world without borax is very hard to determine. For skarns, it is felt that a fairly long near-surface flow path or stationary time allowing boron to separate into a boron-rich fluid was necessary. Such conditions in the Himalayan Plateau appear to account for that area's very long-lasting and frequent high-boron geothermal springs, but that is the only area for which such conditions have been determined (although it appears similar in surface detail to the Puna region). There have been few studies on the types of basalt or their composition connected with borate deposits, or on the types of tuff. If the prior subduction of high-boron sediments or rocks is a factor, there appears to be no method of detecting this.

With the smaller deposits (and Searles Lake), the only essential factor in developing a lacustrine borate deposit is the presence of a long-lasting, fairly normal high-boron geothermal spring (such as the many hundreds known today) adjacent to an arid-climate closed basin with the correct balance of basin leakage or overflow to allow a borate deposit to form. The spring must have lasted long enough for the boron to accumulate and concentrate, often thousands of years. It appears that most borate deposits also occasionally varied from being shallow lakes to playas. Most of the small borate playas that exist today probably would not have survived burial, being too small to have withstood the mud-covering process to bury and protect them.

Springs that lasted for very long periods (i.e., the upper salts at Searles Lake, 40,000 years) could eventually form a large deposit. Searles Lake, however, had very limited brine seepage, so a mixed salt deposit was formed. The more general case would have more leakage (as have almost all of the world's other large saline mineral deposits of halite, potash, trona, and sodium sulfate), and with a high-boron spring, pure colemanite, borax, or mixed Na–Ca borate deposits could have formed. Thus, the requirements for a large borate deposit may well have been (a) a long-lasting high-boron geothermal spring, which required a long volcanic period or a continuing high underlying magma temperature (as at the Puga kernite deposit); (b) a closed basin with an arid climate, ample basin depth (or subsidence) for a thick accumulation of salts, and often broad shallow margins with a deep center; and (c) sufficient seepage and/or periodic flushing to remove the nonborate salts.

References

Alonso, R. N., and Robertson, D. B. (1992). La genesis de kernita en los yacimentos de borax. In *Instituto Recursos Minerales, Universidad Nacional de la Plata,* No. 2, pp. 1–8.

Anovitz, L. M., and Grew, E. S. (1996). An introduction. In *Mineralogy, Petrology and Geochemistry of Boron.* Mineral. Soc. Amer., Reviews in Mineralogy, Vol. 33, Ch. 1, pp. 1–40.

Anovitz, L. M., and Hemingway, B. S. (1996). Thermodynamics of boron minerals: Summary of structural, volumetric and thermochemical data. In *Mineralogy, Petrology and Geochemistry of Boron,* (F. S. Grew and L. M. Anovitz, eds.), Vol. 33, Ch. 5, pp. 181–262, Mineral. Soc. Amer., Reviews in Mineralogy.

Christ, C. L., and Clark, J. R. (1977). A crystal–chemical classification of borate structures with emphasis on hydrated borates. *Phys. Chem. Min.* **2,** 59–87.

Christ, C. L., and Garrels, R. M. (1959). Relations among sodium borate hydrates at the Kramer deposit, Boron, California. *Am. J. Sci.* **257,** 516–528.

Erd, R. C. (1980). The minerals of boron. In *Mellor's Inorganic and Theoretical Chemistry,* Vol. 5, Pt. A, pp. 7–71. Longman, New York.

Foshag, W. F. (1921 April–May). The origin of the colemanite deposits of California. In *Econ. Geol.* **16**(3), 199–214.

Frondel, C., and Morgan, V. (1956). Inderite and gerstleyite from the Kramer Borate District, Kern County, California. *Am. Min.* **41,** 839–843.

Garrett, D. E. (1992). *Natural Soda Ash; Occurrences, Processing and Use.* Van Nostrand Reinhold, New York.

Garrett, D. E. (1995). *Potash: Deposits, Processing, Properties and Uses.* Chapman & Hall, London.

Hawthorne, F. C., Burns, P. C., and Grice, J. D. (1996). The crystal chemistry of boron. In *Boron Mineralogy, Petrology and Geology* (E. S. Grew and L. M. Anovitz, eds.) Vol. 33, Ch. 2, pp. 41–116, Mineral Soc. Amer., MSA Reviews in Mineralogy.

Hay, R. L. (1966). *Zeolites and Zeolitic Reactions in Sedimentary Rocks.* Geol. Soc. America Special Paper 85.

Heller, G. (1986). *A Survey of Structural Types of Borates and Polyborates,* No. 131, pp. 39–98, Topics Current Chemistry.

Helvaci, C. (1996). Personal communication.

Helvaci, C., Stamatakis, M. G., Zagouroglou, C., and Kanaris, J. (1993). Borate minerals and related authigenic silicates in northeastern Mediterranean Late Miocene continental basins. *Explor. Mining Geol.* **2**(2), 171–178.

Ingri, N. (1962). Equilibrium studies on borate polyanions. *Acta Scand.* **16**(2), 439–448.

Ingri, N. (1963). Summary of borate polyion studies. *Svensk. Kem. Tidskr.* **75**(4), 199–210.

Kitano, Y., Okumura, M., and Idogaki, M. (1978). Coprecipitation of borate–boron with calcium carbonate. *Geochem. J.* **2,** 183–189.

Malinko, S. V., Sashkin, D. P., and Yurkina, K. V. (1976). Fedorovskite: A new mineral of the roweite-fedorovskite isomorphic series. *Zap. Vses. Mineral. O-va,* **105**(1), 71–85.

Mellini, M., and Merlino, S. (1977). Hellandite: A new type of silicoborate chain. *Am. Min.* **62**(1–2), 89–99.

Morgan, V. (1980). Boron geochemistry. In *Inorganic and Theoretical Chemistry*, Vol. 5, Sec. A2, p. 112, Longman, New York.

Muessig, S. (1959). Primary borates in playa deposits: Minerals of high hydration. *Econ. Geol.* **54,** 495–501.

Nies, N. P. (1980). "Alkali Metal Borates: Physical and Chemical Properties." In *Inorganic and Theoretical Chemistry,* Vol. 5, Sec. A9, pp. 342–501. Longman, New York.

Nies, N. P., and Campbell, G. W. (1980). Inorganic boron-oxygen chemistry. In *Inorganic and Theoretical Chemistry,* Vol. 5, Ch. 5, pp. 53–231, Longman, London.

Ozol, A. A. (1977). Plate tectonics and the process of volconogenic–sedimentary formation of boron. *Int. Geol. Rev.* **20**(6), 692–698.

Sanford, W. E., and Wood, W. W. (1991 September). Brine evolution and mineral deposition in hydrologically open evaporite basins. *Am. J. Sci.* **291**(7), 687–710.

Chapter 2 | Borax and Sassolite Deposits

2.1 BORAX

2.1.1 Argentina

The Puna region (Fig. 2.1) in the South American high Andes is a major borate province. It includes the northwestern corner of Argentina, the southwestern corner of Bolivia, the southeastern corner of Peru, and the northeastern edge of Chile. The total area encompasses 200,000 km^2, 45,000 km^2 of which are in Argentina. There are numerous playa borate deposits in each country, but only Argentina has buried deposits, with reserves of 100 million metric tons of ore. The region has desert alluvial plains and dry lakes at altitudes of 3500–4000 m, and numerous volcanos up to 6,000 m high. It is bounded by the volcanoes of the Cordillera Occidental on the west and the much lower Cordillera Oriental on the east. The region was formed 15 million years before the present (Mybp), but the borates were deposited much later, first in the Upper Miocene epoch of the Tertiary period (6 $\pm$ 1 Mybp), then in the Lower Pleistocene epoch (1.5 Mybp), and finally in the Quaternary Holocene period, or comparatively recently (Alonso, 1986; Alonso, Jordan, Tabbutt, and Vandervoort, 1991).

2.1.1.1 Loma Blanca

The 14-km-Loma Blanca deposit is located 70 km southeast of the northwestern tip of Argentina, 180 km northwest of Jujuy, and 10 km southwest of Coranzuli, at an elevation of 4150 m. Its age is 6.99 million years (My), and it has 20 million metric tons of reserves at 13.5% B_2O_3. The deposit is in the 250- to 300-m-thick Sijes formation, with 30 m of borates interlayered with greenish or gray tuffites and claystones. Below it are 113 m of green lacustrine sediments, and above it a travertine bed, followed by 107 m of pyroclastic rocks. The deposit contains predominantly borax ($Na_2B_4O_7 \cdot 10H_2O$), lesser amounts of inyoite ($Ca_2B_6O_{11} \cdot 13H_2O$) and ulexite ($NaCaB_5O_9 \cdot 8H_2O$), and minor amounts of colemanite ($Ca_2B_6O_{11} \cdot 5H_2O$; Table 2.1). Some kernite has also been reported (Solis, 1996). The ore zone is quite complex, aligned northeast–southwest and bordered by shear faults (Fig. 2.2). Most of it is nearly horizontal, but it can dip <20° near faults or where mild plastic flow and doming have formed gentle synclines and anticlines. The borates may occur as disseminated crystals

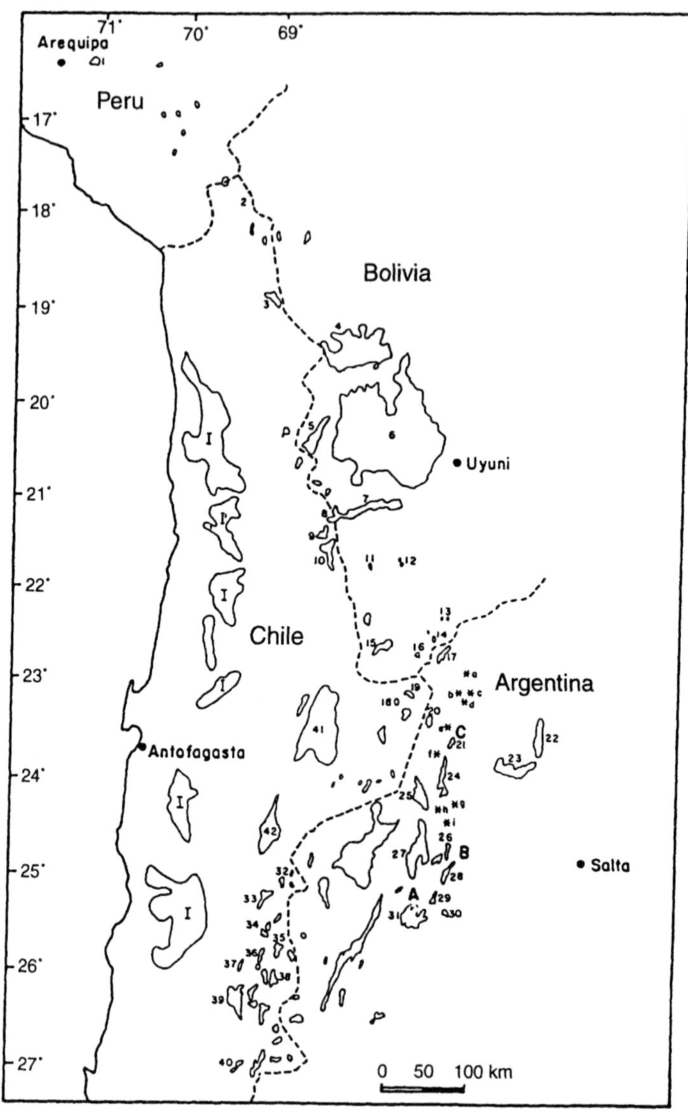

The *Roman* numeral (*I*) indicates nitrate salars with borates in the Pampa del Tamarugal region (North of Chile). *Arabic numbers* indicate the salars with recent borate deposits in Peru, Bolivia, Chile and Argentina. They are: *1* Lagune Salinas; *2* Cosapilla; *3* Chilcaya; *4* Coipasa; *5* Empexa; *6* Uyuni; *7* Chiguana; *8* Ollague; *9* Carcote; *10* Ascotan; *11* Pastos Grandes; *12* Capina; *13* Mamacoma; *14* Curuto; *15* Chalviri; *16* Luriques; *17* Laguna Vilama; *18* Zenobia; *19* Dioloque; *20* Jama; *21* Olaroz; *22* Laguna Guayatayoc; *23* Salinas Grandes; *24* Cauchari; *25* Rincón; *26* Pastos Grandes; *27* Pozuelos; *28* Centenario; *29* Ratones; *30* Diablillos; *31* Hombre Muerto; *32* Aguas Calientes; *33* Pajonales; *34* Aguas Amargas; *35* De la Isla; *36* Aguilar; *37* Infieles; *38* Lagunas Bravas; *39* Pedernales; *40* Maricunga; *41* Atacama; *42* Punta Negra. *Small letters* indicate the places with geysers and hot spring borate deposits such as *a* Coyahuaima; *b* Arituzar; *c* San Marcos; *d* Volcancito; *e* Libertad; *f* Tropapete; *g* Antuco; *h* Socacastro; *i* Blanca Lila. *Capital letters* indicate the main Late Miocene borate deposits: *A* Tincalayu; *B* Sijes; *C* Loma Blanca

Figure 2.1 Distribution of principal borate localities in the Andes. (From Alonso and Viramonte, 1990; reprinted by permission of Springer-Verlag Gmbh & Co. KG.)

Table 2.1

Chemical Analyses of the Different Ore Types at Loma Blanca[a] (wt%)[b]

	Ulexite			Inyoite		Borax
B_2O_3	14.24	12.28	10.75	16.58	23.19	19.49
CaO	6.06	4.80	6.82	6.87	6.82	6.82
Na_2O	3.42	3.29	3.72	2.42	2.36	7.32
Cl	1.02	0.28	0.22	0.29	0.36	0.22
Fe	0.05	0.10	0.15	0.04	0.02	0.15
Insoluble	66.07	60.04	66.38	53.66	34.74	66.38

[a] Each ore type is in calcitic, tuffitic mudstones.
[b] Alonso, Helvaci, Sureda, and Viramonte, 1988.

in claystone, mudstone, and tuffite (Fig. 2.3); as thin layers interbedded with these rocks; as small nodules with radiating, cauliflower, or "cotton ball" texture; or as crystals in vein and vug fillings.

The borax crystals are usually colorless and transparent, but in some places they are green, gray, or pale brown due to fine inclusions. The 0.5–4 cm (most commonly 1–2 cm) euhedral or subhedral crystals are similar to those recently formed in seven Argentine salars (Alonso, 1986). The borax usually occurs as disseminated crystals in the clay–tuffite matrix, with a 2:1 matrix:borax ratio. The 3-m lower and 1-m upper borax beds contain essentially no other borates, but above and below the beds are layers in which borax occurs with ulexite. Also, when borax outcrops or is near the surface, it often has been changed to ulexite or calcite. (There are calcite pseudomorphs of borax.) The four ulexite beds above and below each borax bed, in ascending order, are 3, 1, 0.75 and 0.5 m thick, and contain cauliflower or cotton ball nodules in the clay–tuffite matrix. Some randomly oriented 0.5- to 1-mm fibrous crystals grow on the nodules, and the ratio of ulexite to the matrix is from 1:1 to 2:1, or ~15% B_2O_3. The very soft ulexite is usually gray from clay impurities, although occasionally it is pure white. In the upper part of the deposit some ulexite appears to have been recrystallized, and occasionally it forms pseudomorphs of borax.

Inyoite occurs in 2 m (lower), and 2 or 3 m (upper) thick beds, composed of 0.1- to 0.4-m-thick sub beds of clear, gray, or brown tabular or intergrown <10 cm groups of crystals or nonoriented crystal aggregates. There are some veins of massive inyoite and parallel groups of thin tabular clear, colorless, perfectly shaped euhedral crystals, usually with ulexite.

Colemanite occurs in the lowest part of the deposit as small nodules, with radiating crystals in 2-mm to 3-cm geodes, or continuous <20-cm beds. The crystals in each form are usually colorless and transparent, or occasionally gray. The smaller nodules are spherical, and the larger ones ovoid. Some contain vugs, and others have a core of coarse crystals that occasionally radiate

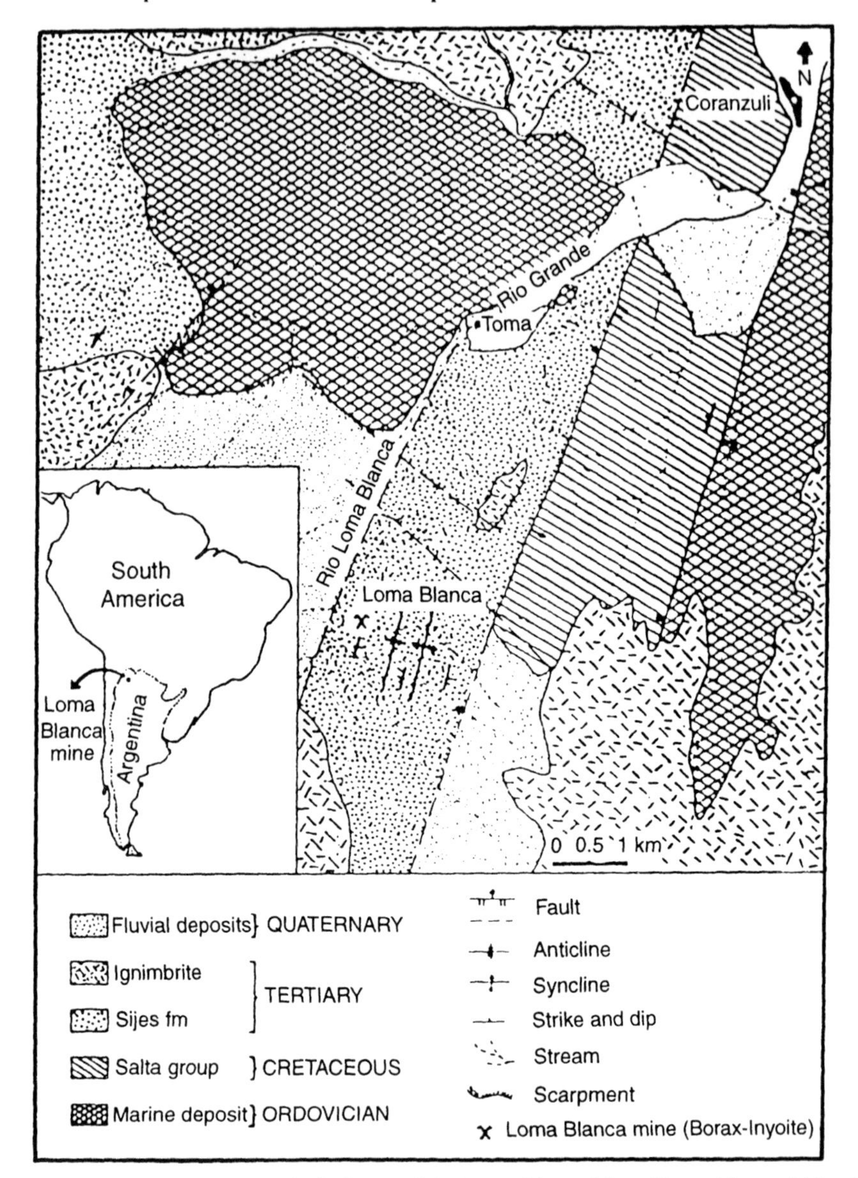

Figure 2.2 Locality and geological map of the Lome Blanca Mine. (From Alonso, Helvaci, Sureda, and Viramonte, 1988, reprinted by permission of Springer-Verlag Gmbh & Co. KG.)

from separate locations in the nodules. Many have occlusions of the clay–tuffite matrix.

Small amounts of teruggite ($Ca_4MgAs_2B_{12}O_{18}·20H_2O$) have been found in the upper part of the deposit as 2–8 cm nodules of fairly pure white powdery

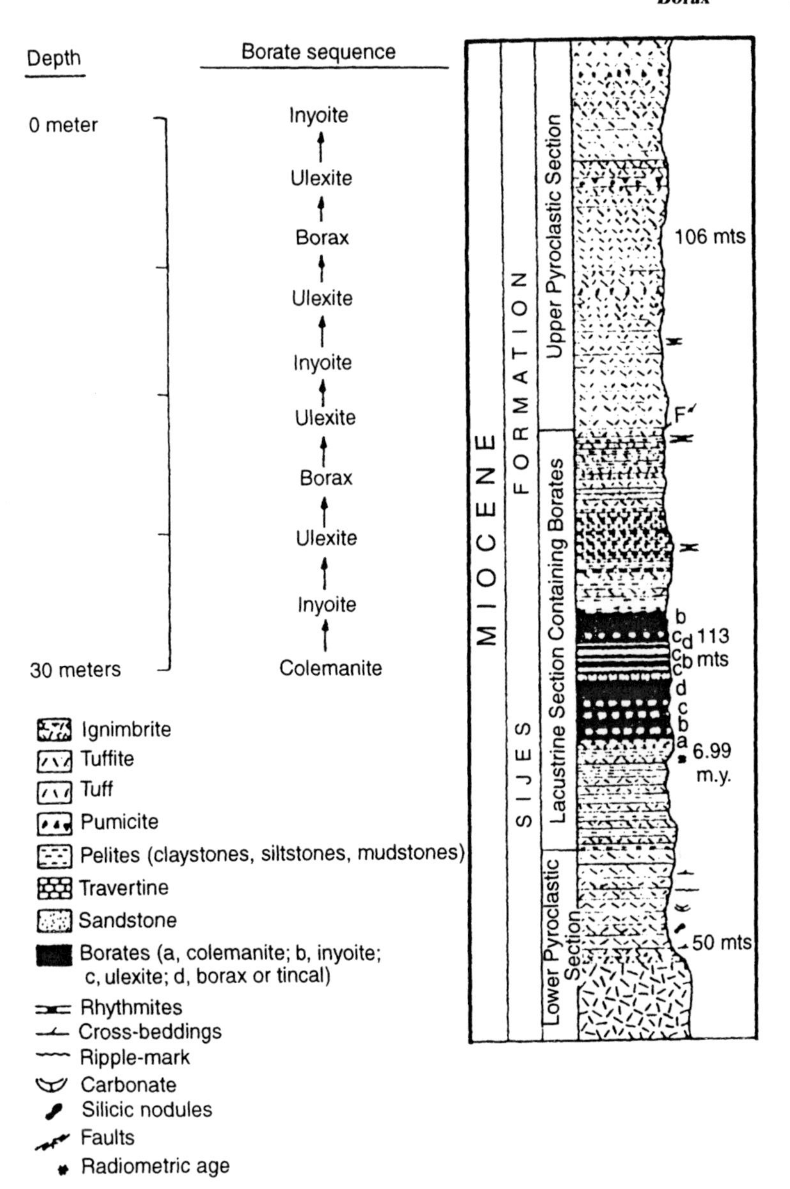

Figure 2.3 Borate sequence and stratigraphy at Loma Blanca. (From Alonso et al., 1988; reprinted by permission of Springer-Verlag Gmbh & Co. KG.)

crystalline aggregates of small euhedral crystals, usually associated with inyoite, ulexite, calcite, aragonite, realgar (AsS), and orpiment (As_2S_3). The latter two arsenic minerals and sulfur are found in the clay, tuff, and borates all through the deposit. Calcite and aragonite are common in the borate zones, and there is essentially no gypsum. Montmorillonite is the dominant clay mineral, and illite is a minor component. Chlorite is present throughout the deposit, but most abundant near or in tuff or tuffites. The borate bedding planes are undisturbed, and there are few pseudomorphs or intergrowths of one borate with another (Alonso, Helvaci, Sureda, and Viramonte, 1988). Before 1982, Loma Blanca was considered an inyoite deposit and only mined sporadically. However, after the discovery of borax at a shallow depth, the mine was more actively developed as an open pit by S. R. Minerals (Barbados), producing borax, ulexite, and inyoite (Alonso, 1986; Solis, 1996).

2.1.1.2 Tincalayu

The Tincalayu deposit, discovered between 1923 and 1927 by Catalano (1930), was first described in the literature in 1929. It is located on a large 10 km long, 2 km wide, and 150 m high (appearing as a low mountain) peninsula extending into the northwestern corner of the Salar de Hombre Muerto (Lonja Negra in Fig. 5.1). It is composed primarily of the 250- to 300-m-thick Sijes Formation, with the basement rock halite (>147 m thick, 10 km long, and 2 km wide; the Halite Member). The Borate Member lies directly on the halite, and is covered by the 25-m Pelitic Member (ferruginous claystones, with lesser amounts of fine sandstones, claystones, and tuffs, along with a few thin gypsum layers).

The entire zone is strongly folded, with areas of the borax covered by fissured basalt. The deposit is 1000 m long (north–south), 100–150 m wide, 10–60 m (average 10–30 m) thick, 40–60 m deep, and quite irregular. It thins at the edges. Folding has increased the central thickness and shortened the deposit, and much of the borax has been fractured and brecciated. The deposit is in the southern part of the peninsula at an altitude of ~4100 m. The area has 60–80 mm of rain per year, and winter temperatures can reach −32°C. Tincalayu's age is 5.86 Mybp, and it contains 10 million metric tons (mt) of 18% B_2O_3 ore. It has been mined as an open pit by Boroquimica Samicaf since the 1950s, who ship the ore by truck and rail 400 km to a processing plant near Salta (Alonso and Gonzales-Barry, 1995; Alonso and Helvaci, 1988).

The ore is primarily borax, with only several limited zones of kernite ($Na_2B_4O_7{\cdot}4H_2O$) and the rare appearance of 15 minor borates. In the lower half of the deposit, pinkish-brown sandy–silty mudstone layers predominate, whereas in the upper half, the impurities are mainly green tuffs (Fig. 2.4). Because of the strong kneading effect caused by folding and faulting there are many different textures in the ore: massive, sacroidal, rhythmic beds; disseminated crystals, druse, veins, geodes, and the like (Alonso, 1986; Alonso and Viramonte, 1990).

The borax occurs in massive beds that are colorless, light green, or red due to clay inclusions or yellow-orange due to arsenic sulfides. There are also some layers of millimeter- to centimeter-sized euhedral crystals in pale green tuffaceous mudstone that form rhythmic beds with other sediments. In some

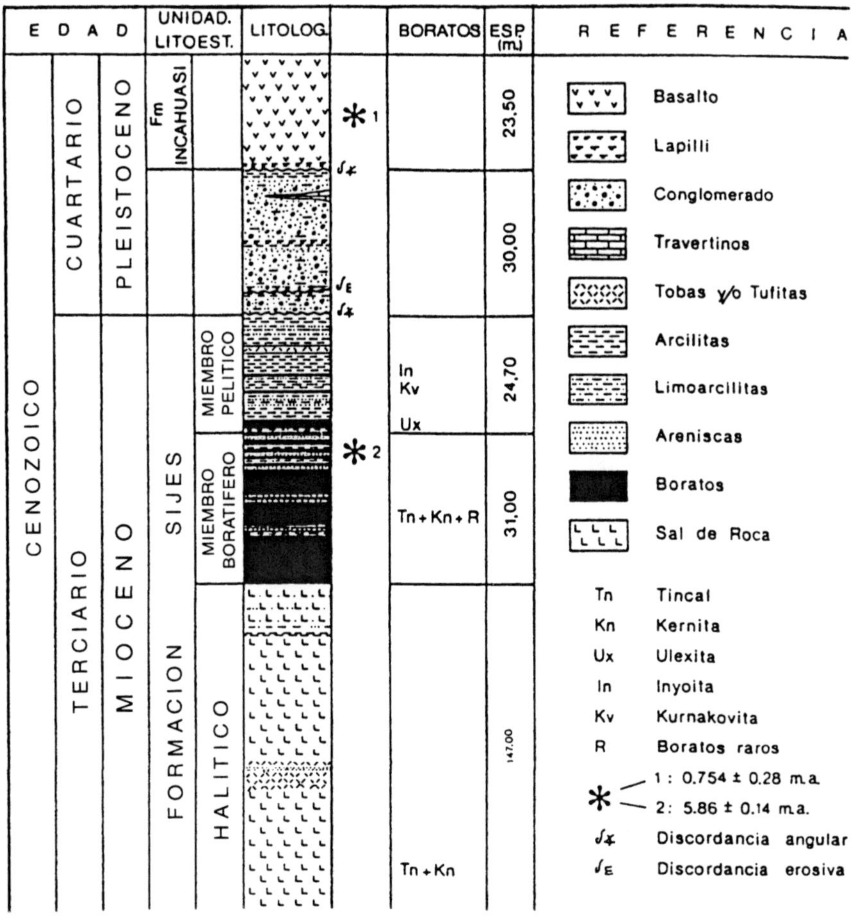

Figure 2.4 Typical stratigraphic column of the Tincalayu borate deposit. (From Alonso and Gonzalez-Barry, 1995; reprinted by permission of the Dr. Alonso.)

areas the beds' sedimentary form and stratification are apparent, but they are usually modified by tectonic movement. Some disseminated crystals of borax have also been found in the underlying halite (Alonso and Gonzales-Barry, 1995). Kernite occurs principally in the lower part of the deposit, forming irregular isolated <10 m lenses completely surrounded by borax, and occasionally <30 cm veins cutting through the borax or, in a few places, the underlying halite. Kernite forms aggregates of clear, transparent acicular crystals <30 cm long and 10 cm wide (Alonso and Robertson, 1992; Hurlbut, Aristarain, and Erd, 1973).

Ulexite occurs in modest amounts in the mud–sand–clay above the borax, usually as disseminated white, fibrous, silky compact mm-sized nodules in zones alternating with barren clay or in layers of irregular, deformed <6 cm nodules. In some areas a thin layer of groups or isolated impure, opaque, reddish-brown euhedral <5 cm (<2 cm average) crystals of inyoite ($Ca_2B_6O_{11}\cdot 13H_2O$) covers the deposit. In other areas, <1 m beds of <10 cm kurnakovite ($Mg_2B_6O_{11}\cdot 15H_2O$) crystals are in the overlying clastic material.

Many other borates occur in very small amounts and in limited locations, such as ezcurrite ($Na_4B_{10}O_{17}\cdot 7H_2O$), whose 7.5 × 1.4 × 1.2 cm nodules of somewhat parallel crystals in fine-grained borax are similar to kernite. Ameghinite ($NaB_3O_5\cdot 2H_2O$) occurs as spheroidal, elongated 2- to 3-cm nodules within borax, but separated from it by tincalonite, and often with ezcurrite and rivadavite. Rivadavite ($Na_6MgB_{24}O_{40}\cdot 22H_2O$) has a similar occurrence, being completely surrounded by borax and in 1- to 9-cm silky spheroidal aggregates. Aristarainite ($Na_2MgB_{12}O_{20}\cdot 10H_2O$) occurs as <1 mm crystals within mixed zones of borax, tincalonite, and kernite. Mcallisterite ($Mg_2B_{12}O_{20}\cdot 15H_2O$) occurs as <1.2 cm crystals within the cores of <8-cm borax nodules or in veins of rivadavite. Inderite ($Mg_2B_6O_{11}\cdot 15H_2O$) outcropped in <1 m beds (Muessig and Allen, 1957), and ginorite ($Ca_2B_{14}O_{23}\cdot 8H_2O$) and strontioginorite [$(Ca, Sr)_2B_{14}O_{23}\cdot 8H_2O$] were found by Hurlbut and Erd (1974) and Aristarain, Rusansky, and Walker (1977), respectively. Probertite ($NaCaB_5O_9\cdot 5H_2O$) occurs as idiomorphic crystals or as white radial aggregates in both borax and kernite. Searlesite ($NaBSi_2O_6\cdot H_2O$) is found in some borax beds as spherical <1 cm nodules composed of radially oriented crystals. Meyerhofferite ($Ca_2B_6O_{11}\cdot 7H_2O$) is a rare transformation product of inyoite, and tincalconite ($Na_2B_4O_7\cdot 5H_2O$) of borax (Alonso and Gonzales-Barry, 1995). Realgar and orpiment appear as yellow or orange intergrowths in the borax, sometimes in high concentrations. Gypsum and anhydrite are present with the ulexite, usually disseminated or in thin layers. Occasional cubes of halite have been found in cavities in the borax, and some thin layers of thenardite (Na_2SO_4) occur in the underlying salt (Alonso, 1986).

2.1.2 China and Tibet

2.1.2.1 Xizang (Tibet) Lakes

There are more than 57 borate-containing Tibetan lakes to the southwest of the Qinghai-Xizang (Tibet) Plateau (Kistler and Helvaci, 1994; Fig. 5.7), and very likely all of the world's borax from antiquity to 1818 came from these deposits. They were fed by thermal springs near rift zones in the Gangdisi-Nianqingtanggula and Himalaya Mountains, and many of the area's hot springs are still high in boron (average 26 ppmB; maximum 471 ppmB). Their discharge mantles usually contain borates, and many flow into closed basins. The region's acidic lava rocks' boron content is also quite high (average 24–39 ppmB; maximum 621 ppmB). As the Indian and Asian plates collided, there was considerable buckling and elevation to form the Himalayan and Tanggula (and other) mountains, as well as the Himalayan plateau. However, there was neither complete subduction of the Indian plate, nor elevation and buckling of the two plates. The result was considerable melting during the past 40 million years. Much of the molten rock rose to form a liquid layer in the upper crust, making this crust about twice as thick as normal. This partially molten zone helps to explain why the Himalayan plateau is so flat, even when ringed by mountains: the soft underlying layer could not support large mountains (Monastersky, 1996). It may also explain why the area has so many high-borate, apparently long-lasting geothermal springs, with the boron presumably having accumulated and concentrated in waters accompanying the molten zone. Perhaps a similar phenomenon may have occurred in the Puna region of the Andes to form its plateau and large number of borate deposits.

A few of the borax lakes in Tibet are listed in Table 2.2 with their brine types and dominant borate minerals. Examples of their B_2O_3 content are Zabuye Caka, 1.45%; Banguo Co, 1.22%; and Yagedong Co, 1.02% (their average pH is 8.7). Many of the Tibetan lakes have a high sodium carbonate content, and when there is more sodium carbonate or bicarbonate than calcium and magnesium, borax is the dominant borate mineral, but when there is not, either ulexite or Mg borates (when the brine is high in $MgCl_2$) are more common. Other salts in the Chinese lakes include halite, trona, natron, mirabilite, thermonatrite, northupite, and hydromagnesite, with the high hydrates predominating in the area's cold climate. Dujiali Lake is said to still be actively mined for limited borax production (Orris, 1995), and Zabuye Caka contains high concentrations of the alkali metals (potassium, lithium, rubidium, and cesium) as well as borax, sodium, chloride, sulfate, and carbonate (Chen, 1986).

2.1.2.1.1 Early Tibetan Production

The locations in Tibet from which borax had been recovered before the 1900s have been given only fragmentary attention, primarily because the "miners" sold the borax in small amounts to local bazaar dealers, who in turn sold it to sheep herders who would carry it over and down from the Himalayas to an Indian market city. There it was sold to a merchant who took it to a port city where it was again sold to an agent of the refiner and/or final marketer. The areas of borax-containing lakes called "Tibet" extended on the northeastern side of the Himalayas and its adjacent ranges for more than 1600 km (1000 mi), starting in the Ladakh region of India and reaching beyond Lhasa in Tibet, attaining altitudes of 4000–4600 m (13,000–15,000 ft). This region may be the largest of the world's three major playa–borates areas.

A 1631 report noted that Ahmedabad, and to a lesser extent Kathiawar, India were gathering points for borax from Tibet (Tavernier, 1889), and a 1640 publication said that it was traded in the bazaars of Gujarat, having come from near the river Jankenekar. (The European translation of Tibetan names

Table 2.2

Examples of Borate Lakes in the Xizang (Tibet) Area[a]

Lake	TDS[b] (g/l)	pH	B_2O_3, (g/l) (B as ppm)	Chemical type	Borate mineral[c]
Nieer-Co	233.99	7.6	4.48 (1128)	Sulfate	Ulexite, kurnakovite
Calaka (II)	107.22	9.5	5.69 (1596)	Carbonate	Borax
Xia-Caka	201.47	8.4	2.98 (770)	Carbonate	Borax
Zabuye-Caka	377.2	9.1	7.8 (1759)	Carbonate	Borax
Laguo-Co	91.29	7.4	2.30 (655)	Sulfate	Ulexite
Guojialin (Dujiali)	220.50	8.8	6.05 (1539)	Carbonate	Borax, tincalconite
Pengyan-Co	340.54	9.0	3.06 (709)	Carbonate	Borax
Banguo-Co (II)	187.02	8.6	2.56 (670)	Carbonate	Borax
Geerkunsha	365		4.63 (1053)	Carbonate	Borax, ulexite
Zhacang-Caka[c] (II)	384.78	8.0	2.14 (480)	Sulfate	Kurnakovite, pinnoite, inderite, ulexite
Da-Qaidam[d]	340.57	7.5	2.58 (598)	Sulfate	Pinnoite, ulexite, inderite, etc.
Xiao-Qaidam[d]	239.53	7.8	1.27 (318)	Sulfate	Pinnoite, ulexite

[a] Sun and Li, 1993.

[b] Total dissolved solids.

[c] Borax $Na_2B_4O_7 \cdot 10H_2O$; inderite $Mg_2B_6O_{11} \cdot 15H_2O$; kurnakovite $Mg_2B_6O_{11} \cdot 15H_2O$; pinnoite $MgB_2O_4 \cdot 3H_2O$; tincalconite $Na_2B_4O_7 \cdot 5H_2O$; ulexite $NaCaB_5O_9 \cdot 8H_2O$.

[d] See Table 5.4 for the brines' analyses.

was often very inaccurate, and later suffered further from the Chinese characters-to-English variability. Few of the early literature names can be recognized on current maps.)

In 1773 a Madras merchant noted that borax was dredged from Necbal lake (Travis and Cocks, 1984). Blane (1787) stated that 30 days north of Betowle (Butwal), Nepal there is

A small valley, surrounded with snowy mountains, in which [there] is a lake about six miles (9.6 km) in circumference. The water . . . is constantly hot, so much so that the hand cannot be held in it for any time. The . . .

> *earth is full of a saline matter in such plenty that after falls of rain or snow it concretes in white flakes upon the surface. ... Upon the banks of this lake, in the winter season when the falls of snow begin, the earth is formed into small reservoirs by raising it into banks about six inches (15 cm) high. When these are filled with snow, the hot water from the lake is thrown upon it, which, together with the water from the melted snow remains in the reservoir to be partly absorbed by the earth, and partly evaporated by the sun. After which there remains ... a cake of sometimes half an inch (1.3 cm) thick of crude borax which is taken up. ... It can only be made in the winter season, because the falls of snow are an indispensable requisite, and also because the saline appearances upon the earth are strongest at that season. When once it has been made upon any spot ... it cannot be made again upon the same place till the snow shall have fallen upon it and dissolved three or four times. After ... the saline efflorescence reappears, ... it is again fit for the operation.*
>
> *The borax ... is transported from hill to hill upon goats, and passes through many different hands before it reaches the plains (of Hindostan). ... When brought down from the hills, it is refined from the earth and gross impurities by boiling and crystallization. ... [Besides being] very hot [the lake water is] ... very foul, ... as [if] it were greasy. ... It boils up in many places, and has a very offensive small. ... After being purified, it [the borax] sells in the market here for about 15 rupees* per *maund. The only mode of information [about the borax] is through some of the wild and unsettled [Tartar] mountaineers, for the place is inaccessible even to the inhabitants of Hindostan, and has never been visited by any of them. ... The cold in winter is ... so intense that everything is frozen up, and ... life can only be preserved by loads of blankets and skins. In the summer ..., the reflection from the sides of the mountains, which are steep and close to each other ... renders the heats insufferable (Blane, 1787, pp. 298–299).*

Another report during the same period notes that

> *Twenty-eight days journey to the north of Nepal, and twenty-five to the west of Lassa, the capital of Thibet, there is a vale about eight miles (12.8 km) broad. In a part of this vale there are two villages, ... the inhabitants of which are wholly employed in digging the borax, which they sell in Thibet and Nepal. ... Near the two ... [villages] there is a pool of a moderate size, and some smaller ones, where the ground is hollow, in which the rainwater collects. In these pools, after the water has been some time detained in them, the borax is formed naturally. The men, wading into the water feel a kind of a pavement under their feet, which is a sure indication that borax is there formed, and there they accordingly dig it. Where there is little water, the layer of borax is thin; and where it is deep, it is thicker. ... Over the latter there is always an inch or two (2–5 cm)*

of soft mud. ... Thus is the borax produced merely by nature, without either boiling or distillation. The water in which it is formed is so bad that the drinking a small quantity of it will occasion swelling of the abdomen, and in a short time death itself. The [crystal] that yields the borax is of a whitish colour, and ... ten days journey farther north, there is another valley named Taprè *where they dig borax, and another still farther called* Cioga. ... *Borax is in the Hindoo and Nepalese languages called* soaga. *If it be not purified, it will easily deliquesce, and in order to preserve it any time, till they have an opportunity of selling it, the people often mix it with earth and butter. (de Rovato, 1787, pp. 471–473)*

Saunders (1806, pp. 406–407; 1789, pp. 96–97), visiting Lake Tengri-Nur in 1783, said

It is about fifteen days journey from [the monastery] Tissoolumboo, and to the northward of it [in another report, northwest]. It is encompassed on all sides by rocky hills, without any brooks or rivulets near at hand. ... Its waters are supplied by springs, which being saltish to the taste, are not used by the natives. The tincal is deposited or formed in the bed of the lake, and those who go to collect it dig it up in large masses, which they afterwards break into small pieces for the convenience of carriage, exposing it to the air to dry. Although tincal has been collected from this lake for a great length of time, the quantity is not perceptibly diminished. ... The cavities made by digging it soon wear out or fill up, [and] it is an opinion with the people that the formation of fresh tincal is going on. They have never yet ... [found] it in dry ground, or high situations, but it is ... [recovered] in the shallowest depths and the borders of the lake, which, deepening gradually from the edges towards the centre, contains too much water to admit of their searching for the tincal conveniently. ... From the deepest parts they [harvest] rock-salt, which is not to be found in shallows or near the bank. The waters of the lake rise and fall very little, being supplied by a constant and unvarying source, neither augmented by the influx of any current, nor diminished by any stream running from it. The lake, I am assured, is at least twenty miles (32 km) in circumference, and standing in a very bleak situation is frozen for a great part of the year. The people employed in collecting these salts are obliged to desert from their labour so early as October on account of the ice. Tincal is used in Thibet for soldering, and to promote the fusion of gold and silver. ... Thibet ... contains [tincal] ... in inexhaustible quantities.

Much farther to the east Strachey (1848, pp. 327, 331, 548, 551) noted,

The salt and borax mines of Gnari, ... are obtained from different spots in the same vicinity, and both worked in the same way by washing the earth taken from the surface of the ground in which they are developed

by natural efflorescence. ... On the north side of the Gangri mountains is a high valley inhabited by shepherds and salt carriers. North [and east?] of that are the salt and borax fields.

Montgomerie (1870, pp. 50, 52–53, 55) reported on

"An area at ... an altitude [of] 15 to 16,000 feet (4600–4900 km) ... with numerous lakes producing salt and borax." Somewhat further, toward Tengri Nur, they "passed numerous other salt and borax lakes. Borax fields were seen at Rooksum and Chak-chaka, and numbers of people were working them. ... At one [lake] ... about 100 men were at work near a camp of some 30 tents. ... The borax generally was said to find its way down to Kumaon [also Babuk], Nepal. ... As to borax, there appears to be any amount of it to be had for the digging, the Lhasa authorities only taking a nominal tax of about 8 annas [or a shilling] for ten sheep or goat loads, probably about 3 maunds or 240 lbs (110 kg)."

Montgomerie (pp. 55, 57) also later reported on seeing the borax lake, Bul-Tso, near lake Tengri-Nur, both of which are north of Lhasa. Other lakes mentioned as Tibetan borax sources in the 1700s and 1800s include Yamdok Cho (Lake Palte) south of Lhasa, which apparently had also been worked for a very long time. Far to the west are Purang-Chaki and Rudok, the latter producing the high-quality "water borax" (Travis and Cocks, 1984).

Cunningham (1870, pp. 140–141) described areas visited in 1845: Bitter Lake, White Lake, and the Salt Lake (Khaori-Talao). He noted that the latter

Lies in the Salt-Covered Plain (Thogji-Chanmo), and is 30 mi (48 km) northwest of Tshomoriri at an altitude of 15,684 ft (4780 m). It is about 2.5 × 5 mi (4 × 8 km) in length and width. The shores of the lake, and the plain have crusts of trona, natron, and borax.

Other reports speak of the winter cooling of warm brine, solar evaporation to crystallize borax, the "evaporation of ill-tasting foul water," and production of "high-quality water borax" (Heyden, 1909). Also, in contrast to the other major borate lakes areas, there are many active thermal springs still feeding the Tibetan lakes, adding to the deposits. The borax deposits in eastern Tibet appear to have originally been the major borax source because of the more difficult Himalayan passes in the West. Most of the western deposits were in or near the Ladakh mountain range in Kashmir, and those such as Puga were not large, but even so, they outlasted those from the east with the help of a sorting and cleaning operation, and for a short period, a refinery at the small town of Jagadhi, India. Border disputes between Tibet and Sikkim, Bhutan, and Assam essentially cut off the easier eastern transport routes in the latter part of the 19th century.

The Tibetan delivery system for borax to India was reported by Robottom (1893, pp. 28–33):

Tincal is brought into the [nearest] town in Tibet and sold at the bazaars. . . . Sheep owners buy it from the small dealers and put it into their saddlebags, placing as much as . . . 30–40 lb (14–18 kg) in each bag, or about equal in weight to the animal that is carrying it. Some few goats are also employed in this trade. When they have received their loads, they are started on their wearisome journey, many of the flocks carrying nothing but tincal. Each sheep-driver carries a distaff and bobbins, and as they travel along, every bit of wool that falls from the sheep, or that sticks to the thorny bushes with which the sheep may come in contact, is carefully collected. The wool thus gathered is spun into yarn or strong thread and then woven into cloth, which in its turn is made into bags. These are covered outside with sheepskins to prevent the tincal from getting wet, and also to protect the woolen bags from getting torn by the thorny bushes. Numbers of the sheep and goats die on the road, and their flesh is always eaten by the drivers. From 800 to 1000 sheep constitute a drove.

The animals are driven from seven to nine miles (11.2–14.4 km) a day, and it takes from six to eight weeks for the journey from the starting point to Moradabad. They travel through a pass of the Himalayan Mountains, about 100 miles (160 km) north of Almira. In one part of the pass there is no grass for a distance of 25 miles (40 km). The young underwood has to be cut down for the leaves to feed the sheep. Some points of the pass are 15,000 ft (4600 m) above sea level. The tincal is sold in the bazaars at Moradabad by the Bootus to native dealers, and it is sent from thence to Calcutta (by ox cart, and then boat down the River Ganges).

There have been many other reports of these early borax caravans, indicating that during the 1700s and 1800s they were fairly common in Tibet and all of the Himalaya passes. Hedin (1909, pp. 356–357) even told of special borax–salt roads throughout the country to accommodate them. Montgomerie (1870, pp. 52–53) reported on yak caravans as well:

Their salt [and borax] was laden on about sixty yaks, each carrying 1½ to 2 maunds (120 to 160 lbs (46–73 kg). . . . Two men were able to manage this large number of yaks as the road was a good one.

The Tibetan borax production reached its peak of 1600 tons in 1885, then declined and stayed in the 900–1300 tons/yr range through 1919. Sheep and goat caravans remained as the transport method over the Himalayas (Brown, 1921).

2.1.3 India

2.1.3.1 Puga Valley, Kashmir

One of the world's oldest commercial borate deposits is in the Puga (literally a hole) Valley of Ladakh, Kashmir (in the Himalayan Mobile belt west of the main Himalayas) at an altitude of 4400 m. Broad runoff aprons surrounding numerous hot springs contain surface deposits of kernite ($Na_2B_4O_7 \cdot 4H_2O$), with a total commercial area (>3mm, <6.4 cm thick) of 0.485 km^2 in 1977 (0.76 km^2 originally). The crude mixture that can be scraped from the surface averages 44% kernite, and ~550 tons of kernite are added each year from the springs. In 1977 there were 5500 metric tons of kernite in the deposit, and <3 m trenches did not find any borates, indicating that the 1 m (3 ft) deposits of the mid-1800s have been removed (Kashkari, 1977).

The Puga Valley is part of the extensive Ladakh borax spring zone, with hot springs extending from Rupshu to Chushul and to Hundes and Mansarower in Tibet. The Puga springs flow through faults in the Paleozoic Puga formation containing both older and younger interbeds of gneiss with quartzites and quartz–mica schists (in places gypsiferous). The shear zones have been intruded by amphibolites and underlain by older biotite–granite or quartz gneiss, often intruded by quartz–tourmaline veins. The springs have feeble fumarolic activity, but profuse hydrogen sulfide and steam accompany some of them. The geothermal gradient is 51.2°C/km (2.81°F/100 ft; Sinha, 1971), and the hot springs' water enters at 5600 m, developing a maximum $\delta^{18}O$ shift of 2% (Navada et al., 1991).

Cunningham (1870, pp. 144–145, 239–240) first described the Puga deposit after a visit in 1845:

> *The hot springs of Puga ... occur in a rivulet called Rulang-Chu for a length of about two miles (3.2 km). The springs vary in strength from gentle bubbling to strong ebullition, and the temperatures vary from 80 to 145°F (27–63°C).*

The river joins the Indus 16 km (10 mi) from Puga.

> *The air in this volcanic area bears the sulphurous smell of hydrogen sulphide, and ... a mixture of borate of soda and other salts is deposited along both banks of the river, along with sulphur as pure transparent crystals.*

Cunningham saw no human activity when he was there in September, but heard that shepherds still came in the summer to collect small quantities of borax. The damp borate was initially light pink to green, but after drying it changed to a dull white. The quantity of borax recovered was 7.3 metric tons/yr (16,000 lbs/yr), and sulfur 730–1500 kg/yr (1600–3200 lb/yr; Cunningham, 1870).

Dr. H. von Schlagintweit (1878, pp. 518–522) investigated the deposits in 1857, describing the hot springs and the narrow river channels where borax occurred in the banks "over a large area" with an average thickness of 1 m (3 ft). The valley's altitude was >4600 m (15,000 ft),

Presenting a strange and striking land formation with a wavy surface with large isolated borax accumulations in the shape of ninepins.

Some of the springs registered temperatures of 54–58°C, and there were also deposits of sulfur and gypsum. The kernite contained impurities of salt, boric acid, ammonium chloride, magnesium sulfate, sodium sulfate, and alum. At that time, kernite harvesting had almost ceased, but caravan leaders told Schlagintweit that Puga had once been the area's main borax source (Travis and Cocks, 1984).

Grabau (1920) confirmed that the deposits had been 1 m thick, and that spring temperatures were 54–72.5°C [Sinha (1971) measured temperatures of 45–82°C, the latter being the area's boiling point]. The yearly average air temperature is 1.3°C; the hot spring zone is 6 km (4 mi) long, 433–451 m (1420–1480 ft) wide; and low cones of kernite and travertine are formed around the discharge vents, some rising 4.6–6.1 m (15–20 ft). The terrain is rough and weathered, and the borates are often covered by sand and dust. The water from the hot springs flows over broad terraces into ponds and marshes and then the river.

The kernite crystallizes in aggregates of clear or white acicular crystals, but in the summer the surface rehydrates to borax, then to a fine white powder of tincalonite. The spring water contains 120–150 ppmB, but quickly concentrates to >1300 ppmB as it flows over the mantel, ponds, and marshes. The water also contains a high content of each of the alkali metals (Table 2.3; Chowdhury, Handa, and Das, 1974; Kashkari, 1977). The deposits are still

Table 2.3

Concentration of the Puga Valley Thermal Springs and Kernite Deposits, ppm[a]

		B_2O_3 (B)	Na	K	Li	Rb	Cs
Average spring water		(135)[b]	588	57	5.9	0.9	10.5
Typical near-spring[e] deposits		—	—	—	103	138	1400
Typical crude kernite, acid soluble		—	—	—	2900	156	990
Typical crude kernite, water soluble		224,000	—	—	2770	95	460
Spring water, Sinha[c]		(114–134)	609	64–70	Ba	Ca	Mg
					20	7.3	2.1
TDS[d]	pH	Cu	Sr	HCO_3	Cl	SO_4	F
2,202	7.9	2.0	.04–0.4	744–887	340–410	128	10–14

[a] Chowdhury, Handa, and Das, 1974.
[b] Bx1,000/TDS = 52–61.
[c] Sinha, 1971.
[d] Total dissolved solids.
[e] Also, Mn 20–200; Ga 10; Ca, Pb, Ag, Ni in "significant amounts."

periodically mined at rates such as the 550 tons reported for 1942, with the final processing done in Jammu, India (Lyday, 1992).

2.1.4 Nepal

Studies by the Royal Society of London in 1787 identified a series of hot lakes in Nepal that crystallized borax near their edges in the winter, and from which borax could be harvested in a fairly pure form. Between 1840 and 1850 other deposits from which borax could be produced were also found. The amount of borax harvested from these lakes, together with that from Puga, was sufficiently large that Britain (who purchased their output) became the world's largest borax consumer during this period (Lyday, 1992).

2.1.5 Turkey

By far the world's largest reserves of borates occur in a 300 km east–west, 150 km north–south, L-shaped area in western Turkey (Fig. 2.5). It is south of the Marmana Sea in western Anatolia and 70–100 km east of the Aegean Sea and the city of Izmir. There are four main borate districts, each having different mineralization, but of a similar age, with sediments deposited from inland lakes during active volcanic periods.

In the northwestern corner of the borate district (Bursa province), the

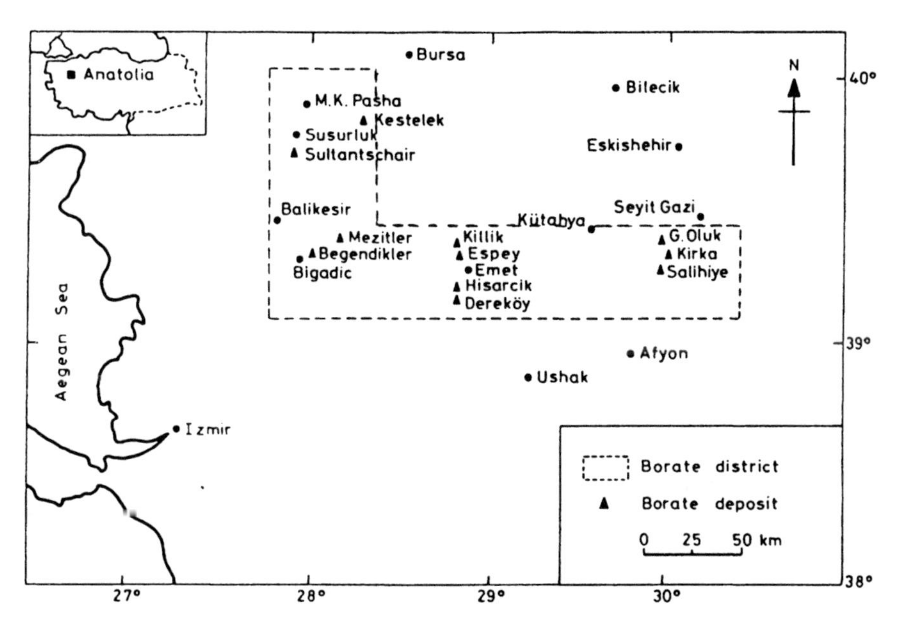

Figure 2.5 Location of the major borate deposits in Turkey. (From Helvaci, 1977; reprinted by permission of the University of Notingham and Dr. Helvaci.)

Sultancayiri (Sultantschair) deposit contained predominantly priceite in a gypsum–clay matrix, whereas nearby (but more easterly), Kestelek has colemanite in more normal clay–marl strata. Bigadic (Balikesir province), in the district's southwestern corner, has both colemanite and ulexite, whereas Emet (Kutuhya province), in the south-central district, is similar to Kestelek. Finally, Kirka (Eskishir province) in the southeastern corner contains primarily borax. Each deposit was formed in the Neogene or Middle Oligocene Tertiary period. All contain considerable tuff, with basalt either above or below the sedimentary sequence and/or extensive basalt (rhyolitic, dacitic, trachytic, or andesitic lava) near them. Limestone and marls are a dominant feature of the strata, with clay or shale as the usual matrix material. Each deposit was formed in closed basins with an arid, temperate climate (inferred by the somewhat associated lignite; Helvaci, 1977). There is a general trend in the depositing brine's alkalinity from the northwest (gypsum matrix, pH < 7) to the southeast (borax, pH ~ 10), and then further east to the Beypazari trona deposit (pH of 10–11). In many deposits there are signs of a playa environment part of the time, with mud cracks, rain indentations, bird tracks, and some borates as nodules, similar to those formed in near-surface muds in present-day playas.

2.1.5.1 Kirka

Kirka, the world's largest borax deposit, is located in western Anatolia 240 km west of Ankara, and in the southeastern corner of the borate district. Its age dates from the Upper Tertiary (Neogene) period, and it is in a 400 m sequence (on a fossiliferous limestone base) of various lake sediments and volcanic tuff capped by basalt (Fig. 2.6). The tuff is from andesitic–trachyandesitic lava, and the basalt is an olivine-rich type. The borate zone ends 130 m below the basalt. It outcrops, dips to 160 m, and its area is in the form of an inverted L. The basin had broad shores and a central deep zone to allow evaporation, concentration, and progressive calcium precipitation from the edges to the center. This resulted in the borax (Fig. 2.7) being surrounded by ulexite, and the ulexite by colemanite. The minor impurities in the various borate minerals (Table 2.4) infer a geothermal source because of their high content of strontium, iron, manganese, zinc, arsenic, and other metals. Also, the concentration of impurities in each of the borate minerals and the magnesian calcite matrix are all similar, and the fairly high content of sulfate and chloride in the borate crystals (relatively low-adsorption ions) indicates fairly high contents of these salts in the crystallizing brine. The chloride impurity increased from 132 to 278 to 351 to 470 in the calcite, colemanite, ulexite, and borax, respectively, demonstrating the increasing brine strength in this crystallization sequence.

Colemanite outcrops were first discovered at Kirka in 1960 on its western and southern edges. However, when the area was more fully drilled, it was found to contain primarily borax, with only secondary colemanite and ulexite. It has >500 million metric tons of 25% B_2O_3 borax, and minor amounts

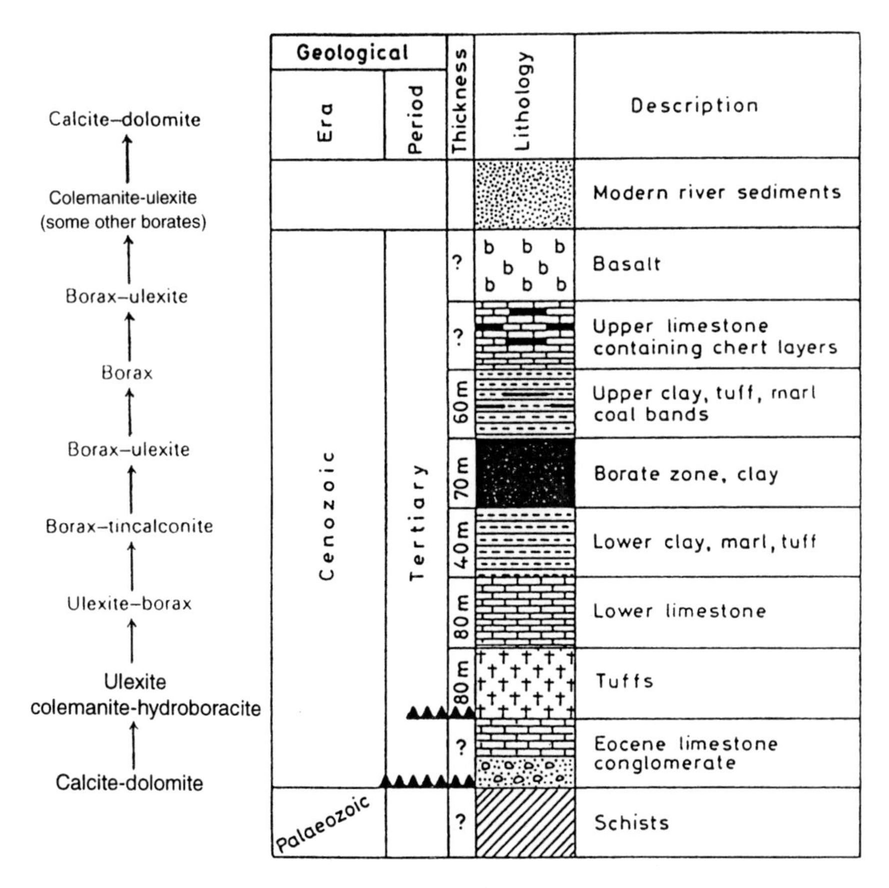

Figure 2.6 General borate sequence and stratagraphic column of the Kirka borax deposit. (From Inan, 1972; reprinted by permission of the Institution of Mining and Metallurgy, Transactions, Section B, Applied Earth Science.)

of hydroboracite ($CaMgB_6O_{11} \cdot 6H_2O$), inderite ($Mg_2B_6O_{11} \cdot 15H_2O$), inyoite ($Ca_2B_6O_{11} \cdot 13H_2O$), kurnakovite ($Mg_2B_6O_{11} \cdot 15H_2O$), meyerhofferite ($Ca_2B_6O_{11} \cdot 7H_2O$), tincalonite ($Na_2B_4O_7 \cdot 5H_2O$), and tunellite ($SrB_6O_{10} \cdot 4H_2O$). Calcite, dolomite, clay (mostly smectite and some illite and erionite), tuff, quartz, biotite, sanidine, albite, anorthoclase, and feldspar are in the matrix (Helvaci, Stamatokis, Zagouroglou, and Kanaris, 1993). The highest borax concentrations are in the central area, and its 1-mm to 1-cm crystals (when cutting across bedding planes <10 m long, 2 m diameter) are light pink, yellowish orange, or gray due to inclusions of clay or other impurities. There are usually no other borate minerals with the borax, but a thin film of tincalonite forms on exposed borax crystals, and there is some conversion to ulexite at clay interfaces.

Ulexite surrounds the borax, predominantly as thick fibrous beds, but there are also cauliflower-like nodules in layers of <2–3 m with crystals of <1–5 cm

Figure 2.7 Typical borax ore from Kirka. (From Etibank, 1994; pictures courtesy of Etibank.)

and occasional columnar clusters, rosettes, and cotton ball masses, including pseudomorphs after borax. Its color is usually white to gray, and it has been transformed to colemanite in some areas. A larger colemanite zone surrounds the ulexite, occurring as both nodules (with some geodes) and beds of colorless, gray or pink crystals. Some inyoite occurs primarily on the southern edge of the deposit above the upper colemanite. Smaller amounts of kurnakovite ($Mg_2B_6O_{11}\cdot15H_2O$) are found in the center, also over the upper colemanite in thin 1–20 cm-long discontinuous layers in clay just above the deposit. Tunellite occurs only in very small amounts in some clay layers near the ulexite (or kernite) deposits as colorless to white <10 cm needle-shaped crystals surrounded by

Table 2.4

Analyses of Pure Samples of Borate Minerals from the Kirka[a] Deposit

	Weight %					Parts per million			
	Borax	Tincalonite	Ulexite (Cole)[b]	Tunellite		Borax	Tincalonite	Ulexite	Tunellite
B_2O_3	40.57	45.95	43.39 (42.4)	54.88	Fe	624	438	730	389
H_2O	30.49	31.32	32.17 (19.0)	18.46	Cl	470	282	351	134
Na_2O	8.07	21.30	7.98 (0.20)	0.14	Al	232	49	381	138
CaO	3.60	0.11	13.52 (19.0)	0.16	Ba	129	130	86	1070
MgO	2.90	0.41	0.79 (6.0)	0.53	Cr	58	7	7	5
SiO_2	2.09	0.06	0.56 (6.9)	0.13	Mn	19	0	7	1
SO_3	0.50	0.54	0.35 (0.11[c])	0.06	Sn	14	10	24	2
SrO	0.24	0.004	1.69 (0.55[c])	27.46	P	11	2	5	8
K_2O	0.05	0	0 (0.28)	0	Ce	2	0	47	108
As_2O_5, ppm	90	10	10 (40[c])	170	Zn	1	4	4	18
CO_2	—	—	— (3.9)	—	Pb	1	2	0	0
Al_2O_3	—	—	— (0.91)	—	Cu	0	0	18	477
					Ni	0	2	4	2
					Br	0	0	1	3
					U	0	0	3	0
					Th	0	0	1	0

[a] Helvaci, 1997.

[b] Colemanite ore (not pure crystals); Inan, Dunham, and Esson, 1973. They quoted typical ulexite ore as containing 34.6% B_2O_3, and dolomitic calcite 0.84% B_2O_3.

[c] Average content in borax, ulexite, colemanite, and calcite; Inal et al., 1973.

borax. The borax, ulexite, and colemanite appear to be primary throughout the deposit, with only a few limited areas where secondary transformations have occurred (Helvaci, 1977; Inan, 1972; Inan, Dunham, and Esson, 1973).

2.1.6 United States

2.1.6.1 Borax Lake, California

This small lake 130 km (80 mi) northeast of San Francisco was the first commercial borate deposit operated in the United States, supplying essentially all of the country's borate needs (~500 tons) from 1864 to 1868. Its value was first recognized in 1856 by Dr. John A. Veatch, who had spent 2 years prospecting in north-central California for borates, analyzing the many boron-containing hot springs, lakes, and playas. By 1864 he had obtained mineral leases, organized the California Borax Co., and started production (Vonsen and Hanna, 1936).

The lake occupies a small depression in a peninsula extending into the southeastern end of the large adjacent freshwater Clear Lake, and has a limited (basalt) drainage basin. The area has had recent volcanic activity, and the remains of an alkaline borax hot spring (Little Sulfur Banks, "still giving off sulfurous gases") are above the lake in a basalt ridge. The region contains many hot "mineral springs, of sulfur and bitter waters" with small to considerable concentrations of boron (92–293 ppmB). About 2.4 km (1.5 mi) away there is a "white powder hill" (Sulfur Banks), a weathered (partly from SO_2) volcanic mound, which was mined for elemental sulfur in 1865 and later (1870) as a rich source of mercury. A hot spring at its base at one time gave off CO_2–H_2S gas, and probably had water similar to that of Little Sulfur Banks (Table 2.5). Borax Lake had a springtime area of 80–120 hectares (200–300 acres) and a depth of 1–1.2 m (3–4 ft), which diminished to 20–24 hectares (50–60 acres) by the end of summer. The rich borax area was ~16 hectares (40 acres) in its center; there were <0.3-m spires of calcareous tufa around the edges, and the lake supported an extensive algae growth. This gave the brine a high organic content, which considerably enhanced the normal "soapy" feel from sodium carbonate and borax.

The central portion of the lake's mud surface was smooth, soft, and plastic, with no crystals in its upper 30–40 cm (12–15 in.), but beneath that the mud began to feel gritty with small, perfectly shaped borax crystals. With increasing depth the crystals' size increased, and the grit became sand. Then, 2–3 cm deeper and 0.6–1 m (24–36 in.) from the surface, there were 0.6-m- (2-ft-) wide, and 10-cm- (4-in.-) thick layers of crystals. They were still separate, but tightly packed, with little or no mud between them. The mats occurred randomly, were not continuous, and may have formed as several layers, one above the other, separated by 10–15 cm (4–6 in.) of mud. Beneath the borax layers the mud became much more dense, and the crystals were larger but

Table 2.5

Several Springtime Brine Analyses of the Borax Lakes, California (ppm)

	Borax Lake[f]						Little Borax Lake	
	Sulfur Banks Hot Spring[a]	Chatard, 1890	Whitehead & Feth, 1961[b]	Vonsen & Hanna, 1936		Ayers, 1882[c]	Whitehead & Feth, 1961	
Na	1,550	1,230	6,140	11,680	11,090	—	1,330	3,390
K	50	87	295	466	544	—	285	731
Mg	0	—	25	107	68	—	8.7	24
Ca	11	—	4.0	9	58	—	6.4	8.0
Sr	—	—	—	—	—	—	—	5.3
R_2O_3	—	—	2	3	—	(NaCl 7,960)	—	—
Cl	741	180	5,760	9,890	11,260	4,830	365	905
CO_3	0	782	2,650	6,890	3,240	13,640	1,220	3,770
HCO_3	3,830	—	1,160	—	2,470	(Na_2CO_3 24,100)	970	626
SO_4	119	224	20	44	49	—	5.8	10
B	720	105	290	432	490[e]	1,492	150	360
SiO_2	86	—	4.3	—	—	($Na_2B_4O_7$ 6,940)	0.9	20
Br	Tr	—	19	12	—	—	—	—
PO_4	—	—	6.6	6	—	6.9	—	—
NO_3	—	—	13	—	—	—	0.8	—
Organics	—	49	—	—	—	—	—	—
TDS	7,620	2,560	16,200	29,500	30,660	39,000	3,850	9,540
Density	—	—	1.010	—	—	—	—	1.006
pH	—	—	9.7	—	—	—	9.7	9.6
Hardness	—	—	112	—	—	—	52	120
Conductivity	—	—	22,100	—	—	—	5,780	12,400
1000 × B/ Total salts	95	41.0	17.9	14.6	16.0	38.3	39.0	37.7

[a] White, 1957; 57°C; also I 1.4[d]; CO_2, and H_2S in gas.

[b] Also I 8[d]; Al 1. ; F 1; NH_4 0.5; Zn 0.2; Cu 0.02; As 0.01; Fe, Mn, Pb, NO_2 0ppm.

[c] Summer analyses.

[d] Veatch in 1857 (Hanks, 1883) also claimed that the brine had a considerable iodine content.

[e] Split between 5[illegible]4 ppm B_4O_7 (141 ppm B) and 1380 ppm B_2O_4 (349 ppm B).

[f] Foshag (1921) stated that the brine's soluble salt content was 18% borax, 62% Na_2CO_3, and 20% NaCl.

less numerous. Finally, a layer of "blue clay" was reached, and the deposit changed considerably. Above it the crystals were <5 cm (2 in.) long, and there also were many smaller crystals. In the blue clay there were only a few scattered crystals, 13–18 cm (5–7 in.) long and 5–10 cm (2–4 in.) wide, few crystals <5 cm long, and no small crystals. The larger crystals often weighed <0.5 kg (1 lb), and the blue clay was so firm that exact molds remained when the crystals were removed, with little clay remaining in or on them. No borax was deeper than 0.3 m into the blue clay. All of the crystals had a greenish color, but it did not persist after they were dissolved and recrystallized (Ayres, 1882).

The borax was recovered from 1.2 m (4 ft) square and 1.2–1.8 m (4–6 ft) deep, steel cofferdams floated on rafts to the desired location, and suddenly dropped. The sharpened lower edges penetrated the upper mud and made a seal in the denser mud. The top of the cofferdam was pounded so that it was just a little above the water level, and the water in it was pumped or bailed out. Then the surface mud was removed by clam shell-type "tubs" and dropped into the preceding hole. The borax-mud was then excavated, loaded into boats, and taken to the shore. At the plant the mixture was agitated into a thin slurry with lake brine and passed through a riffle (sluice) box, just as if the process were a placer gold operation. The larger borax crystals remained behind the steadily decreasing-size riffles [boards 1.3–5 cm (0.5–2 in.) high] positioned at intervals of 10–15 cm (4–6 in.) across the open rectangular sluice box. The clay and smaller crystals flowed over the riffles and out the end of the box, whereas the larger ones stayed behind the riffles and were frequently removed. They were then washed, drained, and dissolved in hot water. The settled, clear leach liquor was next put into small lead-lined tanks to cool and crystallize into a 99.95% product. Occasionally, a cofferdam could yield 400 kg (900 lbs) of borax from the riffle box, the equivalent of ~12% recoverable borax in the mud.

The simplicity of the operation, as well as the low capital and operating cost made it a very attractive process for 4 years until the rich areas of the lake had been mined. Then the operators switched to an attempt to leach the entire mud–crystal mixture with boiling water and crystallize the borax from the settled liquor. For this operation the mud was initially mined by damming off a portion of the lake and digging the mud with the labor of 130 Chinese workers. This was soon changed to the use of a floating steam dredge, and plant processing that included air-drying the mud, followed by a mild "roast." The change was a futile attempt to improve the very poor yields and product purity, but, instead, it further increased the plant, fuel, and labor costs. A second stage of borax crystallization was required, and because it was long before cheap and efficient settling agents were available, a large amount of entrained lake brine entered the plant. The solids left as a thin slurry with most of the dissolved borax still present, and thus the recovery was poor. A product was still made, but in 1868 several holes were drilled in the lake trying to find a stronger borax brine. It was found at a shallow depth in one hole, but they drilled deeper and encountered a dilute artesian aquifer. This water and an unusually wet winter flooded the processing plant, thus closing it permanently. The deposit's production was 12 tons in 1864, 125 tons in 1865, 201 tons in 1866, 220 tons in 1867, and only 32 tons in 1868 (Ayers, 1882; Goodyear, 1890; Irelan, 1888; Simoons, 1954; Ver Planck, 1956; Vonsen and Hanna, 1936).

A reconnaissance in 1934 by Vonsen and Hanna (1936) found the lake almost dry, with only 0.4 hectares (1 acre) of brine, 5-8 cm (2-3 in.) deep (the lowest level since 1861). A thin crust of sodium chloride had crystallized on the dry playa and at the edges of the lake, and under the brine was an expansion-fractured 8- to 10-cm (3- to 4-in.) bed of trona. Later, by the end of winter, all of the surface salts had redissolved. The mud near the (summer) shoreline changed from being moderately firm to a black viscous ooze covered by a thin crust of salts toward the lake's center. At a 0.6-m (2-ft) depth some irregular nodules of trona were found, with small crystals of northupite, numerous 1-mm gaylussite crystals, and some glauberite. In the center of the lake at a depth of 1.5 m (5 ft), a layer of firm blue clay contained abundant borax crystals. One crystal was 10 cm (4 in.) long, and the others were graded to smaller sizes. In test holes near the eastern shore, a thick layer of small borax crystals was found at a depth of 0.5–0.6 m (1.5–2 ft). Occasional crystals of teepleite were also noted. In the 66 years since the lake had been mined, apparently borax had again grown in its muds.

2.1.6.2 Fish Lake Marsh, Nevada

This small playa (Figs. 2.8 and 2.13), like several others, had some surface areas containing only borax, and others with ulexite (Table 2.6). However, here the dominant mineral was borax, and the former processing plant was designed only for borax, even though some ulexite was processed (the average harvested ore contained 31% borax and 45% other soluble salts). The salt-encrusted portion of the playa had an area of 10 km^2 (4 mi^2), and was 5.6 km (3.5 mi) long (north–south) and 1.6 km (1 mi) wide. The entire valley's area was 40 km^2 (15.6 mi^2), and its drainage basin 4000 km^2. It was unusual in also containing 800 hectares (2000–3000 acres) of good farmland, with fresh water from several small mountain creeks (Papke, 1976; Williams, 1883).

The richest borax area was in the northeastern corner, which in addition to surface crusts had clusters of small white borax–trona mounds 15–20 cm (6–8 in.) high. In the mud beneath these mounds were <1-m crystalline layers with a similar borax composition. Further south were 40 hectares (100 acres) covered with dazzling white efflorescent salts, also rich in borax. The crust was moist and flaky, resembling fresh snow, and could even be compressed into balls that would harden and retain their shape. Both areas were inexpensive to harvest and process. South of the white area were 16–20 hectares (40–50 acres) containing efflorescent and cotton ball ulexite (and gypsum), but with

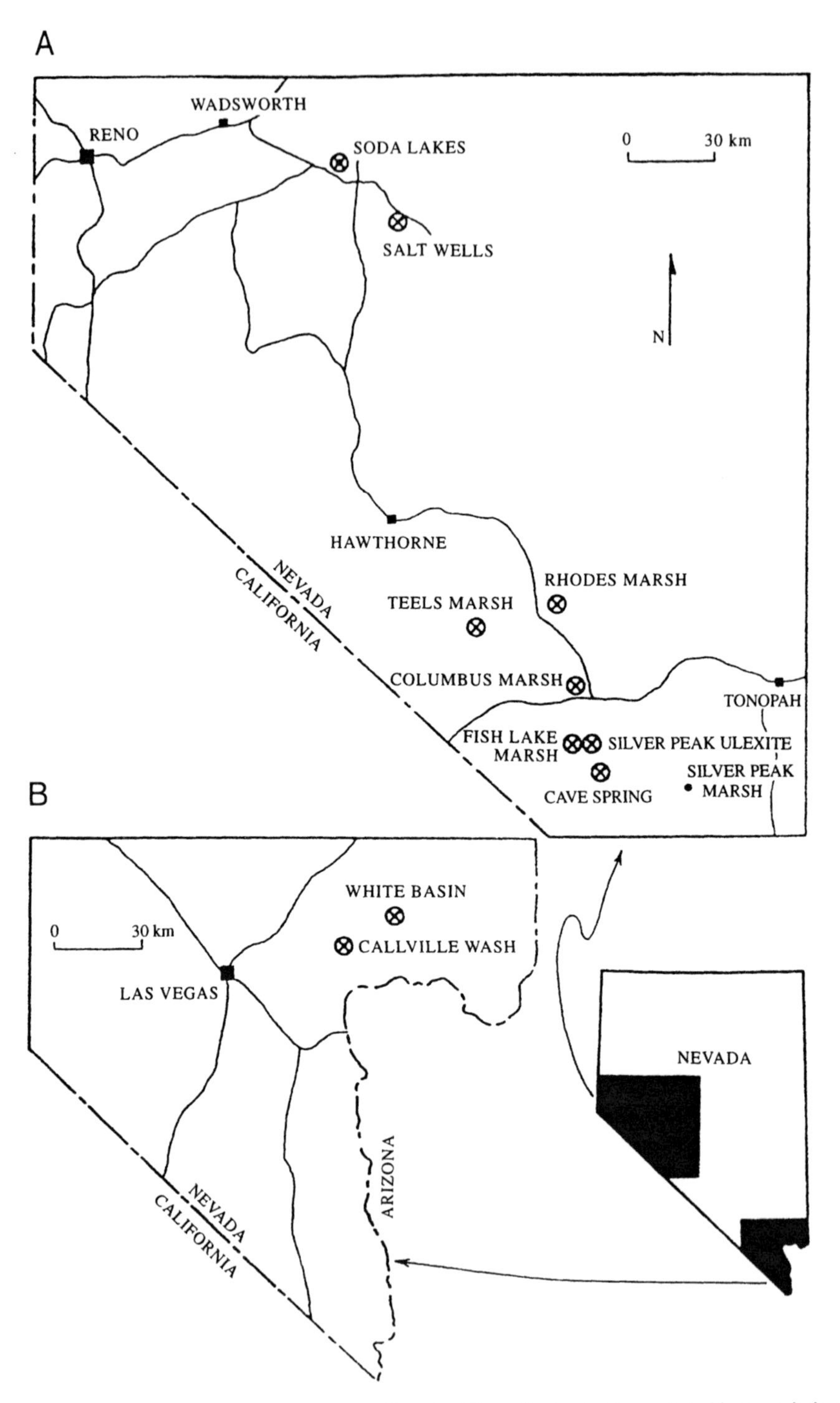

Figure 2.8 Major borax occurrences in Nevada. (From Papke, 1985; reprinted by permission of the Society for Mining, Metallurgy and Exploration, Inc.)

Table 2.6
Examples of Fish Lake, Nevada Brine, and Surface Salt Analyses (wt%)

			Surface salts from areas with:				
Interstitial brine[a]			Typical harvest[b]	High soda ash[c]	Cotton balls[c]	High gypsum[c]	High salt[c]
NaCl	13.1	Na	~16.1	31.97	4.63	2.41	34.18
Na_2SO_4	5.0	Ca	—	—	5.75	18.27	—
KCl	3.5	Cl	~7.8	15.52	1.97	0.82	37.44
Na_2CO_3	2.7	SO_4	~4.6	9.15	0.10	33.14	5.76
$Na_2B_4O_7$	0.75	CO_3	~12.0	23.85	1.31	—	8.28
Total	25.0	B	3.50	0.64	5.17	2.12	0.71
		Al, Fe, Mg	—	—	10.13	2.20	—
		Insols.	~5	3.46	35.4	5.72	4.86
		H_2O	~19.2	12.24	20.9	22.66	6.00

[a] Gale, 1917.
[b] Williams, 1883.
[c] Spurr, 1906.

a grassy, high-clay surface. Even further south were 160–200 hectares (400–500 acres) of lower-grade borax crusts containing either high-soda ash or salt mixtures. Throughout this area were small wet zones where springs had formed areas high in ulexite (Williams, 1883).

The earliest operation on Fish Lake Marsh was in 1873 by the Mott and Piper Co. which produced 2 tons/day of purified borax. Then in 1875 Pacific Coast Borax moved from their more expensive ulexite operation on Columbus Marsh and established a much larger processing plant. Both companies operated profitably for 10 years, but by 1883 the production was in serious decline (Williams, 1885), and in 1890 the borax reserves were essentially exhausted. Crude borax was raked or dug from near the surface and piled into long windrows to drain. It was then shoveled into wagons and moved to a plant stockpile. Pacific Coast Borax located its plant on the slope of a hill on the west side of the playa to take advantage of gravity flow of the processing streams. Ore storage was on a flattened area above the plant, with a stone retaining wall, 4.6 m (15 ft) high and 46 m (150 ft) long, supporting it and serving as a side of the plant. Ore was dumped onto the pile as the wagons entered from one side of the upper shelf and left on the other.

In the plant, six dissolving tanks, 2.7 m (9 ft) in diameter and 2.1 m (7 ft) deep, were on the next lower level adjacent to the ore retaining wall, and 5-cm (2-in.) plank platforms extended from the ore area to partly cover the tanks. This allowed ore to be dumped into them easily, and provided a platform for observation and manual agitation. Steam pipes were mounted in the bottom of each tank, with perforations every 10 cm (4 in.) to allow the steam to heat and partly agitate the slurry. About 1.2 m (4 ft) of ore and 1 m (3 ft) of water were added to a tank, bringing the level to 0.6 m (2 ft) from the top. The steam was then turned on and the mixture brought to a boil for several hours, with some stirring. When the density reached the desired strength, the steam was turned off and the slurry allowed to settle until the liquor was clear, which usually took 10–12 hr. The hot borax solution was then siphoned off and sent to a group of 45 galvanized iron crystallizer tanks sitting on the next level, 9.1 m (30 ft) below. Here the liquor cooled and borax crystallized on the tanks' walls. When cooled, the liquor was drawn off and discarded to the lake. The crystals were then removed, sprayed with water, allowed to drain and dry, and bagged for shipment. The drying and storage building, at the plant's lowest level, was 19.5 m (64 ft) long and 7.3 m (24 ft) wide. It also contained small rail lines to improve the handling efficiency for the solids.

A small amount of fresh water was added to the residual mud in the dissolving tank, and it was heated and leached again. After this leach wash, the mud was dumped into tanks directly below on the crystallizer tank level. There it was analyzed, and if it contained enough boron, it was loaded into a car and run along a track to a "Hickley elevator," which conveyed it to the ore stock pile. Here it was mixed with new ore and leached again. If the boron level was low enough after either the first or second leach, the mud was dropped to the next level (in the drying–storage building), placed in cars, and sent by rail to 15 m (50 ft)-square, 0.76 m (2.5 ft)-deep tailings ponds. After drying in these ponds, the mud formed a borax crust, which was periodically harvested and leached.

A boiler–pumping building was also located on the leach tank level. The boiler was 4.9 m (16 ft) long and 1.4 m (54 in.) in diameter. Two steam-driven pumps were located next to it. One, with a 15-cm (6-in.) cylinder was used to pump water to the leach and crystallizer tanks for washing and general utility use. The other pump had a 10-cm (4-in.) cylinder and was used to supply water from the well to the boiler and for ore leaching. The water well was a pit 6.7 × 5.5 m (22 × 18 ft) long and 5.5 m (18 ft) deep (see Table 4.8). It was estimated that the plant had 460 m (1500 ft) of steam pipes ranging in size from 1.3 to 10 cm ($\frac{1}{2}$ to 4 in.) in diameter between the pumps, leach tanks and crystallizers, and for general utility use (Williams, 1885). In this plant the second ore leach appears to have been primarily a mud-washing step, but also helped to dissolve any ulexite in the ore. A third leach would certainly be for ulexite. The harvested salts from the borax area contained considerable trona, which slowly converts ulexite to borax. Thus, in the later stages of the operation, some ulexite probably was carefully blended with the borax–trona ore to allow the trona to react with it during a three-leach process. This would further increase the borax recovery efficiency, extend the limited ore reserves, and obtain a reasonable recovery from the ulexite. Lakes containing only ulexite had to purchase soda ash for this purpose.

2.1.6.3 Kramer Deposit; Boron, California

The second largest but the oldest and best known of the world's sodium borate deposits is at Boron, California. It is 140 km north northeast of Los Angeles in the western corner of the Mojave Desert. The town of Boron is 5 km south of the deposit, and the Santa Fe Railroad's Kramer station is 10 km to the east. This station was the only point of reference at the time of discovery in 1913, so it became the deposit's name. The larger town of Mojave is 50 km to the west, and the southern terminus of the Sierra Nevada and Tehachapi mountains as they join the Coast Range is a short distance further west. South and southwest 80 km were the Lang and Frazier deposits in the Coast Range, and east 40 km the Four Corners deposit.

The Mojave Desert is an area of relatively moderate relief, and it is part of the Basin and Range Province that encompasses the Death Valley, Nevada, and Mexican borate deposits. It is generally flat, with only scattered hills and low mountains between its broad valleys. Its surface consists largely of sand and gravel carried there by sheet-flood alluvial fans (Gale, 1946). The mountains and basement rocks considerably predate the Middle Miocene Tertiary borate deposit (Siefke, 1991).

The initial discovery of borates occurred when a homesteader, Dr. Otelia Suckow, drilled a water well and found colemanite at a depth of 113 m (370 ft). This resulted in considerable nearby drilling, but only low-grade colemanite and some ulexite were found. In 1924 Dr. Suckow opened a shaft into his colemanite area, and the commercial production of colemanite was initiated. It was not until 1925, however, that borax was discovered by the Pacific Coast Borax Co. in a well 5 km (3 mi) east of the colemanite discovery. They quickly installed a shaft and initiated the Baker mine. Later, Dr. Suckow also found borax and started the Suckow Borax mine 2.4 km (1.5 mi) to the west of Baker. In 1927 the Western Borax Co. located a shaft halfway between the other two, but 0.8 km (0.5 mi) to the south, and all three companies were rapidly in full production. Later, Pacific Coast Borax bought the other properties and gained control of most of the deposit.

The Kramer deposit's age is 19 ± 0.7 My, and it is in a small basin within a much larger Tertiary basin. Beneath and above the lacustrine zone are primarily arkosic (feldspar-rich rock derived from the decomposition of granite) sands and silts with some volcanic rocks and calcite. These sediments were interrupted in the past by a flow of acidic (olivine basalt to latite) lava (the Saddleback Basalt), which formed the base for most of the deposit (Fig. 2.9). It flowed from nearby vents, one of which is 3.2 km (2 mi) northwest of the Suckow mine, and formed an interfingering 60- to 180-m (200- to 600-ft) mass with a gentle syncline on its surface. This, after subsequent earth move-

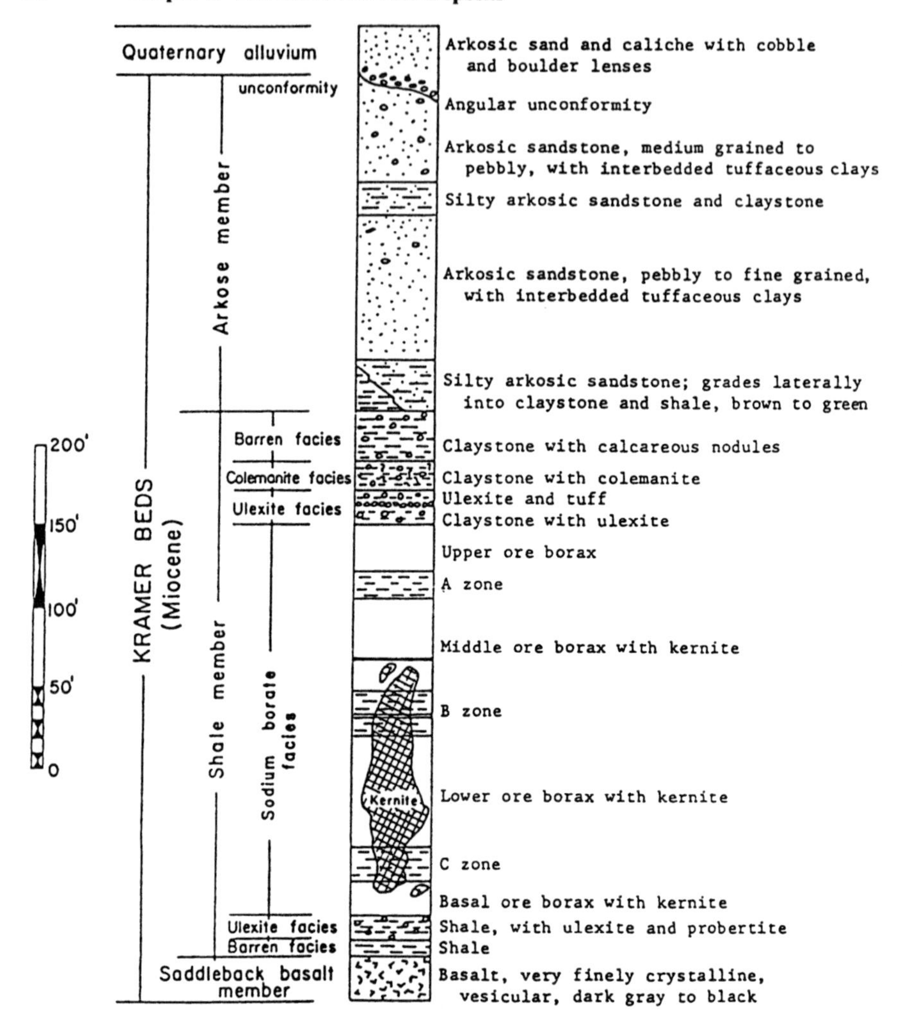

Figure 2.9 Generalized stratigraphic section of the Kramer beds. (From Bernard and Kistler, 1966; reprinted by permission of the Northern Ohio Geological Society.)

ment, allowed it to outcrop nearby (the Saddleback Mountains) and to form the basin for the borates (except in one location on its southern border). Borax hot springs also must have opened then, or shortly thereafter. After the deposit was formed, it subsided, perhaps caused by the east–west Western Borax fault. Then late in the Middle Pliocene epoch, the Mojave block was uplifted, causing most of the present folding and faulting (Siefke, 1991).

By an unusual coincidence, the three initial mines all were near the edge of the borax and formed the apex of a triangle that roughly defined the limits of the deposit (Fig. 2.10). The total borate area (including ulexite and colemanite) is 8 km (5 mi) long (east–west) and <1.6 km (1 mi) wide (north–south). The sodium borate area is 2.4 km (1.5 mi) long (east–west) and >0.8 km (0.5 mi) wide north–south [3.2 km long and 1.6 km wide, Siefke, 1991]. This encompasses 200 hectares (500 acres), with the high-grade borax thickness 17–69 m (56–226 ft), average 23 m (75 ft) [24–76 m (80–250 ft), Obert and Long, 1962]. The borax ($Na_2B_4O_7 \cdot 10H_2O$) was 46–305 m (150–1000 ft) deep and dipped <10–15° south. It had a lenticular form enveloped by ulexite on the top, bottom, and three sides, which in turn was enclosed by colemanite on three sides and the top, extending far beyond the ulexite. The Western Borax fault bounded the borates on the south, and probably had been active and formed a scarp, thus becoming one edge of the basin during the depositional period. In some zones of the deposit, there were scattered, and occasionally large masses of kernite ($Na_2B_4O_7 \cdot 4H_2O$; Gale, 1946).

The upper section of the basalt in some areas contains ulexite and searlesite in cracks and crevices. Also, near the basalt contact colloidal silica has formed an "apple-green siliceous shale" or chert, indicative of shallow water reaction with the lava. Ostracods are also present (Gale, 1946). Bernard and Kistler (1966) categorized and named the various strata above the basalt in the lacustrian zone (see Fig. 2.9). First, there is the 7.6- to 15-m. [(25- to 50-ft; 1- to 15-m) Gale, 1946] thick footwall shale, divided into an initial ~3.7-m (12-ft) barren, and then a ulexite zone. It is a layer of dark greenish to gray, or almost black, slatey shale, thinly laminated and micaceous, and its beds are frequently quite contorted. It locally includes some sandstone or coarse arkosic conglomerates (Obert and Long, 1962). Barnard and Kistler (1966) described the claystone and shale in all of the deposit's beds as being gray-green to dark green, primarily montmorillonitic, and occasionally tuffaceous. In the footwall shale's upper section it contains considerable ulexite in irregular patches, nodules, or veinlets of dense fibrous, pearly-white crystals that are occasionally mixed with structureless, compact, granular probertite. In a few locations some of the ulexite also appears to have been converted to white, hard, almost porcelain-like howlite.

Above the footwall shale are the borax beds, occurring in or interbedded with clay and shale, along with some beds of tuff, arkosic sands, and conglomerates. The clay and shale appear to be ancient shallow lake-bed or playa sediments closely related to the preceding and ongoing volcanic activity during the period of deposition. They are soft and easily eroded, being chiefly thin beds of montmorillonite and illite clays, with some lenticular to tabular discontinuous layers of coarse to fine-grained, clay-filled cemented arkosic sands. They helped preserve the borax from being leached because of their low permeability, as they were deposited with, over, and adjacent to the borate layers. Both the clay and the borates were further protected by the conglomerates that eventually covered them, and by the basin's later sinking and

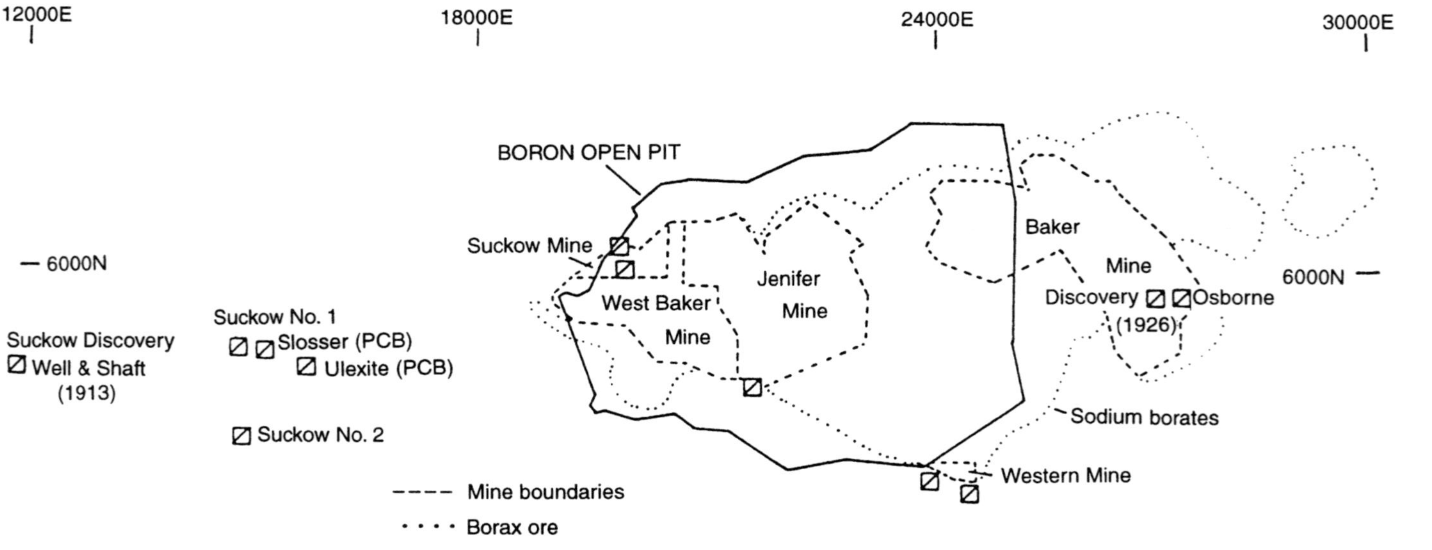

Figure 2.10 Outline of the Kramer (Boron) ore body, and the location of various early operations. (From Siefke, 1991; reprinted by permission of the Society of Economic Geologists, Guidebook Series.)

deforming through fault slippage. Some of the beds undulate and roll, and faults have created vertical displacements of 2–3 cm to 15–23 m. Some are gently tilted and folded into broad arches, locally producing dips of <30° (Fig. 2.11).

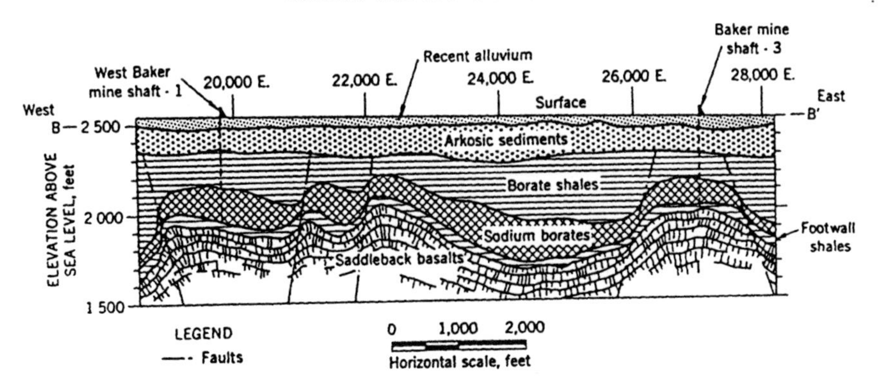

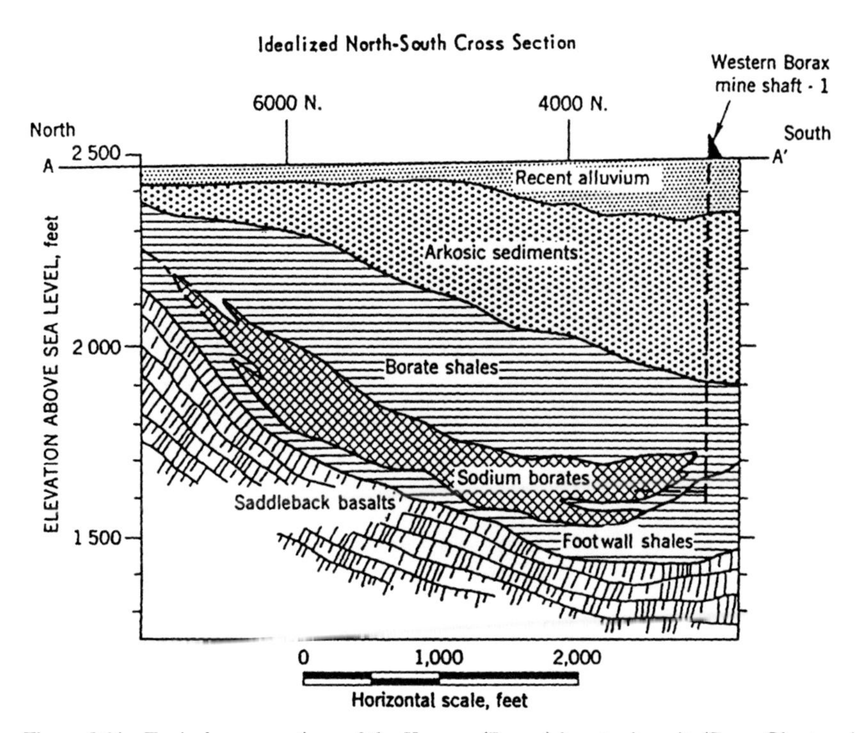

Figure 2.11 Typical cross-sections of the Kramer (Boron) borate deposit. (From Obert and Long, 1962.)

The sodium borates occur in seven different zones, of which four typically contain high-grade ore (>75% borax): Upper, Middle, Lower, and Basal. These zones are separated by three lower grade beds. A, B, and C, respectively, which contain <60% borax as disseminated <0.3- to 2.5-cm (0.13- to 1-in.) crystals in a claystone matrix. Only in the central region are all of the beds present, with the Basal zone being the thinnest and smallest, and the Lower bed the thickest and most extensive. In thickness and $\%B_2O_3$ the beds typically are: Upper ore, 2.1–7.6 m (7–25 ft); A shale, 3.1–9.4 m (10–31 ft); average, 19 ft, 16.6%; Middle ore, 10.7–15.8 m (35–52 ft), average, 29 ft, 29.4%; B shale, 3.7–10.7 m (12–35 ft), average, 14 ft, 6.5%; Lower ore, 3.7–27.4 m (12–90 ft), average, 17 ft, 25.5%; C shale, 3.1–10.7 m (12–35 ft); Basal ore, 3.1–6.1 m (10–20 ft). The 0.3–2.5 cm ($\frac{1}{8}$–1 in.) borax crystals sometimes appear to be graded in size, with the larger ones at the top (Bowser and Dickson, 1966). In many beds there is as little as 20% claystone with the borax, but in other zones the borax crystals are disseminated within the claystone matrix. There are also numerous green to brown 1.3–61 cm (0.5 in.–2 ft) thick, most commonly <7.6 cm (3 in.), interbedded claystone layers with the borax beds. Some pairs have equally thick beds, but others are quite variable. Many of the claystone beds are fractured, with borax-filled crevices, and occasionally they are >30 cm (1 ft) wide and somewhat displaced.

Above the borax beds there are again two zones of shale, the lower one 5.5–11.3 m (average 9.1 m) thick containing ulexite grading upward into colemanite, and the upper zone 9.1–15.2 m (average 10.7 m) consisting of "barren shale." The boundary between the ulexite and borax zones is sharp, and on the sides tuff layers extend smoothly between the borax deposit and the ulexite-containing shales. Gale (1946) described the shale in this zone as fine-grained (colloidal), soft, dark, and putty-like. It dries to a shrunken, light gray or green hard clay, and absorbs and retains water with considerable swelling. Obert and Long (1962) noted that the blue clay was massive, compact and with thin laminations. Arkosic sand interbeds grade into the blue clay, especially in the borders. There are two <0.6-m (2-ft) thick ulexite beds above the borax in the central part of the basin, associated with gray tuff. The ulexite is white with a fibrous texture, and in the upper bed some colemanite has replaced ulexite. Besides the two beds, there are nodules of ulexite, primarily in the lower part of the section, with lesser amounts above the ulexite beds. Ulexite is also a common crack- and vein-filling material in this zone. In the colemanite zone the colemanite occurs as thin beds of glassy to milky white crystals with prominent cleavage planes and frequent nodules in the upper and lateral zones, or as thicker beds with fewer nodules in its lower section. The lower nodules blend into the underlying ulexite zone to form mixtures of both types of nodules. The colemanite zone also blends smoothly into the upper green shale beds without much distinction except for the lack of borates in the upper shale. The upper barren shale is greenish, micaceous, thin bedded, and contains arkosic constituents mingled with tuff (Gale, 1946).

Kernite occurs in some of the borax beds (it made up ~30% of the remaining deposit in 1994; Kistler and Helvaci, 1994), and is never lower than 1 m (2–3 ft) from the deposit's lower edge. Many of the claystone beds near or in the kernite are somewhat broken and distorted, although within individual kernite areas they are still sharp and distinct. Near the edges of the kernite some of the ore is in isolated crystal masses (or pods), and commonly these masses have clear borax crystals and/or claystone rims around them. Borax adjacent to the kernite also may have the pod-type structure. Some of the borax below, and occasionally above the large kernite masses has a finely crystalline "sucrosic" structure as if it dissolved and rapidly recrystallized. This is particularly true of the lower 1.5–3 m (5–10 ft) of the Basal ore, which usually is glassy or sucrosic (Bernard and Kistler, 1966), and there are irregular masses of clay between some kernite areas. The kernite crystals may be either fine-grained or up to 0.1–152 cm (<0.02–5 ft) long, and often they are glass-clear and perfect in shape. Much of the kernite is in the Basal bed, but the Lower bed also contains some in its lower half, and there is some in a few south-central areas of the Middle ore. None has been found in the Upper ore. It occurs primarily in the south-central, far southwestern, and eastern areas of the deposit. Some probertite ($NaCaB_5O_9 \cdot 5H_2O$) is found in the claystone beds in the general area of the kernite as radiating prismatic crystal aggregates or rosettes (Siefke, 1985).

The deepest part of the deposit is at its southern edge, where it terminates rather abruptly 150–460 m north of the east–west Western Borax fault. Its actual edge is near (and north of) the Portal Fault, trending 30° north of west, except for part of a kernite bed which has been displaced more than 610 m (2000 ft) to the west, and now is south of the Portal fault. In the south-central area, the southern side has also been lowered 122 m (400 ft), but the displacement rapidly decreases to the west where the fault branches. There are five major northwest trending faults (and at least three subordinate ones) that affect the central and western part of the deposit. They have caused 0- to 15-m (0- to 50-ft) displacements (appearing to increase with depth), folds, and thinning of the borate beds adjacent to the faults in many areas, particularly on the upper side. Barnard and Kistler (1966) felt that neither plastic flow nor water leaching was a factor in the bed thinning.

Smith (1968), however, suggested that considerable borax had been leached from the deposit. In the northwestern section near the Portal Fault, the claystone bedding over the Upper borax has an obscure structure, and the two overlying ulexite beds are missing. Also, a Middle ore clay–tuff marker bed is at a level that should have had 30 m (100 ft) of borax over it, but does not. These features indicate that as the beds were tilted they were raised sufficiently in this area for groundwater to dissolve the borax. Some surface outcrops of adjacent shale beds may also indicate missing borax. Ore thinning at greater depth, valleys in the upper ore surface, pinching-out, discontinuities, and the presence of Mg-borates may all represent leaching. Sucrosic borax and the fracturing of the kernite and its accompanying shale almost certainly are the result of dissolving and recrystallizing (Smith, 1968).

Specific descriptions of local areas of the deposit help to show the variability of its structure. The Jenifer mine was slightly to the west of the deposit's center, and in that area the total of the three high-grade borax beds was 0–105 m (0–345 ft), average 53 m (175 ft), with each bed 7.6–24 m (25–80 ft) thick. The beds were separated by 3–11 m (10–35 ft) of lower-grade borate-shale. None of the mined beds were continuous over the entire area, and they contained random horizontal 0- to 5-cm layers of clay (Obert and Long, 1962). The Western Borax mine was near the south-central edge of the deposit, and its major ore was kernite found in beds or small angular to rounded crystalline masses in greenish, fine-grained clay. In the upper half of the deposit, the kernite bedding was somewhat obscure, but in the lower half it was distinct though contorted (Obert and Long, 1962). Gale (1946) reported that the ore was 18–34 m (60–110 ft) thick and contained only 10–15% shale, with no indication of significant bed disruption or alteration.

The Suckow or West Baker mine was at the northwestern corner of the deposit, with a total of 61 m (200 ft), average 40 m (130 ft), of high-grade borax (3 beds; Fig. 2.12). The beds were separated and surrounded by olive-green clay and low-grade ore, and the borax was 46–107 m (150–350 ft) deep, sloping 10–15°. The beds were cut by a number of northwest-bearing near-vertical faults, displacing them 0.6–15 m (2–50 ft), but the faults had been recemented and were tight and strong. The mine was dry except for one small artesian spring (Obert and Long, 1962). Gale (1946) described the deposit as being relatively undisturbed; secondary veins of borax cut through the bedding planes. Ulexite veins and nodules were in the over- and underlying borax-clay, some of which appeared to be a probertite replacement.

The Baker mine was near the eastern edge of the deposit, with its borax bed (only one was mined) 110–150 m (360–500 ft) deep and 2.4–35 m (8–115 ft), average 23 m (75 ft) thick. In half of the mine the bed was flat; elsewhere it dipped 17–24°, and its lower section contained both borax and kernite (Obert and Long, 1962). Gale (1946) described the area as a "broad, flat-topped anticlinal nose with a plunge to the southeast (p. 369)." The borax bed was 32 m (105 ft) thick at the No. 2 shaft, "with some interbedded shale (p. 370)," and at a depth of 114 m (375 ft). Much of the ore was well bedded, and some kernite occurred as crystals in the borax bed. Some also had apparently reverted to borax, with a layer of borax around the kernite.

More than 80 minerals have been reported in the deposit, including 18 borates as follows: Only a few grains of 0.1–0.6 mm *garrelsite* ($NaB_3Si_2B_7O_{18}\cdot 2H_2O$) have been found in ulexite in the footwall clay adjacent to

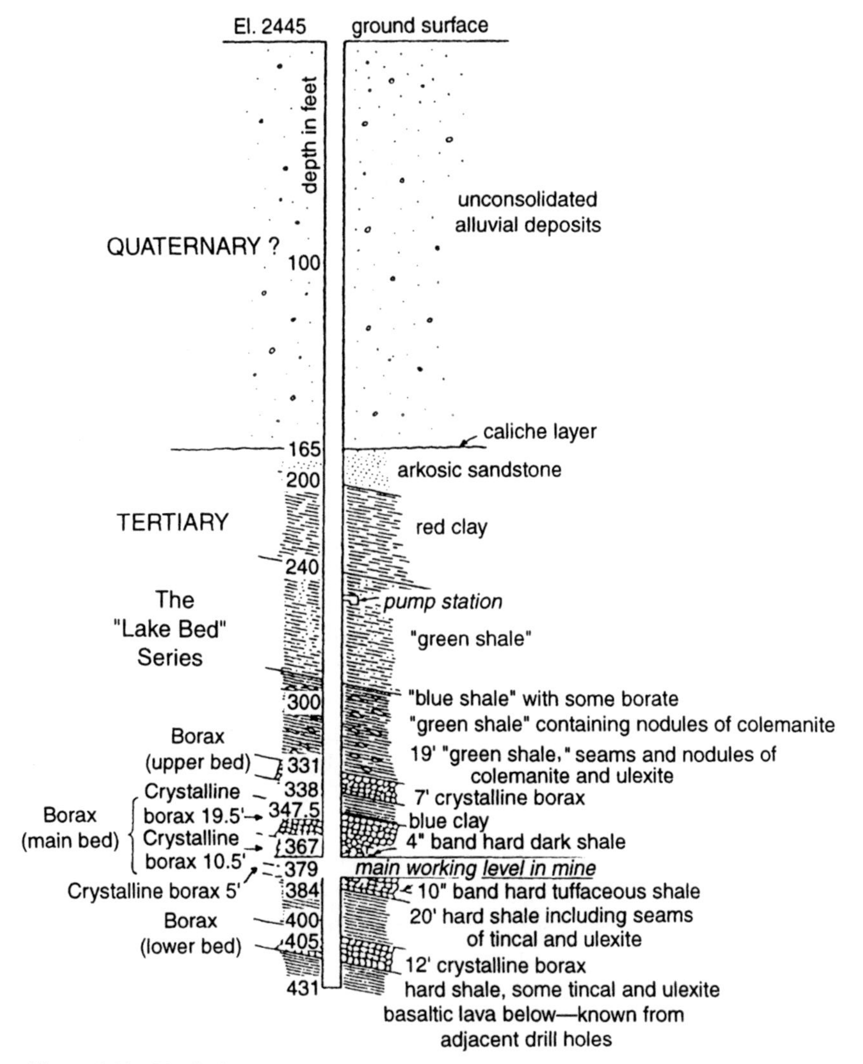

Figure 2.12 Vertical section through the original shaft of the Suckow borax mine. (From Gale, 1946; reprinted by permission of the California Department of Conservation, Division of Mines and Geology.)

the Basal bed. *Howlite* ($Ca_4Si_2B_{10}O_{23}\cdot 5H_2O$) occurs as sparse cauliflower like nodules in the clay or siltstone surrounding the deposit, chiefly in the lower ulexite. In the Western Borax mine it was found in a "water drift," in drill holes, and as thin platy 0.15-mm crystals in a colemanite geode. *Hydroboracite* ($CaMgB_6O_{11}\cdot 6H_2O$) nodules, often hollow with acicular colorless to white crystals, are occasionally found in the claystone immediately above the Upper borax. Hydroboracite resembles ulexite, but is coarser. *Inderite* ($MgB_6O_{11} \cdot 15H_2O$-I, monoclinic) has been found as many single crystals, generally with *kurnakovite* ($Mg_2B_6O_{11} \cdot 15H_2O$-II, triclinic). It occurs as slender, elongated, four-sided, colorless to yellowish glassy prisms with large oblique terminal faces, 2 mm–4 cm in diameter, and often broken in the long dimension. Kurnakovite has been found in large, shapeless 200- to 300-kg masses and as crystal clusters or <30-cm single crystals. It occurs above the borax in the base of the clay–ulexite matrix, and on or close to fault surfaces. It is colorless, or bluish gray.

A single lens of *invoite* ($Ca_2B_6O_{11} \cdot 13H_2O$) was found in the open pit mine near the top of an upthrust (above the borate beds) where movement along a small fault had fractured a thin bed of "satin spar" ulexite. It had both intergrown and tabular colorless to white <2.5-cm crystals. Some inyoite had also been deposited in the soft clay below the ulexite bed, with small amounts of *meyerhofferite* ($Ca_2B_6O_{11} \cdot 7H_2O$) and colemanite. The meyerhofferite was found as massive white and silky fibrous pseudomorphs after inyoite, and some samples contained ulexite. Near the base of the upper ulexite bed the meyerhofferite consisted of small white prismatic crystals disseminated through green claystone, with realgar and natrolite. Minette (1988) noted that *probertite* ($NaCaB_5O_9 \cdot 5H_2O$) is "a widespread but minor (less than 0.5%) constituent of the kernite ores ... and is common in the footwall (p. 315)" clay. It often occurs as 3- to 4-cm globular clusters or rosettes of white, radiating, needle-like crystals. It can also resemble ulexite as beds of parallel fibers, but it is coarser, with 23 × 0.5-mm crystals. In the Western Borax and Suckow mines it was found as doubly terminated crystals in massive borax beds, or as rounded grains in the sucrosic borax of the Basal bed. In the Western Borax mine it also occurred as transparent, platy masses of columnar <6-cm crystals along bedding planes and fractures in the claystone. It is usually colorless and transparent with perfect prismatic cleavage, and often is found with cream-colored masses of *searlesite* ($NaBSi_2O_6 \cdot H_2O$) in the footwall bed.

Searlesite is a common-to-abundant mineral, especially in the footwall ulexite–claystone, and in altered tuff layers of the uppermost colemanite–claystone beds. It also occurs as white radiating crystal groups in the underlying basalt's cracks, and as scales in a buff-colored marl-like rock, or compact masses in a limestone-like rock in the footwall ulexite. Probertite is commonly associated with it. The searlesite is found throughout the deposit as small, thin colorless tabular crystals that are almost square, but with curved sides in altered beds of tuff, often accompanied by *boron-bearing K-feldspar* and zeolites. *Sassolite* (H_3BO_3) has been found in the Western Borax mine's "water drift" as sparse, minute pearly white scaly crystals disseminated in siltstone. In the Baker mine it was found at the contact of the lower claystone and the basalt.

Tincalonite ($Na_2B_4O_7{\cdot}5H_2O$) is fairly common as a dehydration product of borax, which turns to an opaque white rapidly on exposure to Boron's generally hot dry air. *Tunellite* ($SrB_6O_{10}{\cdot}4H_2O$) is not common, occurring primarily as single crystals disseminated in claystone. When found in horizontal strata, it is near fault planes, and above the Upper borax it also is found along vertical faults. Tunellite is generally colorless to pearly grey, but sometimes brownish to violet. When massive, it resembles fine-grained dolomite, with long thin 2 × 13-mm single crystals, or <0.5-cm cubic clusters with a pearly luster and perfect, almost micaceous, cleavage. Large single <4 cm crystals of *barium tunellite* (15% BaO) have been found with realgar and ulexite (Morgan and Erd, 1969).

Many assessory minerals are also present, such as the bright orange-red to blackish-red realgar (AsS) and stibnite (Sb_2O_3). A 0.6- to 6.1-m (2- to 20-ft) interval in the top section of the Upper bed contains relatively large amounts of realgar and stibnite, giving the borax and claystone a red to orange color. These minerals occur fairly consistently between two shale beds (U1 and U2), but are absent when the Upper bed is <4.6 m (15 ft) thick. Sparsely scattered realgar is also crystallized in the Lower borax. Part of the upper half of the Middle ore has a brown coloration, perhaps from realgar, which is found in the claystone immediately above it and all through the deposit. In the borax, realgar exists in finely disseminated filaments and lacy structures, whereas in the claystone it is found as microscopic crystals coloring the clay red or orange. In the upper ulexite bed of the overlying claystone–ulexite unit, realgar forms small masses and sometimes larger crystals. It also occurs as small lenses of essentially pure massive columnar crystals in the B (Middle) low-grade borax bed. Realgar often occurs in the occluded clay in the borax or along crystal boundaries, and in the claystone it can occur with ulexite as a fracture-filling material. It is found as fine-grained masses in the kernite (although less abundantly; Morgan and Erd, 1969). Bowser (1964) called realgar a "most spectacular sulfide mineral, ... with the contrasting brilliant red of the realgar [against] the white borates and green clay [being] one of the most striking features of the deposit (p. 151)."

Stibnite is far less abundant, but in some places it does occur with the realgar. It can be seen in borax as fine-grained parallel and radiating crystals between the much larger borax crystals (Bowser, 1964). Morgan and Erd (1969) noted that stibnite occurred as very small (microscopic) single needle-like shiny black crystals and clusters throughout the deposit, but was only concentrated in a few locations. This includes zones of realgar-rich clay at the Upper borax–claystone contact and in the Middle borax bed. Spherules of stibnite with a radiating structure have been found in a kernite–probertite–clay bed, and minute crystals of iridescent purple stibnite were noted in a core section. In a few vugs in calcite or colemanite, bright yellow orpiment-appearing crystals also have been found.

The alkaline solution-glass alteration sequence from volcanic glass or tuff to zeolites to analcime to boron-bearing K-feldspar to searlesite has been indicated by these minerals being present throughout the deposit and in cavities in the upper surface or near fault zones in the underlying basalt. The zeolites phillipsite, gmelinite, heulandite, herschelite, and clinoptilolite have been reported, and feldspar, analcime and searlesite are common.

2.1.6.4 Saline Valley, California

This small desert basin is located 29 km (18 mi) northeast of Keeler and 120 km north of Searles Lake (Fig. 2.13). It is 48 km long, 32 km wide, bordered on the west by the 3050 m Inyo Mountains and on the east by the 1830 m Panamint Range (Hardie, 1968). It has a 41 km^2 salt or mud-covered surface at an altitude of 329 m (1080 ft). The basin's drainage enters the playa from below after disappearing into the surrounding alluvial fans. The area has 100 mm (4 in.) of rain per year, but periodic thunderstorms occasionally flood the lake's surface. There are many large boulders (<2–3 tons) on the mud flat with long skid marks behind them, presumably moved by windstorms.

The salts in different segments of the lake are quite variable (Hardie, 1968), with each of the four quadrants surrounding the lake having quite different rocks and run-off water. In the northeastern segment with its volcanic rocks, the waters contain Na_2CO_3, Na_2SO_4, and borax; the southeastern segment with its granite has mostly NaCl with some Na_2SO_4 and $MgSO_4$; the southwestern segment with its limestone has Na_2SO_4 and $MgSO_4$; and the northwestern segment with considerable quartz has Na_2SO_4 and some $MgSO_4$. NaCl is a major component and borax a minor one in each quadrant (see Table 5.5; Garrett, 1992). The lake has one area with >0.5 m thick fairly pure halite, and the northeastern section once had 280 hectares (700 acres) of borax-containing effervescent crust on top of the salt. The borax area was analyzed in 1874, and claimed in 1885 by the Conn & Trudo Borax Company. By 1888 they were producing 40 tons/mo, or 400 tons/yr of borax, and the production remained vigorous until 1895. The borax-containing crust was variously said to be 7.6–15 cm (3–6 in.) thick, with 54.4% borax (Yale, 1892); 15–61 cm (6–24 in.) thick in places, with <90% borax (Bailey, 1902); or <46 cm (18 in.) thick (Gale, 1913).

The surface crust was removed and hauled in carts to the plant, where it was shoveled into long hemispherical wrought-iron tanks ("pans") partly filled with boiling water. They were set on arches of stone and fired from beneath with creosote bushes or shrubs growing next to the springs. The mixture was stirred with long poles until the salts were dissolved. Then the fire was extinguished, and the pans were settled for <10 hr until a clear liquor could be withdrawn. This brine was sent to vats to cool and crystallize borax for about 6 days. Then the liquor was pumped out and discarded, as were the residual solids from the leach tanks. The borax was chipped off the vats' walls,

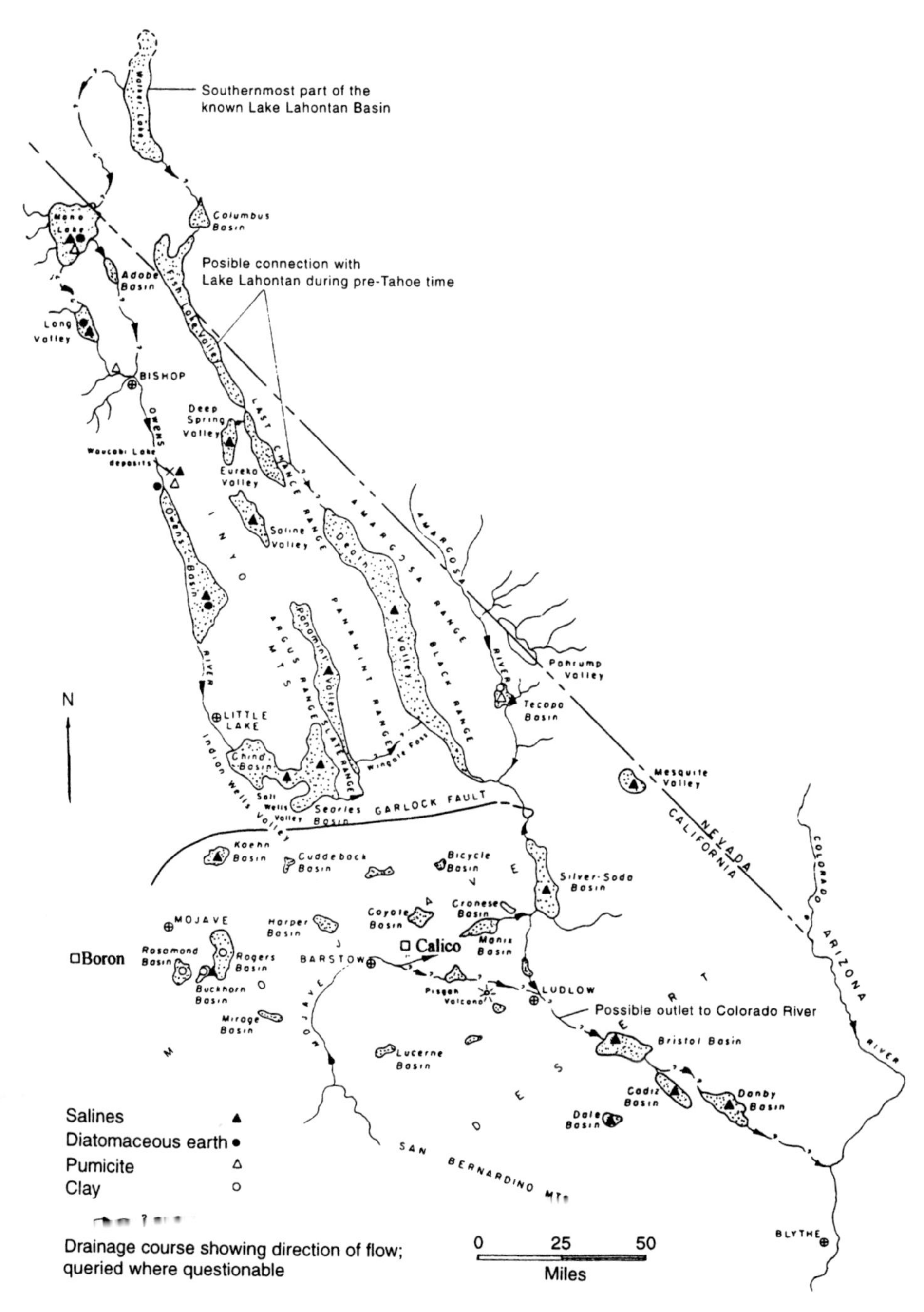

Figure 2.13 California–Nevada borax playas, including Saline Valley, Searles Lake, Death Valley, and Fish Lake Valley. (From Ver Planck, 1956; reprinted by permission of the California Department of Conservation, Division of Mines and Geology.)

removed, drained, dryed, and loaded into sacks for shipment (Gale, 1913). The operation had a staff of 30, and process and drinking water was piped 2.4 km (1½ mi) from a spring in the mountains to the west. The product was hauled by mule-drawn wagons 97 km (60 mi) to the nearest rail station. A second, much smaller plant was located by some warm springs to the northeast of the lake. Conn & Trudo, however, had the distinction of being the last California playa operation to close (in 1907) when the Lila C colemanite mine opened (Yale, 1892).

2.1.6.5 Searles Lake, California

The Searles Valley is ~200 km south-southwest of Los Angeles, near the Sierra Nevada Mountains (see Fig. 2.13) at the northern edge of the Mojave Desert and in the southwestern corner of the Basin and Range Province. The valley is 32 km long, and 16 km wide, with a 1600 km^2 drainage area. The lake is nearly elliptical, 16 km long and 10 km wide, and 63% of its 100 km^2 (40 mi^2) surface area is mud, whereas 37 km^2 is halite. Its elevation is 492–495 m, it has 25–250 mm/yr [average 96.3 mm (3.79 in.)] of rain, and temperatures of −7 to 49°C [annual mean 19.1°C (66.3°F); >38°C (100°F) common in the summer]. During some winters, 2–10 cm of water stand on its surface for a short period. Searles Lake's lowest zone of salts, the Mixed Layer, contains only beds of trona, alone or with other CO_3 and SO_4 salts. The younger, or the Upper and Lower Salts, are complex mixtures of many minerals and are porous, containing the brine used for the large scale production of borax and other salts since 1920. The upper deposits were formed from geothermal spring and runoff waters during the most recent Ice Ages, 40,000–10,000 years before the present (ybp), with the final salts deposited 3500 ybp. The underlying Mixed Layer's age is ~150,000 yr, and the valley started filling with 930 m of sediments 3.2 Mybp. The first scattered soluble salts crystallized 2 Mybp (Smith, 1997). The total reserves of borax (solids and brine) are 181 million metric tons (66.1 million metric tons B_2O_3; Fairchild, Lovejoy, and Moulton, 1997).

Searles Lake once received most of its water from the Owens River. When Owens Lake filled to 67 m above the present surface, it overflowed to China Lake, and when it, in turn, was 12 m above the present playa, it overflowed into Searles Valley to form a lake as high as 200 m above the present surface. From Searles Lake water could overflow to the Panamint Valley, where a >280 m-deep lake could flow into Death Valley. However, apparently only fresh water ever overflowed into these latter two basins. Material balances around the composition, the flow rate of several hot springs above Bishop, the age of the deposit, and the minerals in Owens Lake and Searles Lake indicate reasonably clearly that these springs were the primary source of their boron, potassium, lithium, tungsten, carbonate, and sulfate (Garrett, 1992).

Searles Lake contained three types of borax deposits: mixed surface salts, near-surface crystals in the mud, and brine and crystals in the Upper and Lower Salts (the latter will be reviewed in Chapter 5). Borax was discovered on the surface by Dennis Searles and E. M. Skillings in 1863, but its potential value did not become apparent until after the discovery of ulexite in several Nevada playas. Because borax was far cheaper to recover than ulexite, in 1873 Teels Marsh, and in 1874 Searles Lake began to be mined. Dennis Searles and others claimed four 65-hectare (160-acre) borax patents, including a 120-hectare (300-acre) low area containing the richest borax (in about the center of the western edge of the playa). J. W. Searles then became manager of the San Bernardino Borax Mining Co., which operated profitably until it and another small company were forced to close in 1894 because of the much cheaper borax from the Calico colemanite mines.

Runoff water that flooded the playa in some winters could accumulate <0.3 m (1 ft) deep in the low area and remain for several weeks. This area had 30 cm (12 in.) of permanent salt just under its surface, and after each flooding, and/or ~6 mo of capillary evaporation (the brine level was very near this area's surface), a thick effervescent mixed salt crust was formed. The near-surface brine was "a brown oily, smelly solution," with a 1.24 g/ml density. There was considerable sand in the crust on this near-shore area from the occasional severe spring dust storms. There also was "crystal borax," or "nests" of borax in the underlying mud, occurring "one foot [0.3 m] below some blue mud," with much smaller amounts of $Na_2CO_3 \cdot 10H_2O$, $Na_2SO_4 \cdot 10H_2O$ and NaCl (Robottom, 1893). Some borax crystals were found only 7.6–30 cm (3–12 in.) deep, and the crystal masses were not continuous, occurring as multiple 0.6- to 1-m- (2- to 3-ft) wide clusters, 1–1.3 m (3–4 ft) deep (Hanks, 1893). The <18-cm (7-in.) borax crystals were harvested during exceptionally dry periods, and 700 tons of refined borax were produced in one campaign (Hake, 1889). The mined mud–crystal mixture was screened and washed, and the crystals then quickly leached in hot water to dissolve the nonborax salts. The remaining borax was removed, dried, packaged in 35 kg (76 lb) bags, and sold in that form (Robottom, 1893). Unknown to these early operators, a much larger borax crystal-containing area (~200,000 metric tons) occurs at a slightly greater depth elsewhere in the lake, probably formed when it was the lowest area of the playa.

When the effervescent crusts were >2.5 cm (1 in.) thick, they were loosened by shovels and "swept" into windrows spaced far enough apart to allow a horse cart to travel between them. Then after draining, the salts were shoveled into carts and hauled <3.2 km (2 mi) to the plant, which was on higher ground, up from the shore (Fig. 2.14 is an idealized, out of proportion, but illustrative sketch). A typical harvest salt analyses was 12% borax; 10% Na_2CO_3; 16% Na_2SO_4; 12% NaCl; and 50% sand, water, and clay (Hake, 1889). For most of the plant's operation, the ore averaged ~8% borax on a refined product

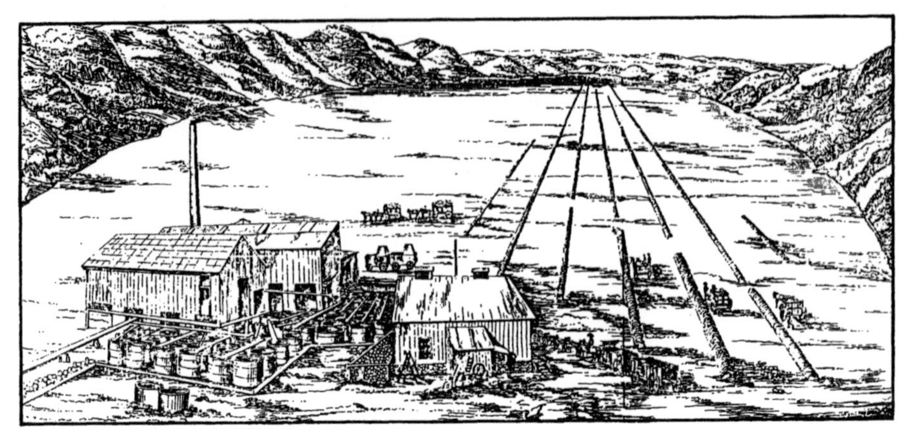

Figure 2.14 Idealized view of the surface borax operation at Searles Lake. (From Hake, 1889; reprinted by permission of the Society of Chemical Industry, London.)

basis, inferring a 67% plant yield. Other reports (Bailey, 1902; Hanks, 1883) stated that 13 tons of ore were required to produce one ton of product.

At the plant the salts could be sent to a ~2000-ton storage pile or processed immediately by first being crushed, and then loaded into one of five 26,000- (7000-gal) leach vessels. These tanks were made of 7.6-cm- (3-in.-) thick Oregon cedar and were 2.1 m (7 ft) deep, with a 1 m^2 (10.5 ft^2) bottom area. They had removable Oregon pine lids and were mounted above the 30 cylindrical galvanized-lined cooling crystallizers (initially wooden with no lining) to allow a gravity flow. The first borax leach used residual liquor from the second leach, if available, and was heated by steam sparging through holes in 3.1-cm ($1\frac{1}{4}$-in.) pipes in the bottom of each tank. The slurry was boiled, with some agitation by long hand-pulled rakes, until the end of the day when it reached a 16–30°Be (1.125–1.26 g/ml) density. The steam was then turned off, the tank settled >8 hr or until the next morning, and the clear liquor decanted through a movable withdrawal pipe into several of the primary crystallizer tanks. There it cooled and crystallized for 5–9 days while the mud from the leach tank was flushed out and discarded. The residual liquor was removed and sent to a large galvanized-lined solar evaporation pan to crystallize high-borax salts (which were recycled) before being flushed to the lake. The borax contained some insolubles, considerable other salts, and was colored by the organics in the brine. Consequently, the crystals were removed from the walls of the crystallizers, shoveled out, and returned to a leach tank with boiling brackish water. There they were heated and hand agitated until dissolved and the density was 18°Be (1.145 g/ml). After settling, the clear solution was sent to a group of nine galvanized-lined recrystallizer tanks, six 14.6 m (48 ft) long and three 11 m (36 ft) long, and all 1.2 m (4 ft) deep and 1.4 m (4.5 ft) wide.

After the brine had cooled to 49°C (120°F) and the borax had crystallized, the liquor was sent to the primary leach tanks, and the crystals were removed from the walls, shoveled onto draining boards and allowed to dry. Over 100 tons/mo of 99.85% borax was produced (Table 2.7), which because of the area's hot summers, became somewhat dehydrated, saving 5% on shipments and bringing a 1¢/lb premium (Hanks, 1883).

Initially, the only building in the plant was for the 1.1 m (42 in.) steam boiler [with a 81 cm (32 in.) flue pipe] used to power pumps and heat the leach liquor. It was first fired by burning creosote bushes (greasewood), but wagons gathering it soon had to travel >16 km (10 mi) and be gone for weeks. In 1887 a switch was made to the more economical crude oil, brought back by wagons 115 km (72 mi) from Mojave, the closest rail connection. Two sources of water were used: (a) drinking and boiler water piped 12 km (7.5 mi) from springs in the Argus Range in the northwest, (b) and 1% Na_2CO_3 brackish artesian (it rose 0.9–3 m [3–10 ft]) water from fourteen 17.8-m (55-ft) wells just upslope from the shore to the north, which was used for ore leaching and other process needs. Fifty Chinese laborers were employed, and there were 65 animals for the hauling. The wages paid per person were 1 foreman, $8.00/day; 4 mechanics, each $4.00/day; 50 laborers, each $2.00/day; 2 teamsters, each $100/mo; and 3 clerks, each $150/mo. The company supplied room and board for all but the clerks, making the total labor cost ~$50,000/yr. The product-haulage mules (two 20 mule teams) and the harvest-haulage mules cost $0.50/day/mule (DeGroot, 1890; Hanks, 1883). The refined borax

Table 2.7
Examples of Borax Product Analyses in the U.S. in 1882 (wt%)[a]

	Searles Lake, California	Teels Marsh, Nevada
Product		
Borax	99.85	98.87
NaCl	Trace	0.50
Na_2SO_4	Trace	0.25
Insolubles	Trace	0.38
Surface water, other	0.15	—
Insolubles		
Silica	—	73.67
Organics	—	6.43
Na_2O	—	5.54
CaO	—	3.89
Fe_2O_3	—	1.43
Water, etc.	—	9.04

[a] Williams, 1883.

product was placed in casks, and 10-short-ton loads were hauled to the railroad at Mojave in strong wagons ("desert schooners"). The round trip took 8 days, costing twice as much to ship the product to the railroad as from there to the East Coast. All of the necessary supplies and equipment were brought back in the wagons. The plant was considered to be well built and operated, and quite efficient for its period (Hake, 1889). When it closed, it was sold to Pacific Coast Borax, and the surface deposits were never worked again.

2.1.6.6 Teels Marsh, Nevada

Teels Marsh (Figs. 2.8 and 2.15) is 100 km west-northwest of Tonopah in Mineral County at an altitude of 1495 m (4904 ft). It is 6.4–9.6 km (4–6 mi) long and 1.6–4 km (1–2.5 mi) wide, and has an area of 16–31 km^2 (6.2–12 mi^2) and a drainage basin of 837–881 km^2 (323–340 mi^2). Its rainfall is 14 cm (21 cm in the mountains), with most of the water sinking into alluvial fans and entering beneath the lake. The climate is dry, with hot summers and cold

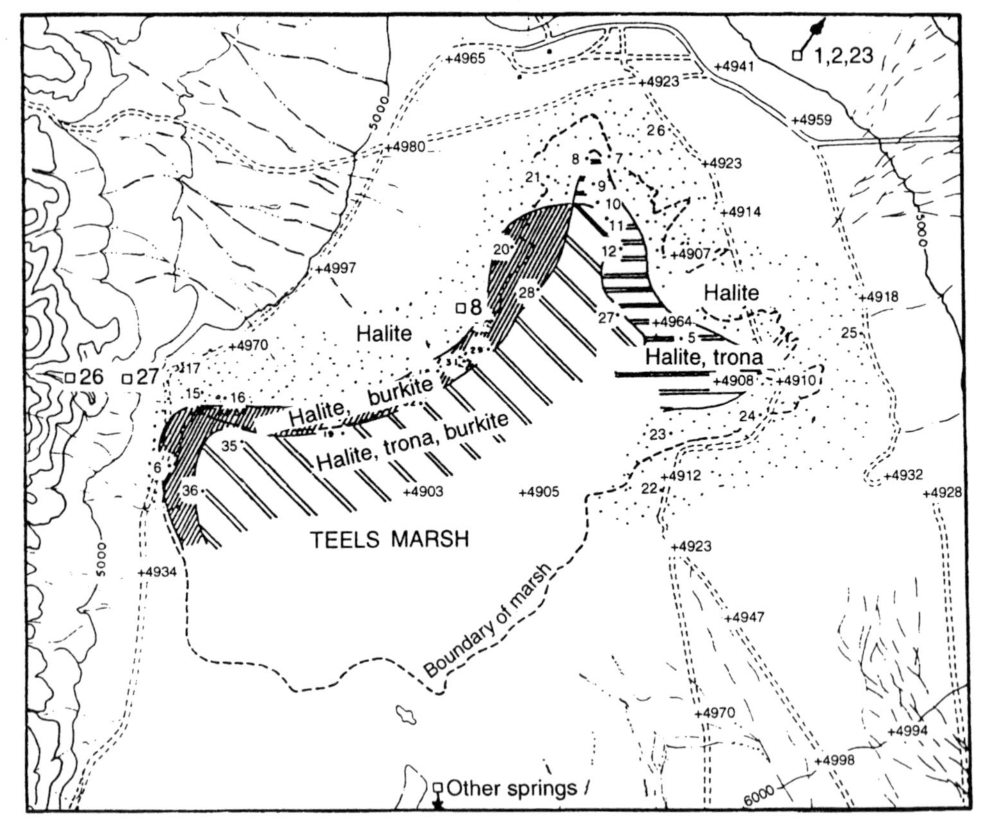

Figure 2.15 Recent distribution of salt crusts on Teels Marsh. (Adapted from Everts, 1969.)

winters, and has a net evaporation rate of 127 cm/yr. Its richest borate area was in the north, and 12 small springs in volcanic rock with carbonate-rich waters, and 3 in a small section to the northeast in sedimentary rocks, high in calcium, magnesium and sulfate flowed into the marsh. The boron content was 5–7 ppm (Table 2.8), and the combined springs and runoff formed a strong carbonate brine under most of the playa (Table 4.8; Everts, 1969; Papke, 1985; Smith and Drever, 1976; Van Denburgh and Glancy, 1970).

Borax ($Na_2B_4O_7 \cdot 10H_2O$) and tincalonite ($Na_2B_4O_7 \cdot 5H_2O$) occurred on the surface as a soft powdery mass (Table 2.9), and under it was 1.3–46 cm (0.5–18 in.) of borax crystals. The salts also contained trona, halite, in some zones burkeite, and various insolubles including searlesite ($NaBSi_2O \cdot H_2O$), giving the marsh's surface the appearance of "a vast bed of white sand resembling sea foam" (Hanks, 1883). The near-surface sediments contained clay, ryolite ash, halite, and calcite; only in higher elevations near the sulfate-containing springs was gypsum present. The upper tuff's age is 1,000 yr, and that of gaylussite at 5.5 m (18 ft) depth 10,760 yr (Hay, 1966). Because the playa maintains a fairly high brine level (0–2 m depth) in the lowest (northern) section, evaporation (both capillary and surface) caused the borate deposit to form rapidly in the summer, and brine cooling had a similar effect in the winter. Thus, when the surface crust was removed, it slowly replenished itself. The upper 3 m of playa sediments were brown and relatively uniformly permeable. Because the incoming waters enter the playa primarily from aquifers, the borax did not become saturated until it reached the surface, causing the deposit to form there (in the summer) or just below (winter).

Salt was recovered from the central part of Teels Marsh in 1867, but borax was not analyzed until 1872 by a young gold and silver prospector, F. M. "Borax" Smith. After difficulty in obtaining a large number of 8-hectare (20-acre) placer claims on the richest borax areas, Smith and his brother obtained financing and built a large processing plant on the west side of the marsh. Production started in 1873 and continued until 1892. There were also several smaller operations on the Marsh. Soon after production started, a price war began with the operators at Little Borax Lake (and later the Nevada ulexite operators), causing the price to drop low enough to close the Little Borax Lake operation, and to give the Teels Marsh plants temporary (in 1878) financial difficulties. Most plants quickly recovered, however, and for ~10 yr Teels Marsh and Searles Lake were the world's major borax producers.

The borax was recovered by raking the "dirty snow or chalk" sandy mixture of surface salts into windrows, with the workers standing on the undisturbed crust because the underlying mud was quite soft (an "unfathomable slime"; Anon., 1892). After draining, the windrows were shoveled into wagons and hauled to the plant on a small hill near the playa. In the Smith operation the borax was slowly shoveled into boiling water in one of nine 2.4-m- (8-ft-) diameter, 2-1-m- (7-ft-) deep iron tanks. They had previously been 30–50%

Table 2.8

Teels Marsh, Nevada Spring and Deep Well Analyses (ppm)[a]

Spring no. (Rock type)	Estimated										
	Na	K	Ca	Mg	Cl	SO_4	CO_3[b]	SiO_2	B[c]	TDS[d]	pH
8, 26, 27	363	16.4	12.4	0.2	244	298	104	69	6.2	1,113	8.43
Playa surface											
NW corner[g]	730	29	53	2	660	480	92	—	12.0	2,142	7.41
N center edge[h]	875	77	26	7	687	103	345	—	30.3	2,366	7.60
1, 2, 23 (sedimentary)	49	2.9	127	66	25	489	73.6	20	9.8	862	7.00
20, 21, 22 (mafic)	24	13.7	17.5	10.6	8.9	17.8	79.5	65	9.7	247	7.33
3, 5, 25, 30 (intermediate volcanic)	483	53	50	20.2	269	290	382	49	6.2	1,602	7.37
28, 29 (felsic)	285	23.5	46	14	176	134	249	25	4.5	957	7.33
Deep playa brine											
2–3 m[e]	11,500	625	—	—	18,000	468	254	128	422	33,300	8.9
3–8 m[f]	64,600	2,630	—	—	68,500	9,960	17,800	122	341	166,000	9.4

[a] Smith and Drever, 1976.
[b] Alkalinity, reported as CO_3.
[c] Estimated by ion balance (missing anions) for the 1976 analyses.
[d] Addition of all ions.
[e] Hay, 1966, SW edge. Also HCO_3 786, PO_4 68, F 28, Br 52, As 0.03, I 0.14.
[f] Hay, 1966, center, SW half. Also HCO_3 5840, PO_4 312, F 154, Br 85, As 2, I 1.9.
[g] Everts, 1969. Also Li 0.7, Cu <20.
[h] Everts, 1969. Also Fe 0.06, Cu <20.

Table 2.9

Teels Marsh Surface Salt Analyses (ppm)[a]

	Halite plus		
	Trona[b]	Trona, burkite[c]	Only halite[d]
Na	73,500	69,800	3.85
Fe	0.09	0.18	tr
Cu	0.19	0.04	0.02
Cl	10,600	10,500	5,170
SO_4	10,800	8,920	2,070
CO_3	9,090	18,600	367
HCO_3	22,300	19,300	660
Total carbonate	34,100	34,500	1,026
Sum of ions	126,290	130,350	12,117
B[e]	7,360	6,090	—
pH	9.14	9.16	8.95

[a] Everts, 1969.
[b] Northeastern corner.
[c] Northern and western sections. The subsurface brine also contained K 3230; Ca 2.7; Mg <400.
[d] Eastern edge, in the center.
[e] Estimated by the missing anions in an ion balance of the reported ions.

filled with water, with steam sparged in the bottom through small holes in 5-cm (2-in.) pipes. There were two 61-cm (24-in.) boilers burning piñon pine from the adjacent mountains or sagebrush from the valleys, and they also operated Cameron & Douglas water pumps. The slurry was manually mixed during continued heating until the brine had a density of 20–30°Be (1.16–1.26 g/ml). The steam was then turned off, and the contents settled overnight. The clear solution was next decanted into 16-gauge galvanized iron crystallizers 1.8 m (6 ft) square at the top and 2.1 m (7 ft) square at the bottom and cooled for several days, with borax crystallizing on the walls 1.3–10 cm (0.5–4 in.) thick. The residual brine was then drained and sent to a shallow wooden 0.2-hectare (0.5-acre) solar pond to crystallize additional salts (which were recycled), and the residual brine was discarded. The borax crystals, knocked and scraped from the empty crystallizers, were usually redissolved and recrystallized to produce a "refined borax" (see Table 2.7; Hanks, 1883). The leached mud was also mixed with water, heated, agitated, and releached to increase the yield before it was discarded. Prior to 1882, when the Carson and Colorado Railroad was completed, the plant's product was hauled by wagon 209 km (130 mi) to the Central Pacific station at Wadsworth (Papke, 1976).

2.1.7 Minor Occurrences

2.1.7.1 Antarctic

Small amounts of <3-cm (usually 0.5- to 1.0-cm) borax crystals and some pseudomorphs of kernite have been found in a Lewis Cliff, Buckley Island supraglacial moraine. The crystals were in <1.4-m borax blocks, with discrete pods of euhedral borax in a matrix of clear, euhedral tabular laths of nahcolite ($NaHCO_3$). The crystals were elongated, punky, white four-sided rhombic prisms, with a typical length-to-width ratio of 10. Trona and glauber salt have also been found in the 1500 m^2 area, and 1 km north was a supraglacial kettle pond with a high boron content. The moraine area was very close to the blue ice margin of an active glacier, but appeared not to have been deposited by recent glacial action. The borax's age was 78,000 yr and its nahcolite matrix 150,000 yr. Small irregular fragments of black, fine-grained limestone were scattered in the nahcolite matrix (Fitzpatrick and Muchs, 1989).

2.1.7.2 Russia

2.1.7.2.1 Mud Volcanoes

Mud volcanoes on the Kertch peninsula near the Sea of Azov contained borax, halite, trona, and fluorides in the extruded mud, and formed efflorescence on the dried sediments and run-off pools and marshes. The mud's temperature was low (Grabau, 1920). The mud's water initially contained 47–1867 ppmB, and the water from 25 other mud volcanoes in Russia had 0.8–2150 ppmB (Morgan, 1980). The Kertch occurrence was commercially worked for a short time, with the fresh mud containing 0.09–0.12% boron, and borax was <42% of the total salts. Dried salt accumulations were harvested and then leached. The gas in the eruptions was carbon dioxide.

2.1.7.2.2 Armenia

A small borax production has been reported from a mixed-salt deposit in the Dzulfa area (Orris, 1995).

2.1.7.3 United States

2.1.7.3.1 New Mexico

The Alkali Flat playa west of White Sands was 1.6 km (1 mi) wide, 8 km (5 mi) long, and was said to contain 4% borax in its surface salts (composed primarily of trona and halite; Northrop, 1959). Borax beds also were reported near Farmington in 1903, but the author (Herrick, 1904) felt that this was unlikely, as was a report of borax at Lake Lucero (Orris, 1995).

2.2 SASSOLITE

2.2.1 Italy and Larderello (also Spelled Lardarello)

The 200-km^2 (77-mi^2) region of steam vents, "soffioni" (soffoni), of Tuscany had been observed for centuries (perhaps 2000 yr; Donato, 1951), but they were in a foreboding area, thought to be that described by Dante in his *Inferno* as the place of ultimate punishment. They had the appearance of white and shifting clouds, hissing and blasting from the ground with a rumbling noise and smell of rotten eggs. Their waters sprayed 2–3 m (8–10 ft) into the air, and during heavy rains the runoff killed fish in adjacent streams for miles. The vapors initially were analyzed as 95.5% superheated steam, 4.3% CO_2, and 0.2% other gases (N_2, H_2S, NH_3, CH_4, H_2 and 0.01% H_3BO_3; Table 2.10), and could be toxic to man and animals when new fissures and fumaroles suddenly developed. The condensed vapors in the atmosphere formed pools (lagoni) containing dilute boric acid, ammonium sulfate and other salts. The boric acid was initially <1 g/1 (175 ppmB), but in older pools it could be <19.3 g/1 (3370 ppmB). Various borates have been found in the mantles and pipes, including sassolite (H_3BO_3), bechilite ($CaB_4O_7 \cdot 4H_2O$), biringuccite ($Na_4B_{10}O_{17} \cdot 3H_2O$), lagonite ($Fe_2B_6O_{12} \cdot 3H_2O$), larderellite ($NH_4B_5O_8 \cdot 2H_2O$), nasinite ($Na_4B_{10}O_{17} \cdot 5H_2O$; ezcurrite), santite ($KB_5O_8 \cdot 4H_2O$, very small clear and colorless crystals; Merlino and Sartori, 1970), and sborgite

Table 2.10
Various Gas Analyses (on a Water-Free Basis) of the Larderello Area Steam Vents (Volume %)[a]

	1842	1868 Lardarello	1868 Serrazzano	1868 Sasso.	Range (Ave.)[b]
CO_2	57.3	90.5	87.9	88.3	92–95 (93.7)
H_2S	1.3	4.2	6.1	5.4	2–2.5 (2.1)
CH_4	—	2.0	1.0	2.6	— (2.5)[c]
N_2	34.8	1.9	2.9	1.6	— (0.7)
H_2	—	1.4	2.1	2.0	—
O_2	6.6	—	—	0.1	—

[a] Hanks, 1883.
[b] Donato, 1951. Also H_2O 95.53%, H_3BO_3 0.78%, NH_3 0.43%. Other factors: temperature: 143–204°C (290–400°F); pressure: 71–390 psi; flow rate/vent: 400–1300 fps; noncondensable gas/1000 lb steam: 2400–3600 ft^3 at 0°C, 760 mm Hg; condensate: 0.3–0.4% H_3BO_3 (520–700 ppm B), 0.1–0.3% NH_3.
[c] Including hydrogen.

($NaB_5O_8 \cdot 5H_2O$). Sulfur, various ammonium salts, thenardite, orpiment, and silicates are also present. Recent deep drilling has indicated that the geothermal field's age is 2.5–3.7 Myr (Pliocene epoch), and that it has had continuous flow ever since. The basement rocks are 2000 m deep, with the thermal energy coming from an old granitic intrusion. Above the base, in ascending order, are quartzite and phyllite, clastic sediments, dolomite and anhydrite, limestone, sandstone, and alluvial rocks. The base temperature is 400°C, and the geothermal gradient 120–150°C/km (6.6–8.2°F/100 ft; Del Moro, Puxeddu, di Brozolo, and Villa, 1982). The fumarolic brine composition is quite different from the area's thermal springs (Bencini and Duchi, 1988).

Boric acid was first analyzed in brine pools in 1777, but it was not until 1808 that crude boric acid was produced on a very small scale. In 1820 Francesco Lardarel began purchasing land and mineral rights and initiated an evaporation–cooling process for the condensed brine. Larger pools were made around the vents, lined with stones, and filled with water. They were 1.2–1.8 m (4–6 ft) deep, depending on the pressure of the steam, and $<$13.7–18.3 m (45–60 ft) in diameter. During excavation, steam was vented through chimney pipes away from the workers to prevent them from being burned. Often the water from an upper pool, after ~24 hr of contact would be sent to the next lower pool, and sometimes to many in series (Fig. 2.16A). When the concentration reached ~2% H_3BO_3, the brine was taken to the plant and the pools refilled with water. In the plant the brine was evaporated to its saturation point, and then settled, cooled, and crystallized. The crystals were removed, washed, dried, packaged in wooden casks, and sold as crude boric acid.

Initially there were no roads, drinking water, local labor or housing at the site, and Lardarel very quickly exhausted the wood supply for his evaporators. In 1827 he conceived of using the steam as a heat source for evaporation. This was done by placing a lead tank, 0.9–1.8 m (3–6 ft) square, on top of a fumarole, with a vent pipe rising from the bottom through the tank. Its bottom was sealed by water in the vent's pool but did not block the brine's circulation. In this manner strong brine was formed as before, but when at the correct density it was placed inside the lead tank to complete the evaporation, which greatly improved the economics of the operation (Robottom, 1893). The economics were further improved in 1840 when a method for drilling into the steam reservoir was developed. This provided a much larger boric acid supply [more 140–230°C steam, with a higher boron content (0.03%)] and inexpensive heat for evaporation. In 1905 this system provided the basis for the first geothermal power plant. Production of boric acid was 75 ton/yr in 1818–1828, 4427 tons in 1828–1838, 680 tons in 1839, and a steady growth to 1800 mt/yr, then 2000 mt/yr in 1860 and 3,000 mt/yr in 1900. From 1840 to 1873 it was the world's leading borate source, and its lower price almost curtailed delivery from Tibet.

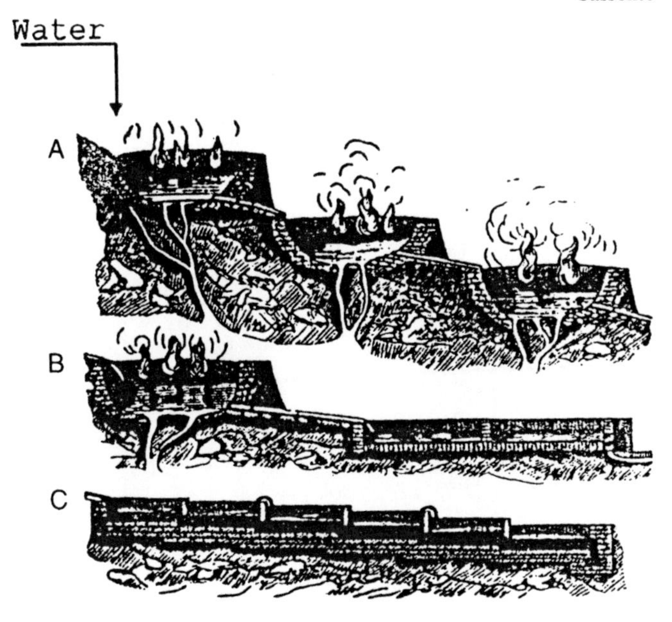

Figure 2.16 Boric acid absorption at Larderello, Italy. A. original natural flow of brine from one lagoon to another; B. original brine flow from the last lagoon to the plant evaporator; C. later arrangement of brine flow from one bore hole-fed lagoon to another. (From Hanks, 1883; reprinted with permission of the California Department of Conservation, Division of Mines and Geology.)

A U.S. consul visited the plant in 1882 noting (Hanks, 1883, pp. 63–72),

> *A shallow pond is dug, and in it an artesian well is bored, which at a small depth invariably strikes a bed of… [steam]. Not content with vapor alone, the boring is carried down till the well gives water. The boring machinery is then withdrawn and water let into the pond.*

The water fed to the pools turned slate-blue after ~24 hr.

> *Its temperature is considerably above 100°C. Dense vapors rise for many yards (m) above the ground, heating the air so much as to render it unpleasant to remain long near them. Efflorescent salts and decomposed rock ejected by the steam lie scattered all around the heated surface of the ground, along with sulfur incrustations and many sulfate … [salts], besides iron pyrites in minute veins in the fragments of rocks.*

When the lagoni water had been evaporated to 1.5–2% H_3BO_3, the brine was sent to the "vasco" (Fig. 2.17A), an 18.3-m- (60-ft-) square settling tank with a tiled roof. Clear brine flowed from it, and next entered a >60-m- (200- to 300-ft-) long building (for rain protection, Fig. 2.17B) containing an inclined "Adrian evaporator," designed by Lardarel and lined with lead. Three parallel

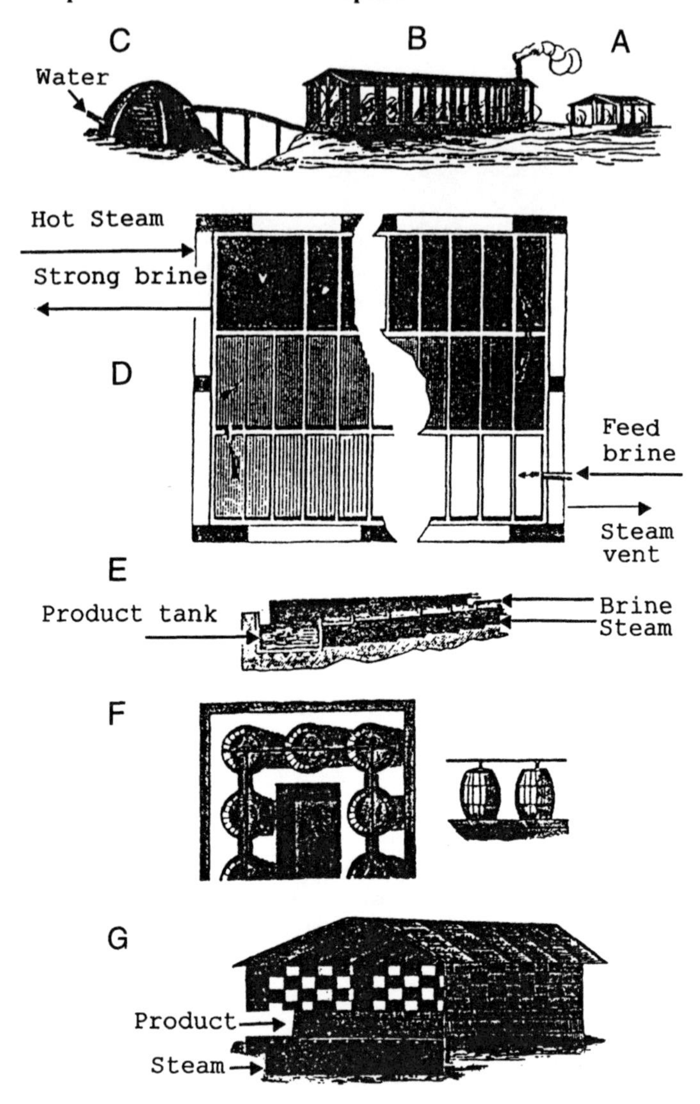

Figure 2.17 Boric acid recovery at Larderello, Italy. A. brine settler; B. plant evaporator building; C. steam generator for plant brine evaporator; D. top view of plant (Adrian) evaporator; E. side view of last evaporator line; F. crystallizer feed arrangement and barrels; G. product drying building. (From Hanks, 1883; reprinted with permission of the California Department of Conservation, Division of Mines and Geology.)

56-cm- (22-in.-) wide shallow pans with 1.8-m- (6-ft-) long subpans, 8.5 mm ($\frac{1}{3}$ in.) below the preceding pan with a 1.3-cm- ($\frac{1}{2}$-in.-) wide and high barrier (Fig. 2.17D), were arranged for the brine to cascade the length of the building. It first flowed along the outer row, then was pumped to the middle series of pans. At the end of that row, it was pumped to the other outer row. The final pan in the diagonal corner from the feed entrance was a deep liquid storage reservoir. Under each row of pans was an arched brick tunnel for the steam, initially generated by sending water to an adjacent steam vent that was totally enclosed (Fig. 2.17C) by a 3-m- (10-ft-) high stone dome. "The imprisoned high pressure steam ... acquires immense power, and thumping loudly against the dome ... seems ready ... to destroy it." The steam was conducted to the passageways under the evaporator, flowed countercurrent to the brine, and finally was vented. During the later operations, steam was obtained directly from drill holes. The entire process took 12 hr compared to the previous 62 hr.

After the feed brine had passed through 50 to 60 subpans it became slightly yellow, and by the end, when the brine contained 15–16% H_3BO_3, it was a bright golden yellow viscous syrup with a distinctive odor. It was periodically sent to a crystallizing house containing a series of large barrels 0.91–1.07 m (3–3.5 ft) in diameter (Fig. 2.17E). Plugs in a wooden trough allowed each to be filled, and the brine slowly cooled and crystallized for 4 days. The barrels were then drained, the liquor recycled, and the 2- to 3-cm- (1-in.) large hexagonal plates of H_3BO_3 with a flaky appearance and a pearly luster were removed. They were placed in large wicker baskets to drain and then emptied on the floor of a drying house (Fig. 2.17F), which was also heated from underneath by steam from the fumaroles. The boric acid was periodically raked and turned until dry, and then it was shipped in 614-kg (1350-lb) casks.

One of the operating difficulties was the scarcity of water, and often in the summer, Larderello's company was almost the only one that could operate. His plants frequently worked only half time, even with water from the mineral baths at Morbo. Production costs in $/ton were: labor, $10.00; packing and shipment, $13.00; administration and overhead, $8.60; maintenance and miscellaneous, $28.40; taxes, $13.00; total, $73/ton. The company employed 40 people, producing 3–3.5 ton/day in the face of especially difficult maintenance (the lagoons could collapse or diminish in steam supply, requiring new wells). Boring was often done in areas having many small steam jets, which could burn the workers or cave in. Corrosion was severe (even in the adjacent towns), and all aspects of the job were demanding and required careful control. Even so, the operation was very successful.

Unfortunately, some ammonium borate and sulfate also crystallized with the boric acid, reducing its quality (Table 2.11). Much of it was sent to Larderello's refinery in Marseilles, but by 1859 it began to be shipped to other European refiners also. By 1890 there were 10 separate boric acid operations in Tuscany, 7 of which belonged to Lardarel. He was an unusually caring and generous owner, and even built weaving factories in the villages to provide work for the wives of the plant workers. His company also developed housing, schools, a church, a music center, and other facilities for its employees, and even provided free health insurance and pensions (Grabau, 1920; Hanks, 1883;

Table 2.11

Examples of the First Crude Boric Acid Produced at Larderello, wt%

	Hanks 1883[a]		Dupont 1910
H_3BO_3	76.49		83.46
$(NH_4)_2SO_4$	8.51		5.30
$MgSO_4$	2.63		7.50
H_3SiO_3	1.20		—
$CaSO_4$	1.02		—
Na_2SO_4	0.92		—
K_2SO_4	0.37		—
		Fe, Al	0.30
$FeSO_4$	0.37		
$Al_2(SO_4)_3$	0.32		
NH_4Cl	0.30		
$H_2SO_4{-}H_3BO_3$[b]	1.31		
H_2O (hydrate)	1.31		1.44
Organics	Trace	+Insoluble	2.00

[a] 1842 analysis. The impurities in the salts were considerably reduced in the later manufacturing period. Other analyses: boric acid 89.0, 84.3; impurities 11.0, 15.7.
[b] Indicated to be a $H_2SO_4{-}H_3BO_3$ double salt.

Kistler and Smith, 1983; Robottom, 1893; Travis and Cocks, 1984; Williams, 1885).

In more recent times, boric acid has only been recovered from the steam condensate from the area's geothermal power plants. The steam's boric acid content was 0.35 g/kg of steam in 1950, and the production of H_3BO_3 was ~2000 metric tons/yr. Exhaust steam from the turbines was condensed, forming a 0.3–0.5% H_3BO_3, 0.1–0.3% NH_3 solution. It was concentrated in conventional evaporators with low-pressure steam to crystallize H_3BO_3, and after centrifuging, washing, and drying, its purity was 95–97%, or 99.5% if recrystallized. Borax, ammonium bicarbonate, and boron carbide were also produced (Donato, 1951).

2.2.2 Venezuela

There are three boron-containing "solfataric" springs in the Chaguarama valley of Venezuela. One of them,

> *Providencia, is a boracic soffoni, like those of Larderello. ... From small basins of hot water steam issues, hissing in white clouds. The waters contain alkaline compounds, ... together with boric acid, ammonium chloride, etc.*

The western spring, Potosi, was a series of active geysers, which periodically erupted,

> *With great vehemence, throwing up a large column of hot water and steam,*

and encrustations of chalcedony and silica formed around each vent. To the east, Buena Esperancia consisted of many small "lagoni," which interrupted the dense tropical forest with white barren round spots containing 50–100°C springs. They erupted much more mildly than at the other two areas, but as their water cooled it formed vast accumulations of sticky amorphous sulfur, which later hardened. The surrounding sandstone had become bleached white, saturated with silica containing considerable cinnabar, and dark areas of marl had small crystals of pyrite in their fissures (Cortese, 1904, pp. 741–742).

2.2.3 Minor Occurrences

Vulcano, in the Lipari Islands near Sicily, emitted abundant boric acid, which reacted with the surrounding rocks and crystallized hydrogen, sodium, and calcium borates in its vents and mantles. Sassolite was also common at Mount Vesuvius, Italy, and was found in fissures after the 1851 eruption (Chamberlin, 1912; Hanks, 1883).

> *Boric acid is a common exhalation accompanying volcanic eruptions, ... [and] is found in notable amounts on Stromboli and Etna as a sublimate in lava cavities and cracks. Acidic magmas give off appreciable amounts of boric acid and fluorine as they cool. Granite magmas also may give off boric acid (Keys, 1910, p. 701).*

Boric acid with lithium and strontium has been noted in the waters of Montecatini, Italy, just as sassolite encrustations are often found in fumaroles of the Shows-Shinzan and Asama volcanoes in Japan (Watanabe, 1967). Morgan (1980) noted that the boric acid content of 16 fumaroles throughout the world varied from 0.6 ppmB (Kilauea Iki, Hawaii) to 1130 ppmB (Suclick) and averaged 67 ppm. Minor amounts of sassolite have been reported from many sources, including efflorescence near water inflow at the Kramer deposit, Siegler, Tuscan Springs, and the Geysers in northern California, Steamboat Springs, Nevada, and Yellowstone National Park (White, 1957). It has been found in colemanite outcroppings in Death Valley and in several Tibetan localities (Smith, Almond, and Sawyer, 1958).

References

Alonso, R. N. (1986). *Occurrences, Stratigraphic Position and Genesis of the Borate Deposits of the Puna Region of Argentina.* Doctoral dissertation, University Nacional de Salta, Department of Natural Sciences, Argentina, 196 p.

Alonso, R. N., and Gonzales-Barry, C. E. (1995). *Geology of the Tincalayu (Salta) Borax Deposit.* 3rd National Conference of Economic Geology, Vol. 3, pp. A21–A36.

Alonso, R. N., and Helvaci, C. (1988). *Mining and Concentration of Borates in Argentina* (Y. Aytekin, ed.), pp. 301–305, Proc. 2nd Int. Min. Proc. Symp., Izmir, Turkey.

Alonso, R. N., Helvaci, C., Sureda, R. J., and Viramonte, J. G. (1988). A new tertiary borax deposit in the Andes. *Mineralium Deposita* **23,** 299–305.

Alonso, R. N., Jordan, J. E., Tabbutt, K. J., and Vandervoort, D. S. (1991, April). Giant evaporite belts of the Neogene Central Andes. *Geology* **19,** 401–404.

Alonso, R. N., and Robertson, D. B. (1992). *La Genesis de Kernita en los Yacimentos de Borax.* Instituto de Recursos Minerales, Universidad Nacional de la Plata, No. 2, pp. 1–8.

Alonso, R. N., and Viramonte, J. G. (1990). Borate deposits in the Andes. In *Stratobound Ore Deposits in the Andes* (L. Fontbote, ed.), pp. 721–732, Springer-Verlag, Berlin.

Anon. (1892). The history of borax in the United States. *Eng. Mining J.* **54,** 247–248.

Aristarain, L. F., Rusansky, J., Walker, M. F. (1977). Ulexita de Sijes, Provencia de Salta. *Obra del Cent. del Museso de la Plata* **4,** 23–47.

Ayers, W. O. (1882, July). "Borax in America. *Popular Sci. Monthly* **21,** 350–361.

Bailey, G. E. (1902). The saline deposits of California. *Calif. State Min. Bureau Bull.* **24,** 49, 63–65.

Barker, J. M., and Lefond, S. J. (1985). Boron and borates: Introduction and exploration techniques. In *Borates: Economic Geology and Production* (J. M. Barker and S. J. Lefond, eds.), pp. 15–34, SME-AIME, New York.

Barnard, R. M., and Kistler, R. B. (1966). Stratigraphic and structural evolution of the Kramer sodium borate ore body, boron, California. *Second Symp. Salt* **1,** 133–150.

Bencini, A., and Duchi, V. (1988). Boron distribution in thermal springs from Tuscany and Northern Latium. *Rend. Soc. Ital. Minera. Petrolog.* **43,** 927–933.

Blane, W. (1787). Some particulars relative to the production of borax. *Philosophical Trans. Royal Soc. London* **77** (Pt. 1, No. 28), 298–299.

Bowser, C. J. (1964). *Geochemistry and Petrology of the Sodium Borates in the Non-Marine Evaporite Environment.* Doctoral dissertation, University of California at Los Angeles, 243 p.

Bowser, C. J., and Dickson, F. W. (1966). Chemical zonation of the borates of Kramer, California. *Second Symp. Salt* **1,** 122–132.

Brown, C. (1921). *Borax: Principal Characteristics,* pp. 21–29, Bulletin of Indian Industry and Labour, No. 12.

Catalano, L. (1930). Resena geologica preliminar de la Puna de Atacama. *Revista Minera* **2**(9), 270–286; **2**(10), 289–305.

Chamberlin, R. T. (1912). The physical setting of the Chilean borate deposits. *J. Geol.* **20,** 766.

Chatard, T. M. (1890). Natural soda. *U.S. Geol. Surv. Bull.* **60,** 27–101.

Chen, Y. (1986). *Hydrochemistry and Evolution of Interstitial Brine in the Zabuye Saline Lake of Tibet,* pp. 176–184. Bulletin of the Institute of Minerals Department Chinese Academy of Geological Science, No. 2.

Chowdhury, A. N., Handa, B. K., and Das, A. K. (1974). High lithium, rubidium and cesium contents of thermal spring water, spring sediments and borax deposits in Puga Valley, Kashmir, India. *Geochem. J.* **8,** 61–65.

Clautice, K. H., and Mowatt, T. C. (1982). *Trona Occurrences Within the Yukon Flats Basin, Alaska.* U.S. Bureau of Mines Open File Report, No. 69–81, 34 p.

Cortese, E. (1904, November 10). A quicksilver deposit. *Eng. and Mining J.,* **78,** 741–742.

Cunningham, A. (1870). *Ladak* (a reprinting of earlier reports), pp. 136–137, 140–141, Sager Publications, New Delhi-I, India.

da Rovato, F. G. (1787). Father Joseph da Rovato's letter to the Royal Society relative to borax (Patna, 1786). Translation *Philosophical Trans. of the Royal Soc. London* **77**(2), 471–473.

DeGroot, H. (1890). *The Searles Borax Marsh,* pp. 534–539. California State Mineral Bureau, Tenth Annual Report.

Del Moro, A., Puxeddu, M., di Brozolo, F. R., and Villa, I. M. (1982). Rb-Sr and K-Ar ages of minerals at temperatures of 300–400°C from deep wells in the Larderello Geothermal Field (Italy). *Contrib. Mineral Petrol.* **81**(4), 340–349.

Donato, G. (1951). Natural steam power plants of Larderello. *Mech. Eng.* **73,** 709–712.

Dupont, F. M. (1910, December). The borax industry. *J. Ind. Eng. Chem.* **2,** 500–503.

Ericksen, G. E., Hosterman, J. W., and St. Amand, P. (1988). Chemistry, mineralogy and origin of the Clay Hill nitrate deposits, Amargosa River Valley. *Chem. Geol.* **67**(1–2), 85–102.

Etibank (1994). *Annual Report.* Sihhiye Cihan Sok. No. 2, 06443 Ankara, Turkey.

Everts, C. H. (1969). *The Evolution of Playa Waters, Teels Marsh, Mineral County, Nevada.* Master's Thesis. Department of Geology and Geophysics, University of Wisconsin, 105 p.

Fairchild, J. L., Lovejoy, M. E., and Moulton, G. F. (1997). *A New Technology for the Soda Ash Deposits Near Trona, California.* 1st International Soda Ash Conference, Rock Springs, WY, June 10–12, 10 p.

Fitzpatrick, J. J., and Muchs, D. R. (1989). Borax in the supraglacial moraine of the Lewis Cliff, Buckley Island Quadrangle: First Antarctic occurrence. *Antarctic J. of the United States* **24**(5), 63–65; Fitzpatrick, J. J. and Muhs, D. R. (1990). *Antarctic Res. Ser.* **50,** 57–69.

Foshag, W. F. (1921). The origin of the colemanite deposits of California. *Econ. Geol.* **16,** 199–214.

Gale, H. S. (1913). *Salt, Borax and Potash in Saline Valley, California,* pp. 416–421, U. S. Geological Survey, Bulletin 540-N, Contributions to Economic Geology for 1912, Pt. 1.

Gale, H. S. (1917). *Potash,* pp. 397–481, U.S. Bureau of Mines Yearbook, Pt. 2.

Gale, H. S. (1946). Geology of the Kramer borate district, Kern County, California. *Calif. J. Mines Geol.* **42**(4), 325–378.

Garrett, D. E. (1992). *Natural Soda Ash: Occurrences, Processing and Use.* Van Nostrand Reinhold, New York, 636 p.

Goodyear, W. A. (1890). *Lake County,* pp. 237–238, Calif. Min. Bureau Rept. 10.

Grabau, A. W. (1920). Principles of Salt Deposition. In *Geology of the Non-Metallic Mineral Deposits* (Vol. 1). Deposits Formed from Volcanic Waters, pp. 337–340, McGraw-Hill, New York.

Hake, C. N. (1889, November 30). An account of borax lake in California. *J. Soc. Chem. Ind.* **35,** 854–857.

Hanks, H. G. (1883). *Report on the Borax Deposits of California and Nevada.* Third Annual Report of the State Mineralogist, California State Mining Bureau, Pt. 2, pp. 5–111.

Hardie, L. A. (1968). The origin of the recent non-marine evaporite deposit of Saline Valley, Inyo Country, California. *Geochim. et Cosmochim. Acta* **32,** 1279–1301.

Hay, R. L. (1963). Zeolitic weathering in Olduvai Gorge, Tanganyika. *Geol. Soc. Am. Bull.* **74,** 1281–1286.

Hay, R. L. (1966). *Zeolites and Zeolitic Reactions in Sedimentary Rocks.* Geological Society of America, Special Paper 85, pp. 32–34.

Hedin, S. (1909). "Journeys in Tibet, 1906–1908," *Geographical J.* **33**(4, April), 353, 356–357, 379, 382, 385, 392.

Helvaci, C. (1977). *Geology, Mineralogy and Geochemistry of the Borate Deposits and Associated Rocks at the Emet Valley, Turkey.* Doctoral Dissertation, University of Nottingham, England, 338 p.

Helvaci, C., and Palmer, M. R. (1997). *The Boron Isotope Geochemistry of the Neogene Borate Deposits of Western Turkey,* Geochim. Cosmochim. Acta, **61**(15), 3161–3169.

Helvaci, C., Stamatakis, M. G., Zagouroglou, C., and Kanaris, J. (1993). Borate minerals and related authigenic silicates in northeastern Mediterranean Late Miocene continental basins. *Explor. Mining Geol.* **2**(2), 171–178.

Herrick, C. L. (1904). Lake Otero, an ancient salt lake basin in southeastern New Mexico. *Am. Geol.* **34,** 186, 189.

Hunt, C. B. (1975). *Death Valley.* Univeristy of California Press, Berkeley, 234 p.

Hurlbut, C., Aristarain, L. F., and Erd, R. C. (1973). Kernite from Tincalayu, Salta, Argentina. *Am. Min.* **58,** 303–313.

Hurlbut, C., and Erd, R. C. (1974). Aristarainite, $Na_2O{\cdot}MgO{\cdot}6B_2O_3{\cdot}10H_2O$, a new mineral from Salta, Argentina. *Am. Min.* **59,** 647–651.

Inan, K. (1972). New borate district, Eskishehir-Kirka Province, Turkey. *Inst. Mining, Metal. Trans. Sect. B,* **81**(789), B163–B165.

Inan, K., Dunham, A. C., and Esson, J. (1973). Mineralogy, chemistry and origin of Kirka borate deposit, Eskishehir Province, Turkey. *Inst. Mining, Metal., Trans. Sect. B,* **82,** B114–B123.

Irelan, W. J. (1888). *Lake County.* Report of the State Mineralogist, California Mineral Bureau Report 8, pp. 324–329.

Jones, B. F. (1965). *The Hydrology and Mineralogy of Deep Springs Lake.* U.S. Geological Survey Professional Paper 520A, 56 p.

Kashkari, R. L. (1977). Replenishment studies of borax deposits of Puga Valley, Ladakh District, Jammu and Kashmir State. *Indian Miner.* **31**(1), 34–39.

Keys, C. R. (1910). Borax deposits in the United States. Transactions, American Institute of Mining Engineers (AIME), **40,** 701–710.

Kistler, R. B. (1996). Personal Communication. Chief Geologist, U.S. Borax Inc., Valencia, California.

Kistler, R. B., and Helvaci, C. (1994). Boron and borates. In *Industrial Minerals and Rocks* (S. J. Lefond, ed.), pp. 171–186, New York, AIME.

Kistler, R. B., and Smith, W. C. (1983). Boron and borates. In *Industrial Minerals and Rocks* (S. J. Lefond, ed.) 5th ed., pp. 533–560, New York, AIME.

Lyday, P. A. (1992). *Boron.* Annual Report, U.S. Bureau of Mines, Washington, D.C., 11 p.

Meighan, C. W., and Haynes, C. V. (1970). The Borax Lake site revisited. *Science* **167**(3922), 1213–1221.

Merlino, S., and Sartori, F. (1970). Santite, a new mineral phase from Larderello, Tuscany. *Contr. Min. Petrol.* **27,** 159–165.

Minette, J. (1988, October). A notable probertite find at Boron, California. *Min. Record* **19**(5), 315–318.

Monastersky, R. (1996, December 7). Tibet reveals its squishy underbelly. *Sci. News* **150**(23), 356.

Montgomerie, T. G. (1870). Narrative report on the trans-himalayan explorations made during 1868. *J. Asiatic Soc. Bengal* **39**(2), 47, 50, 52–53, 55, 57.

Morgan, V. (1980). Boron geochemistry. *Inorg. Theoretical Chem.,* **5**(A), 72–152.

Morgan, V., and Erd, R. C. (1969, September–October). Minerals of the Kramer borate district. *Calif. Div. Mines and Geol. Min. Info. Serv.* **22**(9 and 10, Pts. 1 and 2), 143–153, 165–172.

Muessig, S., and Allen, R. D. (1957). The hydration of kernite. *Am. Min.* **42,** 699–701.

Navada, R., et al. (1991). Isotopic studies of geothermal waters. *Isotopenprasis* **27**(4), 160–162.

Northrop, S. A. (1959). *Minerals of New Mexico,* pp. 148–149, University of New Mexico Press, Albuquerque.

Obert, L., and Long, A. E. (1962). *Underground borate mining, Kern County, California.* U.S. Dept. Interior, Bureau of Mines, Report of Investigation 6110, 67p.

Orris, G. J. (1995). *Borate Deposits.* U.S. Geological Survey Open-File Report 95-842, 57 p.

Palmer, M. R., and Helvaci, C. (1995). The boron isotope geochemistry of the Kirka borate deposit, Western Turkey. *Geochim. et Cosmochim. Acta* **59**(17), 3599–3605.

Papke, K. G. (1976). Borates, evaporites and brines in Nevada Playas. *Nevada Bur. Mines, Geol. Bull.* **87,** 35 p.

Papke, K. G. (1985). Borates in Nevada. In *Borates: Economic Geology and Production* (J. M. Barker and S. J. Lefond, eds.), Ch. 6, pp. 89–99, SME-AIME, New York.

Robottom, A. (1893). The tincal trade, pp. 28–33; Boracic acid, pp. 84–88; Crude borate of Soda, pp. 143–146. *Travels in Search of New Trade Products,* Jarrold & Sons, 10 & 11 Warwick Lane, E. C., London.

Saunders, R. (1789). Some account of the vegetable and mineral productions of Boutan and Thibet. *Phil. Trans. Royal Soc. London* **79**(9), 96–97.

Saunders, R. (1806). Observations botanical, mineralogical and medical. In *An Account of an Embassy to the Court of the Teshoo Lama in Tibet,* pp. 406–407, W. Bulmer and Co., London.

Siefke, J. W. (1985). Geology of the Kramer borate deposit, Boron, California. In *Borates: Economic Geology and Production.* (J. M. Baker and S. J. Lefond, eds.), pp. 159–165, SME-AIME, New York.

Siefke, J. W. (1991). The boron open pit mine at the Kramer borate deposit: The diversity of mineral and energy resources of southern California. *Soc. Econ. Geol. Guidebook Series* **12,** 4–15.

Simoons, F. J. (1954). Borax: Nineteenth century mines and mineral spring reports of Lake County, California. *Calif. J. Mines Geol.* **50,** 300–303.

Sinha, B. P. (1971). *Geochemistry of the Thermal Springs in the Puga Valley, Ladakh,* pp. 264–265. Proceedings of the 58th Indian Science Congress, Part III, Section V, Geology, Geography, Abstract No. 62.

Smith, C. L., and Drever, J. I. (1976). Controls on the chemistry of springs at Teels Marsh, Mineral County, Nevada. *Geochim. et Cosmochim. Acta* **40,** 1081–1093.

Smith, G. I. (1979). *Subsurface Stratigraphy of Late Quaternary Evaporites, Searles Lake, California.* U.S. Geological Survey Professional Paper 1043.

Smith, G. I. 1997. Personal communication.

Smith, G. I., Almond, H., and Sawyer, D. L. (1958, November–December) Sassolite from the Kramer borate district, California. *Am. Min.* **43,** 1068–1078 (Nov., Dec.).

Smith, W. C. (1968). Borax solution at Kramer, California. *Econ. Geol.* **63,** 877–883.

Solis, A. R. (1996). *Summary, Loma Blanca.* Universidad Nacional de Jujuy, Instituto de Investigaciones Tecnologicas Minerase Industriales, 6 p.

Spurr, J. E. (1906). *Ore Deposits of the Silver Peak Quadrangle,* Nevada, pp. 158–159. U.S. Geol. Survey Prof. Paper 55.

Strachey, H. (1848). Narrative of a journey to Cho Lagan, Tibet. *J. Asiatic Soc. Bengal* **17**(2), 327, 331, 548, 551.

Sun, D., and Li, B, (1993) Origins of borates in the saline lakes of China. *Seventh Symp. Salt* **1,** 177–194.

Swihart, G. H., McBay, E. H., Smith, D. H., and Siefke, J. W. (1996). A Boron isotopic study of a mineralogically zoned lacustrine borate deposit: Kramer, California. *Chem. Geol.* (*Isotope Geosci. Sect.*) **127,** 241–250.

Tavernier, J. B. (1889). Travels in India (English translation by Ball). *London,* **2,** 16.

Travis, N. J., and Cocks, E. J. (1984). *The Tincal Trail-A History of Borax,* pp. 24–26. Harrap, London.

Van Denburgh, A. S., and Glancy, P. A. (1970). *Water Resources Appraisal of the Columbus Salt Marsh-Soda Spring Valley Area.* Water Resources-Recon. Series Report 52, 61 p.

Ver Planck, W. E. (1956). History of borax production in the United States. *Calif. J. Mines Geol.* **52**(3), 273–291.

Von Schlagintweit, H. (1878). Borax in the Puga Valley, Ladak, India. *Munchen Acad. Sitzungsb.* **8,** 518–522.

Vonsen, M., and Hanna, G. D. (1936). Borax Lake, California. *Calif. J. Mines Geol.* **32**(1), 99–108.

Wang, J. (1987). A preliminary study of the characteristics and conditions for forming the Anpeng Trona deposit. *Petrol. Explor. Develop.* **5,** 93–99.

Watanabe, J. (1967). Geochemical cycle and concentration of boron in the earth's crust. In *Chemistry of the Earth's Crust* (A. P. Vingradov, ed.), Vol. 2, pp. 167–178, Israel Program for Scientific Translations, Jerusalem.

White, D. E. (1957). Magmatic, connate and metamorphic waters. *Bull. Geol. Soc. Am.* **68,** 1659–1682.

White, D. E., and Waring, G. A. (1961). *A Review of the Chemical Composition of Gases from Volcanic Fumaroles and Igneous Rocks,* pp. C311–C312. U.S. Geological Survey Professional Paper 424-C, Article 261.

Whitehead, H. C., and Feth, J. H. (1961). Recent chemical analyses of waters from several closed basin lakes and their tributaries in the western United States. *Geol. Soc. Am. Bull.* **72**(9), 1421–1425.

Williams, A. (1883). *Borax,* pp. 568–569. U.S. Geological Survey, Mineral Resources for the U.S. for 1882.

Williams, A. (1885). *Borax,* pp. 861–862. U.S. Geol. Survey, Mineral Resources of the U.S. for 1883–1884.

Wise, W. S., and Kleck, W. D. (1988). Sodic clay–zeolite assemblage in basalt at Boron, California. *Clays Clay Min.* **36**(2), 131–136.

Yale, C. G. (1892). *Borax,* pp. 494–506. U.S. Geol. Survey of Mineral Resources in the U.S. for 1889–1890.

Zhang, M. et al. (1989). *Saline Lakes on the Qinghai-Xizang (Tibet) Plateau.* Beijing Science, Technique Press, Beijing.

Zheng, X. et al. (1988). *Salt Lakes In Xizang (Tibet).* Chinese Science Press, Beijing.

Chapter 3 Calcium, Magnesium, or Silicate Buried Deposits

3.1 ARGENTINA

3.1.1 Sijes (Pastos Grande)

The Sijes district has four major borate mining areas: Monte Amarillo, Monte Verde, Esperanza, and Santa Rosa (Fig. 3.1), with six small open pit mines operating in 1994, and reserves of 20 million metric tons (mt) (Kistler and Helvaci, 1994). The deposits' altitude is ~4000 m and its age is 5.7–6.8 million years (My). It contains bedded colemanite [$Ca_2B_6O_{11}{\cdot}5H_2O$; Monte Verde, Esperanza (upper beds)], inyoite [$Ca_2B_6O_{11}{\cdot}13H_2O$; Monte Verde (middle beds)], and hydroboracite [$CaMgB_6O_{11}{\cdot}6H_2O$; Monte Amarillo (lower beds)] within the Sijes Formation. The deposits are located discontinuously over 30 km, south southwest to north northeast, with more than 10 1–4 km subbasins deposited from a chain of lakes, and with thermal springs probably opened by the Quevar volcanic complex to the north during three separate periods. The deposits change laterally from borates to gypsum (or anhydrite) to clastics or pyroclastics and dip to the east. The <1-m (average 0.5-m) borate beds are folded, faulted, and interlayered with claystones, siltstones, sandstones, tuff, tuffites, gypsum, or anhydrite.

The borates occur in many forms: massive, nodular, lenticular, interlayered, and disseminated. Hydroboracite and colemanite are often yellowish-white, and the beds of the former are usually quite pure, but can contain clay or tuff. Small amounts of ulexite ($NaCaB_5O_9{\cdot}8H_2O$), inderite ($Mg_2B_6O_{11}{\cdot}15H_2O$, with inclusions of a bituminous material, Llambias, 1963), meyerhofferite ($Ca_2B_6O_{11}{\cdot}7H_2O$), nobleite ($CaB_6O_{10}{\cdot}4H_2O$), gowerite ($CaB_6O_{10}{\cdot}5H_2O$), and probertite ($NaCaB_5O_9{\cdot}5H_2O$) also have been noted. Bird tracks, fish and fauna fossils, mud cracks, and rain prints are in the adjacent clay (Alonso 1986, 1992b; Alonso and Viramonte, 1990).

3.1.1.1 Esperanza

This area is at an altitude of 3850 m, on the east side of the Sijes Mountains. Its colemanite zone is 37 m thick, 3 km long (north–south), dips 20° east, and is cut by a few faults with 20–100 m displacement. The borate beds, <0.6 m (0.2–0.3 m average, some only 1–10 cm) thick, are interlayered with dark

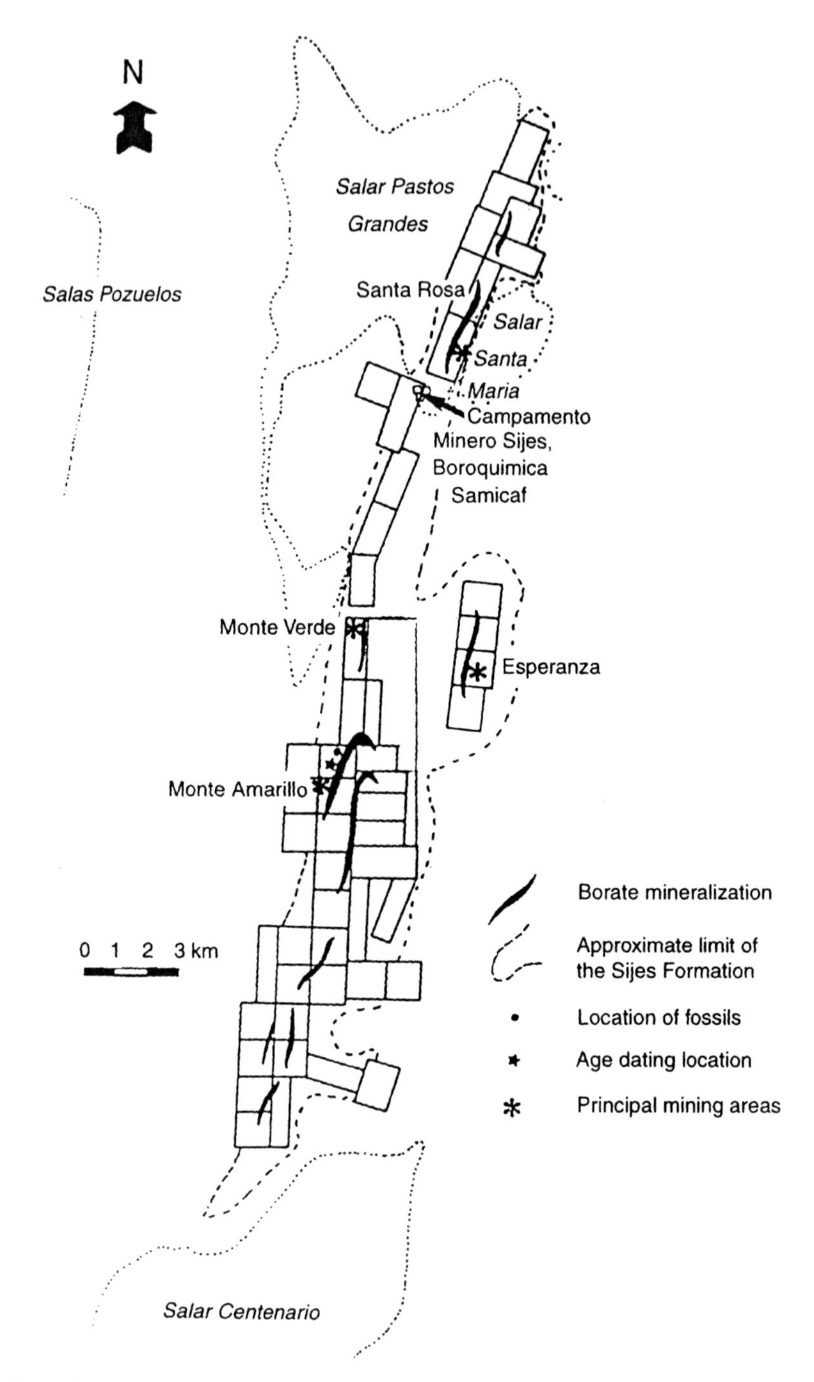

Figure 3.1 Layout of the Sijes Borate Deposit. (From Alonso, 1986; reprinted by permission of the Universidad Nacional de Salta.)

brown (some light green or yellow) fanglomerates, which are conspicuous in outcrops. Thinner layers of gypsum occur near the top and outer edges of the formation. The colemanite is primarily in beds, with a limited zone of tightly packed nodules (deformed into an ovoid shape). The 2 mm–3 cm nodules occasionally are widely dispersed, combined into small aggregates, or in geode form. The colemanite is often black or dark colored (from Fe or Mn impurities), but some thicker beds have clear and transparent crystals, and colemanite can be a vein-filling material. There are some beds of hydroboracite, and much smaller amounts of inyoite and ulexite. The hydroboracite is usually fibrous, found as massive lenses or beds, and can have colemanite in its fractures. Inyoite can form a sparse covering of fibrous crystals perpendicular to beds of hydroboracite. Ulexite occurs as small nodules in the clay over the borates, and some realgar and orpiment are in the borates and clay. Open pit operations with machine excavation have produced 100,000 metric tons of 28 ± 2% B_2O_3 ore (Gonzales-Barry and Alonso, 1987).

3.1.1.2 Monte Verde

This 378-m formation includes a 113-m colemanite and inyoite zone, with 11 m of borates in 0.2- to 2-m beds that outcrop more than 3 km (north–south) and dip 6–35° east. Faults cause 10- to 30-m breaks every 500–1000 m. The beds are interlayered with green and brown claystone, siltstone, tuff, some white sandstone, or occasionally gypsum–mudstone layers. These layers are quite uniform and continuous with large crystals, some containing only colemanite, others pure inyoite, and still others mixtures. There are small zones every ~50 m of hydroboracite, occasionally with gypsum. The borates merge into sandstone in the north, and gypsum and clay in the south. The deposit's structure, in ascending order is as follows: clay and gypsum with colemanite as tightly packed 5-mm to 2-cm nodules, or thin beds; two layers of white tuff; 1 m of fairly rich colemanite (with some inyoite, hydroboracite, and secondary ulexite) interbedded with clay, gypsum, and grey tuff; a layer with no borates; and upper and lower borate beds separated by layers of tuff, clay, gypsum, and some inyoite. The upper bed in most zones has thin layers of colemanite and inyoite in sandy clay, but one zone has a thick inyoite bed with some colemanite and hydroboracite lenses and veins.

Colemanite also occurs as vein-filling crystals and geodes. Inyoite's form is similar to that of colemanite, but thick beds and individual fragile, light yellow to green (usually transparent) crystals predominate, often intermixed with colemanite. Hydroboracite is most often found in beds with colemanite or inyoite as <10% of the borates. Its crystals are yellowish or white, fibrous, large, and tough. Ulexite is primarily in the claystone above the deposit as pure, white, small nodules, veins, or patches, apparently an alteration product.

Arsenic sulfide minerals are common. The reserves contain 3 million mt at 20% B_2O_3 (Alonso, 1992a).

3.2 CHINA

3.2.1 Liaoning

In 1988 there were 55 underground and open pit mines in 112 separate borate deposits (12 contained 98% of the reserves) in the Liaoning area of the Liaodong Peninsula in northeastern China (Fig. 3.2). The ore is szaibelyite (ascharite, $Mg_2B_2O_5 \cdot H_2O$) and szaibelyite–ludwigite [$(Mg, Fe)_2FeBO_5$], with some suanite ($Mg_2B_2O_5$; when pure, 41.5% B_2O_3, 2.7% FeO, 0.17% Fe_2O_3, 2.5% SiO_2, 1.54% F, 0.15% MnO) in fractured magnesian marble as numerous veins filled with borates, magnesite, and magnetite. The deposit has reserves of 44 million mt at 8.4% B_2O_3(5–18% B_2O_3), 64% of the country's total B_2O_3 (Kistler and Helvaci, 1992; Lyday, 1994). Borate production began in 1980, with 100,000 tons/year of ore from an open pit mine (Anon., 1980).

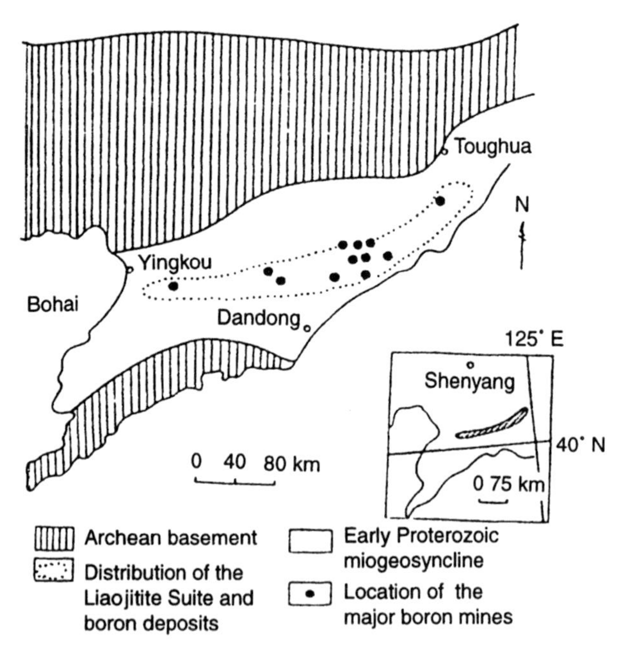

Figure 3.2 Tectonic setting of the Liaojitite Suite and distribution of borate deposits. (From Zhang, 1988; reprinted with kind permission of Elsevier Science-NL, Sara Burgerhartstraat 25, 1055KV Amsterdam, The Netherlands.)

The deposit is a skarn in an Early Proterozoic sial (upper-crust granitic layer) rift basin, 140-km wide and 300- to 400-km-long, which was part of a eugeosynclinal sequence (Fig. 3.3; a trough where volcanism is associated with sedimentary formations). It is highly mineralized, containing B-Fe and Pb-Zn deposits in the eugeosynclinal facies, and pyrite, phosphates, and the world's largest magnesite deposits in miogeosynclinal (no volcanism associated with sedimentation) facies. The deposit is >2 km thick and contains Liaogi Granite as the basement rock. The Boron-Bearing Series (age, 2.05–2.3 billion years bp) is a tourmaline-bearing granite; with albite-rich leptynite, only quartz-rich volcanics and magnesian carbonate layers. It is characterized by its distinct banded structure) and the Turbidite Series. The boron deposits are of two types: ascharite and ascharite-ludwigite. The latter type has much larger reserves than the former, but for ease of processing, in 1988 only the ascharite was mined. Some of the ludwigite deposits contain entirely borates; others are a rich mixture; and the smaller deposits usually are lower in grade. Stillwellite ($REBOSiO_4$) is fairly common, with its rare earths distributed roughly the same as in the basement granite and magnesian marble (Zhang, 1988).

The borates are closely related to the underlying volcanics (Fig. 3.4), with the granite's thickness determining the marble's thickness, and this in turn the ore's thickness. Also the ^{12}C and ^{34}S analyses in the various groups are nearly the same, again indicating a common source. When tourmaline (10.2%

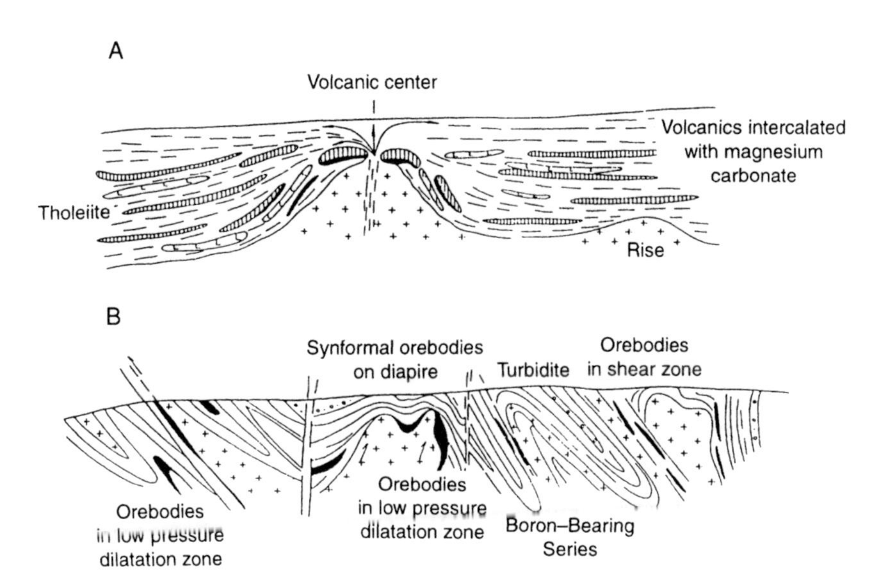

Figure 3.3 Model showing the formation (A) and modification (B) of borate deposits in the Liaodong Peninsula. (From Zhang, 1988; reprinted with kind permission of Elsevier Science-NL, Sara Burgerhartstraat 25, 1055KV Amsterdam, The Netherlands.)

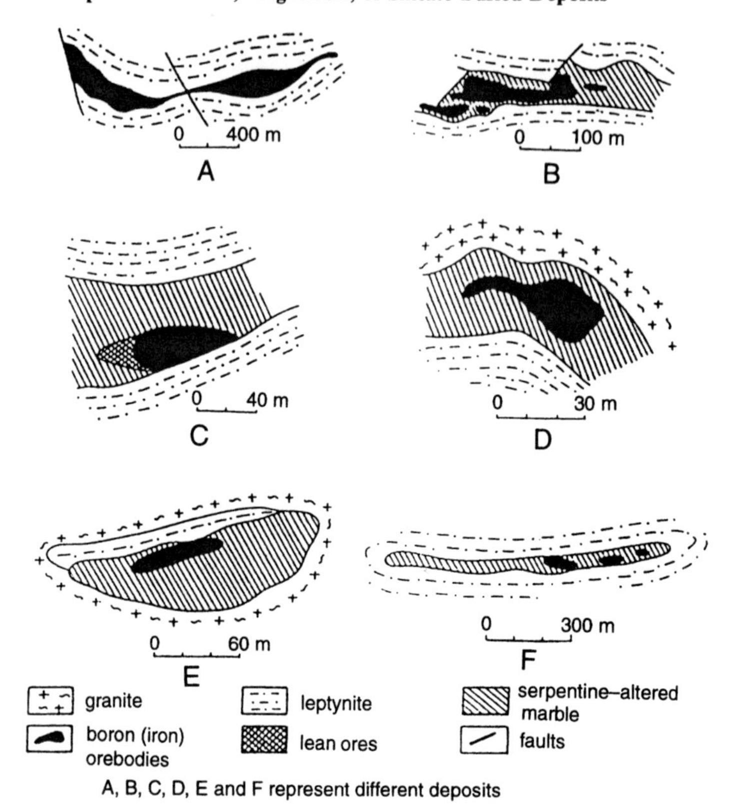

Figure 3.4 Relationship between borate ore bodies and their host rocks. (From Zhang, 1988; reprinted with kind permission of Elsevier Science-NL, Sara Burgerhartstraat 25, 1055KV Amsterdam, The Netherlands.)

B_2O_3) is not present in the volcanics, the other minerals contain smaller amounts of boron: hornblende 570 ppm B_2O_3, biotite 90 ppm, albite 30 ppm, and potassium feldspar, a trace, implying that the granite had been more thoroughly stripped of boron by the accompanying fluids at that location. The ore bodies have gone through major changes since deposition, with more than 40 domes in the area, but the ore is never more than 120 m from its granitic core, indicating that plastic granite from the ancient volcanoes caused the doming. It also frequently altered the rocks at the marble-volcanics contact to serpentine. The ore's distortion has also led to many fabric patterns for the borates (Zhang, 1988). Lixue (1985) earlier described these borate deposits as magnesium skarns formed by intruding magma and its accompanying hydrothermal, boron-bearing solutions. Peng and Palmer (1995) and several others have also examined the deposit. There have been many other articles on various skarns in China, as indicated in the References.

3.3 MEXICO

3.3.1 Magdalena Area

A 60-m (200-ft) outcrop of howlite ($Ca_4Si_2B_{10}O_{23}\cdot5H_2O$) has been noted 13 km southeast of Magdalena in Sonora at *Mesa del Almo,* as discordant veins through 10 m of Tertiary lacustrine sediments. Minor amounts of meyerhofferite ($Ca_2B_6O_{11}\cdot7H_2O$) and priceite ($Ca_4B_{10}O_{19}\cdot7H_2O$) were also detected, along with clinoptilolite, phillipsite, analcime, calcite, quartz, and illite. About 50% of the borate veins are 10–30 cm thick, and interbedded with pale red to green tuffaceous shales, 40% are 1- to 20-cm beds with fine-grained silty sandstone, and 10% are 1- to 80-cm beds with zeolitic or lithic tuffs. The howlite occurs as porcelaneous nodules, or as 2- to 30-cm veins in echelons perpendicular to fractures or bedding planes. Some thin 2.5- to 15-cm lenses form along bedding planes, and 0.9- to 1.8-m veins in fault zones. The subhedral or tabular 2- to 37-μ howlite crystals contain, in ppm: 5560 K, 2859 Fe, 531 As, 361 Ti, 270 Sr, 62 Rb, 48 Ba, 23 Zn, and 18 Pb. The howlite veins dip at 65–75°, and the "ore" averaged 6.3% B_2O_3 in shale, and 8.5% in sandstone and tuff (the veins are 33–43% B_2O_3; Lefond and Barker, 1979, 1985; McAnulty and Hoffer, 1972).

A drilling program by Vitro-U.S. Borax in the Magdalena area found colemanite ($Ca_2B_6O_{11}\cdot5H_2O$) with some howlite in a deposit containing >100 million mt at >10%B_2O_3 (Lefond and Barker, 1985). In a 1991 pilot plant study, it was reported that benificiation and removal of the ore's high arsenic content was difficult (Kistler and Helvaci, 1992). Borates were also found 15 km northwest of *Tubutama*, and 70 km northwest of Magdalena by the Mexican government's trenching and drilling (1300 m of core) program (Fig. 3.5). There were high-boron springs (with borate mantles) in the area, and colemanite outcrops. The estimated reserves were >29.6 million metric tons at 8.3% B_2O_3, including 1.41 million metric tons at 17.7% B_2O_3. Flotation tests on the ore gave concentrates with 32–43% B_2O_3 and 46–96% yields, but neither flotation nor calcination reduced the strontium, arsenic, and other impurities sufficiently to meet U.S. specifications (Arriaga, Pena, and Gomez, 1986).

The colemanite- (with some howlite) bearing zone was 4–21 m wide, >3000 m long, and in some areas <200 m deep. It dipped <30°, with 1-mm to 1-m (usually 1- to 10-cm) beds interlayered with fine- to medium-grained mudstone and sandstone (except to the south where the beds were green clayey tuff). The ore also occurred as lenses, veins, disseminated crystals, nodules, and a fibrous form with gypsum. Some 1- to 10-mm fracture zones contained colemanite, including a small amount that extended through a 5-m underlying shale bed and into fractured limestone. Even lower, colemanite could occur in 1-cm to 1-m beds in fibrous gypsum. Clinoptilolite and celestite often were found with the borates, which were of Late Miocene age, overlying

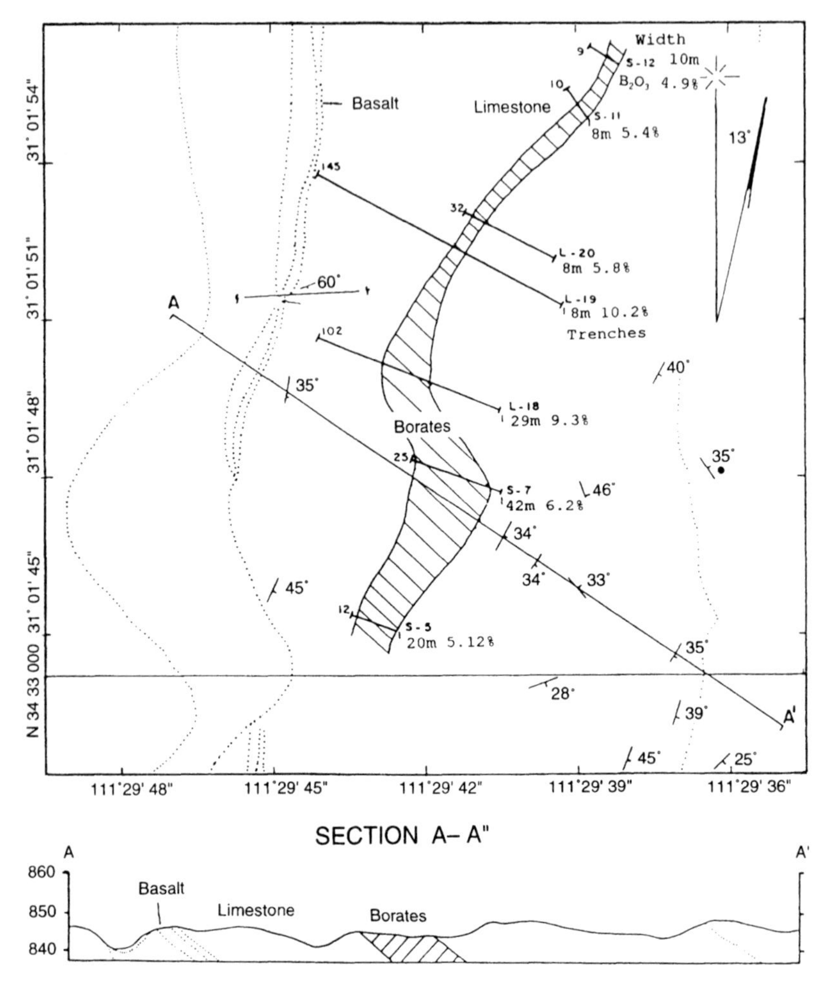

Figure 3.5 Tubutama borate outcrop. (Adapted from Caballero, Menedez, and Rocha, 1980.)

Early Miocene basalt (Lefond and Barker, 1985). Caballero, Melendez, and Rocha (1980) noted that the colemanite's anhedral and subhedral crystals had a granular texture, a vitreous luster, a hardness of 4–4.5, and a 2.42 density. They were colorless to white, transparent to translucent, and contained 50.9% B_2O_3. Some howlite was found throughout the deposit, occasionally as veins or porcelain-like nodules, with a hardness of 3.5, and a 2.58 density. Mcallisterite ($Mg_2B_{12}O_{20}\cdot15H_2O$), wardsmithite ($Ca_5MgB_{24}O_{42}\cdot30H_2O$), and ulexite ($NaCaB_5O_9\cdot8H_2O$) occurred in efflorescence near a local spring in trace amounts.

3.4 RUSSIA

The *Bor (Dalnegorsk)* deposit is in the Primorye Range of eastern Siberia near the Sea of Japan, 400 km (250 mi) northeast of Vladivostok. It produced 95% (80%; Crowe and Ignatiev, 1993) of Russia's 200,000 tons/yr B_2O_3 production in 1990 from 6–12% B_2O_3 ore. The deposit is a datolite ($Ca_2B_2Si_2O_9 \cdot H_2O$) and danburite ($CaB_2Si_2O_8$) skarn formed in Triassic limestones and siltstones, and intruded by two or three stages of magma and hydrothermal solutions. It is 2 km long, <500 m wide, <1000 m deep, and contains 600 million metric tons of 8–10% B_2O_3 ore (30–45% datolite). The deposit is mined (open pit) and processed by a workforce of <4800 (Kistler and Helvaci, 1992).

The deposit was first formed with danburite and minor amounts of datolite crystallized after the intrusion of granitoids and hydrothermal fluids into calcite at a depth of 1–2 km 50–60 My ago. A second intrusion 32–50 My ago by alkali basaltoids and accompanying fluids transformed danburite into datolite (except in one area), and formed new datolite (Malinko, 1992). There are numerous datolite pseudomorphs after danburite, and 14–17% NaCl alkaline brine in voids, indicating that the intruding solution was reactive and at >350°C (Khetchikov, Gnidash, and Ratkin, 1990a, 1990b). Rusinov, Kudrya, Laputina, and Kuzmina (1994) described the deposit as a "rhythmically banded pyroxene–wollastonite skarn formed in three stages (p. 570)." The first stage at 390–410°C originated the rhythmic zoning with pulsating flow, forming danburite. The second stage caused many replacement reactions, including the formation of datolite by the alteration of pyroxene and danburite.

Ratkin, Khetchikov, Gnidash, and Dmitriev (1992) noted that the borates are in the form of nodules, aggregates, spherules, or rhythmically banded massive beds. The deposit contains many other minerals such as native bismuth (Malinko, 1992; Malinko, Dubinchok, and Nosenko, 1992) and considerable lead and zinc (Hamet and Stedra, 1994). The latter visualized the skarn first forming at 600°C, followed by a stage at which galena (PbS), sphalerite (ZnS), datolite and many other minerals were deposited. Then there was a period with 120–350°C brine forming other sulfide minerals, followed by a postore stage precipitating many lower-temperature minerals (i.e., quartz, calcite, fluorite, and zeolites). Carbonaceous black shales in the calcite provided reducing conditions for the native bismuth (Buslaeva, 1994). Crowe and Ignatiev (1993) stated that the datolite is in a central zone with unmineralized zones surrounding it. The deposit is so large because the limestone was steeply dipping and capped by the final, datolite-forming andesite flows that restricted the escape of the mineralized fluids. In the edges of the skarn, danburite also was converted to axinite [$(Ca, Mn, Fe^{+2})_6Al_4Si_8B_2O_{31} \cdot H_2O$]. Other reports on this deposit are listed in the References.

3.5 TURKEY

3.5.1 Bigadic

The Bigadic deposits are in Balikesir Province, 3–15 km north of Bigadic at an elevation of ~400 m, with upper and lower borate zones bounded on the east and west by extensive volcanics (Table 3.1), and on the northwest by the same olivine basalt that is over and under the ore zone. The deposit is in a 3 × 10-km block-faulted basin that dips 10–60°, has a trend northeast–southwest, and was formed during the Neogene epoch (17 Mybp). The lower ore zone is 35–130 m thick, the upper 20–110 m, and in many areas they outcrop. The deposits have average ore grades of 40% and 29% B_2O_3 for the colemanite and ulexite, respectively (Table 3.2), and appear to have been formed in separate subbasins of a larger basin. The ore has had little alteration. Its original stratification is intact, even though the beds are moderately folded and faulted, and some are fault bounded. Ulexite is the dominant mineral at Acep, Arka, Kireclik, Kurtpman, and On Gunevi, although colemanite is the major mineral in the entire area. The two borate zones rarely occur with equal ore tonage in any mine. In general, the upper beds occupy a smaller area in the center of the deposit, and the lower beds are scattered in smaller zones around the periphery. The lowest colemanite bed in the lower borate zone correlates throughout the district, but none of the other beds correlate at all (Helvaci, 1984, 1994; Helvaci and Alaca, 1991).

The lower borate beds alternate with limestone, marl, tuff, claystone, or mudstone, and are over an alternating marl–tuff–limestone sequence (Fig. 3.6). They are under a banded section of claystone and limestone containing some chert. The individual borate beds are 0.2–40 cm thick, and the total thickness of the sub-beds 0–30 m. The upper borate zone is similar to the lower one, except that it contains more limestone and may lie on a waxy claystone. The matrix is tuff, tuffites, clay, and limestone, with minor amounts of anhydrite and gypsum (found only as thin interbeds with colemanite in the lower borate zone of three mines). Each of the borates can occur as nodules, thin or massive beds, and lenses. Colemanite's ($Ca_2B_6O_{11}·5H_2O$) most common ore form is nodules, occurring in many sizes and shapes, from minute clusters to <50-cm ovoid nodules with frequent thin discontinuous layers of clay or tuff. The nodules often have overgrowths, cracks, hollow centers and other imperfections. Fibrous colemanite, massive beds, and thin layers interbedded with clay also are quite common. Individual zones of nodules, crystals, and beds can be up to 5 m thick.

Ulexite deposits are always found above colemanite, even if widely separated, occurring as massive beds, dense or cotton ball nodules, and relatively pure fibrous crystals (with a silky luster) filling cracks and faults. Ulexite is always in or interlayered with clay or tuff, and its <5-cm crystals may be

Table 3.1

Analyses of Several Volcanic Rock Types in the Emet[a] and Bigadic[b] (Lower Numbers) Areas, wt% or ppm

	wt%					ppm[c]			
	Tuff	Dacite	Basalt	Andesite		Tuff	Dacite	Basalt	Andesite
SiO_2	43.93	67.14	48.66	54.25	As	1367	78	15	48
	66–77	65.7	54–61	57.4					
Al_2O_3	12.21	13.58	13.91	14.23	Sr	1077	258	609	621
	11–17	17.1	14–18	15.9		43–1,400	477	430–700	579
CaO	8.54	2.10	12.13	6.45	Ba	544	1293	1056	1113
	1.0–2.9	3.7	5.8–8.0	6.7					
K_2O	7.09	4.61	4.29	5.55	Ce	494	84	135	142
	3.0–5.0	3.8	1.2–3.7	4.3					
MgO	6.06	0.70	3.94	5.25	Zn	129	86	65	78
	1.0–1.9	0.75	2.4–8.5	3.8					
Fe_2O_3	3.08	1.81	4.76	6.06	Ni	84	29	138	176
	0.6–2.1	2.8	3.0–4.2	4.7					
TiO_2	0.63	0.38	1.66	1.84	Cr	52	8	263	266
	0.1–0.5	0.59	0.9–1.1	0.96					
FeO	0.60	0.44	0.72	0.70	Pb	46	59	17	14
	0.1–1.7	0.22	3.1–4.1	1.9					
Na_2O	0.29	1.92	1.84	2.31	Th	35	25	14	15
	0.5–2.6	4.0	1.8–3.0	2.7					
MnO	0.13	0.06	0.16	0.12	U	7	8	11	7
	0.02–0.15	0.08	0.05–0.14	0.12					
CO_2	3.86	0	1.61	0	Cl	359	1001	136	194
SO_3	0.88	0.06	0.07	0.08					
B_2O_3	0.72	0.32	0.24	0	Br	0	4	0	1
	0–0.77	0.23	0	0.45					
H_2O	6.53	4.63	3.32	2.59					
	0.6–7.8	0.74	0.2–1.3	2.41					

[a] Hel-aci, 1977. Clay in the deposit contained 0.60% B_2O_3.

[b] Hel-aci, 1995. Also (numbers, except P_2O_5, are ppm) *Tuff* P_2O_5 0.03–0.23%, Rb 150–210, Y 11–29, Zr 23–180, Nb 15–18, Ni 3–11; *Dacite* P_2O_5 0.19%, Rb 122, Y 28, Zr 246, Nb 14, Ni 6; *Basalt* P_2O_5 0.37–0.75%, Rb 87–137, Y 29–33, Zr 143–258, Nb 10–19, N 7–151; *Andesite* P_2O_5 0.58%, Rb 182, Y 24, Zn 164, Nb 15, Ni 10.

[c] Average of both areas, except Sr.

Table 3.2
Analyses of Colemanite and Ulexite Product from the Bigadic Deposit, wt%

	Early production[a]		1980 Production[b]		1995 Ore[c]	
	Colemanite	Ulexite	Colemanite	Ulexite	Colemanite[d]	Ulexite[e]
B_2O_3	45.50	42.39	40–42	37.52	33–44	32–40
CaO	27.15	13.69	24–25	23.90	25–28	12–16
SiO_2	3.01	1.64	5–6	7.76	3.1–9.5	1.1–12.4
MgO	0.10	0.23	1.5–2.0	3.65	1.1–4.7	1.1–6.3
Na_2O	—	8.68	—	6.07	0.39–11.1	6.2–7.7
Al_2O_3	—	0.01	0.5–1.0	0.41	0.1–0.4	0.2–0.4
Fe_2O_3	0.34	0.18	0.2–0.3	0.25	0.04–0.07	0.01–0.05
K_2O	(CO_2 0.53)	(CO_2 0.08)	—	0.37	—	—
SrO	—	(S″ 0.07)	—	0.33	1.1–2.1	0.5–2.1
As	—	—	0.06–0.12	—	0–0.01	0
H_2O, etc.	23.39	34.28	23.58–28.74	19.74	22–25	30–35

[a] Murdock, 1958. Sales were made on the basis of 43.5% B_2O_3 for colemanite (43.0% minimum when dried at 105°C); 39.5% B_2O_3, with a 39.0% minimum for ulexite. Composite samples were taken from the Acep, Gunevi, and Kireclik mines.
[b] Albayrak and Protopapas, 1985.
[c] Helvaci, 1995.
[d] Also SO_3 0.3–2.4; Sr 1400–11,400 ppm; Li 4–269 ppm; Ba 41–132 ppm; Rb 2 ppm.
[e] Also SO_3 0.02–0.53; Sr 270–11,400 ppm; Ba 38–123 ppm; Li 2–146 ppm; Rb 2 ppm.

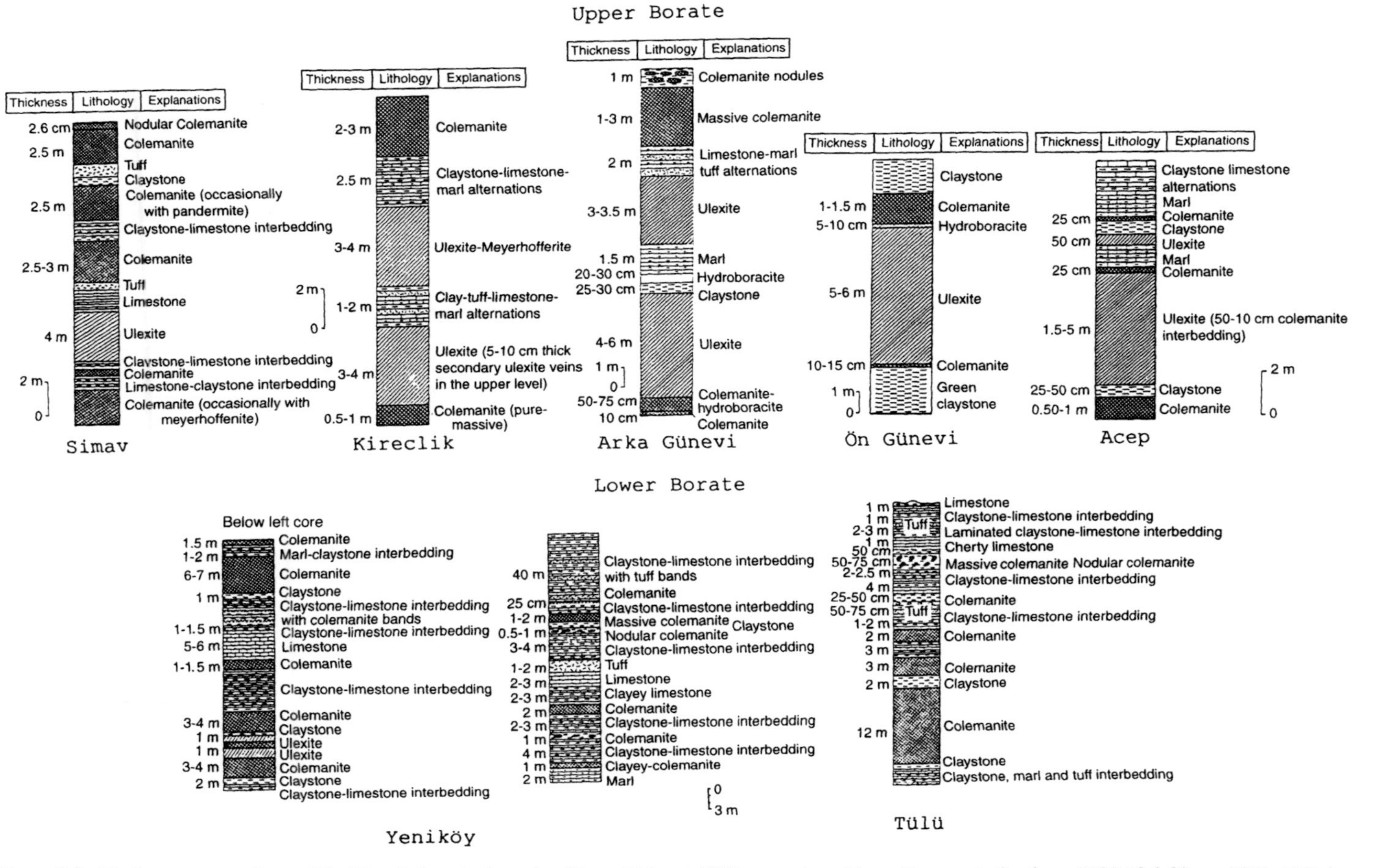

Figure 3.6 Various cross-sections of the Bigadic borate deposits. (From Helvaci, 1995; reproduced from Economic Geology, 1995, Vol. 90, pp. 1251–1253.)

tooth-shaped, banded, or in massive 1.5–5 m layers. Banded ulexite has thin fibrous crystals oriented perpendicularly to the bedding plain in a compact structure that is purer and less friable than the toothed variety. The 3- to 40-cm nodules are usually in 1- to 2-m beds (with some much thicker). There is more ulexite in the center of the deposit, with colemanite extending beyond it. The Bigadic ore has low levels of strontium, magnesium, and sulfate, and much less arsenic than the other Turkish deposits. Its clay minerals belong to the smectite group, predominantly montmorillonite, then chlorite, and some illite (Helvaci and Alaca, 1991). In the center of the deposit, some tuff has been transformed to zeolites (heulandite and clinoptilolite, the latter in commercial purity and quantity), analcime and boron-bearing potassium feldspar, celestite, chabazite, and opal. Current geothermal springs in the area (Table 3.3) contain some boron and deposit extensive travertine mantels (Helvaci, Stamatakis, Zagouraglou, and Kanaris, 1993).

Probertite ($NaCaB_5O_9 \cdot 5H_2O$) is found in <1-m lenses with ulexite in the lower beds and 20- to 30-cm lenses in the upper beds. Its 0.5- to 5-cm fibrous or radiating crystals are occasionally quite abundant. Substantial quantities of hydroboracite ($CaMgB_6O_{11} \cdot 6H_2O$) can also occur locally as nodules or beds, commonly intergrown with colemanite. Inyoite ($Ca_2B_6O_{11} \cdot 13H_2O$) occurs rarely in the upper borate zone at several deposits, but has not been observed at any mine in the lower bed. It is found as crystal aggregates, separate tabular (<2.5-cm) crystals, or very small crystal masses, with cole-

Table 3.3

Typical Analyses of Hot Springs in the Emet[a] and Bigadic[b] Areas, ppm

	Emet area[a]			
	Emet	Hamamkoy	Yenice	Bigadic[b]
Na	47.3	49.5	300	640
Ca	50.5	150	196	29
Mg	24.5	43.8	61.4	14
B	4.8	5.6	26.0	12.4 (10–13)
As	0.08	0.056	0.5	1.62
Sr	0.06	1.10	3.71	0.96
Conductivity	760	1210	3400	—
pH	7.2	7.0	6.9	—
Temperature °C	43.8	48.3	46	90
1000 B/TDS[c]	19	11	22	5

[a] Helvaci, 1977.

[b] Helvaci, 1995. Also HCO_3 1200; SO_4 395; Cl 220; K 71; F 8.2; Li 1.96; W 0.30; Sb 0.12; Ba 0.06; Au 0.02; Ag 0.01; Cd 0.01.

[c] TDS, total dissolved solids; estimated.

manite, ulexite, meyerhofferite, or hydroboracite. Meyerhofferite ($Ca_2B_6O_{11}\cdot7H_2O$) occurs as <10-cm nodules, usually containing uniformly oriented small crystals, with any of the other borates in most of the upper borate beds. Priceite ($Ca_4B_{10}B_{19}\cdot7H_2O$) is occasionally found in the upper borate zone at Avsar, Mezarbasi, and Simav in colemanite, ulexite, and calcite as white chalky <50-cm nodules. Rivadavite ($Na_6MgB_{24}O_{40}\cdot22H_2O$) is rare, found only in tailings, dumps, and ponds as 2- to 3-cm nodules. Some tertschite ($Ca_4B_{10}O_{19}\cdot20H_2O$) has been observed in the upper borate bed at Kurtpman in colemanite, ulexite, and meyerhofferite. Howlite ($Ca_4B_{10}Si_2O_{23}\cdot5H_2O$) is found in both zones in clays alternating with thin colemanite layers at Avsar, Kurtpman, and Simav, or as mixtures of small nodules with colemanite. Tunellite ($SrB_6O_{10}\cdot4H_2O$) occurs in Gunevi and Kireclik's upper borate zone as 1- to 5-cm flattened, colorless, transparent crystals attached to ulexite (or rarely colemanite; Helvaci, 1995; Ozpeker and Inan, 1978).

Borates were first commercially recovered at Bigadic in 1950, and by 1960 the deposit was fully delineated (Bekisoglu, 1962). The mines operating on the lower bed in 1995 were Tuluovasi (Tulu, open pit) and Yenikoy, and on the upper bed were Acep (open pit), Avsar, Gunevi, Iskele, Kireclik, Kurtpman, and Simav. Other closed mines were at Begendikler, Camkoy, Isklar, Mezarbasi, Salmanli (Helvaci, 1995), Borecki (Boreke), Domus, Pinari, Salmanli, Tulu Degirmen (Brown and Jones, 1971), Ankara 2 and 3, and Hannicik (Orris, 1995). Brown and Jones (1971) described the six (2-, 4-, 1.5-, 2-, 4- and 2-m) colemanite beds in one mine (in a 200-m clay–marl zone) as being in the center of a faulted syncline. Near the tops and bottoms of the beds, layers of clay separated thin beds of prismatic colemanite crystals. The colemanite beds were more massive toward the center, occasionally containing little or no clay, and in places, large nodules occurred with compressed clay coatings. Plastic bentonitic clay was above and below the colemanite, and there was a 1-m interval of clay continuing howlite above the top bed (considered too thin to be recovered). The slope of the beds was 40–50°, and they had "contorted bedding due to incompetent folding." A second mine 0.5 km away had two colemanite beds (3.1 and 1.4 m thick), and a third (2 km away) had 2.7-m and 4.5-m (upper) beds. They dipped 20–30°. The upper bed contained massive colemanite with no interlayered clay, and the lower bed had one clay layer. This mine also had thinner borate beds composed of ulexite with a central colemanite layer. When the beds outcropped, they converted entirely to calcite.

Murdock (1958) noted that at the Begendikler mine the ore occurred in 5- to 25-cm lenses within a 1.5- to 1.75-m-thick zone, averaging 50–60% colemanite, and dipping 13° (Fig. 3.7). At Gunevi the borate was 3.3 m thick, averaging 67% colemanite and 33% ulexite. The ulexite was in the center as amorphous-appearing fine crystals, with some colemanite nodules. Kireclik had only 2.5- to 3-m colemanite beds, with many cavities containing large

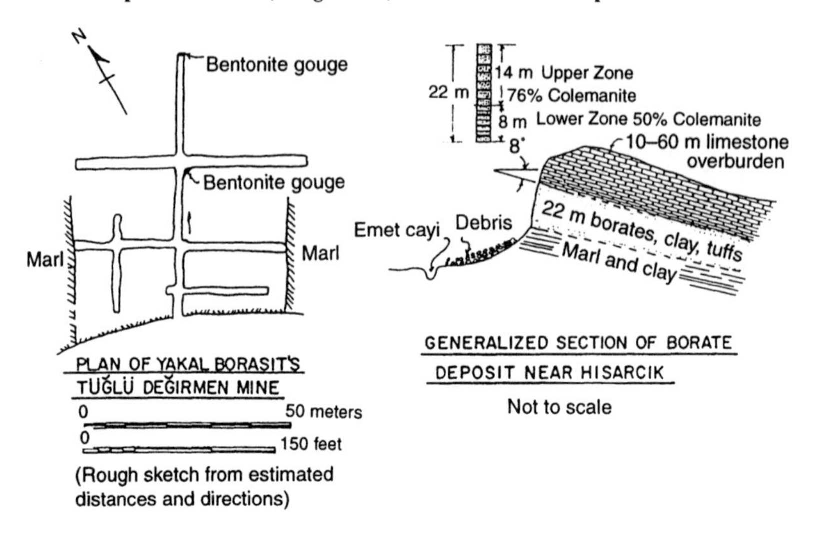

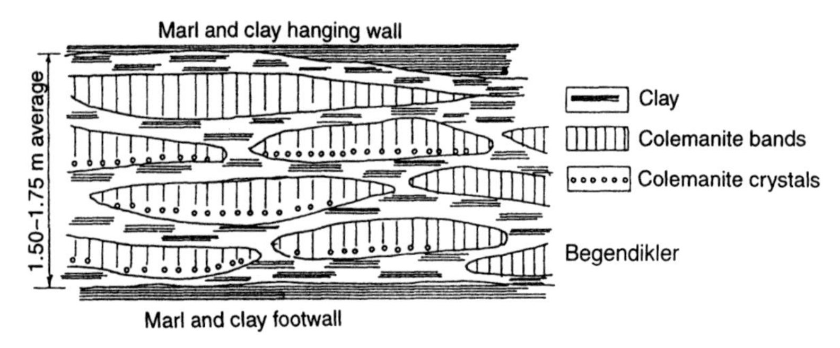

Figure 3.7 Various early Turkish Mines. (From Murdock, 1958.)

crystals. The Acep mine had some 3-m beds consisting primarily of ulexite with limited colemanite at the base, and lower beds of colemanite alone. At Tulu Degirmen the 2- to 2.5-m bed had a large amount of clay or shale partings, as well as nodules of colemanite.

3.5.2 Emet

The Middle Oligocene Emet colemanite ($Ca_2B_6O_{11} \cdot 5H_2O$) region in Kutahya province has numerous deposits, which like Bigadic to the west appear to have been formed in separate, but occasionally interconnected lakes. The 0- to 100-m borate zone (Fig. 3.8) consists of claystone with borates interbedded with limestone, marl, tuff, or clay. The borates outcrop east of the Emet River on the western side of the basin for most of its north–south length, dip 0–20°,

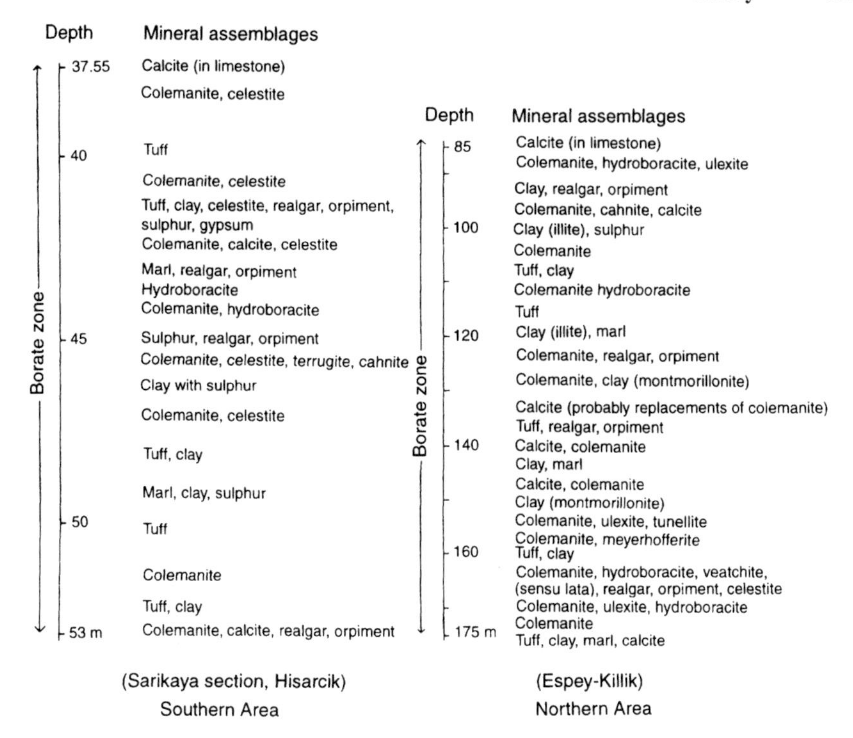

Figure 3.8 Mineral assemblages variation with depth in southern and northern Emet areas. (From Helvaci and Firman, 1976; reprinted by permission of the Institution of Mining and Metallurgy, Transactions, Section B, Applied Earth Science.)

and reach a maximum thickness of 100 m (30 m average, 30 m at Hisarcik) in the center of the Espey and Killik areas. Most of the borate zones are discontinuous, few extending >200 m laterally (Helvaci, 1984). Bekisoglu (1962) and Weiss (1969) stated that the beds dip 5–40°, and are 2 m thick in the north and >20 m in the south, averaging >50% colemanite, with very large nodules. Murdock (1958) noted that Emet's altitude was 1000–1500 m, and that it had greater ore thicknesses than Bigadic, more tuff interlayers, and a limestone capping, silicified at its base. There are numerous faults (some are active, having slipped further during the 1970 Gediz earthquake), thinning, and other dislocations. Brown and Jones (1971) observed that at outcrops there was a 5- to 10-m calcite-altered zone (with a <1-m transition) before unaltered colemanite was reached. There are a few thermal springs west of the Emet River (see Table 3.3).

Colemanite normally occurs as <0.5-m nodules, the smaller ones spherical and the larger ones ovoid. Some have hollow cores, and all consist of small-crystal aggregates, with thin layers of clay and numerous clay inclusions.

Most have formed a compacted 5- to 10-cm halo in their green clay matrix, and their crystals are curved. There are also some areas of bedded colemanite and crystals formed in clusters, disseminated, and found in cavities and fractures.

Many other borates also have been identified: ulexite ($NaCaB_5O_9 \cdot 8H_2O$), meyerhofferite ($Ca_2B_6O_{11} \cdot 7H_2O$), tunellite ($SrB_6O_{11} \cdot 4H_2O$), teruggite ($Ca_4 MgAs_2B_{12}O_{28} \cdot 20H_2O$), cahnite ($Ca_2BAsO_6 \cdot 2H_2O$), hydroboracite ($CaMg B_6O_{11} \cdot 6H_2O$), and veatchite-A ($Sr_4B_{22}O_{37} \cdot 7H_2O$). The meyerhofferite, ulexite, and teruggite usually occur as nodules (teruggite also as white euhedral crystals), whereas it appears that the other borates have formed from colemanite (Helvaci, 1984). Meyerhofferite and ulexite have been found only in the northern basin with colemanite in the lower zone. The nodules of the former contain coarse radiating crystals intergrown with clay, and often have a hollow center filled with delicate acicular crystals. Ulexite's cauliflower-like nodules are soft, and usually gray from clay occlusions. Teruggite occurs only sporadically at one level in the southern basin as white, powdery, 2- to 10-cm nodules containing white euhedral crystals and spherulites of cahnite. Small amounts of tunellite and veatchite have been found only in one deep horizon in the northern area. Hydroboracite is more common, occurring occasionally at three horizons in the northern area and one in the south, primarily at the margins as a colemanite replacement. There is a high arsenic content in all of the deposit (Helvaci, 1977; Helvaci and Firman, 1976).

Montmorillonite (of the Al, Mg, or Al-Mg-Fe type) is the major clay mineral, with illite, chlorite (relatively abundant near tuff layers), and smectite. Clinoptilolite and boron-bearing K-feldspar have been identified, whereas gypsum occurs only in the southern area (Helvaci et al., 1993; Kistler and Helvaci, 1992). Other minerals include calcite ($CaCO_3$), gypsum ($Ca SO_4 \cdot 2H_2O$), celestite ($SrSO_4$), sulfur (S), realgar (AsS), and orpiment (As_2S_3). The volcanic rocks in the drainage basin have an unusually high B_2O_3 content (2400–7200 ppm) as well as other geothermal-type constituents (As, Sr, Ba, etc.; see Table 3.3), and colemanite's impurities (Table 3.4) are also similar to those in the rocks (Helvaci, 1977).

There were four major mines in 1985 (Hamamkoy, Hisarcik, Espey, and Killik), with 85% of the reserves at Espey (former mines were at Goktepe and Derekoy; Orris, 1995). The deposit was discovered in 1956, and that same year commercial operations started at Espey and Killik, first by underground mining, and later with open pits at Hisarcik and Derekoy (Helvaci and Firman, 1976). At Espey, large colemanite nodules were mined in 1971 from a 3-m bed with a 6° dip, and realgar crystals were in the surrounding clay. At Killik, the 1.5- to 2-m bed had 50–60% colemanite as <1- to 2-m spheroidal nodules, often containing some water. At Hisarcik the 22-m ore zone had an 8° slope, the 7- to 8-m lower beds contained

Table 3.4
Average Analyses of Several Pure Colemanite Samples from the Emet Area, wt% or ppm[a]

	wt%			ppm	
	Hisarcik	Espey and Killik		Hisarcik	Espey and Killik
B_2O_3	47.63	47.39	Al	973	2439
CaO	24.19	24.20	Fe	767	1987
H_2O	19.91	20.35	Ce	214	132
SiO_2	2.63	2.29	Cl	132	165
MgO	1.92	1.33	P	119	125
SrO	1.51	1.22	Cu	27	22
Na_2O	0.24	0.19	Ba	27	104
As_2O_5	0.21	0.11	Mn	22	37
K_2O	0.06	0.32	Cr	8	12
SO_3	0.30	0.36	Zn	7	11
			Ni	5	14
			Pb	4	8
			Th	2	2
			Br	0	1

[a] Helvaci, 1977.

25–50% colemanite, and the upper 14-m bed had 75% colemanite. Some realgar was in both zones.

3.5.3 Kestelek

This 17.4-My-old deposit was discovered in 1954 with a 5-m, borate zone, the upper 1.5 m containing tightly packed <1-m nodules, and the lower section having many clay partings (Fig. 3.9; Albayrak and Protopapas, 1985). The borates occurred as nodules, masses of crystals, beds, or thin layers of fibrous and euhedral crystals. Transparent euhedral crystals of colemanite also filled cracks and cavities in some of the shale beds (Kistler and Helvaci, 1992). Smaller colemanite nodules were spherical or ellipsoidal, whereas the larger ones were ovoidal (egg-shaped), some with hollow centers and thin discontinuous layers of clay (Table 3.5). The matrix contained clay (predominantly smectite, with illite and chlorite), marl, limestone, tuff, dolomite, quartz, and clinoptilolite.

The deposit contained a small amount of ulexite ($NaCaB_5O_9 \cdot 8H_2O$), primarily as large, cauliflower-like nodules, with some fibrous beds, rosettes, and crack-filling crystals. The nodules were white to gray and very soft. Probertite

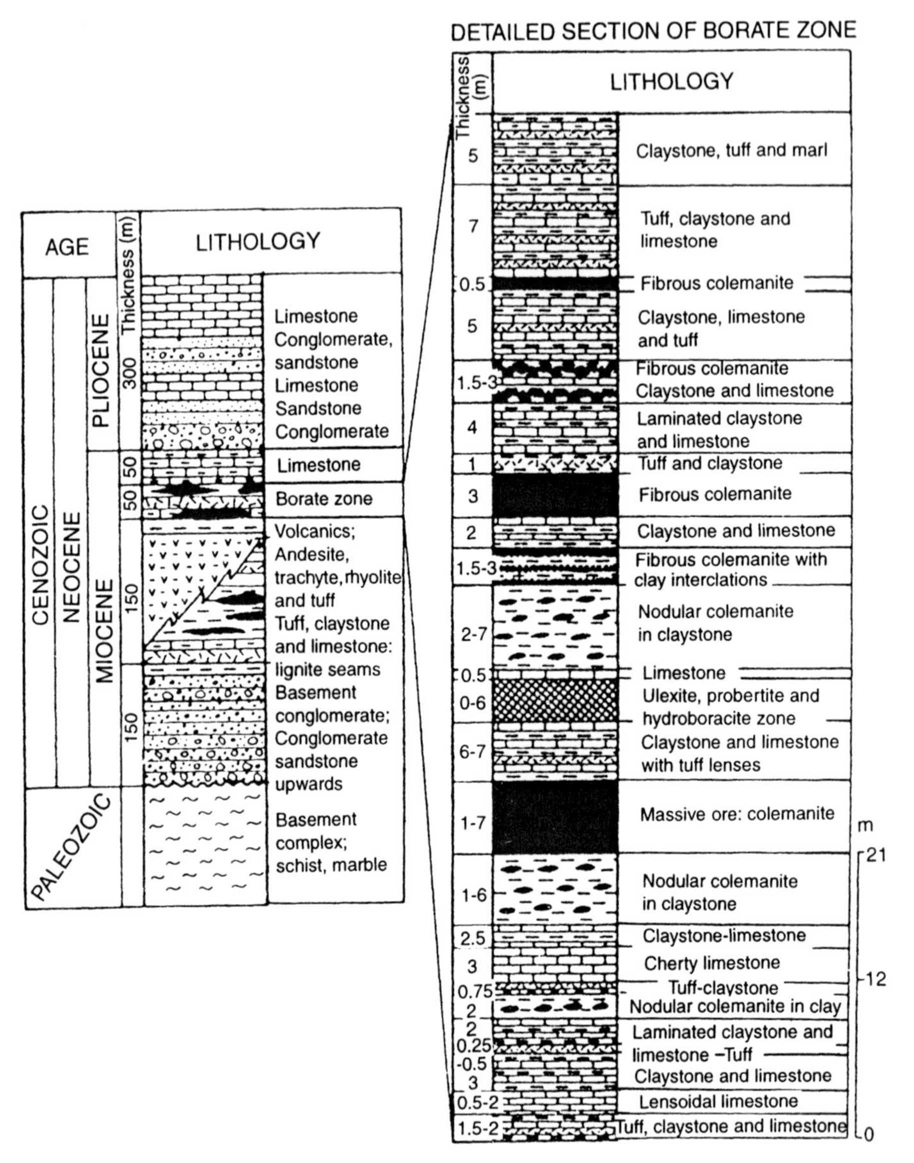

Figure 3.9 Generalized columnar section of the Kestelek area. (From Helvaci, 1994; reprinted by kind permission of the author.)

($NaCaB_5O_9{\cdot}5H_2O$) occurred only in a few deeper areas, always with ulexite as light yellow to white 5-mm to 5-cm radiating or fibrous crystals. Small amounts of hydroboracite ($CaMgB_6O_{11}{\cdot}6H_2O$) are found in all of the Turkish borate deposits, and here it was only in a few borate zones, usually as thin

Table 3.5
Chemical Analyses of Random Samples of Pure Kestelek and Sultancayiri Borate Ore, wt%[a]

	Kestelek		Sultancayiri	
	Colemanite	Clay	Priceite	Howlite
B_2O_3	41.35	2.10	46.49	40.69
CaO	32.05	6.65	34.50	29.98
SiO_2	3.66	17.27	0.21	15.56
Al_2O_3	0.70	26.50	0.04	0.08
MgO	0.59	9.76	0.12	0.19
Na_2O	0.48	1.27	0.80	0.92
Fe_2O_3	0.48	4.48	0.04	0.05
SrO	0.36	0.06	—	—
Li_2O	0.12	0.10	0.03	0.02
K_2O	0.03	0.60	0.05	0.10
As	0.01	0.013	—	—
SO_3	0	0.43	0.57	1.49
LOI[b]	20.17	27.77	18.09	11.48

[a] Helvaci, 1994.
[b] Loss on ignition (primarily H_2O and CO_2).

layers in adjacent clay beds. It formed white, light gray, or yellow radiating needle-shaped conical crystals, with colemanite or ulexite. The deposit has reserves of 10 million metric tons of high-grade ore (Helvaci, 1994).

3.5.4 Sultancayiri

Beginning in 1865 the Aziziye mine near Susurluk became the first borate operation in Turkey, with priceite ($Ca_4B_{10}O_{19}\cdot 7H_2O$) nodules recovered from a gypsum quarry (Robottom, 1893). The Sultancayiri mine, 1 km away, was discovered in 1887, and until 1954 when the ore in the district was exhausted and borate mining ceased, it was Turkey's sole (initially) and major borate deposit. The priceite outcropped, had an 8–10° dip, and was formed 20 My ago. The average 1- to 1.5-m (0.2- to 6-m) borate zone was generally discontinuous, with a 20% content of nodules. They weighed 10 g to 1 metric ton and contained 44–50% B_2O_3 (see Table 3.5). Their outer crust was hard like fine-grained marble, and the crystals were white, compact, and small, as magnesite or chalk. They could be removed easily from the dark gypsum matrix (Dupont, 1910).

The initial 1500-m mining area was connected to a second 1200 × 60-m mineralized zone. A 93-m vertical shaft serviced the mine, and the 1-m ore zone was nearly horizontal. A clay layer occurred at the top and bottom of the ore in the gypsum strata, and the ore was occasionally interbedded with claystone, limestone, and tuff. There was also some howlite, and minor amounts of boron-bearing K-feldspar, clinoptilolite, illite, chlorite, calcite, opal-CT, sanidine, anorthoclase, quartz, and celestite (Helvaci *et al.,* 1993; Kistler and Helvaci, 1992). The howlite ($Ca_4Si_2B_{10}O_{23}·5H_2O$) occurred as small, compact, dense, and structureless nodules that looked chalk-like and earthy. They usually were in thin priceite and colemanite bands in a clay bed, although some were imbedded in priceite and colemanite nodules. The total priceite production was 1 million metric tons of 42% B_2O_3 (Brown and Jones, 1971). Product was sent to the nearby railroad by an aerial ropeway (Bekisoglu, 1962).

A more detailed study by Orti, Helvaci, Rosell, and Gundogan (1997) noted that the borates occur in the central, deepest area of the basin as both beds and nodules, thus appearing to have formed as normal deposits from an evaporating borate-containing lake–playa. However, priceite's complex double pentaborate polymer structure implies that it is a secondary mineral and/or that there was a prolonged crystallization period. It is unlikely that it could have formed in the rapid dynamics of an evaporating lake. In this deposit, its secondary origin is also indicated by its low $\delta^{11}B$ (−20.0; Palmer and Helvaci, 1997), as if it were converted from a moderately low $\delta^{11}B$ mineral such as colemanite (−11.1 here), in which not all of the boron in the conversion liquor crystallized. Thus colemanite might have been the originally deposited borate, which much later was transformed to priceite from intruding high-silica, alkaline geothermal water. This intrusion also would have heated the formation, converting gypsum to anhydrite [because, otherwise, the burial depth of the formation was much too shallow (estimated at 200–250 m) for this conversion (normally >1000 m are required)]. The high-silica, alkaline nature of the waters is suggested by the presence of opal, chalcedony, chert, zeolites, and boron-bearing feldspar, none of which could have formed in the low pH (6.4–7) of gypsum-depositing waters. It appears that some priceite later was converted to howlite because of its even lower $\delta^{11}B$ (−24.5). Geothermal springs of this type are still in the area.

3.6 UNITED STATES

3.6.1 Calico District, California

3.6.1.1 Borate (Old Borate)

Colemanite was discovered in the Calico Mountains in 1883, 16–24 km (10–15 mi) north of Daggett, 13 km (8 mi) from Yermo, and 8 km (5 mi) east of the famous Calico silver mining camp. Its elevation was 910 m (3000 ft), and its age 13–17 My (Smith, 1985). The Coleman Company purchased the best claims in the area, and by 1888 was nearing full production at the largest, the Borate deposit. However, in that year they went bankrupt, and the Pacific Coast Borax Company acquired their borate properties. They continued with the Borate operation, becoming the company's major mine until 1907. When it closed, scavenging recovery continued by the Neel Borate Company and others from 1907 to 1924, with the mine subleased while it still had functional inclined shafts, drifts, and crosscuts. Ultimately, 1800 m (6000 ft) of stoping had been done, and a 200- to 300-m-long, 18- to 21-m-high mine dump followed the outcrop line (Wendel, 1978).

The beds at Borate outcropped for 2.4 km (1.5 mi) along the northeastern slope of the rhyolite-based Calico (Red) Mountain (Ver Planck, 1956, 1957). The colemanite also outcropped periodically over 10–16 km, starting on the northern and western sides of the Calico Mountain, passing down a canyon westward, and becoming mixed with sandy sediments, gypsum, and clay further west. At the outcrops, much of the colemanite had been converted to calcite. The best borate deposits were in the eastern part of this area (Wright, Stewart, Gay, and Hazenbush, 1953), and they soon became honeycombed with mine shafts and drifts. The colemanite was in many disconnected segments that often were difficult to locate. The two beds appeared to be limbs of an anticlinal fold, with one dipping to the north and the other to the south (the dip was always >15–20°). The southern limb contained by far the most ore. Crawford (1894) noted that the colemanite beds were highly folded and faulted, and intruded by dikes of liparite, hornblende, and andesite breccias. Yale (1905) also commented on the discontinuous nature of the borate veins, occurring as "pockets which may develop into very large deposits."

Colemanite in the Calico district was in the Tertiary (Middle Miocene) Barstow Formation as two beds (five; Piper, 1985) separated by a 12- to 15-m (40- to 50-ft) interval. The beds occurred as lenses in soft shale and were 2.1–3 m (7–10 ft) thick (0.6–4.6 m; Piper, 1985), and contained large, glassy colemanite crystals. The colemanite was beige to "snowy white, and composed of radiating crystals of singular beauty. ... The beds vary in color from red to green and gray, and have somewhat the appearance of indurated mud (p. 60)." The average ore had 15–20% B_2O_3 (Bailey, 1902). There were also some disseminated colemanite crystals or 10- to 20-cm nodules (with occasional celestite) randomly distributed in silty shales, thin interbeds of limestone, or dark brown calcitic shale. Some of the nodules were geodes with well-formed crystals; a few appeared to have two generations of growth; and others had celestite or selenite in the center. Small colemanite veins also filled cracks and cavities in the adjacent strata (Keys, 1910).

Beneath the colemanite were 15.2 m of boron-rich shale containing some thin colemanite layers or crack-filling veins, and sporadic 1.3-cm porcelaneous nodules of howlite. Over the borates was a bed of greenish gray silty shale, yellow tan limey shale and limestone. The rocks surrounding the colemanite were thin-bedded greenish-gray calcareous shales and sandstones, tuff, or yellowish-tan coarse sediments with some chert (Storms, 1893). Park (1991) noted that between the ore zones were calcareous concretions containing aquatic and terrestrial arthropods, zeolites (phillipsite, epistilbite, clinoptilolite, and chabazite), clay (montmorillonite, ailed, and smectite), sulfates (alunite, alunogen, celestite, kainite, and rozenite), carbonates, and iron-rich sulfides and oxides. Foshag (1921) was struck by the lack of metamorphism in the borate ore zone, even though the core of the Calico Mountains was formed by a later rhyolite flow. Foshag (1921, 1922) observed that some of the chalky outcrops had <1-m zones of flat tabular crystalline howlite. Howlite also occurred as thin shiny plates on some colemanite crystals. Occasionally ulexite was found in cracks in the lower andesitic tuff as fine needles in radiating masses, and bakerite ($Ca_8B_{10}Si_6O_{35}{\cdot}5H_2O$) was found in some vughs and cracks. There were also some younger stratified layers of gypsum (Piper, 1985).

3.6.1.2 American Borax Company

This mine at the foot of the north slope of Lead Mountain between Yermo and Barstow was most active between 1904 and 1907 (Wright *et al.,* 1953). The outcrop was 3.2 km long and 4.5–6 m (15–20 ft) wide at the surface exposure. In 1905 its production rate was second to that of the Borate mine at "a carload of boric acid every five or six days." Its ore contained 7–30% B_2O_3 (average 10–12%), and was claimed to be enough for a 50-year operation (Yale, 1905). The mine had two 1.5-m (5-ft) (4.5- to 6-m; Wright, 1953) borate beds dipping 70–75°, and consisting of disseminated colemanite in blue clay ("borate mud"). These beds were separated by 24 m (80 ft) of sandy shale containing uneconomic amounts of colemanite (Keys, 1910). The mine was initially an open pit, and later underground with a 60-m shaft at first (Bailey, 1902), then 120 m (400 ft) of inclined shafts connected by 300 m (1000 ft) of drifts for the overhand stopes. Ver Planck (1956) noted that the mine was first operated in 1894, and called the Columbian. It was sold in 1899 to the Columbian Mining and Chemical Company, then again in 1901 to the Standard Sanitary Company, which operated it as an affiliate. The *Blumberg* mine was 3.2 km (2 mi) south of the Borate, operating on a 15-m (50-ft) "mud" colemanite bed (Bailey, 1902). The *Cave Springs* borax operation held deposits of "mud" (perhaps with trona, halite, and salt cake) south of the Avawatz Mountains, but the property was never commercialized (Bailey, 1902).

3.6.1.3 Centennial

This mine was located high on the slope of the Calico Mountain, and only produced 600–700 tons of ore. The bed under the 75-m outcrop had large glassy colemanite crystals in black shale, also described "as irregular lumps in a chalky white howlite (?) matrix, distributed along a single zone in the shale (p. 224)." The bed dipped 20° northeast and had a strike of north 57° east. The mine had several small open pits with limited underground operations (Wright *et al.*, 1953).

3.6.1.4 Columbia Mining and Chemical Company

The two mines operated by this company were 7.2–8 km south (the Humphries, or Gem mine) and 9.6 km northwest (the Columbia mine) of Daggett, and were first mined in 1897 or 1898. Wright *et al.* (1953) described the Gem's ore as containing 8–18% B_2O_3 (with some railcar loads shipped at 30% B_2O_3), and occurring in shaley beds with subordinate sandstone or limestone. These low-grade, finely disseminated colemanite shales had a distinctive outcrop color and thus were easily recognized by prospectors. The mine had a 30-m shaft with drifts at the 15- and 30-m levels. The *Oasis Mining and Oil Company* was sinking a shaft on its borax claims 2.4 km northwest of Marion in 1902 (Bailey, 1902), but ore was never mined. The *Owens* mines were located 8 km northwest of Daggett, 9.6 km southwest of Calico, and "the borate beds had been exposed by the uplift of a mass of eruptive rock parallel to the Calico range." The ore was 12–15 m thick and had a 70° dip. A 60-m vertical shaft was sunk into the deposit (Crawford, 1894).

3.6.1.5 Palm Borate Company (American Board of Promoters)

This mine was 10.4 km from Daggett, 0.8–2.4 km south of the main colemanite zone, with gray shale exposures on the east side of Calico Mountain (Yale, 1905). It was another of the four operations on low-grade borate shales or "muds" with 5–8% B_2O_3 ore: American, Columbian, Palm, and Western. Their processing plant was completed in 1907 (as Borax Properties Inc.), but operated only 10 days when a drop in boric acid prices (caused by the opening of the Lila C mine) resulted in its bankruptcy (Wright *et al.*, 1953). The mines of *Stevens and Greer* were located intermittently over an 8 km distance, starting 8 km east of Calico along the southern side of the range. Only small open cuts and drifts were made, but in Garfield Canyon a 12-m bed was exposed for 300 m. The deposits farther east were quite irregular. They were much disturbed and often had a synclinal fold. Their "mud" deposits covered a stretched-out area of >0.25 hectares (1 acre; Crawford, 1894, 1896). The *Union Borax Company* in 1919–1920 sank a 195-m (640-ft) shaft ~100 m (several 100 ft) south of the Borate ore zone. Neither the shaft nor a north-trending drift encountered commercial ore, although colemanite and howlite could be seen in their ore dump (Wright *et al.*, 1953).

3.6.1.6 Western Mineral Co. (Bartlett)

This company was the first to process low-grade "borate muds" containing 7–20% (average 10%) B_2O_3. Their mine was near (southeast of) Calico, with a 12- to 18-m (average 7.5- to 9-m), >60-m-deep bed of bluish-black, gray or red clay-like "mud" that looked like a fine-grained shale or sandstone (Bailey, 1902). It dipped 30° northwest, with a northeastern strike, and contained stringers of howlite, celestite, and gypsum. It was mined by an inclined shaft with drifts in both directions at 15-m intervals. A total investment of $140,000 produced <500 tons of 95% boric acid before the plant was closed in 1907 (Wright *et al.*, 1953).

3.6.1.7 Hector (Fort Cady) Area

In a drill hole near the Santa Fe Railroad's Hector station, 56 km (35 mi) east of Barstow, colemanite was found in a 58-m (191-ft) interval (Table 3.6). It is of Miocene or Pliocene age, occurring as <0.05-mm, 0.1- to 0.8-mm (the most common), and 0.5- to 1.6-mm ("spherulitic") crystals, generally in a tuffaceous clay or anhydrite rock. It is pale yellowish-brown to orange or gray, and usually occurs in <60-cm (2-ft) beds, but also is present as veins and lenses. There are some secondary veins of clear colemanite or anhydrite and a few <3-mm howlite blebs, generally associated with inclusions or layers

Table 3.6

Colemanite Occurrence in the Hector Area[a]

Depth (ft)	Interval (ft)	Estimated colemanite content (wt%)	Depth (ft)	Interval (ft)	Estimated colemanite content (wt%)
1335.8	0.4	20	1401.8	0.1	30
1347.3	0.5	75	1406.1	4.3	5
1353.7	0.4	50	1411.0	1.0	50–70
1354.3	0.7	10	1432.8	2.8	20
1354.6	0.3	80	1436.0	1.7	15
1355.3	0.3	75	1442.9	2.8	10
1356.3	0.1	75	1455.3	5.0	15
1358.1	1.8	20	1457.7	0.7	40–50
1361.7	0.7	75	1461.4	0.4	50
1362.7	0.5	10	1462.0	0.9	70–80
1370.3	2.3	15	1474.7	0.5	50–70
1380.6	1.4	50–70	1510.5	3.5	70–90
1384.9	0.5	50–70	1511.0	0.5	90
1385.2	0.3	5–7	1511.5	0.5	60
1386.8	1.0	50–70	1512.0	0.5	10
1387.5	0.5	15	1512.6	0.6	60–80
1398.6	0.6	80	1525.9	3.0	7
1401.7	0.5	50–70			

[a] Madsen, 1970.

of clay. Some celestite is present as a calcite or colemanite replacement (Madsen, 1970).

3.6.2 Chetco Priceite Deposit, Oregon

This small deposit of priceite ($Ca_4B_{10}O_{19}{\cdot}7H_2O$) is located 19 km (12 mi) northwest of Brookings, Oregon, which is 48 km (30 mi) north of Crescent City, California. It was worked commercially from 1891 to 1892, although limited quantities were mined and shipped to San Francisco as a silver polishing compound in 1871–1872. The deposit outcropped about 150 m from the Pacific Ocean, 6 m above Lone Ranch (Cresswell) Creek, where it was first noted by the ranch owner on a freshly caved bluff. Because priceite looked and felt like chalk, it became widely used by carpenters and coopers at the local fisheries. The rancher attempted to dig a 23-m (75-ft) shaft to the ore from above, but gave up when he encountered considerable water. The property was then sold several times, eventually to Pacific Coast Borax. Samples were analyzed, identified as a new mineral, and named after the analyst (Staples, 1948). The priceite was friable and turned to dust when exposed to air for long periods, but it was hard, strong, and quite pure and uniform when mined. Its rhombic platelet crystals were generally milky white and very small. The nodules were easily broken with a pick (Gale, 1921a).

Some of the priceite filled seams and cavities in the associated serpentine rock as hard white beds, but most was found as nodules completely encased by a "soft green clay, talc or black slate." These nodules weighed up to 200 kg (450 lbs), were fairly uniform, and often touched each other. Branching off the main ore zone were 9- to 0.1-kg (20-lb pea-gravel) nodules, some of which had a crust of aragonite (Staples, 1948). In another description, the nodules were said to be scattered in a black clay matrix, many in continuous or intersecting lines. When the clay was dug from under the larger nodules, they dropped cleanly away from the dense, polished black clay. The clay was much compressed, and the miners could easily feel this change with their picks. Each nodule also had feeder veins flowing trickles of brine into its back or bottom.

The early miners in 1871–1872 described the stratigraphy in descending order: soft green clay or talc with streaks of white waxy steatite (serpentine decomposes to such masses); highly fractured black "slate"; slate with decomposed serpentine in the fractures; slate with hard priceite filling all seams and cavities; a layer with priceite filling the seams in decomposed serpentine (which resembled a soft blue clay with green and white veins); and finally, priceite nodules, some with their upper half mixed with blue steatite, but usually pure milky white (Gale, 1921a).

Pacific Coast Borax unsuccessfully tried to mine the deposit, then leased it to others, with about 40 miners holding leases at the peak of production.

The largest entry tunnel was 146 m (480 ft) long, encountering 4.5 m of priceite at 46 m (150 ft), then 23 m (75 ft) of barren rock, and finally priceite again. The ore was broken, put in 45-kg (100-lb) bags, and shipped from a new wharf near the deposit to San Francisco for processing. The miners received $23/ton for the ore, but litigation among the leasees stopped operations. One miner had shipped 580 tons, operating on the bank opposite the discovery outcrop, and he had blocked out 3600–5400 tons of additional ore, which could have lasted 2–3 years.

No further work was done on the property, and the tunnels caved or were flooded once the operation ceased. A second outcrop had been sighted 6.4 km to the north (Staples, 1948), and a third 1.6 km south of the mining area (Shaffer and Baxter, 1975). They were in a line parallel to a nearby rhyolite dike containing considerable graphite in its 16 km (10 mi) of exposed length. Both the priceite and rhyolite contained abnormal amounts of Cu and Sr, and the priceite had more Fe and Mg than found elsewhere (Staples, 1948).

3.6.3 Coastal Range, California (Los Angeles Area)

3.6.3.1 Frazier Mountain

The Frazier Mountain borate district was on the south end of Mount Pinos in the northeastern corner of Ventura County. It was 85 km (50 mi) northwest of Los Angeles in the Coastal Range, near the southern end of the Sierra Nevada mountains. Colemanite ($Ca_2B_5O_{11}{\cdot}5H_2O$) was discovered in 1898, and production began at the Frazier mine in 1899. The company was sold to the Stauffer Chemical Company in 1900. In 1905, 75 men were employed, and the production was 90 metric tons (100 tons) per month. During 1902, mining started at the nearby Columbus mine of the Columbus Borax Company, and both operated until 1907 (when the Lila C dropped prices). The Frazier mine had shipped 23,000 metric tons and the Columbus mine 7000–8000 metric tons of hand-sorted 35–45% B_2O_3 product, the two representing a combined value of $1,000,000. In 1907 work started on the Russell mine between the other two, and it shipped ore from 1911 to 1913 (Ver Planck, 1956). Transportation from the mines at first was by mule teams, then by a new traction engine, which Stauffer used to haul 136-metric-ton loads 96 km (60 mi) to the railroad first at Lancaster, and later at Bakersfield at a rate of 90–180 tons/month (Bailey, 1902; Evans and Vredenburgh, 1982; Yale, 1905).

The borate bed's age is 15 My (Smith, 1985), and it is in a matrix of shale and limestone (Member 4 of the middle Tertiary Plush Ranch Formation) within 180 m (600 ft) of basalt. The borate-containing shale had a typical fine-grained sedimentary structure, and the colemanite beds were very irregular [average 1.2-m (4-ft) thick, 34% B_2O_3], made up of large milky-white (even glassy in some areas) but mostly grayish or black crystals. Often the crystals were without definite arrangement or had a radial structure. There was also colemanite vein filling in most of the adjacent limestone cracks and cavities, as well as "needle ore," consisting of "stringer veins of distinct cross-fibrous structure [needles], generally formed in thin bands between evenly bedded shale layers adjacent to the main colemanite beds (Gale, 1914)." The deposits could be nearly flat (at the crest of an anticlinal fold), or more commonly have a near-vertical dip. In the Russell mine, the ore zone was bounded by a slip-fault, and all of the deposits were highly folded and faulted. The outcrops seldom contained colemanite, but were characterized by both limestone ridges (either massive or with a rough, porous travertine-like appearance) and by a high content of (later crystallized) gypsum, such as selenite crystals. Little gypsum was in or near the colemanite, occurring only as a later intrusion in thin stringers following bedding planes and in fractures. Gypsum currently is depositing in the mine openings. The outcrops also contained "buttons" of flattened spheroidal disks of unknown composition (Bailey, 1902; Gale, 1913a, 1914).

The basalt above and below the shale–limestone–colemanite beds contained some zeolites, analcime, feldspar, and chalcedony geodes. The companion beds of shale and limestone appeared to be in shear zones because they were shattered and cemented by colemanite, with some shale crushed and crumpled. Some large and irregular masses of priceite ($Ca_4B_{10}O_{19}{\cdot}7H_2O$) were reported (such as outcrops that were "almost entirely priceite"), but most of the priceite occurred in shale layers above, below, and in the basalt. Some howlite ($Ca_4Si_2B_{10}O_{23}{\cdot}5H_2O$) had also been found (Gale, 1913b). The Frazier mine's ore-bearing thin bedded gypsiferous shale outcrop was 12 m thick at the mine opening, within a 36-m-thick bed of weathered limestone, and with basalt both above and below it. The colemanite bed had whitish or grayish crystals, 1.5–1.8 m (5–6 ft) thick in one tunnel, but it varied considerably in thickness. Heavy timbering was required to support the entries. An upper entry with an 11-m (35-ft) incline had a 21-m (70-ft) horizontal drift in 4.3 m (14 ft) or ore. A lower entry was horizontal for 49 m (160 ft), inclined at 45° for 85 m (280 ft), and then level to the ore. The main haulage tunnel, was 0.4 km (0.25 mi) long at a still lower level, with the first 180–210 m (600–700 ft) in basalt and then ore. It had branched horizontal tunnels, extensive drifts, and crosscuts to mine the stopes. A lower 520-m (1700-ft) tunnel did not find ore (Gale, 1913b).

The Russell mine was on the south flank of an (east–west) anticlinal fold of the borate bed. It had a 61-m (200-ft) main shaft in basalt, and a 107-m (350-ft) tunnel at its base extending to the ore. The mine then followed the ore with a 60° dip and a north 75° east strike to the 76-m level with a haulage winze. The basalt near the ore contained considerable zeolites, and the ore was in a >76-m-thick shale–limestone bed. The ore was discontinuous, thickening [<15 m (50 ft)] and thinning, with the highest grade adjacent to a fault and some "needle ore" filling the fractures. The large, glassy colemanite crystals varied from white to dark, and near the fault were spotted or blotchy. There were some cavities in the ore, and much of it was mingled with fractured limestone. However, it still had a typical sedimentary banded appearance. In one crosscut in basalt a "massive vug of clear white ... priceite was observed." It was completely enclosed by the basalt, hard and compact, with a chalky-white uniform grain size. The hand-sorted first grade colemanite ore contained 42.5% B_2O_3, and the second grade 29% B_2O_3 (Gale, 1913b).

The Columbus mine had shipped only a small amount of product before 1907 when it first closed. It was sold to the National Borax Company in 1912, and an extensive plant was built, including a new mine shaft. The main entry was 61 m (200 ft) in basalt to the ore, and then a winze adit followed the ore to lower levels where there were extensive underground workings. The ore-bearing limestone was 3 m thick at its outcrop and 18 m (60 ft) when intersected by the entry tunnel. Then it dipped 55° north with a north 70° east strike. The lower section of the mine flooded in 1912, and the entire operation was abandoned the following year. Many prospect tunnels were made in this general area, but little additional colemanite tonnage was mined, except for 130 metric tons in Bitter Creek Canyon (Gale, 1913b).

3.6.3.2 Lang (Tick Canyon; Sterling Borax Company)

In 1907 colemanite was discovered in Tick Canyon in the Coastal Range, 8 km (5 mi) north of the Southern Pacific Railroad's Lang station and 48 km (30 mi) northeast of Los Angeles. In 1908 the Sterling Borax Company (owned by Stauffer Chemical, American Borax, and two borax processing companies) started mining. The colemanite was sorted underground and shipped to Lang by narrow gauge rail. As the mine grew deeper and borax prices fell, the company was sold in 1921 to Pacific Coast Borax who operated it until 1923, and dismantled it in 1926 (Ver Planck, 1956).

The 4.9-m (16-ft) colemanite bed outcropped for 305 m (1000 ft) in Tick Canyon. The bed was near the top of the Oligocene Vasquez Formation in fine-grained and thinly bedded red and brown sandstone, mudstone, and purplish silty shale. Its ore was richer than at Calico and its age was 20 My (Smith, 1985). It had a strike of north 75° west, and a dip of 70–80° south. The ore zone was deformed into folds modified by many cross- and longitudinal faults. The colemanite bed was <9.1 m (30 ft) thick (average 1.8–3 m), with subbeds that "alternated with layers of black carbonaceous shale ... some ... quite bituminous" (Eakle, 1911, p. 180) and 305 m (1000 ft) long, enclosed by and intermixed with shale. Thinner colemanite beds and several thin howlite-bearing layers were also present but not mined. Noncommercial colemanite also occurred for 3.2 km (2 mi) northeast of the mining area. About 0.4 km east they were offset 305 m by a fault (Gay and Hoffman, 1954). Near the mine "the strata are abruptly upturned against a great basic dyke" (Keys,

1910, p. 694). Also, the colemanite was "in cleaved masses, and very often in columnar bands" (Foshag, 1921, p. 103). There was shale under the colemanite, but on top there were "heavy bedded sandstones ... with efflorescence of white alkali salts along their seams and bedding planes" (Eakle, 1911, p. 180).

The colemanite was predominantly a massive bed, with 2-mm to 2-cm crystals, a glassy luster, and generally a gray color due to included mud. Specimens with a divergent columnar structure were common, and single crystals comparatively rare. In the beds some clusters of crystals were thickly grown together and attached to a base of massive colemanite. Some colemanite occurred as nodules, and as a pure white crack and cavity filler (Pemberton, 1968). There was abundant plant life in the colemanite, indicating that the deposit had once been a freshwater marsh. The colemanite's crystallographic properties were slightly different from that of other colemanite (it was called "neocolemanite"). It tended to cleave so strongly that most of the crystals from the mine were cleaved fragments (Eakle, 1911). The ore averaged 33–35% B_2O_3, typically 36.10% B_2O_3, 23.74% CaO, 13.97% SiO_2, 5.14% CO_2, 2.36% MgO, 1.32% Fe_2O_3, 1.27% Al_2O_3, and 16.10% H_2O (Dupont, 1910).

Considerable ulexite was found near the base of the colemanite bed at the 75-m (250-ft) level. It was bedded, massive, fibrous, and compact, with the fibers lying in all directions, giving a satiny luster and a botryoidal surface. The ulexite occurred "in irregular masses more or less lens-like and surrounded by thin layers of clay," producing an analysis of 43.12% B_2O_3, 14.14% CaO, 7.05% Na_2O, and 35.68% H_2O (Foshag, 1921, p. 210). Probertite occurred in very small quantities as lenticular nodules in the shale matrix, with 0.5- to 1.5-mm crystals as slender radiating prisms or occasional rosettes. Small amounts of veatchite ($Sr_4B_{22}O_{37}·7H_2O$-I), paraveatchite ($Sr_4B_{22}O_{37}·7H_2O$-II), and howlite ($Ca_4Si_2B_{10}O_{23}·5H_2O$) were found with colemanite in limestone or marl. The veatchite occurred as pearly cross-fibered seams in the colemanite. Some "howlite nodules are imbedded in the colemanite and form 'augen' [lenticular eye-shaped masses] in the strata" (Foshag, 1921, p. 211). Small amounts of howlite were also found as $<$1-m nodules with a cauliflower-like surface. They were white, gray, or black, but porcelain-like and white inside. Some nodules had 2-mm howlite crystals on their surface as rosettes [in the tailing piles, along with bakerite ($Ca_8B_{10}Si_6O_{35}·5H_2O$), pyrite ($FeS_2$), realgar (AsS), arsenopyrite (FeAsS), stibnite (Sb_2S_3), tremolite ("mountain leather," $Ca_2Mg_2Si_8O_{23}·H_2O$) and celestite ($SrSO_4$); Pemberton, 1968]. Calcite was rare in the deposit (Foshag, 1921).

3.6.3.3 Four Corners (Kramer Junction)

This colemanite deposit is located in the Mojave Desert, 11 km (7 mi) east of the Kramer deposit, 30 km north of the Coast Range, and 100 km north of Los Angeles. The colemanite is found in two areas: the Sunray Mid-Continent Oil Company (quite small and low grade), and Tenneco Mining's Rho.

Their age is 18–20 My (Smith, 1985), and their reserves, as millions of metric tons of ore and percentages of B_2O_3: 12 at 17%, 24 at 14%, 47 at 9%, and 11 at 5%. The ore has a high arsenic content (834–1425 ppm). The Rho deposit has two 0- to 53-m (0- to 175-ft) colemanite beds separated by 7.6–15 m (25–50 ft) of shale, claystone, and sandstone (Fig. 3.10), and is 140–380 m (458–1250 ft) deep. It was discovered in 1957 by a U.S. Geological Survey drilling program in search of borates, and further delineated by private drilling. The colemanite zone is in the 98-m (320-ft) Miocene Tropico Group of greenish silty claystone, shale, and silty sandstone. The formation is a broad syncline, with the colemanite in the 0–15° dip south limb. It occurs as 0- to 6-mm crystals disseminated in montmorillonite–claystone, or as 2.5- to 10-mm (average 6-mm) beds interlayered with silty claystone and sandstone and containing 0.25–27% B_2O_3. Accompanying minerals are calcite, quartz, feldspar, biotite, analcime, heulandite, iron oxide (<2.5-cm nodules and bedding plane coatings), realgar, and orpiment.

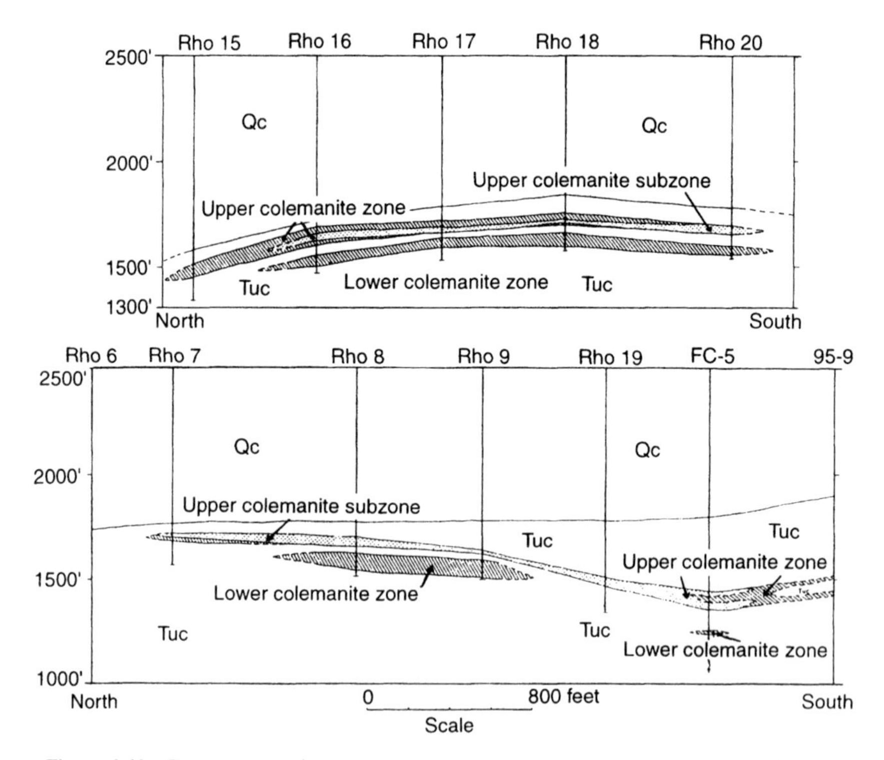

Figure 3.10 Two cross-sections of the Four Corners, California Rho colemanite deposit. (From Evans and Anderson, 1976; Used with permission of the California Department of Conservation, Division of Mines and Geology.)

The Sunray deposit is 3.2 km northwest of Rho, with only three drill cores. The adjacent strata dips 20–55°, and the top of the colemanite is at 140–164 m (458–538 ft). The colemanite zone is 12–30 m (40–98 ft) thick, with 3.7% B_2O_3. Wet scrubbing and desliming tests reduced the As to 1040 ppm on 14.3% B_2O_3, 1400 ppmAs Rho ore. The final B_2O_3 content was 42.6% and the colemanite recovery 61.5%. With calcining, the arsenic increased to 1720 ppm, the B_2O_3 to 54.1%, and the calcining yield was 54.1% (Evans and Anderson, 1976).

3.6.4 Death Valley, California

3.6.4.1 General

The >23 borate deposits in the Death Valley region are mainly in the Black Mountain tectonic block (Fig. 3.11; Table 3.7), bounded by the Death Valley, Furnace Creek, and Sheephead faults. The area is characterized by Tertiary rocks and has had considerable tectonic activity. The borate deposits' age is 6 My (Smith, 1985). They are a few kilometers west and somewhat follow the Furnace Creek fault. The deposits are in the Furnace Creek Formation as a lacustrine sequence within 10 m ("several tens of feet") to 150 m (500 ft) of the base of the formation, above the basal unit (conglomerates containing Paleozoic and/or Artist Drive detritus, and sometimes minor amounts of borates). The borate zone consists of clays, tuffaceous mudstone, shale, sandstone, zones of tuff, limestone and basalt, and borates interbedded with these rocks (Barker and Wilson, 1976). A gypsiferous unit sometimes overlies it, having minor amounts of borates interlayered with fibrous gypsum and mudstone. Some areas have been folded into a broad syncline, and all of the deposits have been cut by steeply dipping faults with some displacement. The deposits are bounded and/or cut by faults, and the beds may dip <45° (Evans, Taylor, and Rapp, 1976). Basalt flows, sills, and dykes are quite numerous, sometimes occurring in the same horizon as the borates. Altered basalts occasionally also contain secondary borates (Norman and Johnson, 1980). The Death Valley borate area is a narrow 3.7 × 36-km band (83 km to include Shoshone), very much like the multideposit Bigadic and Emet borate areas of Turkey. Each has closely related but distinct borate deposits, as if it is part of a larger sedimentary basin with many slightly different hot springs and drainage areas.

The general borate sequence in many deposits starts with colemanite, which surrounds the other borates (if any) and extends beyond them. Ulexite and usually probertite then form as central cores. If probertite is present, the ulexite usually surrounds it. The interior of most deposits contains unaltered bedding planes and a uniform depositional pattern. There are few, if any, alteration indicators (pseudomorphs, slump structures, etc.) for the bulk of

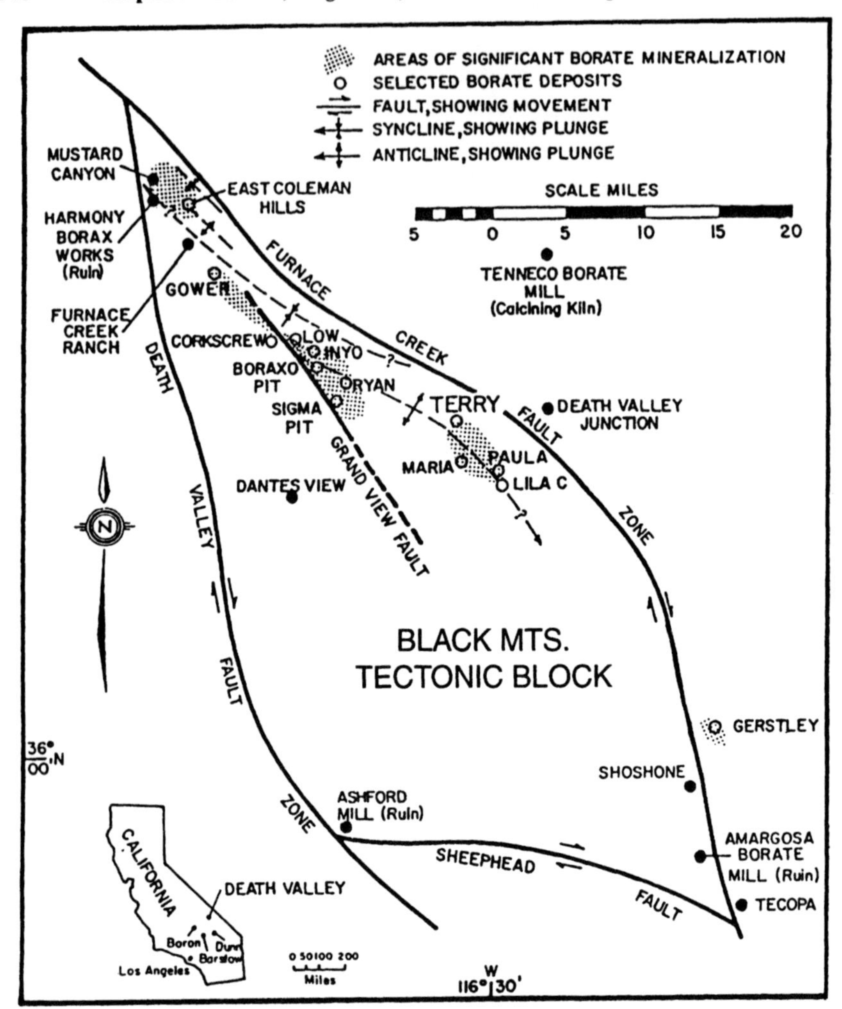

Figure 3.11 Structural interpretation of the Death Valley region. (From Barker, 1980; Used with permission of the California Department of Conservation, Division of Mines and Geology.)

the deposits (Barker and Barker, 1985). Considerable alteration has occurred on the top, around the edges, and intrusively into most deposits, forming more than 26 different borate minerals.

3.6.4.2 Biddy McCarthy (Upper Biddy)

This mine was one of the Ryan cluster (which includes Lower Biddy, Grand View, Lizzie V. Oakley, Payed Out, and Widow Mines; see Fig. 3.9), 32 km (20 mi) southwest of Death Valley Junction. Each mine shipped their ore by

Table 3.7

Borate Mines in Death Valley[a]

Deposit	Inferred reserves, 1000 mt ore[b]	Grade, % B_2O_3[c]	Deposit	Inferred reserves, 1000 mt ore[b]	Grade, % B_2O_3[c]
Biddy McCarthy, Lower	45 C	25 C	Grand View	78 C	14 C
Biddy McCarthy, Upper	260 C 5 UP	18 C 32 UP	Inyo	1200 C 620 UP	22 C 24 UP
Billie I	8860 C 2950 UP	22 C 27 UP	Lila C	0	—
Billie II	2730 C 910 UP	22 C 27 UP	Lizzie V. Oakley	152 C	15 C
Boraxo	154 C 405 UP	20 C 28 UP	Maria	640 C	24 C
Corkscrew	345 C 110 UP	26 C 25 UP	Monte Blanco	863 C 1810 UP	22 C 18 UP
DeBely	45 C	16 C	Played Out	23 C	15 C
East Coleman	—	—	Sigma	4020 C 7800 UP	20 C 25 UP
Gerstley I	121 U	27 U	Widow No. 3	680 C 1140 UP	18 C 23 UP
Gerstley II	770 C 140 U	22 C 28 U	Widow No. 7	815 C	180
Gower Gulch	—	—	Total[d,e]	21,680 16,011 UP	20 C 26 UP

[a] Evans, Taylor, and Rapp, 1976.

[b] Estimated remaining ore as of 1976. There has been some mining since then.

[c] C = colemanite; UP = ulexite + probertite.

[d] Total B_2O_3, million mt: 4.34 mt C; 4.16 mt UP. Wendel (1978) estimated 5.06 mt C; 3.50 mt UP. Orris (1995) reported other mines called Low and Paula.

[e] Estimates of other areas: Muddy Mountains 1.02 million mt @ 14% B_2O_3 = 0.14 C (Wendel 0.39). Four Corners: 85 million mt @ 10.8% B_2O_3 = 9.18 mt B_2O_3. Hector 120 million mt ore @ 6% B_2O_3 = 7.2 mt B_2O_3. Boron tailings: mined ore 6.4 million mt @ 10% B_2O_3 = 0.64 mt B_2O_3 (75% ulexite). Slimes: 12.7 million mt @ 10.7% B_2O_3 = 1.36 mt B_2O_3.

gasoline motor trains on a "baby gauge" [rails 0.45 (18 in.)–0.6 m apart] line to Ryan's main storage bins. The borate ore occurred in thin-bedded, light-colored shales over a thick section of coarse sandstone and tuff, which forms conspicuous bluffs below the mines. The ore was capped by basalt, which formed the crest of the ridge behind the mines. In each of the area's mines, the colemanite zone was <30 m (100 ft) thick. All were considerably faulted, and faults often bounded the deposits.

At the Upper Biddy the 6.1- to 30-m borate beds were at an elevation of 884 m (2900 ft) and dipped 45° to the northeast. There were some thin shale layers in the colemanite beds and <2% ulexite–probertite in the ore (Evans *et al.*, 1976). The purest ore was in the upper 6.1–21 m (20–70 ft) of the bed nearest the capping, and it had prismatic 0- to 5-cm (0- to 2-in.) crystals, or radiating plumose structures. Under the main beds, colemanite also was found in fractures or mingled with shattered shale. The deposit initially was mined by a glory hole (a limited-size, cone-shaped open pit), with the chute from its base 58 m (190 ft) from the entry (Hamilton, 1921). Mining was then by underground stoping, entirely by hand drilling (as the colemanite was very friable), and the ore was segregated underground into first- or second-class grades. About 150 men were employed, producing 45 metric tons/day of first-class and 140–180 metric tons/day of second-class ore. It was hand-trammed out of the mine in small rail cars through a 64-m (210-ft) tunnel, and loaded into storage bins. The main haulage was later mechanized, extended 550 m (1800 ft) through a hill to Ryan, and then also used by the Grandview, Oakley, and Widow mines. Foshag (1931) found probertite in the mine's dump with a somewhat laminated, flattened radial or compact-radial structure and a greasy luster. Occasionally, it was in the form of nodules.

3.6.4.3 Biddy, Lower

This mine was 0.4 km (0.25 mi) northwest of Upper Biddy, and its colemanite bed dipped 40° to the southwest, with a north 50° west strike. The ore was interbedded with shale and conglomerate, and contained no ulexite or probertite (Evans *et al.*, 1976; "important amounts of ulexite," Foshag, 1931). There were several 1.8- to 3.0-m (6- to 10-ft) ore beds, also first mined by a glory hole, then by tunnels driven into the ore at various levels. The ore was hoisted to the surface in an inclined 20-hp tramway, placed in bins, and later taken to Ryan. A Fairbanks–Morse semidiesel engine drove a Sullivan compressor to supply air for the miner's drills (Hamilton, 1921).

3.6.4.4 Billie

This deposit is in the Upper Furnace Creek Wash in an area folded and faulted along a northwest-trending axis. The ore dips 20–30° southeast with a north 40–50° west strike and consists of colemanite and ulexite–probertite in beds interlayered with and surrounded by calcitic mudstones and shales (Figs. 3.12 and 3.13). The deposit is 1130 m (3700 ft) long and 46–53 m (150–175 ft) thick, with an average width of 220 m (700 ft) and a depth of 46–400 m (150–1300 ft). Its ore reserves are 2,700,000 metric tons of ulexite–probertite at 27% B_2O_3 and 11,000,000 metric tons of colemanite at

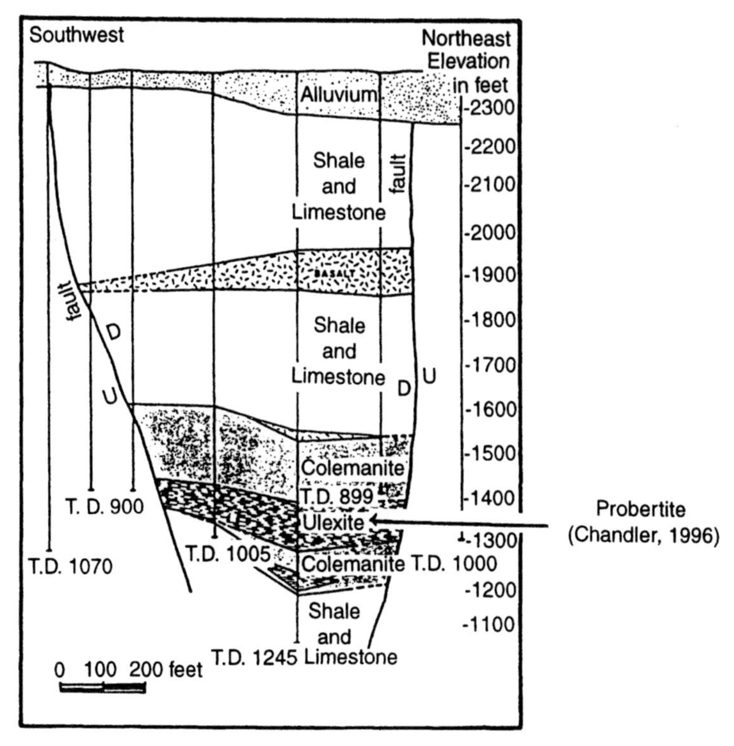

Figure 3.12 Idealized end view of the Billie borate deposit. (From Garrett, 1985; reproduced by permission of the Society for Mining, Metallurgy and Exploration, Inc.)

21% B_2O_3. The colemanite extends well beyond, below, and over the other borates, and it appears that some of the lower probertite was altered to colemanite. Severe earth movement has shattered, rubblized, and contorted (it was later lightly cemented) the Furnace Creek mudstones surrounding the deposit, and also fractured and distorted the borates. There are no major faults in the ore zone, but the southwestern and northeastern edges are bounded by faults (Kistler and Helvaci, 1994; Lyday, 1994, 1995; Norman and Johnson, 1980). Chandler (1996) noted that the orebody had been dropped by two faults on its sides, causing folding and faulting to occur in the ore at ~51-m intervals, with normally ~1-m movement and each block being slightly lower than its predecessor. Projected shoreline features of a series of basins can be traced from the Billie to the Boraxo to the Sigma–White Monster mines, and the residues of ancient geothermal mantles appear to be present. The original bedding planes of the ore have been preserved, and the mudstone beds are sometimes continuous over large areas. The deposit has many vugs, or crystal cavities, and what may be partially dissolved probertite fibers occasionally

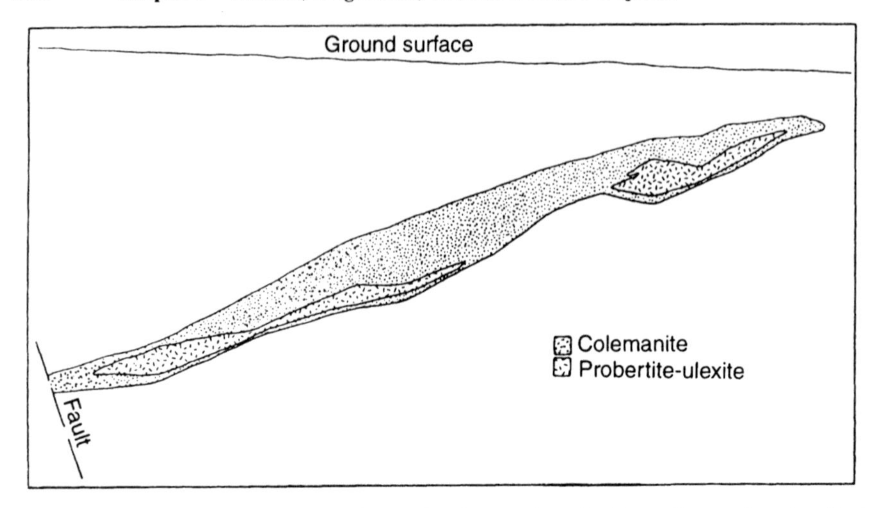

Figure 3.13 Idealized side view of the Billie borate deposit. (From Grace, Douglas, Pool, and Lattanzi, 1985; reproduced by permission of the Society for Mining, Metallurgy and Exploration, Inc.)

have been observed penetrating crystals of colemanite in these vugs. Near the lower edge of the deposit's center is altered basalt.

Most of the colemanite is of a lathe shape or equant habit, although some occurs as radiating crystals (often near probertite) with a modest sodium content. The probertite contains a small amount of strontium, and veatchite ($Sr_4B_{22}O_{37}\cdot 7H_2O$) was found in some colemanite vugs. Limited amounts of howlite occur in the upper part of the deposit near the colemanite–probertite contact. Anhydrite can replace colemanite, having a similar appearance, with a tan, gray, or black color. Celestite ($SrSO_4$) is found in some vugs and with calcite. Strontianite ($SrCO_3$) also has been observed (Chandler, 1996).

3.6.4.5 Boraxo (Thompson)

This deposit is 4 km (2.5 mi) west of Ryan at an elevation of 732 m (2400 ft). It is 850 m (2800 ft) long, 91–210 m (300–700 ft) wide, and 0–40 m [0–130 ft, average 12–14 m (40–45 ft)] thick. It dipped 5–60° (average 40°; the adjacent strata dipped 20–30°) and terminated at a depth of 240 m (800 ft). It was on the south limb of an anticline, 61–610 m above the base of the Furnace Creek Formation, cut on the north near the outcrop by the Pit fault. There were two 0.3- to 1.5-m (1- to 5-ft) ore beds interlayered with limey shale, mudstone, siltstone, or sandstone, and deformed locally by minor folds. Generally, the outer zone of the deposit was calcite-bearing colemanite, and the core a mixture of ulexite and probertite. The average ore grades were colemanite

20% B_2O_3 and ulexite–probertite 28% B_2O_3. In areas with the three borates, the probertite was at the lower edge grading upward to ulexite, and above it, colemanite. However, colemanite occurred alone for much of the deposit, comprising about 50% of the ore (40% probertite). The borates were in a calcareous blue-gray shale with white limestone laminations and scattered limestone, basalt, tuff, quartzite, or shale pebbles. There was some fracturing and mixing of the shale with the borates, increasing (with contortion) at the margins, and some fractures and vugs were filled with colemanite.

Most of the colemanite was white, although some was clear, golden-brown (from strontium), or gray (from clay). Probertite had large radiating honey-colored crystals. Ulexite was generally white, columnar massive, or crystals in thin veins. There were minor amounts of massive white priceite. Small amounts of hydroboracite were found in vugs with more than 75% of small colemanite crystals, and white-to-cream 3 mm × 2.5-cm ($\frac{1}{8}$ × 1 in.) radial crystals occurred in 6.3-cm (2.5-in.) mounds, scattered among similar colemanite mounds with 1.3-cm ($\frac{1}{2}$-in.) crystals. Hydroboracite also occurred as 1-cm ($\frac{3}{8}$-in.) fan-shaped crystals, each with a top 1.6 cm ($\frac{5}{8}$ in.) wide, and several fans often connected. Tunellite occasionally accompanied hydroboracite, but was more common as microcrystals with colemanite or adjacent clay, with occasional sharp tabular 1.3-cm ($\frac{1}{2}$-in.) crystals (Minette and Wilber, 1973). Accompanying minerals included white to sky-blue celestite and white calcite. The arsenic content of the ore was quite low (typical of Death Valley; Barker and Wilson, 1976).

The deposit was first claimed in 1915 and mined through three 6.1-m (20-ft) shafts (mid-1920s to 1930; briefly 1960s) at three levels, with several small stopes between the levels (Evans *et al.,* 1976; Minette and Wilber, 1973). It was converted to an open pit in 1970, with two adjoining pits and a third added in 1974. When the pit became too deep, it again was mined by underground operations. At that time it was 122 m (400 ft) deep, 910 m (3000 ft) long, and 305 m (1000 ft) wide. Drilling under the proposed dump area discovered a new deposit, the Inyo (Smith and Walters, 1980).

3.6.4.6 *Corkscrew, DeBely*

The Corkscrew mine was opened by two adits driven into a 3-m- (10-ft-) thick colemanite bed (with 24% ulexite and probertite) during 1953–1955. The bed outcropped, was 55 m long, had a 50° dip, and there was basalt under the ore as well as tuffaceous mudstone and sandstone above it (Evans *et al.,* 1976). It produced 14,000 metric tons of ~40% B_2O_3 colemanite (used as a forest fire retardant, a practice discontinued in 1962), which depleted the deposit. The DeBely open pit mine also was operated in the mid-1950s to produce fire retardant colemanite. Its ore existed in complex folds and faults, and contained no ulexite. It too was totally mined out (Wendel, 1978).

3.6.4.7 Gerstley

This deposit is near Shoshone, 64 km (40 mi) southeast of Ryan and the other Death Valley deposits. It was discovered in about 1920 as a few small colemanite nodules in a 0.9- (3-ft) clay–shale outcrop at an altitude of 700 m (2300 ft). It had three separate ore zones, created by a fault on the southwest side splitting it into two disconnected sections separated by 610 m (2000 ft). A second fault split the western section, but without much displacement (Fig. 3.14), and other faults caused considerable displacement within the ore. The beds dipped 25°, contained high-grade ore, and where folding or faulting had occurred, they were variously quoted as <18 m (60 ft) and <46 m (150 ft) thick. Colemanite was the dominant mineral (85%), with 15% ulexite and minor amounts of probertite. Part of the colemanite was in massive beds and interlayered (often 2–5 cm thick) with white to orange limestone clay and basaltic sandstone or conglomerate, and part was in irregular fragments and nodules. Colemanite also could occur as radial crystal aggregates and as a filling material in fractures and vugs in both the ore and surrounding layers. Each of the beds had been fragmented and somewhat mixed in the borate zone. The amount of limestone increased near the borates in the form of both beds and nodules. The light-brown to grayish-green clay–shale matrix in places contained beds of tuff, and in one area pillow-shaped basalt lay directly under the borates. A pebble bed of Cambrian limestone and quartzite was impregnated with colemanite next to the deposit.

Ulexite occurred either with the colemanite or in pure massive beds in deeper zones. There was also ulexite in the shale and sandstone above the ore. Some ulexite had either a solid core of colemanite crystals, or colemanite nodules scattered through the ulexite layers. Also, zones of the ore ranged from pure ulexite to mixtures of ulexite and colemanite to pure colemanite. Some probertite occurred in the deep zones (Barker and Wilson, 1976).

Production started in 1924 after the deposit was purchased by Pacific Coast Borax, and was shut down in 1936 (Ver Planck, 1956). The deposit was only 3.2–4.8 km (2–3 mi) from the Tonopah and Tidewater Railroad (Noble, 1926). A shaft was sunk near the discovery outcrop, and at its base a tunnel was driven both into the ore and to the surface of the alluvial slope facing the Amargosa Valley. This became the entry and main haulageway for the mine. Where the entry intersected the deposit, the ore was 6.1–9.1 m (20–30 ft) thick, and there was a 0.6-m (2-ft) ulexite bed on top of the colemanite. Small amounts of ore were mined long after the mine's official closing date. Its remaining reserves are 240,000 metric tons of B_2O_3 (Kistler and Helvaci, 1994).

3.6.4.8 Grand View Mine

This mine was located 1.6 km south of Ryan on the east side of Furnace Creek Canyon at an elevation of 900 m (2950 ft). It had several 1.8- to

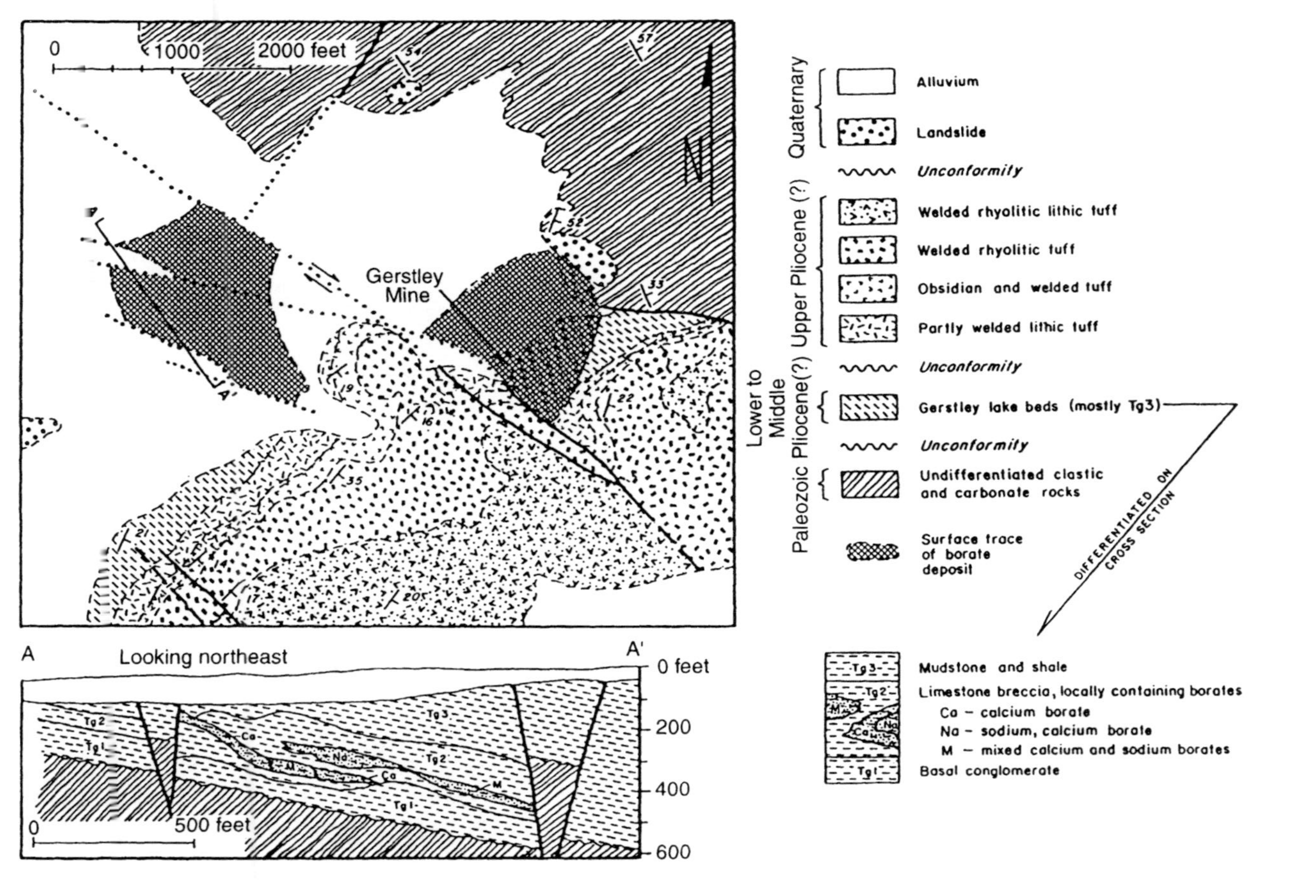

Figure 3.14 Geologic map and cross-section of the Gerstley borate deposit. (From Barker and Wilson, 1976; reprinted courtesy of the Nevada Bureau of Mines and Geology.)

6.1-m (6- to 20-ft) colemanite beds (with no ulexite), with a northeast dip and a northwest and southeast strike. The beds extended from the surface to a flat fault at 90 m and were mined by a series of crosscut tunnels at several elevations, with the main haulage track at the lowest level. The ore was hauled to storage bins on the surface, and then to the main storage at Ryan (Hamilton, 1921).

3.6.4.9 Lila C

This deposit had three parallel 1.8- to 5.5-m (6- to 18-ft) colemanite beds with 40% B_2O_3 ore, each separated by ~7.6 m (25 ft) of barren rock, and traceable in outcrops for 760 m (2500 ft). Colemanite had been discovered in Death Valley near Furnace Creek in 1882, but this was the first mine, started in 1907. As Pacific Coast Borax's deposit at Calico began to exhaust its reserves, in 1903 (some initial work in 1899) development work began on the Lila C mine in Death Valley. About 910 m (3000 ft) of inclined shafts and drifts were developed to enter the ore body, and in 1905 construction was initiated to serve the mine by the Tonopah and Tidewater Railroad. The route passed close to the Lila C, and an 11.2-km (7-mi) rail spur connected it to Death Valley Junction. The mine then opened, with a capacity of 27,000 metric tons/year. Borate at Calico (with less than 1-year's reserves) closed, and the borax price dropped from 6.5–7¢/lb to 4.5–5.5¢/lb in an attempt to expand the market (Gale, 1912; Ver Planck, 1956). This resulted in the closing of many higher-cost borate operations.

The upper bed was 610 m (2000 ft) long, 1.8–5.5 m (6–18 ft) thick, 91 m (300 ft) wide, and dipped at a steep angle. The central bed, 7.6 m (25 ft) below the upper, was <1.5 m (15 ft) thick. The colemanite occurred in the beds as large clear glassy crystals, some with radiating or plumose structures. It also filled cracks and cavities with 0- to 1.3-cm- (0- to 0.5-in.-) long, narrow prismatic crystals or radiating bunches. The massive ore showed a distinct cleavage when shattered and could contain clay or shale inclusions. The entire area was highly faulted and distorted, the beds' strike north 30° west, the dip 45° northeast in the north, with a gradual decline to 20° in the south. The strike curved, following the crest of the ridge above the mine. The ore's thickness was irregular, often due to transverse slips, but after minor offsets could be located easily. The beds were also locally crumpled and irregular. The strata surrounding the ore was light-colored, evenly bedded shales, including sandy layers and thin clay beds interlayered with the ore. Under the deposit were beds of coarse, friable sandstone or tuff, and above it dark gray vesicular lava, some of which cut the borate beds as dykes.

Mining was primarily from two of the beds, of which the central was the most productive. Each was mined by large stopes when possible (depending on their dip and distortion), with the ore dropped to the main entry and haulage level. Initially the beds had been worked from outcrop pits. There were four inclined shafts into the mine, with the main haulage shaft 1070 m (3500 ft) long. The mined ore was hand sorted to remove the highest grade colemanite, and the rest was sent to a processing mill on the property. About 2400 metric tons/month were shipped, and the operation shut down for the two hottest months of the year (Gale, 1912). The mine had 100 employees, and tentatively closed in 1915 as the new Ryan mines were developed, but from 1919 to 1925 an additional 91,000 metric tons were mined (Travis and Cocks, 1984).

3.6.4.10 Inyo Claims, Maria

The Inyo deposit, northwest of the Boraxo pit and bounded by three faults, contains 34% ulexite and probertite and has a 45° dip. It is 370 m wide, 270 m long, and 46–240 m deep (Evans *et al.,* 1976). Its reserves are 1,200,000 metric tons of colemanite, 600,000 metric tons of ulexite–probertite, and 400,000 metric tons of B_2O_3 (Wendel, 1978). Maria's dominant borate was colemanite (>99%), although it had a small core of ulexite. A fault partly bounded the east side, but it was nearly oval shaped and <11.6 m (38 ft) thick, with a 10° dip. The colemanite was in beds, laminated and bounded by a blue-gray calcareous shale. There was also 5–20% of gray to white interbedded limestone, and some beds contained greenish-black basaltic sandstone. The beds had been fractured, highly contorted, and somewhat mixed. The colemanite was gray to white and commonly occurred as small radial clusters in massive irregular fragments, and as a fracture- or vug-filling material in the ore zone and the shale and basalt breccia below it. Thin continuous layers of colemanite also existed in the unbroken shale above the ore, along with some ulexite. Bladed >5-cm (2-in.) golden-brown colemanite crystals were found near the fault line (Barker and Wilson, 1976).

3.6.4.11 Monte Blanco, Oakley

The Monte Blanco outcrop is 0.6 km northwest of Corkscrew, with steeply dipping colemanite beds interlayered with shale. Its reserves are, in million mt: ore 2.68, colemanite 0.18 B_2O_3, and ulexite 0.32 B_2O_3 (Wendel, 1978). Oakley was located south of Grand View and 2 km (1.25 mi) south of Ryan at an elevation of 850 m (2800 ft). Its colemanite beds dipped 45–50° northeast, had a northwest–southeast strike, and were mined by crosscut tunnels at different elevations in the ore. The upper (main) workings were lower than the Widow railroad, so the mine had its own spur track to Ryan (Hamilton, 1921).

3.6.4.12 Played Out, Sigma

Played Out was 3.2 km (2 mi) north of Ryan at an elevation of 884 m (2900 ft). Its several 3- to 6.3-m (10- to 25-ft) colemanite and ulexite beds dipped 40° northeast and had a north 50° west strike. A 240-m (800-ft) entry did not intersect the ore, but ended in a 43-m (140-ft) rise to the mine's sixth, or main, haulage level. Ore was mined from all six levels, with the first encountering 91 m (300 ft) of 6.1-m- (20-ft-) thick ore, and the sixth 61 m of a 6.1-m bed. The colemanite and ulexite were mined separately, and all of the ore dropped to the haulage level and taken to storage bins. From there it was screened on a grizzly, and the high-grade ore removed (by hand sorting; Hamilton, 1921). The Sigma deposit is located 0.6 km south of the Boraxo open pit, and also had been mined as an open pit intermittently since 1975. In 1977 80,000 metric tons of mixed colemanite and ulexite were mined, and the pit was extended (in 1978) to the southwestern corner of the White Monster mine, exposing 30 m of colemanite and ulexite (Wendel, 1978). Kistler (1996) noted that the deposit has 3–3.4 m (10–11 ft) of colemanite in its upper surface and 37 m (120 ft) of ulexite below that (reserves of ~170,000 metric tons of B_2O_3).

3.6.4.13 Terry

This small, shallow colemanite deposit (Fig. 3.15) was located 11 km (7 mi) west of Death Valley Junction and 3.2 km south of the midpoint on the

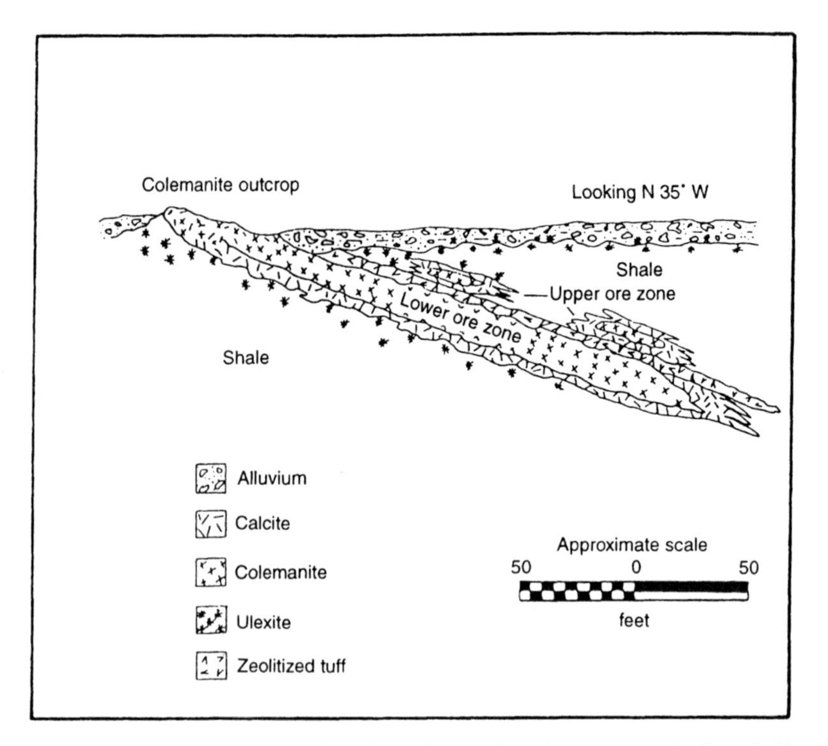

Figure 3.15 Down dip cross-section, Terry borate deposit, Amargosa Valley, California. (From Barker, 1980; Used with permission of the California Department of Conservation, Division of Mines and Geology.)

highway to Ryan. It had a nearly circular outline of 61 × 73 m (200 × 240 ft) and was 2 m (6 ft) thick, 18 m (60 ft) deep, with a domed top, flat bottom, and 27% B_2O_3 ore. The deposit was discovered in 1972, and totally mined out (15,000 tons of ore) with an open pit in 4 months during 1975 by Tenneco. It was one of the few borate occurrences on the northeastern limb of the Ryan anticline, dipping 19° northeast, with a strike of north 36° west. The ore was enclosed by pale olive-green or dusky-yellow calcareous shale of the Furnace Creek Formation and interbedded with basalt, sandstone, conglomerate, tuff, and gypsum. The ore was interfingered with the enclosing beds, and there were laminations, mud cracks, ripple marks, and rip-up clasts in the shale, indicating a variably shallow lake-to-playa environment. Tuff layers between ore zones had been altered somewhat equally to clinoptilolite, chabasite, and phillipsite, with very little analcime and potassium feldspar.

The colemanite was translucent, light brown, 5- to 30-cm spherical aggregates with acicular to bladed crystals that tended to interlock into a poorly packed massive deposit, with some crystal growth between the aggregates. A few nodules were also present with hollow, geode-like interiors. The colemanite near the edges (particularly the upper edge) had been altered to a friable coarse-grained limestone. Colemanite was also crystallized along fracture planes in the ore. There were only minor amounts of hydroboracite ($CaMgB_6O_{11} \cdot 6H_2O$) and ulexite ($NaCaB_5O_9 \cdot 8H_2O$), with the former occurring intermittently throughout the deposit as crystal sprays and vug fillings, but only making up less than 1% of the ore. Its 2- to 10-cm crystals had a long, spherical, radiating acicular habit with a silky luster. Locally, 1- to 10-mm bladed colemanite coated the hydroboracite. Microscopic hydroboracite with colemanite and secondary calcite were found in the weathered edges of the deposit (Countryman, 1977). Flattened, radiating aggregates of acicular ulexite crystals were occasionally found in the parting planes of shales above the ore body. Ulexite also formed thin fibrous aggregates filling fractures 2–3 m below the ore body. Some 4- to 6-mm ulexite cotton balls were in the overlying shale–alluvium contact. Gypsum also was found as a fracture-filling material (Barker, 1980).

3.6.4.14 Widow Mines, White Monster

There were seven different deposits in the Widow mines, located on a ridge east of Furnace Creek Wash at an elevation of 1032 m (3386 ft), and with the high degree of faulting, all were somewhat different. Most of the ore was colemanite, but the central portion of the ore bodies was made up of massive ulexite with some probertite. The latter was found both in colemanite and embedded in soft bentonitic clay (Foshag, 1931). In Widow No. 3 (Fig. 3.16) the ore body was 91 m (300 ft) long and 6.1–61 m (20–200 ft) thick, and had a 40° southeast dip and a north 60° east strike. It had a 430-m (1400-ft) entrance tunnel to the base of the ore, ending with a 30.5-m (100-ft) rise to the No. 2

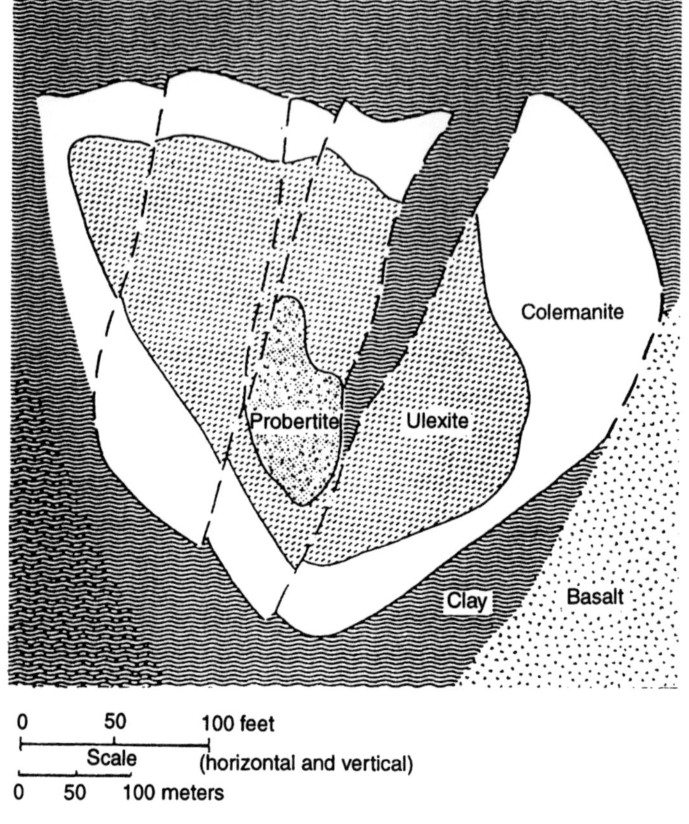

Figure 3.16 Cross-section of Widow 3 mine, Death Valley, California. (From Kistler and Smith, 1983; reproduced by permission of the Society for Mining, Metallurgy and Exploration, Inc.)

level and a 210-m (700-ft) drift in the ore. At its end was another 16.8-m (55-ft) rise to the No. 1 level, which had a 91-m (300-ft) crosscut to the ore, and was connected to a glory hole at the surface. All of the ore was dropped to the lower entry, taken to the surface, and loaded on 10-car, gasoline-driven ore trains for the 3.2-km haul to Ryan. At the adjacent Widow No. 7, the upper level serviced the 152-m- (500-ft-) long, 15- to 18-m- (50- to 60-ft-) thick ore body. About 21 m (70 ft) below this level the deposit was 91–152 m long and 30.5 m thick. Most mining was done by hand drills in order to be more selective, but development work was done with air drills. About 205 men were employed in the Widow mines, producing 70–90 metric tons/day of first-class and 270 metric tons/day of second-class ore (Hamilton, 1921). The White Monster deposit was 68–84 m (223–275 ft) thick and 7–76 m (23–248 ft) deep (Fig. 3.17). Reserves of the adjacent Sigma–White Monster deposits are 3.8

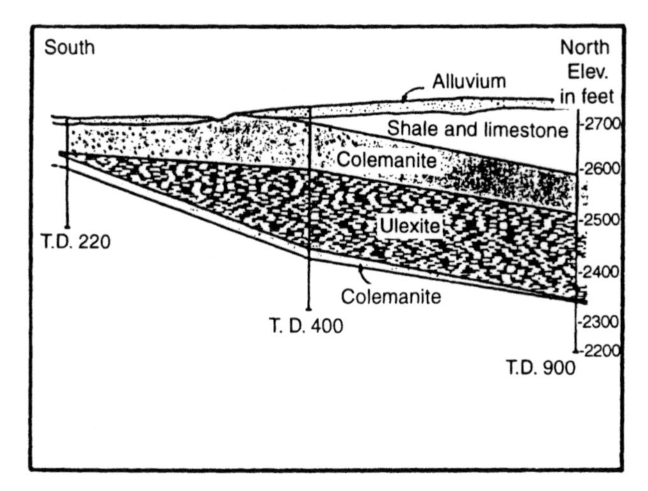

Figure 3.17 White Monster borate deposit. (From Grace *et al.*, 1985; reproduced by permission of the Society for Mining, Metallurgy and Exploration, Inc.)

million metric tons of colemanite and 7.8 million metric tons of ulexite–probertite (Wendel, 1978).

3.6.5 Muddy Mountains, Nevada

3.6.5.1 Lovell Wash (Callville Wash, Anniversary Mine)

This deposit was discovered in 1921 and sold to the West End Chemical Company as the Anniversary Mine (Papke, 1976). The deposit is 18 km southwest of the similar deposit found the previous year in the White Basin (see Fig. 2.8). It is located at an elevation of 610 m (2000 ft), 9.6 km (6 mi) south of Muddy Creek and 38 km [24 mi; 56 km according to Castor, 1993; 72 km (45 mi) by road] east of Las Vegas. Its age is 12–17 My (13–16 My; Smith, 1985), and the deposit is lens shaped with only minor structural distortion. The colemanite was in massive beds, but its interlayering with shale diluted its grade (average 20% B_2O_3, As-free; Noble, 1922).

The single 2.4- to 5.8-m (8- to 18-ft; 2.5- to 7-m according to Castor, 1993) colemanite bed (Table 3.8, Fig. 3.18) was in the Middle Miocene Horse Spring Formation, with 8 m of ridge-forming limestone and marl interbedded with some gypsum at its base, then soft calcareous shales interbedded with limestone and colemanite, with volcanic tuff (3 m of interbedded gypsum and limestone; Castor, 1993) on top. The colemanite was in the north limb of a syncline, with an east–west strike, a 45–65° dip, and about a 910-m (3000-ft) exposed outcrop. The borate ore occurred as compact, essentially pure beds of colemanite, and as nodule layers interbedded with calcareous shale (some nodules contained considerable ulexite). The <3-cm colemanite crystals oc-

Table 3.8
Typical Analysis from the Anniversary Mine in Nevada

Ore,[a] wt%		Pure Colemanite,[b] ppm			
B_2O_3	23.2	Mg	10,700	Cr	5.4
CaO	22.0	Sr	>3300	Cu	<3.8
MgO	9.4	Fe	900	La	<3.8
SiO_2[c]	14.6	Na	600	Mo	<2
R_2O_3	10.1	K	600	Ni	<2
CO_3	19.1	Al	<200	W	<1
SO_4	1.6	Li	172	Sb	<0.2
Colemanite	42.0	Ba	<56		
Dolomite	27.0	As	28		
Silicates[c]	28.5	Y	13.6		
Gypsum	2.5	V	10.2		

[a] Papke, 1976.
[b] Castor, 1993.
[c] Clay and tuff.

curred in several forms: irregular anhedral, anhedral mosaic, columnar, or plumose. The colemanite beds were continuous, but varied in thickness, and some were a meshwork of crystals with interstitial marl. Locally, colemanite filled veins or breccia, and the beds graded into other sediments at the edges (Papke, 1976).

The adjacent calcareous, tuffaceous shaley mudstone with some interbeds of limestone also contained a few calcite nodules locally called "egg shells" or "goose eggs," with a paper thin, multiple-shell structure, or travertine-like appearance. Gale (1921b) described the limestone within and around the colemanite as being in "layers [with] a concretionary or botryoidal upper surface, ... characteristic ... of spring deposits. The colemanite ... can be distinguished from the enclosing beds by faint shades of yellow, green, and pink exhibited by the shale interbedded with [it] (p. 527)." Alonso and Viramonte (1993) stated that the deposit's calcite was travertine, with spring emission vents and bird prints changing laterally into colemanite. In the lower zone were sparse beds of zeolitized green tuff. Castor (1993) found no ulexite in recent borate exposures. The small amount of gypsum occurred as fibrous, columnar beds and veins or as finely crystalline masses. The marl was generally dolomitic, with montmorillonite and occasional hectorite, as well as some celestite. Stromatolites formed from blue-green algae do not grow in very saline water, and yet the domed or underwater form were quite common in the colemanite and adjacent beds. The colemanite had a moderate strontium content and other geothermal-type impurities. The marl and limestone had few impurities, but did contain adsorbed lithium, arsenic, and tungsten.

The mine operated from 1922 to 1927 and shipped ~180,000 metric tons

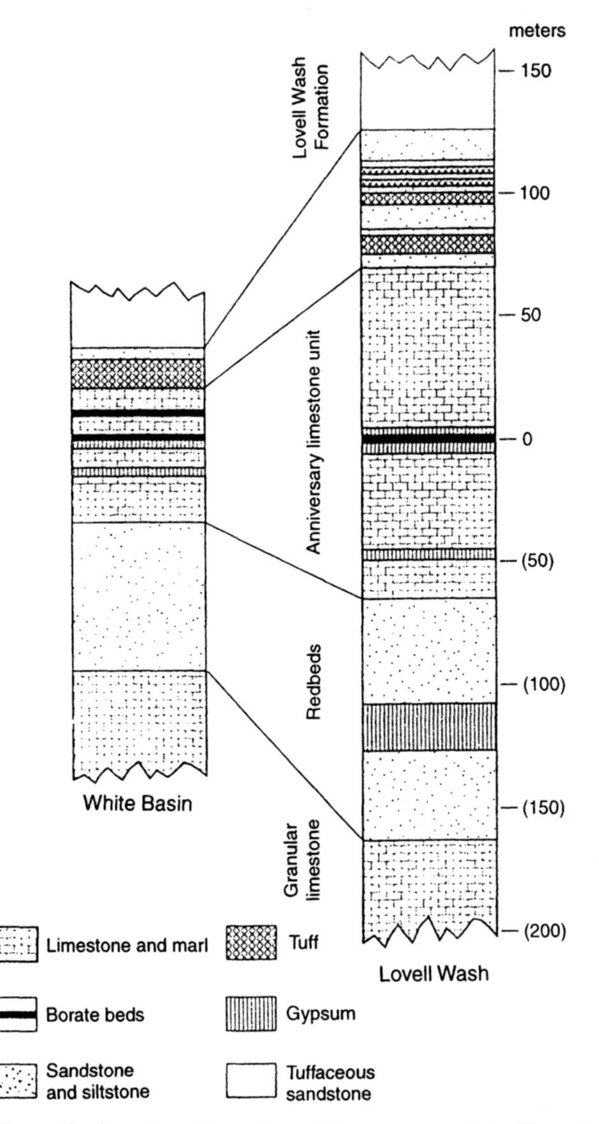

Figure 3.18 Generalized stratigraphic sections of the upper part of the Horse Spring Formation in the Lovell Wash and White Basin areas. (From Castor, 1993; reprinted courtesy of the Nevada Bureau of Mines and Geology.)

of ore to an adjacent processing plant, initially sending 27 metric tons/day of product by tractor-drawn wagons 48 km (30 mi) to the railroad at Dike, Nevada. Later, the mine recovered 160 metric tons/day, and the calciner plant produced 59 metric tons/day of a 40% B_2O_3 product. The mining claims in

1993 were held by the American Borax Company, which further drilled the ore body, finding 3.2 million metric tons of 26% B_2O_3, low-arsenic colelmanite (Castor, 1993; or 1.4 million metric tons of ore and 0.35 million metric tons of B_2O_3 according to Wendel, 1978).

3.6.5.2 White Basin

This deposit was 50 km northeast of Las Vegas, covered 3 km^2, and was bordered on the east, west, and north by high ridges of the Muddy Range. The colemanite outcrop contained considerable paper shale or paper limestone, as well as "globular forms with shells in concentric layers like an onion, called goose eggs by the prospectors (Castor, 1993, p. 12)" (as at Lovell Wash), and which contained some ulexite. Most of the colemanite was in massive beds, but they had been considerably disturbed by faulting and folding (Noble, 1922). The colemanite was discovered in 1920 in two 0.9-m (3-ft) beds dipping gently to the north. It was interbedded with shale, limestone, and volcanic ash, and there were some colemanite nodules (a few with some ulexite; Papke, 1976). The thickness of the borate zone was only about half that at Lovell Wash, and the borates were in the upper part of this zone (rather than the center as at Lovell Wash), fairly near the surface. The borate beds had shallow dips and were quite faulted. The colemanite varied from finely to very coarsely crystalline, and as separate to coalescent crystalline masses with radial to plumose structures. In these clusters the crystals terminated in, and were separated by, marl, and some had diamond-shaped cross-sections. There were some colemanite nodules with vuggy centers and inwardly terminating crystals that filled some cracks and vugs. There was very little ulexite (usually finely fibrous and a colemanite replacement), some celestite, and considerable finely banded crustiform travertine (Castor, 1993).

The initial White Basin claims were sold to Pacific Coast Borax, and later adjacent claims 2.4 km to the east to the American Borax Company. The latter company mined from an inclined shaft with <100 m of drifts from 1922 to 1924, but then lost in litigation with Pacific Coast Borax, who prevailed and shut down the operation permanently. Most of the mining had been done in the western part of the district in a 600-m area. A mill also was located at the site, and according to the size of the tailings pile, ~18,000 metric tons of ore were mined and processed (Castor, 1993).

3.7 YUGOSLAVIA

3.7.1 Jarandol Basin, Serbia

Colemanite ($Ca_2B_6O_{11} \cdot 5H_2O$) and howlite ($Ca_4Si_2B_{10}O_{23} \cdot 5H_2O$) have been reported in commercial-grade deposits in the western Anatolia district of

Serbia, Yugoslavia (Obradovic, Stamatakis, Anicic, and Economou, 1992). The minerals occur in a chain of sedimentary basins extending along the Balkan peninsula, with Jarandol perhaps having the most borates. It is of Miocene epoch, and the area also has magnesite and coal separated by analcime-rich tuff, clay, and marls. The 100-km^2 basin was once twice that large, containing a lake with deep and narrow southern shores and extensive shallow northern margins. There are several boron-rich (up to 2000 ppmB) hot springs in the basin. In one 0.4-km^2 area a <12-m (average 4-m) colemanite bed with subsidiary howlite was interlayered with dark-gray claystone and chert, and bounded by faults. The beds were brecciated, and euhedral transparent colemanite crystals a few centimeters in size also filled fractures in the matrix or underlying beds. The howlite occurred as nodules with margins containing some dark clayey occlusions or <2-cm wormlike bodies.

At a second site there were two borate layers interbedded with tuffaceous rocks. The 2-m upper bed and its surface outcrop contained primarily howlite, but increased in colemanite with depth to 60 m. The 2.5-m lower bed was 150 m deep, and in both beds the colemanite occurred as coarse crystals or fine-grained aggregates. The howlite's botryoidal nodules had very small crystals and were imbedded in clay. Occasionally, howlite could be seen replacing colemanite in the crystals at both sites. Both minerals sometimes contained black inclusions of kerogen as well as thin greenish clay films. At times mixtures of howlite and colemanite nodules surrounded or intruded into fine-grained disseminated colemanite zones. In the outer part of the basin, colemanite and howlite occurred in fine-grained clastic sediments as lens-shaped masses, often with bitumen, calcite, and gypsum. Pure ulexite ($NaCaB_5O_9 \cdot 8H_2O$) and colemanite were usually found in the central section, occasionally along with some tincalonite ($Na_2B_4O_7 \cdot 5H_2O$; Obradovic and Vasic, 1990). Malinko *et al.* (1995) noted the occurrence of a new mineral, studenitsite ($NaCa_2B_9O_{16} \cdot 4H_2O$), in "a borate deposit 280 km south of Belgrade (p. 57)." It was associated with colemanite, howlite, ulexite, and pentahydroborite ($CaB_2O_4 \cdot 5H_2O$) as lenses in shale–marl tuffaceous rocks. The studenitsite was found "as intergrowths of flattened wedge-shaped <5-mm transparent colorless to light-yellow crystals with hatched faces (p. 57)"

3.7.2 Other Areas, Raska

Other deposits in Yugoslavia's chain of basins were said to contain ulexite and borax, analcime, and traces of other borates such as searlesite ($NaBSi_2O_6 \cdot H_2O$) and lueneburgite [$Mg_3(PO_4)_2B_2O_3 \cdot 8H_2O$]. They were found as fissure fillings in magnesite near Jarandol or as <3-mm lueneburgite crystals in cracks of brecciated magnesite (Obradovic, Stamatakis, Anicic, and Economou, 1992). In Bosnia's Lopari–Sibosnica valley and at Kremera in Serbia, searlesite also has been found, in the latter occurring as small ball-shaped grains in magnesite or dolomite. Some layers contained <40% searlesite (average 1.6% B_2O_3). In Bosnia elliptical-shaped searlesite occurred in fractured marl–shale interlayered with tuff or tuffaceous sandstone (Lyday, 1992). Orris (1995) reported similar searlesite at Valjevo–Mionica.

Papezik and Fond (1975) also reported two 10-m and 3-m lenses of howlite and 20- to 30-cm beds near *Raska,* 150 km south of Belgrade. It was white, fine-grained, and interbedded with Tertiary volcanics, clastics, carbonates, and marls. It was also found as scattered nodules in adjacent conglomerates, sandstones, and tuffites. In 1996 this area was announced as "a new boron deposit estimated to contain ~10 million metric tons of reserves" (Anon., 1996).

3.8 MINOR OCCURRENCES

3.8.1 Albania, Greece, and Samos Island

Borosilicate occurrences have been reported in an area of volcanic and ultrabasic rocks in Albania (Tershana and Marishta, 1984). Scattered colemanite with some ulexite also occurs as occasional thin layers, nodules, lenses, or fissure fillings in the saline–alkaline tuffaceous and clay sediments of the late Miocene Karlovassi basin in the Samos Island. The borate layers alternate with claystone, and there are gypsum and celestite beds above the colemanite. The deposit outcrops in a small lens of predominantly colemanite in calcite, with limited ulexite, celestite, dolomite and smectite. Occasionally, calcite has replaced the borates, usually in a spongy form. Colemanite's subhedral to anhedral crystals have well-developed cleavage; there are a few radiating aggregates; and colemanite can be cemented by calcite. The ulexite occurs as small aligned needle-like crystals, usually as inclusions in colemanite. Celestite has formed as thin fissure fillings in the massive colemanite (Stamatakis and Economou, 1991).

3.8.2 Iran and Italy

In Iran the Kara Gol deposit lies between Tebriz and Zencan, with ulexite and hydroboracite extensively deposited (along with gypsum) in a fracture zone (like a hydrothermal metallic deposit), and has supported a small borate operation (Helvaci, 1996). Ulexite and tourmaline-rich granites also have been found in an area near Sirjan, and near Sabzevar there were borates in thin beds of limestone, marl, and clays. Ulexite occurred near Ashin (West Kerman) in south central Iran, discovered in 1871, but there had been only limited recovery by 1916. Ulexite once was mined at Tonkar (Orris, 1995). In 1945 and 1965 borax production was reported at ~100 tons/year (Lyday, 1992; Travis and Cocks, 1984). Ginorite ($Ca_3B_{14}O_{23}{\cdot}8H_2O$) veins in sandstone have been reported (with calcite) at Sasso Pissano, Tuscany, Italy (Allen and Kramer, 1957). Datolite ($Ca_2B_2Si_2O_9 \cdot H_2O$) also has been noted in highly mineralized serpentinites at seven locations in the lower Manubiola Valley, Berceto, Parma (Fabrizio, 1993).

3.8.3 North Korea, Romania, Russia, and South Africa

Minor production of ludwigite ($[Mg, Fe]_4Fe_2B_2O_{10}$), datolite ($Ca_2B_2Si_2O_9 \cdot H_2O$), and kotoite ($MgB_2O_6$) in "pipe-like bodies" in North Korea at Hol-Kol, Kol-don Raitakuri, and Yakutia, southwest of Pyongyang in the 1950s has been reported (Kistler and Helvaci, 1992). The Baita Bihor district of the Black Cris in Romania contains ludwigite, ascharite, and alteration products in the Superior area (Stoici, 1975; Stoicovici and Stoici, 1972). In Russia a quartz–tourmaline pegmatite deposit in the Kamchatka region has 2.25% B_2O_3 ore, with 200,000 tons of B_2O_3 reserves. Some 3–15% B_2O_3 datolite ore has been upgraded by beneficiation to 18% B_2O_3 (or leached directly), and commercially recovered in the North Caucasus (Kerchen and Jaman peninsulas; Lyday, 1992). A complex deposit of 5% B_2O_3 azoproite [$(Mg, Fe)_2(Fe, Ti, Mg)BO_5$] was reported on the western shore of Lake Baikal with 350,000 tons of B_2O_3 reserves (Matterson, 1980). Many other complex borate ores have been reported in Russia, such as datolite in the Soganlug talus near Tbilissi and in the volcanics of the Karadag in the Crimea. Ludwigite has been reported in the polar region of Yakutia. See the References for others. In South Africa, datolite ($Ca_2B_2Si_2O_9 \cdot H_2O$) has been found in the Wessels Mine (Gutzmer and Cairncross, 1993), and also in the hydrothermal fluid intrusion into fault zones of sedimentary manganese ores in the Kalahari manganese field (Gutzmer and Beukes, 1995).

3.8.4 United States

3.8.4.1 Cave Springs, Nevada

Searlesite ($NaBSi_2O_6 \cdot H_2O$) and ulexite ($NaCaB_5O_9 \cdot 8H_2O$) have been found in a series of thinly bedded marls, shales, sandstones, and volcanic tuffs in the Silver Peak Range within the drainage basin of the Fish Lake Valley. The searlesite was found in marl containing mica and beads of clear glassy opal as minute pockets of large granular crystals, and as vein-filling in the marl. The transparent, glassy <2-mm crystals occasionally had a milky coating or carried small beads of opal. The marl could contain <50% searlesite (Foshag, 1935). Searlesite also has been found in many alkaline basins such as Ash Meadows (Orris, 1995).

3.8.4.2 Death Valley, California

Ginorite ($Ca_2B_{14}O_{23} \cdot 8H_2O$) and sassolite ($H_3BO_3$) have been found in soft white to pale yellowish-brown efflorescе in weathered basalt within 50 cm of the surface near the head of Twenty Mule Team Canyon (the Mott open-cut prospect). An underlying altered basalt contained veins of colemanite ($Ca_2B_6O_{11}·5H_2O$) and priceite ($Ca_4B_{10}O_{19}·7H_2O$) in limestone and shale. The ginorite occurred as 1- to 2-mm white pellets imbedded in a pale yellowish-brown matrix of sassolite and clay, and the sassolite as <1-mm pearly plates (Allen and Kramer, 1957). At this and three other nearby locations, gowerite ($CaB_6O_{10}·5H_2O$) also was identified, along with ulexite ($NaCaB_5O_9·8H_2O$), meyerhofferite ($Ca_2B_6O_{11}·7H_2O$), and hydroboracite ($CaMgB_6O_{11}·6H_2O$) in the altered outcrops of <5-cm colemanite and/or priceite beds in fragmented olivine basalt. The basalt was above and adjacent to a colemanite–priceite deposit.

Gowerite was only in exposed zones where the colemanite and priceite had been completely altered, and in the Mott location some of it was in <10-mm globular clusters with radiating small crystals near colemanite remnants. It was with crusts and clusters of subhedral meyerhofferite crystals and some hydroboracite. It also occurred as small clusters in decomposed basalt, like the much more abundant ginorite and sassolite nearer the surface. At the Hard Scramble site the gowerite was in compact 1- to 10-mm globular clusters within the altered basalt, or on vein-like aggregates of meyerhofferite (formed from priceite). Ulexite, ginorite, and gypsum occurred with the gowerite, with ulexite being the most common borate alteration product and gowerite, hydroboracite, and sassolite the rarest. Ulexite occurred throughout the weathered zone, but was concentrated in the upper part. Gypsum, thenardite, and some limonite were also often present (Erd, McAllister, and Almond, 1959).

In a later study by Erd, MacAllister, and Vlisidis (1961), nobleite ($CaB_6O_{10}·4H_2O$), meyerhofferite ($Ca_2B_6O_{11}·7H_2O$), and mcallisterite ($Mg_2B_{12}O_{20}·15H_2O$) were found in these same four locations and three others nearby. All were in altered olivine basalt with veins of colemanite and/or priceite, and with most of the aforementioned borates. The nobleite generally formed crusts on the colemanite or priceite in an open network of tabular crystals. It was also found on weathered colemanite and priceite in older mines. The mcallisterite occurred sparingly in a pale yellowish-brown matrix of sassolite and clay, with small white rods of ginorite. Its 1- to 2-mm (occasional clusters <1–2 cm) crystals looked like rounded pellets (Shaller, Vlisidis, and Mrose, 1965).

Bakerite ($Ca_8B_{10}Si_6O_{35}·5H_2O$) and priceite ($Ca_4B_{10}O_{19}·7H_2O$) have been found in Corkscrew Canyon, with the former occurring as irregular veins in altered basalt accompanied by the zeolites thomsonite and natrolite (containing traces of Pb, Sn, Ag, and Mn). The priceite was in white veins in the basalt and appeared to have been formed from colemanite (Kramer and Allen, 1956). In the Corkscrew mine 3-mm crystals of nobleite were found, and colemanite pseudomorphs after inyoite (Muehle, 1974). In the Mount Blanco district on Furnace Creek, a prospect tunnel was called a "jewelry shop" by

Shaller (1916) because of the beautiful colemanite crystals throughout most of its length. The large sparkling crystals had the appearance of a geode center. Inyoite, priceite, and meyerhofferite were also present, with the inyoite as large crystals with perfect branches on a massive center. In places it had been altered to meyerhofferite, often as a rosette on an inyoite base. The 2- to 3-mm long, 0.3- to 0.5-m-thick crystals had a prismatic habit.

3.9 SPECIAL BORATES OR FORMATIONS

A very large number of borates occur where they may be considered as only low-grade deposits or as a minor component of other rocks or deposits. Borates can be found in many magmatic rocks or formations such as granites, rhyolite, pegmatites, porphyries, serpentines, and even some lavas (Grew, 1996; London, Morgan, and Wolf, 1996). They also may be found in many hydrothermal fluid formations such as metallic ore deposits or skarns (Slack, 1996). Of course, they are common in the mantles or deposits from borate-containing springs or where there has been borate leach-reprecipitation reactions. Finally, in many situations boron may substitute for other elements to a variable extent in other minerals or be adsorbed on them. Some of the more important of these borate types are reviewed in the following sections.

3.9.1 Skarns

Utekhin (1965) noted that there were 95 known occurrences of boron mineral skarns in dolomite or Mg-rich rocks, and 50 in limestone (many others have been discovered since that date). The boron mineralization in the former is characterized by Mg and Fe borates, such as the ludwigite [$(Mg, Fe)_4Fe_2B_2O_{10}$] group. The borates in the latter are predominantly Ca-borosilicates (Table 3.9). The skarns date from the Precambrian to the Cenozoic eras, and may produce a commercial borate deposit, but more often borates are only a small percentage of the rock. Watanabe (1967) theorized that in the later stage of magmatic movement, volatiles such as boron, fluorine, and water are concentrated in a residual solution with the magma. In skarns these solutions then intrude with or after the magma into carbonate formations at temperatures of 300–600°C (judging by the minerals formed). The borate solutions react with the carbonates and adjacent rocks and crystallize. Morgan (1980) observed that the borates usually were deposited after the magma intrusion, often in the limbs of synclines or anticlines where the metamorphosed rocks were heavily fractured. Sometimes the skarn was formed in several stages, with the magma first intruding and being altered, then the accompanying fluids depositing borosilicates, and finally lower temperature minerals being formed. The borates commonly were deposited in the carbonate formation as pockets,

Table 3.9

Common Borate Minerals in Various Types of Skarns[a]

Limestone skarns	Magnesian skarns	Other skarns, general
Axinite	Azoproite	Dumortierite[b]
Cahnite	Fluoborite	Grandidierite
Calciborite	Harkerite	Hambergite[b]
Danburite	Hulsite	Jeremejevite[b]
Datolite	Jimboite	Rhodizite[b]
Frolovite	Kotoite	Stillwellite
Korzhinskite	Kurchatovite	Sussexite
Nifontovite	Ludwigite	Vesuvianite
Nordenskioldine[b]	Orthopinakiolite	
Olshanskyite	Pinakiolite	
Pentahydroborite	Sakhaite	
Sibirskite	Serendibite	
Szaibelyite (ascharite)	Sinhalite	
Tourmaline[b]	Suanite	
Uralborite	Szaibelyite (ascharite)	
Vimsite	Tourmaline[b]	
	Vonsenite	
	Warwickite	
	Wightmanite	

[a] Morgan, 1980.
[b] Often found in the magma of the skarn.

veins, and lenses. With the increased importance of skarns for large-scale borate production, there has been a correspondingly large number of articles on skarn deposits, theorizing on their origin, developing prospecting guides, and summarizing other studies (see the References).

3.9.2 Metal Deposits

Many metallic ore deposits contain borates, such as the *tin* borates in northwestern Tasmania, with economic amounts of B, Cu, Zn, W, and Be, as more than 43 different minerals, including szaibelyite, fluoborite, Sn-ludwigite, warwickite, axinite, tourmaline, and other borates (Kwak and Nicholson, 1988). Similar Sn-B skarn deposits occur in Alaska, East Siberia, and elsewhere. Borates often occur with *lead–zinc* deposits, such as ulexite ($NaCaB_5O_9 \cdot 8H_2O$) and howlite ($Ca_4Si_2B_{10}O_{23} \cdot 5H_2O$) in the Gays River, Nova Scotia deposit (Papezik and Fong, 1975) and many others. Perhaps the most common metallic deposit with borates is *iron ore.* For instance, cahnite ($Ca_2BAs O_6 \cdot 2H_2O$) has been reported in several mines, such as the Kladeborg skarn mine in the Arendal, Norway, as well as an eastern Siberia contact-metasomatic iron ore deposit (Malinko, 1966). Vonsenite [$(Fe, Mg)_2FeBO_5$] is also common, such as at the magnetite skarn deposit at Jayville, New York, which contains <4.80% B_2O_3. Skarn formations also may occur in *manganese* carbonate, such as at the Kaso mine in Japan where jimboite ($Mn_3B_2O_6$) has been formed in the contact aureole of a granitic intrusion (Watanabe, 1967). The boron content of braunite ($Mn_7O_9SiO_3$) from mines around the world averaged 1.05% B_2O_3 ($Mn_7O_9BO_3$; Wasserstein, 1942). *Gold and copper* also may contain borates, as that found in the Hol Kol mine in Pyongyang, North Korea. Ludwigite [$(Fe, Mg)Mg_3Fe_2B_2O_{10}$], warwickite [$(Mg, Fe)_3TiB_2O_8$], kotoite ($Mg_3B_2O_6$), szaibelyite ($Mg_2B_2O_5 \cdot H_2O$) and suanite ($Mg_2B_2O_5$) all have been identified. Many other articles on borates in metallic deposits are listed in the References.

3.9.3 Tourmaline

By far the most common borate (but not found as a commercial ore) is that from the family of tourmalines. They form in many environments, including magma (Henry and Dutrow, 1996) and ore-forming hydrothermal fluids (Slack, 1996), and as secondary minerals. Collectively, they are "the most abundant mineralogical sink for boron" because boron is one of their essential elements, averaging ~3% B (9.7% B_2O_3). They are insoluble in water at low temperatures, slightly more at high temperatures, and much more in hot alkaline solutions. They are thermally stable up to the melting points of most crustal rocks (>700°C) and mechanically very strong (London *et al.*, 1996). As tourmaline is being formed, it is quite responsive to its chemical environment, forming a wide variety of compositions of the formula $XY_3Z_6T_6O_{18}(BO_3)_3W_4$, where T is usually Si with minor or no other elements; W is usually OH but can include O and F (there are 31 anions in the tourmaline molecule); Z is usually Al but can be replaced completely or in part by Mg, Fe^{+3}, Cr^{+3}, and V^{+3}; X is most commonly Na and Ca with minor K; and Y can be any mono, di, and trivalent metal. The structure can accommodate many trace elements, particularly in the elbaite form (Henry and Dutrow, 1996). When tourmalines are found in a granite or pegmatite body they are more concentrated at the margins or in the surrounding rock, and their composition may change appreciably from the edge toward the center. Only a small fraction of the tourmalines are of gem quality, but when they are, they often occur in miarolitic cavities within granitic pegmatites, and usually as the elbaite or liddicoatite forms. When the gem is a light color, it usually is because of the prior removal of Fe and Ti, two strong chromophors (London *et al.,* 1996).

When hydrothermal metal deposits are associated with magma (and not a skarn), there is often considerable tourmaline. Perhaps the best known of such deposits is tin, with or without tungsten or other metals, in greisen veins and replacement formations. Tourmaline is also a common constituent of granite-related breccia pipes and porphyries (and their wall rock) that contain copper, gold, molybdium, and other metals, which formed from hydrothermal solutions. It can be a minor-to-abundant gangue material in many stratabound metal deposits, which occasionally contain >15–20% tourmaline. Such clusters are called tourmalinites, and may have formed by many mechanisms, but replacement reactions with other borosilicates is the most common source under high temperature, alkaline conditions. The close association of tourmaline with metal deposits, especially the magnesium tourmalines, is so great that it has been suggested for use as a prospecting tool (examining tourmaline grains in stream sediments; Slack, 1996).

3.9.4 Boron-Bearing Potassium Feldspar

Another of the fairly common boron-containing minerals is substituted potassium feldspar ($KAlSi_3O_8$–$KBSi_3O_8$). In strongly alkaline solutions, glass-type rocks (tuff, basalt, etc.) can be altered to a sequence of smectite-zeolites-analcime-K feldspar–searlesite. Boron is not appreciably taken up by the earlier minerals and enters the K feldspar only at the final stage of the sequence. If sufficient boron is present in highly alkaline solutions, it can form searlesite and perhaps other borosilicates if the concentrations, reaction time, temperature, and other conditions are favorable (Hay, 1966). An example of this sequence is in Samos Island, where boron-bearing K feldspar makes up >30% of the rocks in the center of its basin and contains 1500–2500 ppmB. The K-feldspar is accompanied by clinoptilolite and analcime (both with a boron content of 10–280 ppm) and occurs as masses of very small crystals. In other highly alkaline Greek basins, the K-feldspars contain 500–5000 ppmB, usually >1000 ppm (Stamatakis, 1989). Several areas of boron-bearing K-feldspars in the United States have been described by Sheppard and Gude (1973). They believed that the boron replacement of aluminum took place very slowly from an interstitial, strong carbonate–borate brine long after deposition, and that the accompanying zeolites usually contained <100 (average 73) ppmB, whereas the K-feldspars had more than 1000 (average 1800) ppmB. The boron altered the monoclinic cell dimensions of the feldspar, and in the lab Al can be completely replaced by boron.

3.9.5 Boron-Bearing Albite (Reedmergnerite)

Boron may also replace the aluminum in albite ($NaAlSi_3O_8$–$NaBSi_3O_8$, also called reedmergnerite). For example, 0.5 × 0.4 × 0.3-mm crystals of reedmergnerite have been found at >610-m (2000-ft) depths in dolomitic rock (near oil shale) of the Green River Formation at Duchesne, Utah. It was with leucospenite, nacholite, and shortite, and there were lesser amounts of eitelite ($Na_2Mg[CO_3]_2$), searlesite ($NaBSi_2O_6 \cdot H_2O$), garrelsite ($NaBa_3Si_2B_7O_{18} \cdot 2H_2O$), and analcime. It also can be made synthetically in the laboratory (Appleman and Clark, 1965; Milton, Axelrod, and Grimaldi, 1955; Milton, Chao, Axelrod, and Grimaldi, 1960). Reedmergnerite with boron-bearing K-feldspar occurs in alkaline silicic rocks in Russia (Martin, 1971).

3.9.6 High-Boron Clays

It has been commonly observed (e.g., Walker, 1975) that clays settled in seawater adsorb considerable boron, particularly illite. Goldschmidt (1954) collected bottom muds from various oceans, finding 50–500 ppmB. Others found 450–500 ppm. Still others found marine shales with 180–320 ppmB and nonmarine shales 5–10 ppmB. Clays associated with Zeichstein potash deposits contained <1500 ppmB. Reynolds (1965) estimated that pure Precambrian illite averaged 422 ± 24 ppmB, and post-Precambrian illite 459 ± 10 ppmB. Frederickson and Reynolds (1960) attempted to determine the adsorption exponent of illite, kaolinite, and chlorite in mixtures of marine clays, and found that pure illite averaged 450 ppmB. Studies from a seawater-diluted estuary (Adams, Haynes, and Walker, 1965) estimated that illite contacted by normal seawater would contain 430 ppmB, and with >50% dilution:

$$\text{Boron in illite} = 358 \times (\text{dilution factor} = y) + 92. \tag{1}$$

If the Zeichstein data were also considered, the formula might read:

$$B = 358y^{0.32} + 92 \text{ for more concentrated seawater.} \tag{2}$$

Lerman (1966) ran four 28-day boron adsorption tests on illite and synthetic seawater and found the following:

$$\log(\text{moles B adsorbed/gram of illite}) = 3.72 + 0.37\,(\log \text{moles B/kg of final solution}) \tag{3}$$

Other data were analyzed that correlated with the constants 4.01 and 0.39, respectively, in formula (3). When put on a simpler basis, the final boron content of the illite (as ppm) from his study was roughly as follows:

$$B_{illite} = 358y^{0.14} + 92 \tag{4}$$

Considerable other data on boron adsorption by clays have been reported, as noted in the References.

References

Adams, T. D., Haynes, J. R., and Walker, C. J. (1965). Boron in holocene illites of the Dovey estuary, Wales, and its relationship to paleosalinity cyclotherns. *Sedimentology* **4**(3), 189–195.

Ageeva, L. I., and Aver'yanov, G. S. (1981). Formation time of the borosilicate skarns of Pamir, USSR. *Dokl. Akad. Nauk Tadzh. SSR* **24**(5), 308–310.

Agyei, E. K., and McMullen, C. C. (1968, August). A study of the isotopic abundance of boron from various sources. *Can. J. Earth Sci.* **5**(4, Pt. 1), 921–927.

Albayrak, F. A., and Protopapas, T. E. (1985). Borate Deposits of Turkey. In *Borates: Economic Geology and Production* (J. M. Barker and S. J. Lefond, eds.), pp. 71–85, Soc. Min. Eng., AIMMPE, New York.

Aleksandrov, S. M. (1975). Geochemical aspects of B-Sn ore formation in Alaskan deposits. *Geochem. Int.* **12,** 139–150.

Aleksandrov, S. M., and Malinko, S. V. (1975). Geochemical peculiarities of endogenic and hypergenic alterations of carbonatoborates. *Geokhimiya,* No. 1, pp. 3–16.

Allen, R. D., and Kramer, H. (1957). Ginorite and sassolite from Death Valley, California. *Am. Min.* **42,** 56–61.

Alonso, R. N. (1986). *Occurrences, Stratigraphic Position and Genesis of the Borate Deposits of the Puna Region of Argentina.* Doctoral dissertation, Univ. Nacional de Salta, Dept. of Nat. Sciences, Argentina, 196 p.

Alonso, R. N. (1992a, September). *Geologia de la Mina Monte Verde (Colemanita–Inyoita), Salta Republica Argentina,* pp. 215–225, 4th Congreso Nacional de Geologia Economica, 1st Cong. Latino Amer. Geol. Econ., Cordoba, Argentina.

Alonso, R. N. (1992b). Estratigrafia del Cenozoico de la Cuenca de Pastos Grandes (Puna Saltena) con enfasis en la Formacion Sijes y sus Boratos. *Revista de la Asociacion Geologica Argentina* **47**(2), 189–199.

Alonso, R. N., and Viramonte, J. G. (1990). Borate deposits in the Andes. In *Strabound Ore Deposits in the Andes* (Fontbote *et al.,* eds.), pp. 721–732, Springer-Verlag, New York.

Alonso, R. N., and Viramonte, J. G. (1993). La cuestion genetica de los Boratos de la Puna. *12th Cong. Geol. Argentino, 2nd Cong. de Explor. Hydrocarburos* **5,** 187–194.

Anon. (1980, August 1). Production begins at Chinese boron mine. *Mining J.* **295** (7563), 95.

Anon. (1996). Boron reserves discovered. *Industrial Minerals,* No. 340, p. 61.

Applemen, D. E., and Clark, J. R. (1965). Crystal structure of reedmergnerite, a boron albite, and its relation to feldspar crystal chemistry. *Am. Min.* **50,** 1827–1850.

Arriaga, H. M., Pena, L. R., and Gomez, A. C. (1986). Resultados de la evaluacion del deposito de boratos de area Tubutama, Sonora. *Geoimet* **141,** 41–60.

Axelrod, J. M., and Grimaldi, F. S. (1954). New minerals in the Green River formation. *Geol. Soc. Am. Bull., Abs.* **65,** 1286–1287.

Badalov, S. T., and Turesebekov, A. (1966). *Comparative Mineralogical–Genetic Characteristic of Datolite.* Zap. Uzb. Otd. Vses. Mineral. Obshchest., Akad. Nauk Uzb. SSR, No. 18, Zh., Geol., Vol. 1967, Abs. No. 4V224, No. 18, pp. 31–37.

Badalov, S. T., Turesebekov, A., Malinko, S. V. (1972). Distribution characteristics of boron in rocks and minerals of the Almalyk ore region (Uzbek, SSR). *Uzb. Geol. Zh.* **16**(1), 78–82.

Bailey, G. E. (1902). Borates. *The Saline Deposits of California, California State Mining Bureau, Bull. 24,* Pt. 2, pp. 33–90.

Barker, C. E. (1980, August). Geology of the Terry borate deposit. *California Geology* **33**(8), 181–189.

Barker, C. E., and Barker, J. M. (1985). Borate deposits, Death Valley, California. In *Borates: Economic Geology and Production* (J. M. Barker and S. J. Lefond, eds.), pp. 103–135, Soc. Min. Eng., AIMMPE, New York.

Barker, J. M., and Wilson, J. L. (1976). Borate deposits in the Death Valley Region. *Guidebook: Las Vegas to Death Valley and Return, Nevada Bur. Mines, Geol.,* Rept. 26, pp. 22–33.

Bekisoglu, K. A. (1962, June–July). Boron deposits in Turkey. *Turkish Econ. Rev.* **3**(4), 12–34.

Blokhina, N. A. (1974). Boron mineralization in skarns of gold sulfide deposits of the Tarorsk Group in Central Tadzhikistan. *Dokl. Akad. Nauk Tadzh. SSR* **17**(8), 47–50.

Brown, W. W., and Jones, K. D. (1971). Borate deposits of Turkey. In *Geology and History of Turkey,* (A. S. Campbell, ed.), pp. 483–492, Petroleum Exploration Society of Libya, Tripoli.

Buslaeva, E. Y. (1994). Halogen-substituted hydrocarbons in carbonaceous black shales. *Geokhimiya* **7,** 1130–1131.

Caballero, A. G., Melendez, H. A., and Rocha, L. P. (1980). "Estudio geologico del depositos de boratos del area de Tubutama, Sonora. *Seminario Interro Sobre Exploration Geologico-Minera, Consejo de Recurces Minerales, Mexico* **3,** 367–402.

Castor, S. B. (1993). Borates in the Muddy Mountains, Clark County, Nevada. *Nevada Bureau of Mines and Geology, Bull. 107.*

Chandler, D. C. (1996, January–February). The Billie Mine, Death Valley, California. *Mineralogical Record* **27,** 35–40.

Chernevskii, N. N. (1973). Pegmatite and skarn boron minerals in alkaline gabbroic rocks in Tien Shan. *Tr. Tashk. Politekh, Inst.* **95,** 25–26.

Chernykh, I. D., Kovalevskaya, O. M., and Kovalevskii, A. L. (1969). Biogeochemical Prospecting for Boron Deposits. In *Biogeokhim. Pooiski Rud. Mestorozhd., Dokl. Sekts., 5th Mtg., 1966,* (A. L. Kovalevskii, ed.), pp. 221–233, Buryat. Inst. Estestv, Nauk, Akad. Nauk SSSR, Ulan-Ude.

Chernyshov, A. V. (1984). Effect of harmful impurities in the Dal-negorsk deposit ores on the density of the prospecting network points. *Rozved. Okhr. Nedr* **11,** 24–26.

Couch, E. L., and Grim, R. E. (1968). Boron fixation by illites. *Clays Clay Min.* **16,** 249–256.

Countryman, R. L. (1977). Hydroboracite from the Amargosa Desert, eastern California. *Mineralogical Record* **8,** 503–504.

Crawford, J. J. (1894, 1896). Borax, pp. 34–35, Twelfth (and Thirteenth) Report of State Mineralogist, Calif. Div. Mines, Geol.

Crowe, D. E., and Ignatiev, A. V. (1993, October). Geology and geochemistry of the Dalnegorsk borosilicate skarn deposit, Far East Russia. *Geol. Soc. America, Abs.* **25**(6), 403.

Dorofeev, A. V. (1974). Mineralogy, distribution, and origin of boron mineralization in skarn deposits, as illustrated by one of the Siberian regions. In *Vopr. Rudonosn. Yakutii* (G. N. Gamyanin, ed.), pp. 156–167, Izd. Fil. Akad. Nauk SSSR.

Dupont, F. M. (1910, December). The borax industry. *J. Ind. Eng. Chem.* **2,** 500–503.

Eakle, A. S. (1911, June 28). Neocolmanite, a variety of colemanite, and howlite from Lang, Los Angeles County, California. *Univ. of Calif. Publications, Bull. Dept. Geol.* **6**(9), 179–189.

Embrey, P. G. (1960). Cahnite from Capo di Bove, Rome. *Rome Min. Mag.* **32,** 666–668.

Erd, R. C., McAllister, J. F., and Almond, H. (1959). Gowerite, a new hydrous calcium borate from the Death Valley region, California. *Am. Min.* **44**(9–10), 911–919.

Erd, R. C., McAllister, J. F., and Vlisidis, A. C. (1961). Nobleite, another new hydrous calcium borate from the Death Valley region, California. *Am. Min.* **46,** 560–571.

Evans, J. R., and Anderson, T. P. (1976). *Colemanite Deposits Near Kramer Junction, San Bernardino County, California. Calif. Div. Mines and Geology, Spec. Publ. 50.*

Evans, J. R., Taylor, G. C., and Rapp, J. S. (1976). *Mines and Mineral Deposits of Death Valley National Monument, California.* Calif. Div. Mines Geol., Spec. Rept. 125.

Evans, J. R., and Vredenburgh, L. M. (1982). Colemanite deposits near Stauffer, Ventura County, California. *Geol. Mineral Wealth Calif. Transverse Ranges* (Mason Hill Volume), pp. 495–499, So. Coast Geol. Soc.

Fabrizio, A. (1993). Minerals of serpentinites from the Lower Manubiola Valley, Berceto, Parma, Italy. *Riv. Min. Ital.* **2,** 67–78.

Fang, J. H., and Newnham, R. E. (1965). The crystal structure of sinhalite. *Min. Mag.* **35,** 196–199.

Finley, H. O., Eberle, A. R., and Rodden, C. J. (1962). Isotopic boron composition of certain boron minerals. *Geochim. et Cosmochim. Acta* **26,** 911–914.

Foshag, W. F. (1921). The origin of the colemanite deposits of California. *Econ. Geol.* **16,** 199–214.

Foshag, W. F. (1922). Calico Hills, San Bernardino County, California. *Am. Min.* **7,** 208, 209.

Foshag, W. F. (1931). Probertite from Ryan, Inyo County, California. *Am. Min.* **16,** 338–341.

Foshag, W. F. (1935). Searlesite from Esmeralda County, Nevada. *Am. Min.* **19,** 268–274.

Frederkson, A. F., and Reynolds, R. C. (1960). Geochemical method for determining paleosalinity. In *Clays and Clay Minerals* (A. Swineford, ed.), Proc. 8th Nat. Conf. Clays, Pergamon Press, New York. Vol. 8, pp. 203–213.

Gale, H. S. (1912). *The Lila C. Borax Mine at Ryan, California,* pp. 861–866, U.S. Geol. Survey, Mineral Resources.

Gale, H. S. (1913a). Borate deposits in Ventura County, California. *U.S. Geol. Survey, Bull. 540-0,* pp. 434–456.

Gale, H. S. (1913b). *The Origin of Colemanite Deposits,* pp. 3–9, U.S. Geol. Survey, Prof. Paper No. 85.

Gale, H. S. (1914). *The Origin of Colemanite Deposits.* U.S. Geol. Survey, Shorter Contributions to General Geology, 1913, Prof. Paper 85.

Gale, H. S. (1921a, December 24). *Priceite, the Borate Mineral in Curry County, Oregon,* pp. 895–898, Mining and Scientific Press.

Gale, H. S. (1921b). The Callville Wash colemanite deposit. *Eng. Mining J.* **112**(14), 524–530.

Garrett, C. R. (1985). *Development of a Roadheader Mining System at American Borate Company,* pp. 886–901, Rapid Excavation and Tunnelling Conference, Proc. 2, Soc. Min. Eng., AIMMPE, New York.

Gay, J. E., and Hoffman, S. R. (1954). Mines and mineral resources of Los Angeles County, California. *J. Mines Geol.* **50,** 506–508.

Goldschmidt, V. M. (1954). *Geochemistry,* pp. 280–291, Clarendon Press, Oxford.

Gonzalez-Barry, C. E., and Alonso, R. N. (1987). El deposito neoterciario de boratos esperanza, salta. *Decimo Cong. Geol. Argentino, San Miguel de Tucuman* **2,** 63–66.

Grace K. A., Douglas, R. F., Pool, T. C., and Lattanzi, C. R. (1985). Valuation of borate resources in Death Valley: A case study. In *Borates; Economic Geology and Production,* (J. M. Barker and S. J. Lefond, eds.), pp. 237–251, Soc. Min. Eng., AIMMPE, New York.

Grew, E. S. (1996). Borosilicates (exclusive of tourmaline) and boron in rock-forming minerals in metamorphic environments. In *Boron: Mineralogy, Petrology and Geochemistry* (E. S. Grew and L. M. Anovitz, eds.), Vol. 33, Ch. 9, pp. 387–502, Mineral. Soc. Am., Reviews in Mineralogy.

Gutzmer, J., and Beukes, N. J. (1995). Fault-controlled metasomatic alteration of Early Proterozoic sedimentary manganese ores in the Kalahari Manganese Field, South Africa. *Econ. Geol.* **90**(4), 823–844.

Gutzmer, J., and Cairncross, B. (1993). Recent discoveries from the Wessels Mine, South Africa. *Min. Record* **24**(5), 365–368.

Hamet, M., and Stedra, V. (1994). The Pb-Zn-B deposit of Dalnegorsk, East Siberia. *Riv. Mineral. Ital.* **3,** 259–270.

Hamilton, F. (1921). *Borax,* pp. 274–277, 17th Report of the State Mineralogist for 1920, California Dept. of Mines and Mineral Resources.

Hay, R. L. (1966). *Zeolites and Zeolite Reactions in Sedimentary Rocks,* pp. 32–34, Geol. Soc. Amer., Spec. Paper 85.

Helvaci, C. (1977). *Geology, Mineralogy and Geochemistry of the Borate Deposits and Associated Rocks at the Emet Valley, Turkey.* University of Nottingham, England, 338 p.

Helvaci, C. (1984). Geology and genesis of the Emet borate deposits, Western Turkey. *27 Int. Geol. Congr., Moscow* (Sect. 13–16) **7,** 262–263.

Helvaci, C. (1994). Mineral assemblages and formation of the Kestelek and Sultancayiri borate deposits. *Proc. 29th Int. Geol. Congr.,* Pt. A, pp. 245–264.

Helvaci, C. (1995). Stratigraphy, mineralogy and genesis of the Bigadic borate deposits, Western Turkey. *Econ. Geol.* **90,** 1237–1260.

Helvaci, C. (1996). Article in preparation on the petrography of the Sultancayiri deposit, personal communication.

Helvaci, C., and Alaca, O. (1991). Geology and mineralogy of the Bigadic borate deposits and vicinity. *Min. Res. Expl. Bull.* **113,** 31–63.

Helvaci, C., and Firman, R. J. (1976). Geological setting and mineralogy of the Emet borate deposits, Turkey. *Trans., Sect. B, Inst. Mining Metal.* **85,** B142–B152.

Helvaci, C., Stamatakis, M. G., Zagouraglou, C., and Kanaris, J. (1993). Borate minerals and related authigenic silicates in northeastern Mediterranean Late Miocene continental basins. *Explor. Mining Geol.* **2**(2), 171–178.

Henry, D. J., and Dutrow, B. L. (1996). Metamorphic tourmaline and its petrologic applications. (E. S. Grew and L. M. Anovitz, eds.), Vol. 33, Ch. 10, pp. 503–558. In *Boron: Mineralogy, Petrology and Geochemistry,* Mineral. Soc. Am., Reviews in Mineralogy.

Honea, R. M., and Beck, F. R. (1962, January–February). Chambersite, a new mineral. *Am. Min.* **47**(1–2), 665–671.

Huang, D., and Peng, C. (1985). Mineralogy study of vonsenite from Shenkeng, Lianping, Guangdong. *Bull. Yichang Inst. Geol. Min. Res., Chinese Acad. Geol. Sci.* **9,** 25–31.

Keren, R., Gast, R. G., and Bar-Yosef, B. (1981). pH-dependent boron adsorption by Na-montmorillonite. *Soil Sci. Sco. Am. J.* **45,** 45–48.

Keren, R., and O'Connor, G. A. (1982). Effect of exchangeable ions and ionic strength on boron adsorption by montmorillonite and illite. *Clays Clay Min.* **30**(5), 341–346.

Keys, C. R. (1910). Borax deposits in the United States. *Trans. AIME* **40,** 694–699.

Khetchikov, L. N., Gnidash, N. V., and Ratkin, V. V. (1990a). Evolution of the mineral-forming medium based on data from studies of pseudomorphs after danburite crystals in voids in the Dalnegorsk borosilicate deposit. *Dokl. Akad. Nauk. SSSR (Mineral)* **315**(6), 1466–1469.

Khetchikov, L. N., Gnidash, N. V., and Ratkin, V. V. (1990). Formation conditions of crystals of datolite, quartz and danburite in the Dal'negorsk borosilicate deposit (based on thermobarogeochemical data). *Min. Zh.* **12**(2), 78–84.

Kistler, R. B. (1996). Personal communication. Chief Geol., U.S. Borax Inc., Valencia, Calif.

Kistler, R. B., and Helvaci, C. (1994, 1993, 1992). Boron and borates. In *Industrial Minerals and Rocks* (S. J. Lefond, ed.), pp. 171–186, Soc. Min. Eng., AIME.

Kistler, R. B., and Smith, W. C. (1983). Boron and borates. In *Industrial Minerals and Rocks* (S. J. Lefond, ed.), 5th ed., pp. 533–560, Soc. Min. Eng., AIME.

Kramer, H., and Allen, R. D. (1956). A restudy of bakerite, priceite and veatchite. *Am. Min.* **41,** 689–700.

Kulikov, I. V., Blinov, V. A., and Cherepivskaya, G. E. (1981). Apophyllite, datolite and xonotlite in ores of the Tyrny-Auz Deposit. *Izv. Vyssh. Uchebn. Zaved., Geol. Razved.* **24**(5), 139–143.

Kwak, T. A., and Nicholson, M. (1988). Szaibelyite and fluoborite from the St. Dizier Sn-Borate Skarn Deposit, NW Tasmania, Australia. *Australian Min. Mag.* **52,** 713–716.

Lefond, S. J., and Barker, J. M. (1979). A borate and zeolite occurrence near Magdalena, Sonora, Mexico. *Sci. Communications* **74**(8), 1883–1888.

Lefond, S. J., and Barker, J. M. (1985). The borates and zeolites of northcentral Sonora, Mexico. In *Borates: Economic Geology and Production* (J. M. Barker and S. J. Lefond, eds.), pp. 177–195, Soc. Min. Eng., AIMMPE, New York.

Leonard, B. F., and Vlisidis, A. C. (1961). Vonsenite at the Jayville Magnetite Deposit, St. Lawrence County, New York. *Am. Min.* **46,** 798–811.

Lerman, A. (1966). Boron in clays and estimation of paleosalinities. *Sedimentology* **6**(4), 267–286.

Lisitsyn, A. E. (1969). Exploration criteria in prospecting for boron skarn deposits. *Sov. Geol.* **12**(9), 62–69.

Lisitsyn, A. E. (1972). Role of acidity–alkalinity of hydrothermal solutions in the formation of borosilicates. *USSR, Geokhimiya,* No. 11, pp. 1389–1392.

Lisitsyn, A. E., and Malinko, S. V. (1969). Physicochemical conditions of borosilicate and borate formation in calcic skarns. *USSR Geol. Rud. Mestorozhd.* **11**(4), 34–45.

Lisitsyn, A. E., and Rudiev, V. V. (1978). Some physicochemical conditions for the formation of borosicilates in skarns. In *Teor. Prakt. Termobarogeokhim., 5th Mtg., 1976.* (N. P. Ermakov, ed.), pp. 139–142, Izd. Nauka, Moscow, USSR.

Lisitsyn, A. E., and Vladimirskaya, N. I. (1972). Distribution of borosilicate mineralization in calcareous skarns. *USSR Geol. Rud. Mestorozhd.* **14**(4), 100–104.

Lisitsyn, A. E., Malinko, S. V., Rudnev, V. V., and Fitsev, B. P. (1982). Diverse origin of boron ore mineralization in the Korshunovskoe Iron Ore Deposit. *Geol. Rudn. Mestorozhd.* **24**(2), 14–19.

Lixue, J. (1985). The geologicial characteristics and evolution of Houxianyo boron deposit of the Early Proterozoic, Liaoning Province. *Liaoning Prov. Geol. and Miner. Resour., Liaoning, China, Abs.,* p. 101.

Llambias, H. (1963). Sobre inclusions halladas en cristales inderita, borax y topacio de la Argentina. *Assoc. Geol. Argent. Rev.* **18**(3–4), 129–138.

London, D., Morgan, G. B., and Wolf, M. B. (1996). Boron in granitic rocks and their contact aureoles. In *Boron: Mineralogy, Petrology and Geochemistry* (E. S. Grew and L. M. Anovitz, eds.), Vol. 33, Ch. 7, pp. 299–330, Mineral. Soc. Amer., Reviews in Mineralogy, Washington, D.C.

Lyday, P. A. (1991, 1992, 1994, 1995). *Boron.* Annual Report, U.S. Bureau of Mines, Washington, D.C.

Lyday, P. A. (1992). History of boron production and processing. *Ind. Min.* (*London*) **303,** 19, 21, 23–25, 31, 33, 34, 37.

Lyubofeev, V. N. (1966). Effects of the medium and the nature of magmatic processes on the formation of endogenic boron mineralizations. *Tr., Krasnodar. Filial Vses. Neftegazov. Nauch.-Issled. Inst.,* No. 17, pp. 400–403.

Madsen, B. M. (1970). Core logs of three test holes in Cenozoic Lake deposits near Hector, California. *U.S. Geol. Survey, Bull. 1296.*

Majmundar, H. (1985, August). Borate mining history in Death Valley. *Calif. Geol.* **38**(8), 171–177.

Malinko, S. V. (1966). First find of cahnite in the U.S.S.R. *Dokl. Acad. Sci., Earth Sci. Sect.* **166,** 695–697.

Malinko, S. V. (1972). Skarns: Typomorphic association of boron minerals related to calcareous skarns. In *Tipomorfizm Miner. Ego Prakt. Znachenie* (F. V. Chukhrov, ed.), pp. 202–205, Nedra, Moscow, USSR.

Malinko, S. V. (1985). Origin of endogenic boron deposits according to typomorphic properties of boron minerals. *USSR Min. Zh.* **7**(1), 36–45.

Malinko, S. V. (1992). Origin of unique localizations of borosilicate ores in the Dalnegorsk deposit of Primore area. *Min. Zh.* **14**(5), 3–12.

Malinko, S. V., Dubinchuk, V. T., and Nosenko, N. A. (1992). Native bismuth in datolite ores of the Dalnegorsk boron deposit. *Mineral Zh.* **14**(1), 42–52.

Malinko, S. V., and Nosenko, N. A. (1990). Genesis of datolite of the Dal'negorsk deposit and the formation of borosiclicate ores. *Probl. Genetich. i Prikl, Mineral., M.*, pp. 54–72.

Malinko, S. V., and Savinskii, I. D. (1986). Source of matter for endogenic boron ores based on the method of sample recognition. *Sov. Geol.* **4,** 59–66.

Malinko, S. V., Anichich, S., Ioksimovich, D., and Lisitsyn, A. E. (1995). Studenitsite $NaCa_2[B_9O_{14}(OH)_4]\cdot 2H_2O$ the New Borate from Serbia, Yugoslavia. *Zap. Vseross. Min. O-va.* **124**(3), 57–64.

Malinko, S. V., Lisitsyn, A. E., and Shergina, Y. P. (1994). Isotope geochemistry of boron-skarn ore mineralization in active continental margins. *Zap. Vseross. Min. O-va.* **123**(4), 10–20.

Malinko, S. V., Yamnova, N. A., Pushcharovskii, D. Y., Lisitsin, A. E., Rudnev, V. V., and Yurkina, K. V. (1986). Iron-rich warwickite from the Taiga Ore Deposit (Southern Yakutia). *Zap. Vses. Min. O-va.* **115**(6), 713–719.

Malinko, S. V., Lisitsyn, A. E., Naumova, I. S., and Purusova, S. P. (1981). Strontiohilgardite from the Korchunovskoe iron ore deposit. *Zap. Vses. Min. O-va.* **110**(5), 588–593.

Malinko, S. V., Yamnova, N. A., Pushcharovskii, D. Y., Lisitsin, A. E. Rudnev, V. V., Yurkina, K. V., Naumova, I. S., and Purusova, S. P. (1979). Physicochemical parameters of the formation of commercial borosilicate mineralization. In *Osnovn. Parametry Prir. Protsessov Endog. Rudoobraz., Mtg., 1977.* (V. A. Kuznetsov, ed.), pp. 98–106, Izd. Nauka, Sib. Otd., Novosibirsk.

Martin, R. F. (1971, January–February). Disordered authigenic feldspars in the series $KAlSi_3O_8$–$KBSi_3O_8$ from southern California. *Am. Min.* **56,** 281–291.

Martirosyan, R. A., and Babaev, N. I. (1968). Prospecting for boron-containing formations. *Uch. Zap. Azerb. Gos. Univ., Ser. Geol.-Geogr. Nauk* **5,** 94–97.

Matterson, K. J. (1980). Borate ore discovery, mining and beneficiation. In *Inorganic and Theoretical Chemistry*, pp. 152–169, Longman, New York.

Maurice, J. (1966). Boron geochemistry. *Ann. Agron.* (*France*) **17**(4), 367–402

McAnulty, W. N., and Hoffer, J. M. (1972). A new howlite occurrence in Sonora, Mexico. *Bol. Soc. Geologica Mexicana* **33**(1), 21–24.

Mel'nitskii, V. V. (1980). Possible genetic relation of a datolite skarn deposit to the intrusion of grantic rocks. *Khim. Prom-st., Gornokhim Syr'ya*, No. 5, pp. 1–3.

Milton, C. A., Axelrod, J. M., and Grimaldi, F. S. (1955). New mineral, garrelsite $[(Ba_{.65}Ca_{.29}Mg_{.06})_4H_6Si_2B_6O_{20}]$ from the Green River Formation, Utah. *Geol. Soc. of Am. Bull.* **66,** 1597.

Milton, C. A., Chao, E. C., Axelrod, J. M., and Grimaldi, F. S. (1960). Reedmergnerite, the boron analogue of albite, from the Green River Formation, Utah. *Am. Min.* **45,** 188–199.

Minette, J. W., and Wilber, D. P. (1973). Hydroboracite from the Thompson Mine, Death Valley. *Mineralogist Record* **4**(1), 21–23.

Moore, P. B., and Araki, T. (1974). Pinakiolite, warwickite, wightmanite: Crystal chemistry of complex 3Å wallpaper structures. *Am. Min.* **59,** 985–1004; Roweite, its atomic arrangement, pp. 60–65.

Moore, P. B., and Ghose, S. (1971, September–October). A novel face-sharing octohedral trimer in the crystal structure of seamanite. *Am. Min.* **56,** 1527–1538.

Morgan, V. (1980). Boron Geochemistry. In *Inorganic and Theoretical Chemistry*, pp. 72–152, Longman, New York.

Muehle, G. (1974). Pseudomorphs from the Corkscrew mine, Death Valley, California. *Mineralogical Record* **5**(4), 174–177.

Murdock, T. G. (1958). The boron industry of Turkey. *U.S. Bureau of Mines, Special Supplement No. 53, Mineral Trade Notes* **46**(5), 47.

Noble, L. F. (1922). Colemanite in Clark County, Nevada. *U.S. Geol. Survey, Bull. 735,* Pt. I, pp. 23–39.

Noble, L. F. (1926, October). Note on a colemanite deposit near Shoshone, California. *U.S. Geol. Survey, Bull. 785,* pp. 63–73.

Norman, J. C., and Johnson, F. C. (1980). The Billie borate ore body, Death Valley, California. In *Geology and Mineral Wealth of the California Desert* (D. L. Fife and A. R. Brown, eds.), pp. 268–275, South Coast Geol. Soc., Santa Ana.

Nosenko, N. A., Ratkin, V. V., Logvenchev, P. I., Polokhov, V. P., and Pustov, Y. K. (1990). Dal'negorsk borosilicate deposit: Product of polychronous manifestation of skarn processes. *Dokl. Akad. Nauk SSSR (Geol.)* **312**(1), 178–182.

Obradovic, J., and Vasic, N. (1990). Mineral deposits in Miocene lacustrine and Devonian shallow-marine facies in Yugoslavia. *Spec. Publ. Int. Assoc. Sediment.* **11,** 147–156.

Obradovic, J., Stamatakis, M. G., Anicic, S., and Economou, G. S. (1992). Borate and borosilicate deposits in the Miocene Jarandol Basin, Serbia, Yugoslavia. *Econ. Geol.* **87,** 2169–2174.

Orris, G. J. (1995). *Borate deposits.* U.S. Geol. Survey Open-File Rept. 95–842.

Orti, F., Helvaci, C., Rosell, L., and Gundogan, I. (1997). Sulfate–borate relations in an evaporitic lacustrine environment: The Sultancayir gypsum member (Miocene, Western Anatolia). *Sedimentology,* in press.

Otroshchenko, V. D., Kiempert, S. Y., and Khorvat, V. A. (1966). *Accessory Datolite From Skarns and Veins in the Chatkalo-Kuraminsk Mountains,* pp. 38–41, Geol. Rudonos. Pritashkent. Raiona, Akad. Nauk Uzb. SSR, Inst. Geol. Geofiz.

Ozol, A. A. (1976). Basic characteristics of boron geochemistry and formation conditions for volcanic-sedimentary type boron deposits. *USSR, Litol. Polezn. Iskop.,* No. 3, pp. 60–74.

Ozpeker, I., and Inan, K. (1978). Relations of observed mineral assemblages to the evolution of borate deposits in western Anatolia. *Turk. Jeol. Kurumu, Bull.* **21**(1), 1–10.

Palache, C., and Bauer, L. H. (1972). Cahnite, a new boroarsenate of calcium from Frankline, New Jersey. *Am. Min.* **12,** 149–153.

Papezik, V. S., and Fong, C. C. (1975). Howlite and ulexite from the carboniferous gypsum and anhydrite beds in western Newfoundland. *Can. Min.* **13,** 370–376.

Papke, K. G. (1976). Evaporites and brines in Nevada playas. *Nevada Bur. Mines, Geol., Bull.* **87,** 35.

Park, L. E. (1991). *Geochemical and Paleoenvironmental Analysis of Laustrine Arthropod-Bearing Concretions of the Barstow Formation, Southern California.* Master's thesis, University of Arizona, Tuscon.

Pemberton, H. E. (1968, May). The minerals of the Sterling Borax Mine, Los Angeles County, California. *Min. Explorer* **3**(1), 1–10.

Peng, Q. M., and Palmer, M. R. (1995). The Palaeoproterozic boron deposits in eastern Liaoning, China: A metamorphosed evaporite. *Precambrian Research* **72,** 185–197.

Pertsev, N. N. (1968). Inderite and inyoite in the oxidation zone of a skarn deposit. *Miner. Syr'e,* No. 18, pp. 116–120.

Pertsev, N. N., and Dorofeev, A. B. (1971). Danburite mineralization in the calcite veins of magnesian skarn deposits. *Tr. Mineral. Muz., Akad. Nauk SSSR,* No. 20, pp. 120–127.

Peters, T. A., and Peters, J. J. (1978). Famous mineral localities: Paterson, New Jersey. *Min. Rec.* **9**(3), 157–166, 171–179.

Piper, J. R. (1985). Borate deposits of the Calico–Daggett area, California. In *Borates: Economic Geology and Production* (J. M. Barker and S. J. Lefond, eds.), pp. 147–155, Soc. Min. Eng., AIMMPE.

Ratkin, V. V., Khetchikov, L. N., Gnidash, N. V., and Dmitriev, V. E. (1992). Role of colloids and paleohydrothermal cavities in the formation of rhythically banded ores of the Dalnegorsk borosilicate deposit. *Dokl. Akad. Nauk. (Petrogr.)* **325**(6), 1214–1217.

Reynolds, R. C. (1965). The concentration of boron in Precambrian seas. *Geochim. et Cosmochim. Acta* **29**(1), 1–16.

Robottom, A. (1893). The tincal trade, pp. 30–33; Boracic acid, pp. 84–88; Crude borate of soda, pp. 143–146, 173–175; Boracite, pp. 159–163, *Travels in Search of New Trade Products,* Jarrold & Sons, London.

Rudashevskii, N. S. (1969). Datolite and tourmaline from a Siberian deposit. *Zap. Vses. Mineral. Obshchest.* **98**(2), 200–206.

Rusinov, V. L., Kudrya, P. F., Laputina, I. P., and Kuzmina, O. V. (1994). Periodic metosomatic zonation in pyroxene–wollastonite skarns. *Petrologiya* **2**(6), 570–586.

Saet, Yu. E. (1969a). Biogenic migration of boron as theoretical basis for biogeochemical prospecting for boron deposits. In *Biogeokhim. Poiski Rud. Mestorozhd., Dokl. Sekts.*, 5th ed., pp. 204–220, Mtg. 1966, Ed. Kovalevskii, A. L., Buryat Ulan-Ude.

Saet, Yu. E. (1969b). Geochemical principles of complex prospecting for boron-based on the secondary dispersion halo. *Sov. Geol.* **12**(2), 96–109.

Saet, Yu. E., Burenkov, E. K., Gorshenin, A. D., Igumnov, N. Y., and Nesvizhskaya, N. I. (1972). Geochemical criteria for developing, appraising, and interpreting exogenous characteristic anomalies in an endogenous boron deposit. In *Litogeokhimicheskie Metody Poiskakh Skrytogo Orudeneniya* (L. N. Ovchinnikov, ed.), pp. 48–50, Akad. Nauk SSSR, Inst. Mineral., Geokhim. Kristallokhim. Redk. Elem., Moscow.

Semenov, Yu. V., Malinko, S. V., Kiseleva, I. A., and Khodakovskii, I. L. (1987). Thermodynamic analysis of the formation conditions of endogenic calcium borosilicates and borates. *USSR, Geokhimiya,* No. 8, pp. 1182–1190.

Shaffer, L. L., and Baxter, R. P. (1975). Oregon borax; Twenty mule team: Rose Valley history. *Oregon Hist. Q.* **73**(3), 228–244.

Shaller, W. T. (1916). Inyoite and meyerhofferite: Two new calcium borates. *U.S. Geol. Survey Bull. 610,* Ser. 3, pp. 35–55.

Shaller, W. T., Vlisidis, A. C., and Mrose, M. E. (1965). Macallisterite, $2MgO \cdot 6B_2O_3 \cdot 15H_2O$: A new hydrous magnesium borate mineral from the Death Valley region, Inyo County, California. *Am. Min.* **50,** 629–640.

Shcherbinin, V. M., and Chernyshov, A. V. (1974). Problem of the effect of structural–lithological conditions on the localization of borosilicates in calcareous skarns. *USSR, Tr. Dal'nevost. Politekh. Inst.* **58,** 101–108.

Sheppard, R. A., and Gude, A. J. (1973, July–August). Boron-bearing potassium feldspar of authigenic origin in closed basin deposits. *J. Res. U.S. Geol. Surv.* **1**(4), 377–382.

Shima, M. (1963, July). Geochemical study of boron-isotopes. *Geochim. et Cosmochim. Acta* **27**(7), 911–913.

Slack, J. F. (1996). Tourmaline associations with hydrothermal ore deposits. In *Boron: Mineralogy, Petrology and Geochemistry* (E. S. Grew and L. M. Anovitz, eds.), Vol. 33, Ch. 11, pp. 559–644. Mineral. Soc. Am., Reviews in Mineralogy.

Smith, G. I. (1985). Borate deposits in the United States: Dissimilar in form, similar in geologic setting. In *Borates, Economic Geology and Production* (J. M. Barker and S. J. Lefond, eds.), pp. 37–51, Soc. Min. Eng., AIMMPE, New York.

Smith, P. R., and Walters, R. A. (1980, February). Production of colemanite at American Borate Corp.'s plant near Lathrop Wells, Nevada. *Mining Eng.* **32**(2), 199–204.

Stamatakis, M. G. (1989). A boron-bearing potassium feldspar in volcanic ash and tuffaceous rocks from Miocene lake deposits, Samos Island, Greece. *Am. Min.* **74,** 230–235.

Stamatakis, M. C., and Economou, G. S. (1991). A colemanite and ulexite occurrence in a Late Miocene saline–alkaline lake of West Samos Island, Greece. *Econ. Geol.* **86,** 166–172.

Staples, L. W. (1948, May). The occurrence of priceite in Oregon. *Northwest Sci.* **22,** pp. 69–77.

Stoici, S. D. (1975). The paragenesis of ludwigite–ascharite from Baita Bihor (Romania), *Freiberg Forschungsh.* **C308,** 99–108.

Stoicovici, E., and Stoici, S. (1972). Boron ores from the superior area of the Black Cris (Baita, Bihor). *Stud. Univ. Babes-Bolyai Ser. Geol.-Min.* **17**(1), 3–10.

Storms, W. H. (1893). *The Calico Mining District.* California Min. Bur., Rept. 11, pp. 345–348.

Tershana, A., and Marishta, S. (1984). The borosilicate occurrence in volcanic and ultrabasic rocks. *Sul. Shkencave Gjeol.* **3**(1), 73–80.

Tilley, C. E. (1951). The zoned contact-skarns of the Broadford area, Skye: A study of boron–fluorine metasomatism in dolomites. *Min. Mag.* **29**(214), 621–672.

Travis, N. J., and Cocks, E. J. (1984). *The Tincal Trail,* Harrap, London.

Utekhin, G. M. (1965, December). The problem of distribution features and conditions of boron concentration in skarns. *Econ. Geol.* **60**(8), 1750.

Ver Planck, W. E. (1956). History of borax production in the United States. *Calif. J. Mines Geol.* **52**(3), 273–291.

Ver Planck, W. E. (1957). Boron. *Mineral Commodities of California, Calif. Div. Of Mines, Bull. 176,* pp. 87–94.

Vladimirskaya, N. I. (1973). Genetic features of borosilicates based upon studies of gas–liquid inclusions. *USSR, Zap. Vses. Mineral. Obshchest.* **102**(4), 394–401.

Volodin, P. K. (1971a). *Geologic Formation Conditions and Distribution Characteristics of Borosilicate Mineralization in Calcareous Skarns and Skarnoids,* pp. 41–71, Mineral., Geokhim. Genezis Rud. Mestorozhd. Tadsh., Donish, Dushanbe, USSR.

Volodin, P. K. (1971b). *Lithological Prospecting Criteria for Skarn-Type Deposits of Datolite and Danburite Ores,* pp. 72–85, Mineral., Geokhim. Genezis rud. Mestorozhd. Tadzh., Donish, Dushanbe, USSR.

Walker, C. T. (1975). *Geochemistry of boron,* Dowden, Hutchinson & Ross, Stroudsburg, PA.

Wasserstein, B. (1942). On the presence of boron in braunite and manganese ores. *Econ. Geol.* **38,** 389–398.

Watanabe, T. (1953, September). Suanite, a new magnesium borate mineral from Hol Kol, Suan, North Korea. *Mineralogical J.* **1**(1), 54–62.

Watanabe, T. (1967). Geochemical cycle and concentration of boron in the earth's crust. In *Chemistry of the Earth's Crust,* V. 2 (A. P. Vingradov, ed.), pp. 167–178, Israel Program for Scientific Translations, Ltd., Jerusalem.

Watanabe, T., and Ito, J. (1954, May). Paigeite (ferroludwigite) from the Kamaishi iron mine, Iwate Prefecture, Japan. *Mineralogical J.* **1**(2), 84–88.

Weiss, A. (1969). The Emet colemaite deposits in Kutahya Province, Anatolia (Turkey). *Aufschluss* **20**(9), 243–245.

Wendel, C. (1978). *Special Report on Borate Resources.* Mining and Minerals Div., U.S. National Park Service, Washington, D.C.

Wright, L. A., Stewart, R. M., Gay, T. E., and Hazenbush, C. C. (1953). Mines and mineral deposits of San Bernardino County, California. *Calif. J. Mines Geol.* **49**(1–2), 175–177, 220–225.

Yakovleva, A. K. (1986). Boron mineralization of the Pechenga and Allarechensk nickel-containing regions. In *Nov. Dannye Mineral. Magmat. Metamorf. Kompleksov Kol'sk. Poluostrova* (I. V. Bel'Kov, ed.), pp. 31–37, Akad. Nauk SSSR, Kol'sk, Fil., Apatity, USSR.

Yale, C. G. (1905). *Review of the Borax Industry During 1904,* p. 1021, U.S. Geol. Surv., Min. Res. U.S.

Yin, G. (1992). Zacangchaka Ascharite Deposit in Tibet and its industrial utilization. *Feijinshukuang* **2,** 4–6.

You, C. F., Spivack, A. J., Greskes, J. M., Martin, J. B., and Davisson, M. L. (1996). Boron contents and isotopic compositions in pore waters. *Marine Geol.* **129**(3–4), 351–361.

Zaritskii, P. V. (1973). Mineralogy and genesis of accessory boron mineralization. *USSR, Mineral. Sb. (Lvov)* **27**(4), 354–361.

Zhabin, A. G., and Malinko, S. V. (1995). 11-year cycle of solar activity as seen in numeral aggregates of Dalnegorsky borate skarn. *Zap. Useross. Mineral. D-va.* **124**(4), 111–115.

Zhang, Q. (1988). Early Proterozoic tectonic styles and associated mineral deposits of the North China Platform. *Precambrian Res.* **39,** 1–29.

Zhao, A., Zhang, X., Xue, C., and Zhou, J. (1991). Study on the mineralogical features and formative conditions of prehinite and datolite from Tieshan iron mineral deposit, Daye, Hubei. *Diqiu Kexue* **16**(6), 681–686.

Zhao, Y., and Li, D. (1987). Metasomatic phenomena in the granite contact zones of the Gejiu Tin Deposit, Yunnan Province, China. *Zhongguo Dizhi Kexueyuan Yuanbao* **16,** 237–252.

Zheng, G., and Jin, S. (1991). Geological characteristics and genesis of the Heping boron deposit in Changxing County, Zhejiang Province. *Kuangchuang Dizhi* **10**(2), 187–192.

Zheng, M. (1987). Types of borate deposits and their prospecting in China. *Zhongguo Dizhi Kexeyuan Kuangchan Dizhi Yanjiuso Sokan* **20,** 47–54.

Chapter 4 Calcium or Magnesium Surface (Playa or Mantle) Deposits

4.1 ARGENTINA

4.1.1 General

There are about 40 playas in the 885-km Puna region (see Fig. 2.1) of the Andes that contain commercial quantities of borates, and in 1994 14 were being mined, including 6 in Argentina (Kistler and Helvaci, 1994). The discovery of ulexite in the Jesuit ruins of Antofalla indicates that borates may have been mined and used for precious metal working since 1700. However, the first recorded borate leases were in 1876, and the first article on borates was in 1880. The salar borate deposits were formed during the Quaternary period and are concentrated in the northeastern part of the country, an area formed about 15 My ago during the Neogene uplift of the central Andes, producing a large high altitude (≥3000 m) desert plain. It is bordered on the west by the Andean volcanic arc, a chain of tall (<6000 m) volcanoes in Chile and Peru, and on the east by the Eastern Cordillera. This range is lower, but sufficient in size to enclose most of the area.

The climate is cold (reaching −30°C) and dry, with intense solar radiation. There is great diurnal temperature variation, winter snows on peaks higher than 5500 m, and limited and infrequent rain (5–30 cm/year) that can come in sudden downpours. This has resulted in extensive weathering and large alluvial fans. These conditions appear to have remained fairly constant since the area formed, but with occasional periods of renewed vulcanism. About 6–7 My ago, volcanic eruptions accompanied by geothermal springs (and perhaps more rainfall) formed a large halite deposit and three buried borate deposits: Tincalayu, Sijes, and Loma Blanca. Within the last million years other borate springs opened to form the modern borate playas, depositing borates, halite, gypsum (and some calcite), caliche, and travertine (Alonso, Jordon, Tabbutt, and Vandervoort, 1991; Vandervoort, Jordan, Zeitler, and Alonso, 1995).

The most common salar borate is ulexite, occurring in two forms: (a) "papas" (potatoes, cotton balls) or nodules in silt, clay, tuff, or sand with a thin crust and clean white interior, (b) "mud," "barras," "bars," "bancos," or bedded ulexite, usually interlayered with silt and clay. Both forms are variable in area and vertical extent, with the ore zone often 0.1–2 m thick

Table 4.1

Ulexite Analyses from Several Argentine Salars, wt% (Hand Cleaned and Air Dried)[a]

				Pastos Grandes		Salina Grande		
						Tres Morros		
	Antuco	Hombre Diablillos	Elsa Muerto	Pastos Mine[b]	Grandes	Mined ore	Calcined	Nino Muerto
Ulexite								
B_2O_3	32.25	30.64	39.80	34.80 28.15[f]	31.05	36.90	50.60	24.78
CaO	10.04	9.34	12.02	12.64 16.95[f]	9.24	12.10	17.10	8.07
Na[c]	7.88	8.70	9.10	4.54	7.67	9.03	12.12	4.51
H_2O	24.25	30.02	32.92	30.64 35.43[f]	28.01	33.20	10.16	27.58
Impurities								
NaCl	10.03	6.01	5.13	—	17.35	5.08	7.43	29.81
$CaSO_4$	0.63	0.53	0.87	2.30	1.98	1.43[d,e]	—	4.45
MgO	0.42	(4.15 $CaCO_3$)	—	—	—	0.36[d]	—	4.92
Fe_2O_3	2.41	1.34	—	—	1.10	1.15	1.33	0.70
Sand, silica, alumina	0.19	8.44	0.27	11.01 14.24[f]	3.60	3.03	3.24	1.26
Bed thickness (m)	1	0.8–1	—	0.05	1	0.8	—	1
Bed area (hectares)	few	3513	—	—	—	—	—	160

[a] Reichert, 1909.
[b] Alonso and Menegatti, 1990; not a salar deposit, but 1.6 Mybp terraces.
[c] In ulexite only.
[d] Buttgenbach, 1901. Also 0.80% $CaCO_3$ and 0.35% Al_2O_3.
[e] Listed as Na_2SO_4.
[f] Inyoite.

(average 0.5 m), starting within 0–3 m (average 0.2–0.4 m) of the surface. The ulexite can also be under <30 cm of halite, or in surface crusts and efflorescence. Nodules vary in size from 0.1 to 30 cm (usually 5–10 cm), and in a few salars there is a zonation from travertine to ulexite barras to papas to disseminated crystals, and in others from gypsum to borates to halite. In mining, the papas are preferred due to their purity when cleaned and air dried (often >28% B_2O_3; Table 4.1; ~20% as mined, with <50% water). Both varieties contain insolubles, chlorides, and sulfates. Borax is also present in seven salars, usually as small amounts of large well-shaped crystals in a grayish-green or reddish-brown clay–silt matrix. They can be as long as 25 cm, but are usually 1–2 cm long. Recently deposited playa inyoite occurs only in the Lagunita salar.

In most salars there are remnants of old thermal springs, but only one is still active. Many salars also have fault lines along one side through which the borate-bearing hot spring water rose, depositing travertine and ulexite in the strata near the fault and in the playa. At Cauchari the fault line is in the sandy zone of an alluvial fan, whereas at Ratones and Centenario the faults are in the basin's sediments. The hot springs appear to be the sole source of the borates in these salars (and not basin rock leaching) because (a) adjacent salars in identical settings only have borates if hot springs were present; (b) the drainage network is often very limited; and (c) the area of borate deposition may be upslope in the basin (Alonso, 1986; Alonso and Viramonte, 1990).

In most salars there is brine just below the surface, usually saturated (or nearly so) with salt, with a density of ~1.21 g/ml and containing SO_4, K, Mg, B_4O_7, Ca and Li, in that order of abundance, with an average pH of 7.25 (Nicolli, 1981; Nicolli, Suriano, Kimsa, Brodtkorb, 1980). Usually, the excess Ca over the HCO_3 and CO_3 is only little more than the B_4O_7, so the precipitation of calcite and ulexite (and perhaps gypsum) can deplete the calcium, allowing some borax to form.

4.1.2 Salars

4.1.2.1 Acazoque, Antofalla, Antuco

Acazoque's small basin is in an alluvial plain 10 km south of San Antonio los Cobres. Under a thin caliche or travertine cover there is a 20-cm layer of light-brown clay containing small irregular clumps and millimeter-size borax crystals, as well as a little sodium sulfate. When the surface dries, some of the borax converts to ulexite. In Antofalla there are flowing high-NaCl, low borate hot springs at its central-western edge which form thick halite crusts (Alonso, 1986). Antuco is a few hectares in size at an altitude of 4000 m, and contains 1 m of alternating layers of bedded and nodular ulexite. The ore (see Table 4.1) is contaminated with insolubles and halite, but was briefly mined at the turn of the century (Reichert, 1909) and from 1940 to 1949 (Orris,

1995). The salar has an active hot spring that deposits ulexite in its mantle and the adjacent alluvial gravel (Alonso, 1986; Alonso and Viramonte, 1993).

4.1.2.2 Cauchari

This 50-km-long and 3- to 5-km-wide salar at 3900-m elevation is separated from the Salar Olaroz to the north by an alluvial fan. There are borate-containing travertine and tufa cones along its western margin, and an estimated 3 million tons of ulexite extending down slope. The salar in many places has a caliche or travertine surface on top of a layer of sand, with clay under that. It has a generally low water table, and once <1.5-m beds of ulexite with <20-cm nodules. The major ulexite areas were to the north (the Mascota mine) and in the central-western area (the Porvenir and Siberia mines). The ulexite was often in brown sand, which was easy to remove, but it did have a moderate salt content. Travertine often lay beneath the ulexite layer (at Porvenir, 1 m thick). The ulexite was 2 m thick in the Nueva Siberia claim and was mined 8 months of the year. It was dug out, washed, air-dried, and then calcined in rotary furnaces. At the Phoenix deposit 120–140 cm of ulexite in wet clayey sand occurred at the surface or a 50–130 cm depth. Then there was fairly pure sand to the water table. Farther into the salar there were 10 cm of surface efflorescence salts, then 60 cm of ulexite, 80 cm of sandy clay, 1 m of clayey sand, and the water table at 1.5 m (Alonso, 1986; Alonso and Gutierrez, 1984; Muessig, 1966; Reichert, 1909).

Borax has been found in the salar's central and south-central areas as 10-cm to 2-m beds of 3- to 5-cm crystals, or as irregular aggregates in salt-containing red silt, sand, and clayey mud. Near the Inundada mine it occurred in a 50-cm layer 1–1.5 m below the surface, with large, <30 cm crystals. The borax layer was covered by reddish-brown sand with some ulexite nodules. In the Porvenir area the borax zone started at 1.5 m, also under a mud–nodule layer. The borax was in 0.5 m of black, smelly clay, with some sodium sulfate, but no salt or gypsum. A wide travertine and caliche platform was on the nearby shore, with one vent in the salar (Alonso, 1988).

A novel mining operation took place on the salar from 1983 to 1985 by Boroquimica Samicaf. The entire ulexite layer was mined by large-tire excavating machines, trucked to the plant, spread to dry, and when dry sent through rotary tromels. The dry clay and sand fell through the screens, whereas the larger ulexite particles came out the end. Feed nodules had 28.6% B_2O_3 and left at 32.7% B_2O_3; bedded ulexite entered at 16% B_2O_3 and left at 22.5% B_2O_3. In 1988 only the Inundada mine of the Gavenda Mining Company was operating (Alonso, 1988), with some production by Borax SA in 1994 and 1995 (Lyday, 1995).

4.1.2.3 Celti, Centenario

Celti is located a few kilometers south of Turi Lari. It is 2 km long and contains only a limited ulexite deposit, which has been sporadically mined

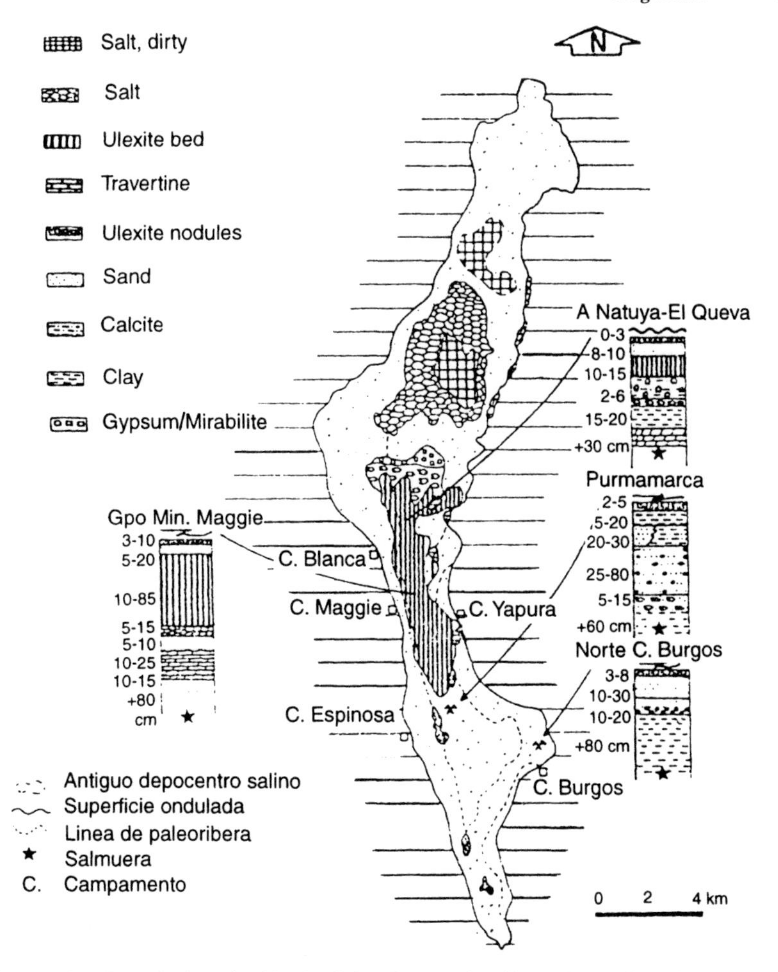

Figure 4.1 Minerals deposited in the Salar Centenario. (From Battaglia and Alonso, 1992; reprinted by permission of Dr. Alonso.)

(Alonso, 1988). The 94-km^2 Salar Centenario is located 170 km west of Salta, with a 1000-km^2 drainage basin and at an altitude of 3850 m. It is long (32 km) and narrow (2.3–6 km), bounded by mountains on its north, east, and west sides, and to the south by a ridge formed by the coalescence of two aluvial fans (separating it from Ratones). Its salts are divided into two quite different zones (Fig. 4.1): In the north there is essentially only NaCl, whereas in the south there are large deposits of ulexite. The north half of the salar has streams originating in predominantly volcanic rock, while the south half receives high-$CaSO_4$ runoff water from sedimentary rocks, which with high-borate geothermal spring water has formed the ulexite deposits. A small area of glauber salt (mirabilite, $Na_2SO_4{\cdot}10H_2O$) and larger areas of gypsum also have formed between the salar's two halves.

In the Maggie mine area the structure from the surface downward consists of a flat to slightly undulating surface with 2–10 cm of a whitish-gray clay–salt crust (Fig. 4.2); 5–20 cm of dry, fine, light-brown sand; a 10- to 85-cm bed of solid, fairly pure (<5% insolubles) and dry ulexite; 5–15 cm of nodular, fused, fairly dry ulexite with few insolubles; 5–10 cm of fine brown-reddish sand; 10–25 cm of massive, pure, compact-to-porous, yellowish, reddish, or dark-gray travertine having crevices filled with green aragonite, thickest to the east and always under the ulexite; 10–15 cm of hard, porous, light-brown caliche with limited interbeds of fine sand (under most of the salar's area); and >80 cm (to the limit of the 1- to 1.5-m-deep prospect holes) of fine, brown, wet sand, with its upper section cemented by carbonates.

The two types of borates (beds or nodules) varied from east to west, with predominantly 30–75-cm beds in the east (near the springs). To the west the beds thinned to extinction, and nodules increased until they were the only ulexite present. Over the entire deposit the ulexite averaged 85% beds and 15% nodules. Near the Purmarca mine an 18-cm layer of borax was found in a reddish calcitic clay under reddish plastic clay containing ulexite nodules. In the area between the halite and ulexite, gypsum crystals and selenite rosettes are common in the surface crust and mud, and small crystals and thin beds of clear, massive (like ice) glauber salt occur in the clay. The 3 × 4-km sodium chloride surface in the north is relatively smooth, white, and polygonated in the area that floods each winter, having a zone within it, and another to the north that is slightly higher, floods less often, and has a rough hard surface with a high clay content. The salt depth is up to 50 m, and the reserves are 700,000 metric tons of bedded and 100,000 metric tons of nodular ulexite. Ulexite was being recovered in 1992 at the Maggie and La Argentina mines (Alonso, 1988; Battaglia and Alonso, 1992).

4.1.2.4 Diablillos

This 3513-hectare salar is west of Hombre Muerto and Ratones at an altitude of 3900 m. It has a smooth surface, an 0.8- to 1.5-m ulexite layer, primarily as beds with some nodules. The brine level is near the surface (Reichert, 1909). The entire drainage basin is in volcanic rocks, and ulexite was deposited on all of the surface and in the upper sediments. The most extensive deposits (estimated at 2.5 million tons) are upslope from the rest of the basin and adjacent to a major east–west fault line (perpendicular to the region's general north–south fault trend). Below the ulexite is a 10-cm bed of caliche or travertine, with occasional layers of silica. No halite or gypsum has been deposited, but some 1- to 2-cm borax crystals were found in the southeastern corner near ulexite in a 1-m reddish clay layer. Small

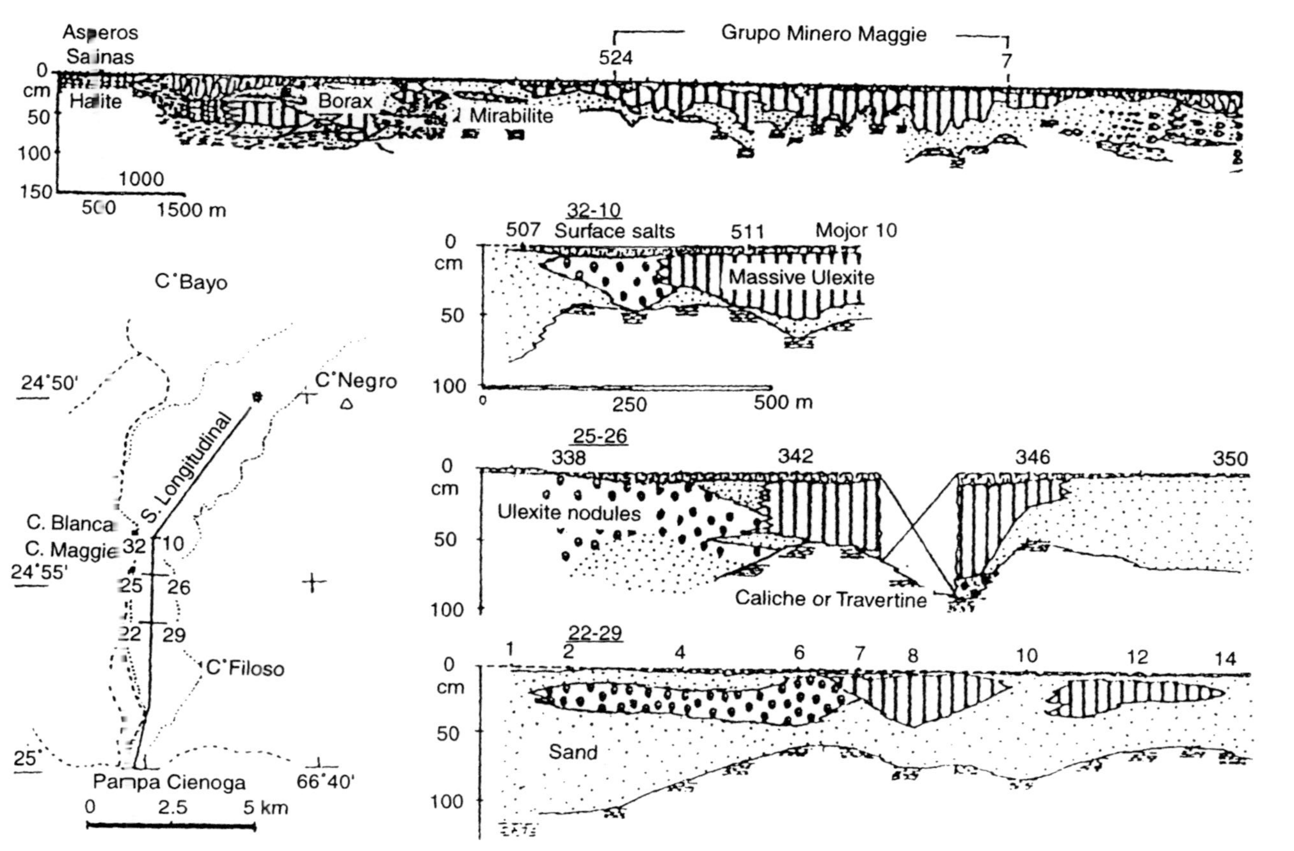

Figure 4.2 Cross-sections of the Salar Centenario. (From Battaglia and Alonso, 1992; reprinted by permission of Dr. Alonso.)

amounts of hydroboracite were once reported in the southeastern corner, but have not been found again. There has been limited ulexite (see Table 4.1) recovery from the deposit, although Borax SA claimed some production in 1994 and 1995 (Alonso, 1986; Alonso, 1988; Alonso and Gutierrez, 1984).

4.1.2.5 Guayatoyac

This playa is the farthest east of the Argentine Puna-region borate salars at a 3960-m elevation. It has estimated 30–35% B_2O_3 ulexite reserves of 620,000 tons. Muessig (1966) noted that the Rio Alumbrio with its former hot springs fed into this salar, and that its ulexite is primarily along its southeastern border under a 20- to 50-cm muddy-brown sand cover. One drill hole revealed 20–25 cm of sand, then 3 cm of a reddish-brown plastic clay, and finally 35 cm of bedded ulexite interlayered (7- to 8-cm beds) with green clay. Below that was more green clay and the water table. The deposits were mined in the early 1900s (as the Salta, Tucuman, Cordoba, and Rosario mines), and occasionally since then (i.e., the Baratoyoc mine). The ore as mined contained 23% B_2O_3 (which increased to 30–32% B_2O_3 after drying) and 6% Cl (Alonso, 1988).

4.1.2.6 Hombre Muerto, Jama

Hombre Muerto (see Figs. 2.1 and 5.1) is the second largest salar (after Arizaro) in the Argentine Puna region. It has both surface and subsurface ulexite deposits (see Table 4.1), but it also has large brine reserves. Because it has been commercialized as a brine deposit, it is discussed later in Chapter 5. The 25-km^2 Jama salar is at an altitude of 3700 m, and its principal ulexite deposit is in its northern end. One drill hole disclosed a mud–saline surface crust, under that 30 cm of gypsum sand, then a 4-cm bed of ulexite, next 40 cm of green clay with some glauber salt, and finally a dark plastic clay with considerable organic matter. The water table was at 1.15 m, and the estimated ulexite reserves are 150,000 metric tons. Nearby are three minor basins: Ana Laguna, Xilon, and Lake Mucar, each containing ~100 metric tons of ulexite. The area has had limited mining (Alonso, 1988; Orris, 1995).

4.1.2.7 Lagunita

This playa is only 100 m long and 50 m wide at an altitude of 4150 m. It is located on the eastern side of the Coyahuaima volcanic complex near the Coyambuyo volcano and the town of Coranzuli, 65 km southeast from the junction of Argentina, Bolivia, and Chile. It has a 5-cm white saline efflorescent crust of salt and ulexite, with some gypsum and calcite on its 60 × 30-m central area, underlain in descending order, by 10 cm of sand, tuff, and mud containing cauliflower-like nodules of ulexite; 10–15 cm of inyoite crystals with minor amounts of ulexite and calcite in a sand–clay–mud mixture; 5 cm of travertine; and sand and gravel to 0.5 m below the water table. The inyoite occurs as disseminated crystals in the mud–tuffite and as <1-cm (generally 0.5- to 1-cm) euhedral crystals in aggregates forming a discontinuous bed. The crystals are clear, colorless, gray or light brown, often perfectly shaped, and usually near both ulexite and calcite. The ulexite nodules are pure white to gray (from clay incorporated in the clusters), very soft, and commonly associated with calcite (which formed at all stages throughout the salar). Travertine also is present on the slopes draining to the playa (Helvaci and Alonso, 1994).

4.1.2.8 Olaroz, Lina Lari

Olaroz is north of Cauchari in the same broad valley, but separated by an alluvial fan. Its main borate area was in its east-central border, with the ulexite found in <1.2-m beds covered by a muddy, salty sand layer. It had a <26% B_2O_3 content and was low in chlorides at the San Nicholas mine of the Industrias Quimicas Baradero Company. Lina Lari is located 40 km west of Turi Lari, and contains small amounts of ulexite and borax. The borax occurs as 1- to 2-cm anhedral crystals in a plastic clay. There has been very little mining activity (Alonso, 1988).

4.1.2.9 Pastos Grandes

This salar is located toward the southern end of the Argentine Puna borate region at an altitude of 3700 m. It has an extensive playa surface as well as a shallow perennial lake fed by the Pastos Grandes River from the north and the Ochaqui-Sijes River from the east. There is a chain of volcanic mountains higher than 6000 m to the north, and both low (in the south) and high mountain ranges on its other sides. The salar has an irregular shape, and its ulexite occurred as both beds and nodules. The ore was ~1 m thick at the La Paz mine and contained considerable halite (Reichert, 1909). Buttgenbach (1901) stated that the bed was 1.5 m thick with 1.13% NaCl.

The salar also has appreciable deposits of ulexite and inyoite in its early Pleisocene (1.6 ± 0.7-My-old) hot spring terraces. They have outcropped over a 150-km^2 area (the Blanca Lila Formation) with, in descending order, 0.3 m of sandy caliche; 5 m of mud–clay interlayered with caliche (and bird and vegetive remains); 3 m of finely laminated light-green to grayish mud–clay; 1.6 m of ulexite and inyoite beds interlayered with light-brown clay–mud; 12 m of sand, clay, and tuff; and then the late Miocene Sijes Formation.

In the north the Andina mine recovered some ulexite from these Pleistocene terraces in 0.5- to 1.2-m beds overlain by travertine, sandstone, and caliche. The individual ulexite layers were 0.1- to 0.2- (<0.5-) m thick and quite pure (see Table 4.1; with a low salt content), and dull reddish inyoite could be intermixed or in separate beds. The currently active Blanca Lila geysers and hot springs are nearby. The Elsa terrace mine was west of the Sijes mining camp, having a 1.6-m borate ore zone containing 5-cm ulexite beds interlayered with equally thick clay beds. Inyoite usually occurred in thin layers next to the ulexite, and as <5-cm crystalline aggregates or rosettes. Both minerals appeared to be primary, but there was some secondary inyoite at the edges of the deposits. The borates were most common in the bedded form, but in some areas there were small ulexite nodules (Alonso, 1986, 1992; Alonso and Menegatti, 1990).

Alonso (1988) noted that some of the borates in the salar floor were also derived from the much older Sijes Formation's leached outcrops. In the eastern border near the Santa Rosa and La Paz mines, 1–1.2 m of excellent quality bedded ulexite had formed, and the leached residue of borate outcrops was apparent. Similar ulexite occurred in the north from the Blanca Lila deposit, and ulexite in the adjacent small *Salar Juanita-Rita* was formed from leaching Monte Amarillo and Sorpresa outcrops.

4.1.2.10 Pozuelos

This 84-km^2 salar is located to the west of Pastos Grandes. In its northeastern corner the sediments contain, in descending order 60 cm of red sand, 20 cm of bedded ulexite, 50 cm of green plastic clay, and disseminated ulexite in fetid organic clay. In the south near the eastern border is a 50-cm bed of nodular ulexite having <5-cm ulexite–clay layers alternating with reddish sand. Underneath is dark-gray to black sand. In the east at a depth of 1.5 m bedded ulexite has been found, covered by dark red clay–mud and interlaced with black clay. There are travertine remains on the salar's border along a fault line. Both areas (and San Mateo) have been sporadically mined by the Espinoza Mining Company (Alonso, 1988).

4.1.2.11 Ratones

This salar is near, and northwest of the much larger Hombre Muerto. It is at an altitude of 3900 m, 8 km wide at its southern end, extends in a north–south direction, and has the Ratones volcano on its southern border. It has a mountain island with a borate-containing bay fairly high on its east slope near the remains of many extinct hot springs (Alonso, 1986). The salar's surface is smooth in places and rough with dirty salt in others, remains wet for much of the year, and contains pure salt in a band from the south to the north. There are 0.5-m (1-m according to Reichert, 1909) ulexite beds in a 1-km^2 area, which are near the surface but in places covered by halite pure enough to be mined as table salt. The extinct thermal springs have a <12-m diameter, and yellow-greenish porous travertine rings with ulexite in the pores. The salar was mined by the Rio Sal Company, and has ulexite reserves of 100,000 metric tons (Alonso, 1988; Alonso and Gutierrez, 1984).

4.1.2.12 Rincon

This salar is north of Pocitos, bordered by a row of volcanoes to the south, the Guayaos Mountain range to the north, and alluvial plains on the sides.

Most of its surface is covered by up to 60-m-deep halite, and in some areas ulexite and borax are found beneath a 20- to 30-cm salt layer. The ulexite occurs as <50-cm (average 30-cm) beds or nodules (the nodules predominate) heavily contaminated with chloride and sulfates (Alonso, 1988). Reichert (1909) described one area as having a 5-cm surface layer of saline efflorescence and caliche on top of 25 cm of white-red marly clay containing fine silky needles of ulexite. Under that was 50 cm of "bancos" ulexite, then 30 cm of wet clay followed by clayey quicksand and the water table. Borax has been found as anhedral, euhedral, and succrodial crystals (called "pork rind" or "corn kernel") along the northeastern border of the salar. The crystals vary widely in size, but the euhedral ones usually are about 0.5 cm long. They are often well-shaped, or irregular aggregates with some halite, found in red silt and sand. They have been almost completely mined out at the Carolina Mine of Boroquimica Samicaf (Alonso, 1988; Muessig, 1966).

4.1.2.13 Salinas Grandes (Puna de Jujuy)

This salar is located south of, and in the same basin as Lake Guayatayoc at an elevation of 3300 m, the lowest salar in the Argentine Puna region. It extends in a northeast–southwest direction and was mined in several areas by a Belgian company during the early mining period. The 30-cm (average) ulexite zone extends from the surface to a depth of <1 m, having nodules on the surface and in the upper zone and bedded ulexite below. A reddish plastic clay is under the ulexite (Alonso, 1988). The Tres Morros "boratera" was north of the town of Moreno and had high-quality, low-salt, 0.8-m (<1-m) beds of nodular ulexite. The ulexite was air-dried, screened to remove adhering clay, taken to the plant, and dehydrated in rotary calciners. The 160-hectare Nine Muerto deposit had 1-m-thick ore in beds and nodules (see Table 4.1; Reichert, 1909).

Buttgenbach (1901) described the salar's ulexite beds as 10–70 cm thick and 0–1.5 m (average 0.6 m) deep. Below the ulexite was a bluish clay that turned gray on drying. The 5- to 15-cm nodules were tightly packed in a sandy clay matrix. Each bed occupied several square meters, and the surrounding area could be barren or have two or three clusters joined together. There could be several beds above each other (the lowest containing the largest and purest nodules), and layers of hard, crystalline, high-gypsum caliche a few centimeters thick sometimes lay between the multiple beds. Some areas of the surface had <1-cm^3 kidney-shaped ulexite nodules, and there always were beds of larger nodules under them. Salt was present throughout the salar (and with the ulexite), was thickest in the center, and usually could easily be leached from the product. Much of the ore was found under the brine level, so it was allowed to dry for 5–20 days, becoming harder and losing 10–20% of its weight. It was then shaken in a basket, causing it to lose most of the adhering clay and all but about 2% of the NaCl. The ulexite reserves were estimated as 500,000 metric tons.

4.1.2.14 Turi Lari

This small salar is different from the others in that it contains considerable carbonates, and its clay has a higher content of lithium (0.14% Li) and arsenic (0.10% As). The clay consists of 70% montmorillonite and dolomite with some calcite, while the silt has 10% each of feldspar, calcite, and dolomite. The salar is in an isolated basin, has a large drainage area, is near extensive Quaternary lava flows, and has extinct mineral springs around it (some with ulexite in their mantles). It has two borate zones: (a) the near-surface (60-cm) zone contains fine ulexite (in limited amounts) and small borax crystals and (b) the next 30 cm has perfectly euhedral <2-cm unoriented prismatic borax crystals in a green plastic bentonitic clay (Muessig, 1958b, 1966). Alonso (1988) described the borax crystals as 1–3 cm long, occurring in the top 90 cm and averaging a mud to borax ratio of 1. In the areas where the salar was completely dry, there was only ulexite. Layers of caliche lie under the borate zone, and in 1988 the deposit was mined by Industrias Quimicas Baradero. Llambias (1963) reported that the borax contained small inclusions of a bituminous material.

4.1.2.15 Laguna Vilama

This small playa is in rugged terrain near the border with Bolivia and has ulexite in both its northern and southern ends. The northern deposit had been excavated at the Pirquitas mine (Boratera Vilama I, II), but the ulexite was of poor quality (Alonso, 1988).

4.1.3 Geysers, Springs, and Their Mantles

Seventeen Argentine borate geysers and/or springs have been found, usually in the vicinity of the borate salars (Fig. 4.3; geysers are artesian thermal springs that form a cone-shaped mantle, whereas springs form a tabular mantle). They might be isolated (e.g., the Tropapete geyser and Lari hot spring) or grouped (e.g., Arituzar, with 8 geysers and hot springs) in a small area. Most of the vents are extinct, or give off only gas (usually CO_2, sometimes H_2S), or have nonborate water. Only Antuco is still forming a borate deposit. The geyser's or hot springs' surface mantles usually contain ulexite, but pinnoite ($MgB_2O_4 \cdot 3H_2O$) has been noted in one of the geysers (Socacastro), and in a few cases borax (e.g., Tropapete, Coyahuaima) is found with the ulexite. Iron–manganese muds are occasionally present, and native silver has been found near the Libertad borate spring, antimonite at Coyahuaima, copper sulfates at Daniel, and anomalous amounts of arsenic and antimony in most cases (Alonso, 1986; Alonso and Viramonte, 1990).

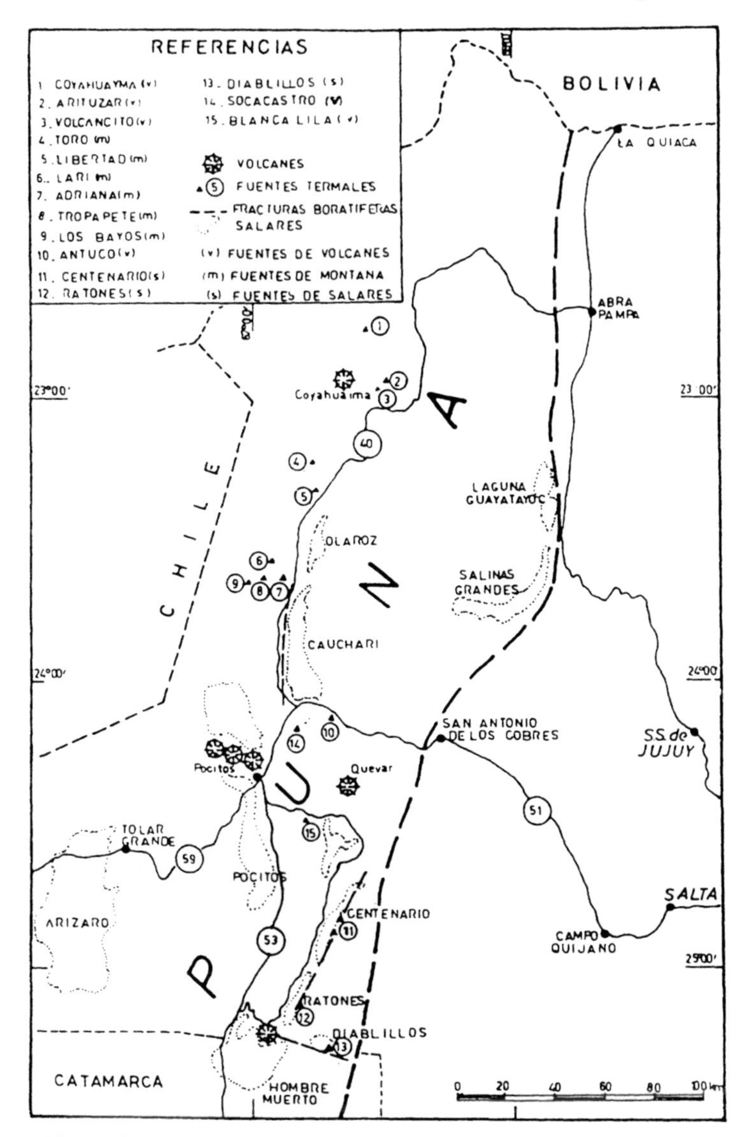

Figure 4.3 Various Argentine borate geysers and hot spring mantles. (From Alonso and Gutierrez, 1984; reprinted by kind permission of Dr. Alonso.)

4.1.3.1 Alumbrio Deposits

These 11 springs are in an area a few square kilometers in size at an altitude of 4200 m, 15 km northwest of Coranzuli. They are "among the most spectacular borate spring deposits ... in the world (Muessig, 1966, p. 153)." Each mantle contained 200–4000 tons of almost pure ulexite. The water from the springs eventually became low in boron and high in calcium bicarbonate, forming calcareous tufa in one cone and irregular thin calcite beds over most of the ulexite in the others. One spring was flowing in 1955 with "59°F (15°C) somewhat salty water," considerable gas (CO_2 ?), and a rough crust of calcite coated its fissure. No borate deposits are near, but the runoff is to the Rio Alumbrio and the Laguna de Guayatayoc with its thin, sparse ulexite deposits (Muessig, 1966).

4.1.3.2 Antuco

The Antuco hot springs are 15 km south of Cauchari, near the volcano Quevar in a group that once covered 4–5 hectares. They formed 50-cm ulexite deposits in an area of volcanic rocks that now have been mined out (Norman and Santini, 1985). One spring is still active and currently deposits ulexite from 17–44°C, high-boron and potassium water containing 151 ppmLi (Alonso, 1986; Viramonte, Alonso, Gutierrez, and Arganaraz, 1984).

4.1.3.3 Coyahuaima (Coyaguaima)

This group of three separate dormant springs is at an altitude of 4600 m, 30 km north of Alumbrio, 80 km west of Abra Pampa, and on the north flank of the Coyaguaima volcanic mass. Each spring has a cone and apron of ferruginous calcareous tufa, with "fluffy" wet ulexite (1–3% halite) both above and below the tufa. Small disseminated borax crystals and irregular masses are in the upper part of the eastern deposit, comprising 3% of this 2-m mantle. The deposit had 10,000 tons of borates in 1955, which supplied the Edith Mine (Muessig, 1966).

4.1.3.4 Volcancito

This spring has a small tabular ulexite deposit with three prominent ulexite–tufa cones in a flat valley of the Rio Blanco 6 km west of the Rio Alumbrio deposits. It no longer flows, but noisy bubbling can be heard beneath each cone, giving them the "little volcano" name. The upper and largest cone has no discernable vent and is formed by ulexite capped by ferruginous calcareous tufa. The middle cone is 20 m to the east and has a similar cap, including fragments that block the 0.5-m vent. When the cap was removed, cold water (hot in 1924) could be seen bubbling vigorously 30 cm below the cone's lip. The lower cone is primarily ulexite and has a 30-cm vent with gas issuing from it. Halite coats the ferruginous tufa in the vent. The ulexite deposit area is 200 × 40 m and <2 m thick, quite pure, porous, saturated with water, and capped by a thin bed of tufa in the upper and central sections. Drainage joins the Rio Alumbrio, with some ulexite efflorescence downstream (Muessig, 1966). The Alejandra mine operated on the deposit (Alonso and Viramonte, 1990). Among other springs, the Archibarca Ravine area (three zones) and the small Libertad mantle have been mined out, and the Ojo de Aqua mantle has been briefly mined (Orris, 1995).

4.2 BOLIVIA

Bolivia has a number of borate-containing salars that have been mined, such as the Salars de Capina Sur, Challviri, Chiguana, Coipasa, Empexa, Laguani, Llipi-Llipi, Luriques, Pastos Grandes, Sacabaya, and Uyuni. Others, such as Chiguana, Cuevitas, Curuto, Laco Poopo, and Mahama Coma also have been suggested as potential borate sources (Muessig, 1966; Orris, 1995). Llipi-Llipi has been estimated to contain 1.2 million tons of ulexite, and the total of all the salars is 20 million tons of B_2O_3 (Leiser, 1996). The Compania Minera Tierra Ltd. in 1988 had ulexite concessions on Capina, Chalviri, Chiguana, Empexa, Leguani, and Pastos Grandes totaling 4800 hectares with claimed 20–37% B_2O_3 ulexite reserves of more than 13 million tons.

4.2.1 Salars de Uyuni and Coipasa

The 9000-km^2 (10,500-km^2 according to Leiser, 1996) Salar de Uyuni may be the largest salt playa on earth, and the adjacent Salar de Coipasa is also very large (see Fig. 2.1). Its altitude is 3653 m, its basin 10,0000 yrs old, and the surface salts 3520 ± 600 yrs old, as are the smaller playas nearby (Erickson, Vine, and Ballon, 1978; Kunaz, 1980; Rettig, Jones, and Risacher, 1980). The area's average rainfall of 20 cm/year (40–50 cm/year, mostly in December–February; Leiser, 1996) appears to have been fairly constant for the basin's life. Uyuni's depth averages 121 m; its longest dimension is 120 km; and much of its surface has a hard, smooth, porous (20–30%), 0.1–20 m (average 3- to 6-m-) thick salt crust. It has 11 salt beds separated by layers of mud and sand, and the total brine in the playa's voids is 13 km^3 (Leiser, 1996).

The salar floods each year to form a 0- to 75-cm-deep lake for 3 or 4 months, and brine is always within 5–20 cm of the surface. It is usually saturated with salt and contains 5.5 million tons of Li (8.9 million tons; Anon., 1992b), 110 million tons of potassium, and 3.2 million tons of boron (Table 4.2). The B-Li brine source primarily is many still-active geothermal springs (i.e., at Calamarca, Challapata, and Oruro). Its major river, the Rio Grande, starts with 11 ppmB, and by the time it is through the salar's delta system, it has a boron concentration of <520 ppm (Lulzaga, 1978). In 1996 the area was said to have the capacity to recover about 5000 tons/month of ulexite, with most of the mining being done by local Indians under a government concession (Leiser, 1996). There are many smaller borate salars at a higher elevation (4000–4500 m) and to the south of Uyuni. One of the largest of these is the

Table 4.2

Average Analyses of Various Bolivian Borate Brines, wt% or ppm as noted

	Salar de Uyuni			Salar de Empexa[a]	Salar de Coipasa[b,c]	Pastos Grandes Salar[d]		
						Hot Springs		Subsurface brine
	1	2	7			Edge[e]	Playa	
Na	8.2	8.72	—	5.4	7.51	317	0.448	7.72
K	0.66	0.72	1.10	0.27	1.10	43	0.051	0.891
Mg	0.64	0.65	—	0.68	1.36	4.5	67 ppm	0.174
Ca, ppm	—	463	—	209	156	10.2	212	1440
Li, ppm	321	349	540	172	350	5.2	69	1800
Sr, ppm	—	14	—	—	17	—	—	—
Cl	14.8	15.71	—	9.7	15.10	560	0.876	15.67
SO_4	1.08	0.85	—	2.8	2.46	36.5	0.014	0.932
HCO_3, ppm	—	333	—	347	747	106	539	608
B, ppm	187	204	525	176	786	2.0	26	376
Br, ppm	—	49	—	—	142	—	—	—
F, ppm	—	10	—	—	33	—	—	—
SiO_2, ppm	—	7	—	—	10	39	37	7.1
pH	—	7.25	7.3	—	7.23	6.55	6.30	7.14
Density	—	1.21	1.19	—	1.231	1.001	1.013	1.194

[a] Ericksen, 1993. Total dissolved solids (TDS) = 19.3.

[b] Rettig, Jones, and Risacher, 1980.

[c] Adjacent to Uyuni.

[d] Risacher and Eugster, 1979. This salar is slightly southwest of Uyuni, in Bolivia. There is also a Pastos Grandes in Argentina.

[e] All analyses as ppm.

100-km^2 Salar de Pastos Grandes, with its boron source being 15 high-discharge hot springs.

4.3 CHILE

4.3.1 Chilean Nitrate Deposits

The Chilean nitrate deposits contain a number of borate minerals: colemanite, ginorite, hydroboracite, iquiqueite, kaliborite, probertite, and ulexite. By-product boric acid has been produced periodically since the 1880s from these unique and puzzling soluble salt formations. The nitrate deposits occur in the northern coastal desert region of Chile over a north–south range of 700 km and an east–west distance of 10–150 km as an almost continuous formation of sodium nitrate and other soluble salts. They are the only substantial concentration of sodium nitrate (6–10% $NaNO_3$) found anywhere in the world, and the only solid-phase deposit containing soluble iodate (0.04–0.08%), perchlorate (0.02–0.04%), and chromate ions. This nitrate (or "caliche") deposit contains an estimated 1 billion tons of sodium nitrate and 100 million tons of boric acid, among its other salts (6–15% Na_2SO_4, 6–10% NaCl, 0.4–1% K, and 0.3–1% B_4O_7). The ore grade varies considerably over the vast deposit, but is usually most concentrated 1–5 m below the surface.

It has been suggested (Garrett, 1985) that the more common ions in the deposit came from higher-elevation salars, thermal springs, and the leaching of the adjacent terrain. Catalytic nitrogen oxidation (from the air) produced the nitrates and the higher oxidation salts, which were captured by the area's heavy daily winter fogs. The salts were transported to the lower valleys through faults and underground aquifers, where the high water tables and (rare) rain, together with the extremely hygroscopic nature of the nitrates, resulted in a capillary movement of the solutions to the surface, allowing evaporation to occur. Periodic partial leaching of the top meter or so of soil from the rains, followed by renewed capillary activity, continued to thoroughly mix, evaporate, and spread the deposit. Wind action, however, was the principal means of redistributing the salts to cover the terrain so completely in this extremely arid desert. To recover product from the caliche, it is first leached. Then by sequential steps of cooling, solar evaporation, recooling, chlorination, and other processing, all of the products can be recovered, including boric acid by a salting-out process.

4.3.2 Salars; Ascotan

Chile also has a large number of borate-containing salars, and five (Ascotan, Aguas Calientes Sur, Atacama, Quisquiro, and Surire) had commercial borate operations in 1994, producing 31,000 metric tons/year of ulexite (Kistler and Helvaci, 1994). Ascotan (Fig. 4.4) is 16 km long and 4.8 km wide at an altitude of 3960 m, with a 173-km^2 surface and a 1090-km^2 drainage area. It has a surface of gypsum sand (common in the Chilean borate salars), a shallow lagoon, brine beneath the surface (Table 4.3), and high volcanos rising from its shores. The railroad from Antofagasta to Bolivia reaches its highest elevation at the salar. Its surface crusts are rich in ulexite, and much of its buried ulexite is in a matrix of diatomaceous earth. In some areas the ulexite is comparatively pure, but in others it alternates with layers of halite and mud (containing glauberite and gypsum; Chamberlin, 1912; Muessig, 1966).

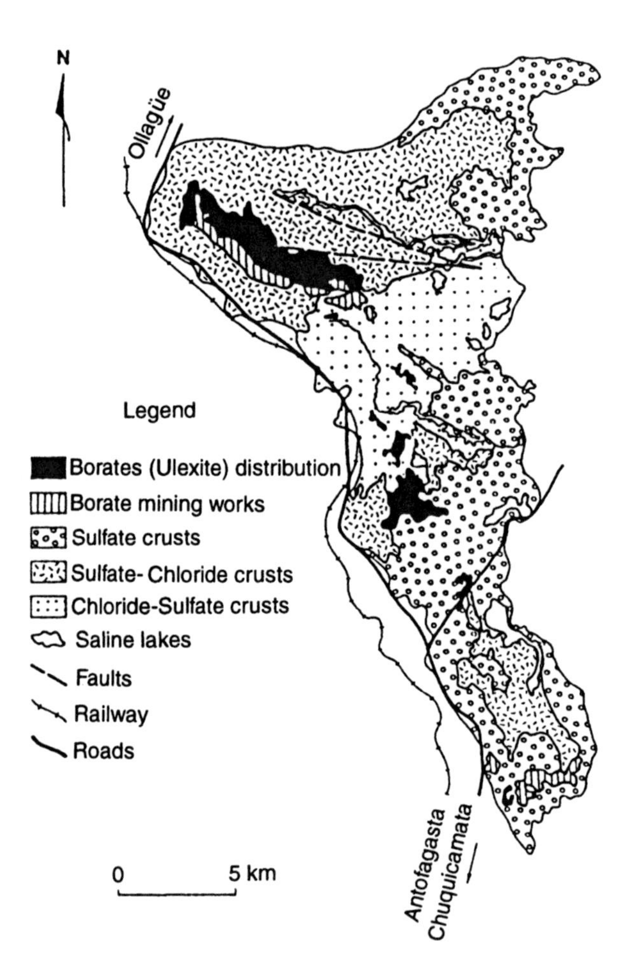

Figure 4.4 Geologic map of the Ascotan Salar. (From Vila, 1990; reprinted by permission of Springer-Verlag Gmbh & Co. KG.)

Table 4.3

Chemical Analyses of Various Boron-Rich Brines from Northern Chile, mg/l[a]

Salar	pH[b]	Dissolved solids	Na	K	Mg	Ca	Li	Cl	SO_4	HCO_3	B
Aguas Calientes[b]	7.7	81,436	25,460	1,183	1,361	2,538	152	46,690	3,154	0	474
Ascoton[c]	7.8	153,600	45,000	3,500	5,125	920	186	70,000	25,000	2,900	783
		47,022	13,870	1,670	827	1,195	82	24,000	4,693	0	595
Atacama	6.6	370,000	91,100	23,600	9,650	450	1,570	189,500	15,900	230	440
		310,000	85,800	13,000	6,350	1,100	940	163,900	8,540	280	360
		190,000	45,100	9,000	5,330	900	520	83,780	18,170	240	360
		73,000	18,220	4,220	1,810	360	290	36,750	3,430	320	100
		62,000	14,840	2,900	1,930	1,080	190	27,500	7,900	100	88
		40,100	10,280	1,690	750	1,160	130	20,300	2,160	92	61
Bellavista	10.4	170,300	50,000	5,403	3,665	5,935	85	100,600	2,720	178	225
Pintados											
Huasco[b]	6.0	150,100	38,000	10,000	1,750	840	130	83,600	13,600	—	2,200
Lugunas	6.8	390,000	126,800	14,280	3,630	110	412	176,600	47,770	406	979
Punta Negra[b]	7.1	271,900	86,000	10,000	2,620	2,080	320	164,500	4,480	—	2,230
Pujsa	8.6	89,298	28,500	1,295	653	375	137	27,660	28,110	0	675
San Martin	—	102,138	28,160	2,614	6,252	1,566	187	60,050	2,490	625	426
Surire[b]	7.5	167,200	54,000	8,700	1,250	750	340	79,800	20,300	90	1,820
Hot Springs[d]	7.8	4,357	1,210	200	28	135	8.3	1,905	534	150	47
River[e]	7.7	152	23	2.8	4.3	1.0	0.1	22	20	33	0.7
Soil[f]	—	—	2.2	0.3	1.8	13.5	65 ppm	1.2	22.9	3.1CO_3	3.1

[a] Ericksen, Vine, and Ballon, 1978.
[b] Vila, 1990.
[c] Ericksen, 1993.
[d] Salas, 1972. also SiO_2 129, NO_3 7.4, Bx 1000/TDS = 11.
[e] Salas, 1972. Also SiO_2 45, Bx 1000/TDS = 4.6.
[f] Salas, 1972. South-center of playa, average 6.1m depth. Also As 158 ppm.

Because of Ascotan's location adjacent to the railroad, it was the first Puna salar mined (1852), but there was not a significant production until the 1880s. By 1900 it was among the world's largest producers, reaching a peak in 1913 of 38,000 tons. It then began to decline, becoming quite small by the 1920s, and almost totally stopped from 1967 to 1978. Ascotan is an Indian word meaning "a place where a dog does not like to live," and in the initial operation during the winter the workers waded ankle deep in ice water, and most afternoons a cold wind developed that engulfed the salar in a cloud of sand, ulexite and gypsum dust.

The laborers ("peones") first removed a 2.5- to 13-cm (1- to 5-in.) hard surface crust ("costra") containing gypsum and dirty salt, and then excavated a 0.5- to 1-m trench to uncover a 0.1- to 1-m layer of ulexite. The ulexite nodules [weighing 0.5–9 kg (1–20 lbs)] were removed with picks and shovels, hand sorted, and spread on "canachas" to dry for a month or two. When dry (Table 4.4), they were broken by stone rollers drawn by cattle. The small particles were removed by hand and discarded, and the larger ones originally were shipped in bags to Antofagasta to be calcined. However, by about 1907 they were calcined at the salar in reverberatory ovens (open pans heated from the top). The borate was shoveled into the furnace bed, stirred frequently with long rakes, and heated by burning a woody fungus ("yareta") brought to the plant by indian contractors. The water content of the calcined ulexite could be reduced to only 20% before the mass fused, partly because of a

Table 4.4

Analyses of Ulexite from the Salars de Ascotan and Mariungo, wt%

	Salar de Ascotan[a]			Salar de Mariungo[b]				
				Surface cotton balls		Surface	Subsurface beds, grade	
	Air dried	Air dried and calcined	Dupont 1910	As mined	Calcined	efflorescence	High	Low
B_2O_3	32.2	39.8	38.04	25.4	53.4	22.5	19.6	17.1
CaO	13.1	15.9	12.34	9.7	23.4	17.4	15.3	11.7
Na_2O	6.2	7.5	15.91	12.4		7.2	11.8	17.5
Mgo	0.6	0.7	0.37	—	—	—	—	—
NaCl	3.1	4.0	Cl 9.64	1.4	—	2.3	—	2.3
SO_4	2.1	2.2	0.84	—	—	12.3	9.2	11.8
R_2O_3	0.1	0.2	0.24	—	—	—	—	—
SiO_3	3.0	3.3	4.95	—	—	—	—	—
Insol.	—	—	—	—	—	10.7	10.3	2.2
Water	39.6	26.4	19.86	51.1	23.2	27.6	33.8	37.4

[a] Anon, 1909.
[b] Fonseca, 1874.

small borax content. The calcine was then removed, cooled, bagged, sent to Antofagasta, and shipped to Europe (Anon., 1909; Robottom, 1893). In 1979 ulexite harvesting was reinitiated on the salar, with much of Chile's estimated 30,000 tons output that year coming from Ascotan (Kistler and Smith, 1983).

4.3.2.1 Atacama, Maricunga

Atacama is a large high-altitude playa in northern Chile that has had some ulexite production in its Tambillo, Tilomonte, and Tilopozo areas (Orris, 1995). However, its most important commercialization is as a brine deposit, so it is discussed later in Chapter 5. Maricunga is located northeast of Copiapo, a town on the western slope of the Andes in a valley formed by the volcanoes Toro, Azufre, and Tres-Cruces. Its altitude is 3800 m, its length 50–55 km, and its width 10–12 km. It has travertine, tufa cones, and terraces along its margins (Muessig, 1966), and even small alum deposits. Ulexite has been mined periodically since at least 1874, with crude ore shipments sent to Antofagasta for export. It can flood in the winter, but the water dries quickly (i.e., 6 cm evaporated in 1873 in 3 days). Surface crusts and efflorescence (called harinas) of ulexite covered most of the playa, and in the mud nodules occurred in a sandy clay with halite. The richest ore was cotton balls on the surface ("bolones"; see Table 4.4) or in the lake's mud at depths of 1–2 m. The surface ulexite was less pure, but easier to harvest. There was also deeper bedded ulexite ("masas") occurring as alternating layers with salt or clay, often 5–10 cm thick (Fonseca, 1874).

4.3.2.2 Pedernales, Pintados, Cariquima, Quillagua

Pedernales, 50 km north of the town of El Savador, has a porous salt mass (containing brine) extending to a depth of 13–28 m. A drilling program indicated brine reserves of lithium, potassium, and borax (Lyday, 1991). Its subsurface ulexite deposits have been worked periodically since the late 1800s, and Minero del Boro has produced ulexite concentrates from these salars since 1984 (Lyday, 1992).

4.3.2.3 Surire (Chilcaya)

This salar, 135 m east-southeast from the port city of Arica at an altitude of 4250 m, has several ~6000-m volcanoes near its basin and one "small peak," Cerro Oquecollo, in its center. It has a perennial lake with hot springs underneath still forming ulexite. Its ulexite zone starts 6–14 cm beneath the surface as nodules and as <1-m (average 30-cm) beds. Production of ulexite started on a small scale in thc late 1800s, but was operated only sporadically until 1986 when Quiborax mined 1100 tons of 28% B_2O_3 ulexite.

In mining, the surface muds and salts are first removed by a small front-end loader. Then the ulexite is excavated by hand labor, placed beside the trenches to dry, and finally loaded into trucks. The salar is estimated to contain

34.5 million tons of ulexite having >14.5% B_2O_3 (1.5 million tons of 35% B_2O_3 ulexite; Leiser, 1996), and significant amounts of boron, potassium, and lithium in its brine (Alonso, 1986; Anon., 1992a; Kistler and Helvaci, 1994; Leiser, 1996; Lyday, 1990).

4.3.2.4 Other Salars

The Salar de Zenobia has hot springs and geysers along its margins. Several other salars have been briefly mined for ulexite, and still others are considered as potential borate deposits, including Aguas Calientes, Aguilar, Aixa and Joya, Carcote, Cosapilla, Diologue, Infieles, Pajonales, Punta Negra (Muessig, 1966), Huasco, La Isla, Las Tias, and Tamarugal (the Chug-chug, Joya and Quebrada de Barrera mines; Orris, 1995).

4.4 CHINA

4.4.1 Tibet

In the borate lakes of Xizang (Tibet) more than 37 ionic components have been found, and 27 minerals. Boron is a major element in the brine of many lakes, existing in concentrations of 32–1440 ppmB (average 542 ppmB). Other average values are Na 6.2 wt%, K 0.66%, Mg 0.47%, Ca 159 ppm, Cl 9.2%, SO_4 2.8%, CO_3 0.12%, HCO_3 948 ppm, Li 320 ppm, F 113 ppm, Sr 6.7 ppm, Rb 5.4 ppm, Cs 1.9 ppm, and As 1.7 ppm. Many lakes have geothermal springs flowing into them, and large mantels containing borates. Borax is by far the dominant borate mineral in the high-carbonate lakes, and in the other lakes ulexite predominates, with borax, inderite, inyoite, kurnakovite, and pinnoite. The kurnakovite usually occurs as 0.3- to 0.5-mm grains compacted in porous strata below the playa surface, and above a clay layer. The pinnoite often combines with carbonates to form a hard surface crust, and sometimes replaces kurnakovite. Ulexite can occur as 4-mm nodules lying directly on a sandy playa surface, and may replace kurnakovite or pinnoite. Many of the lakes are 5600–20,000 years old, with the borate formed more recently (Yang and Zheng, 1985). Because the earliest recovery of borates from these lakes was in the form of borax, further detail is given in the chapter on Borax. There is also some borate recovery from brines, so it is described in the chapter on Brines.

4.5 PERU

4.5.1 Laguna Salinas

This playa is in southern Peru 80 km east of Arequipa at an altitude 4300 m, and is 14 km (east–west) by 8 km (north–south) in size, somewhat surrounded by high mountains (5400–5800 m). It becomes partly flooded (50 cm deep) after the winter rains (December–March) and dries rapidly to form surface crusts containing ulexite ($NaCaB_5O_9 \cdot 8H_2O$), halite, glauber salt and thenardite. Beneath the surface are irregular 10- to 100-cm beds or lenses of ulexite, and in one area, inyoite. Ulexite was first recovered in 1883 by hand digging 1-m-wide, 2- to 3-m-long trenches as deep as the bottom of the ore zone, then removing the ulexite nodules from the mud. Twelve pan dryers were used to increase the grade of the harvested ulexite to about 32–36% B_2O_3, using the local yareta bush as fuel. From 1901 to 1905, 5000 tons were produced, but from 1926 to 1974, only a few hundred tons of air-dried ulexite were sold each year, and the total from the beginning to 1930 was 100,000 tons. Boratos del Peru mined the Laguna in 1975 with backhoes and produced 4500 metric tons of ulexite. There are still reserves estimated at 10 million tons of 32% ulexite (Leiser, 1996; Lyday, 1984).

The playa's structure from the surface downward is 5–15 cm of white surface salts (halite and thenardite), <15 cm of white volcanic ash, 1.2 m of black sandy mud, 20 cm of black mud with some white nodules and lenses of ulexite, 20–50 cm of green mud with more ulexite, and 0.25–1.3 m (average 0.46 m) of the main ulexite bed (Table 4.5), undulating, but fairly continuous

Table 4.5
Analysis of Low- and High-Grade Ulexite from the Laguna Salinas, Peru, wt%[a]

Constituent	Low-grade ulexite	Main ulexite ore	Hand-picked ulexite
B_2O_3	4.17	18.6	38.5
CaO	21.7	19.7	14.0
Na_2O	2.66	5.9	9.5
SiO_2	18.4	10.3	7.9
Cl	1.3	3.2	3.2
SO_3	29.8	20.3	2.0
MgO	2.6	1.8	1.8
K_2O	1.22	1.24	1.2
Al_2O_3	3.14	1.39	0.72
Fe_2O_3	1.74	0.78	0.45
SrO	0.17	0.22	0.15
TiO_2	0.17	0.07	0.03
MnO	0.05	0.04	0.02
Li (ppm)	150	130	120
Cr_2O_3 (ppm)	40	20	10
As (ppm)	86	30	0.9
LOI (427°C)	10.0	17.1	24.4

[a] Norman and Santini, 1985.

over the laguna. The depth to it averages 1.4 m; 0.6 m on the eastern side and 2.9 m on the western side. On the eastern side <15 cm of inyoite ($Ca_2B_6O_{11} \cdot 13H_2O$) occurs in a discontinuous bed of crystalline aggregates under the ulexite (Kistler and Helvaci, 1992; Norman and Santini, 1985). Muessig (1958a) noted that the inyoite occurs only near the Tusca hot springs, primarily as fibrous crystalline aggregates in a discontinuous compact layer with a few small nodules in the upper mud matrix. The 1-hectare (few-acres) bed is 1 m deep in black mud. It appears to be primary, since there are no pseudomorphs, and its inclusions are the same as those of the matrix mud.

Jochamowitz (1907) characterized the mining pits as penetrating: 10–14 cm of surface crust with NaCl, Na_2SO_4, and fine sand; 6 cm of gravel; 20–50 cm of fine sand with some layers of ulexite; 0–2 cm of barren fine sand; and 10–100 cm (average 40 cm) of the main ore body. The ore zone had a low permeability, so ground water rose only slowly from below, and when there were surface pools of brine, there was no ulexite underneath. Two types of ulexite were present: The "silky" type (sedoso) was fairly pure and had a fibrous and brilliant silky luster, and "ordinary" type was less pure. The mining pits (piques) quickly filled with brine from the upper more permeable zone, so small barriers were left between each pit to minimize brine inflow during digging. The pits were laid out in long, straight rows to facilitate hauling the hand-sorted, air-dried ulexite to the plant (oficina). Workers (peons) received 300 centavos/metric ton of ulexite. At the plant each dryer handled 4.25 metric tons/day, taking the 30% B_2O_3 ore to 52% and leaving 12% H_2O in the product. Each dryer burned 3.4 metric tons of yareta/day, for which the laborers were paid 440 centavos/metric ton. The high resin content of this rounded bush with extensive roots made it burn well, but if not used within 20 days it began to rot. The product was hauled to Arequipa by llamas, mules, or donkeys, costing 1400 centavos/metric ton, but a scarcity of "muleteers" limited the production to 150 metric tons/month.

At *Chillicoipa,* near the Bolivia–Chile border, a ulexite mantle is being formed by an active thermal spring (Norman and Santini, 1985). It has had some mining, and the adjacent Laguna Blanca also contains some ulexite (Orris, 1995).

4.6 UNITED STATES

4.6.1 Alvord Valley, Oregon

Alvord Valley is a closed basin 113 km (70 mi) long and 13 km (8 mi) wide in southeastern Oregon. It includes the Alvord Desert in the north, Alvord Lake in the central part, and a small lake and stream in the south. It is bordered on the east by the White Horse and Trout Creek mountains, and on the west by the Pueblo and Steens mountains. Alvord Lake is only ~1 m (few ft) deep, and its area varies with the season and rainfall, averaging 5.6 km (3.5 mi) long and 3.2 km (2 mi) wide. During periods of heavy precipitation, it overflows into the Alvord Desert to the north. In summer the lake is quite small and has alkali crusts on its surface (Libbey, 1960).

There are numerous springs within the valley, some of which discharge water near the boiling point (Table 4.6). The largest (No. 2), flows ~57 liters/sec (900 gal/min) at 36°C (97°F) and has formed a deep pool named Hot (or Borax) Lake, having a diameter of 251 m (275 yd). It is adjacent to the former Rose Valley Borax Company's plant. On the western side of Alvord Desert hot springs (No. 1) in a group discharge ~8.5 liters/sec (135 gal/min) at 76°C (168°F). Approximately 11 km (7 mi) northeast of the north end of Alvord Desert Hickey Springs (No. 5) discharges water at ~100°C through several

Table 4.6

Hot Spring Analyses in the Alvord Valley[a]

Spring no.[b]	Location	Temperature, °F (°C)	Flow rate (gpm)	Total solids,[c] (wt%)	B_2O_3, (ppm)	Bx 1000 / (TDS)
1	At main vent	167 (75)	135	0.298	106	35.6
2	100 m from vent, bottom	97 (36)	900	0.170	61	35.9
	of Hot Lake 1981–1992 av.[d]	15–37°C	—	0.1764	15.5	8.8
3	Pool north of line of vents	122 (50)	—	0.155	31	20.0
4	Pool south of line of vents	198 (92)	—	0.15	56	37.3
5	Side of 9-ft diameter pool (Hickey Springs)	198 (92)	>100	0.168	33	19.6
6	SE vent	136 (58)	45	0.083	4.4	5.3
7	SE vent	95 (35)	30	0.015	1	6.7

[a] Libbey, 1960.

[b] See text for the spring location

[c] Spring 1 contained, as % of the total solids: >10% Si, Na; 1–10% K, B, Ca; 0.01–0.1% Li, Fe, Mg, V; 0.001–0.01% Mo, Ba; <0.001% Ti, Cu.

[d] Michels, 1991. Also, in ppm: Na 476, K 28, Ca 15, As 1.35, Cl 277, HCO_3 403, SO_4 337, SiO_2 203, F 8.1. Air temperature, −1.5 to 19.3°C. Surface elevation, 1244 m. Lake's yearly elevation change, 1.2 cm.

vents. To the southeast up Trout Creek Valley, there are also a few small low-boron thermal springs (Nos. 6 and 7).

In 1899, south of Alvord Lake, borax prospectors observed the white saline crust formed by the evaporation of hot springs water. They interested capitalists in purchasing 809 hectares (2000 acres) of the richest borate portion near the lake [the marsh extended for more than 4000 hectares (10,000 acres)], and erected a plant to produce boric acid. Chinese laborers first scraped about 5 cm (several inches; a new, thinner crust would quickly reform) of the crust into windrows, then loaded it into wagons for the haul to the planet. The harvested salts (a borax–ulexite mixture with 5–20% B_2O_3; Dennis, 1902) were dumped each morning into a 23,000- or 30,000-liters (6000- or 8000-gal) leach tank filled with 36°C Hot Lake water. The tanks were then brought to a boil by burning sagebrush, with sulfuric acid added to form boric acid (but with no excess, to minimize corrosion on the mild steel tanks). Calcium chloride was also added in the winter to prevent glauber salt from crystallizing. The tank was stirred from 3 to 5 PM with long-handled rakes by three laborers, and then settled overnight. In the morning the clear liquor was drained into some of the 24 4500-liter (1200-gal) galvanized "crystallizers" [with "hanger pipes" supporting numerous 0.6 × 20-cm × 1.5-m ($\frac{1}{4}$ × 8-in. × 4-ft) galvanized plates] and allowed to cool for 4–6 days. After the tanks were drained, laborers knocked the boric acid crystals from the hanger plates and tank sides with wooden mallets. The crystals were then shoveled into a wooden trough, washed, stacked to drain and dry in the sun, and bagged.

The residual mud from the dissolving tanks was washed, resettled, and sent to tailings piles. The wash water was used for the next ore-leaching batch, and the end liquor from the crystallizers was also recycled until it became saturated with Na_2SO_4 and NaCl. For the first few years, as the high-boron side of the playa was processed, there was some recycling, but as lower-grade ore was used both the yields and recycling were greatly reduced. This, coupled with longer distances to transport harvest salts and gather sagebrush, and constantly diminishing boric acid prices (because of the new colemanite mines), made the process uneconomical, and forced it to close in 1907.

The operation was conducted by a staff of 25–30 Chinese laborers: 4–6 collected the ulexite (three wagon loads/day, as well as extra quantities stockpiled for the winter); 3 worked at each dissolver and its crystallizers, tending the fires, agitating the tanks, removing crystals and bagging the product; 3 gathered sagebrush for the fires (and extra for winter); the others were assigned various duties (e.g., the cook spent only 1 hour before each meal preparing predominantly rice and dried fish). The staff lived in small adobe buildings near the plant and worked 7 days/week, all year except for a 2-week Chinese New Year vacation. The plant produced <4.5 metric tons/day, 1600 metric tons/year initially, and 400 tons/year later. The product was 96–98% H_3BO_3, and considered to be quite pure (Shaffer and Baxter, 1975).

After the boric acid was sacked, it was hauled to Winnemucca, Nevada by 16- to 24-mule teams. The operation was initially called the Twenty Mule Team Borax Company, but they did not register the name, so later when Pacific Coast Borax did, the Alvord operation was forced by the courts to change its name. Because of the prolific wild roses near the springs the operation became the Rose Valley Borax Company. Their two four-wagon 25-metric-ton boric acid units employed 20–24 mules in the winter and 16–18 in the summer for the 16-day, 240-km (150-mi)-each-way trip.

4.6.1.1 Hot (or Borax) Lake

Michels (1991) reported on Hot Lake as a potential geothermal power source, noting that the lake level was fairly constant during the year, and that its high HCO_3, B, F, and As content were typical of geothermal borate springs. It should have crystallized only borax (and not ulexite), but the other springs must have had a low-carbonate, high-calcium content. The lake is fed from the bottom of its conical basin at a depth of at least 26 m and overflows its built-up mantle in several areas 8 m above the general elevation of the valley. The lake is inhabited by the Borax Lake chub, a cyprinodont minnow unique to this location, and probably a highly adaptive survivor of the ancient Lake Lahontan.

4.6.1.2 China Lake, California

This very small playa (also called Airport Dry Lake) is one of the chain of depressions that periodically overflowed and fed Searles Lake (see Fig. 2.19). Despite the fact that its brines are strongly alkaline, cotton-ball ulexite was reported in brown sandy clay 1 m deep at the interface of coarse (upper) and fine (lower) sediments (Bowser and Dickson, 1966).

4.6.2 Columbus Marsh, Nevada

Borates were first noted on this playa (see Fig. 2.8) in 1865, but only after 1871 when William Troop identified ulexite did commercial operations begin, with the shipment of 0.77 metric tons (1700 lb) of crude product. Four companies were soon producing on the playa, and the largest, Pacific Borax Company, had a mill in the town of Columbus to make refined borax. The lake ultimately supported nine borate operations, with eight scattered around the playa's edges and one near its center. One of these, the Preservaline Company, produced 230 metric tons/year of ulexite with a staff of 20, and then converted it in New York to a preservative for meats and other foods (Yale, 1892). The town of Columbus in 1875 had a population of 1000, and even a newspaper, the *Borax Miner*. However, in 1875 the Pacific Borax Company moved 19 km (12 mi) to the south to the more profitable borax (not ulexite) deposit at the Fish Lake playa. By 1881 the population had dropped to 100, and even when the Carson and Colorado (Tonopah and Goldfield) Railroad was built along its east edge in 1882, the lower freight rates [i.e., $80/ton in 1873 for the 257-km (160-mi) mule-team and railroad haul to San Francisco; $12.50/ton in 1882] still could not revive the operations. Only 220 metric tons of ulexite were shipped in 1892, and none after 1908 (Hicks, 1916; Papke, 1976; Ver Planck, 1956).

The 80- to 100-km^2 (30- to 40-mi^2) Columbus Marsh is roughly elliptical, 14–16 km (9–10 mi) long (north–south) and 5–11 km (3–7 mi) wide (east–west) (the range of estimates is from different authors), at an altitude of 1373 m (4504 ft), with a 960-km^2 (370-mi^2) drainage area and a central surface that has a somewhat lumpy sand–mud texture. It is adjacent (to the north) to the 24-km- (15-mi-) long and 2- to 5-km- (1- to 3-mi-) wide Fish Lake Valley. Ulexite-containing salt crusts occurred predominantly around the Marsh's edges in isolated zones, whereas in its north-center there was sufficiently pure surface salt for its commercial recovery (starting in 1864). The ulexite (and some borax) formed as thin layers in an efflorescent crust (Table 4.7) after each rainy season (Williams, 1883). Under this crust 3- to 15-cm [1- to 6-in.; Papke, 1976; in some places 30- to 60-cm (1- to 2-ft) thick; Ayers, 1882] was a <0.6-m (2-ft) layer of a brown sandy clay mixture containing cotton-ball ulexite. It often was closely packed and could be easily removed by hand. When washed, the nodules were relatively pure, with "beautiful interiors of pearly-white masses of satiny crystals." There was one high-borax area of 200–240 hectares (500–600 acres) on the playa with soda ash (trona)

Table 4.7

Shallow Brine and Surface Salt Samples from Columbus Marsh, ppm[a]

			Water soluble salts (including brine) in surface sediments. Areas of		
	Brine 0–76 cm depth	Well water 4.6–20 m (15–66 ft)	High borate	High carbonate	High sulfate
Na est.	73,600	1,951	18,600	59,400	55,800
Ca	—	—	1,780	0	8,990
K	3,900	188	1,250	880	1,060
Cl	113,200	2,830	22,300	72,400	73,600
CO_3	—	44	600	7,430	0
SO_4	12,700	380	2,900	5,470	38,800
B_4O_7	800	86	16,900	3,450	2,080
SiO_3	—	—	1,000	5,730	420
Total	205,500	5,500	65,300	154,800	182,600

[a] Hicks, 1916. The lack of ion balance is presumably caused by the estimated values for Na.

Table 4.8

Deep Well Water[a] in Several Nevada Borate-Containing Playas[b] and Death Valley,[c] ppm

Brine type	Columbus Marsh		Rhodes Marsh		Teels Marsh		Death Valley[c]	Fish Lake Marsh
	CO_3	SO_4, Cl	CO_3[d]	SO_4	CO_3	SO_4, Cl	SO_4, Cl	CO_3
Na	1,046	1,035 1,951[e]	1,647	210	224	295	88,300	310
K	73	188[e]	58	—	—	—	3,300	—
Ca	21.2	156	10.7	41	30	21	—	12.5
Mg	19	29	1.5	4	11	0	120	3.5
Fe	0.54	—	0.1	—	—	—	—	—
As	0.20	—	0.62	—	—	—	—	—
Cl	1,005	1,174 2,830[e]	1,562	83	147	294	112,700	173
SO_4	289	971 380[e]	1,364	358	85	218	35,700	344
HCO_3	409	119	166	100	377	59	—	194
CO_3	75	5 44[e]	259	0	0	3	—	0
B	9.8	24[e]	62	—	—	1	300	2
SiO_2	35	—	95	—	—	—	—	—
F	2.8	—	24	—	—	—	—	—
PO_4	0.6	—	2.2	—	—	—	—	—
TDS	3,050	5,500[e]	5,798	—	—	950	241,200	—
pH	8.6	7.8	9.0	7.6	8.2	8.5	—	7.8
T, °C	19.5	13	15.7	30	17	19	—	21
Spec. Cond.	5,345	6,200	5,360	1,410	1,400	1,700	—	1,585

[a] Averaging about 50-m depth in the alluvial sands near the playa.
[b] Van Denburgh and Glanay, 1970.
[c] Hanks, 1883; surface brine.
[d] Data averaged with analysis of Jones, 1966.
[e] Hicks, 1916. Water depth 4.6–20 m.

extending from the surface to 38–61 cm (15–24 in.; Ayres, 1882, p. 355). Table 4.8 lists the analyses of water wells near the playa.

After the initial mining period, by 1914 a new <15-cm (6-in.) surface crust had formed in some areas, but with a considerably reduced borate content (Gale, 1914; Hicks, 1916). Bowser and Dickson (1966) described the surface as covered with "phreatophyte (formed from groundwater) mounds" with minor efflorescent crusts between them. Ulexite was in a 2- to 3-cm (thicker under the mounds) zone 10 cm below the surface throughout the area, occurring in clayey–silty coarse sands. There was a distinct plane of demarcation between the ulexite zone and the much finer sediments below it.

4.6.3 Death Valley and Amargosa Valley, California

4.6.3.1 Eagle Borax Works

Borates were first noted in Death Valley in 1873, and they were widely publicized. Many mining claims were located, but the first borax production did not start until 1882 when the Eagle Borax Company began operations. It was located a few miles north of Bennett Wells in the southern, lowest part of the valley, 32 km southwest of Furnace Creek at an elevation of −130 m (−427 ft). The deposit was discovered by Isidore Daunet in an accidental manner as he and prospector friends tried to cross the valley. They had worked at the Panamint mine in the 3367 m-high range to the west of Death Valley, but became discouraged and decided to try their luck in Arizona. Being young and vigorous they embarked on a direct route over the mountains and into Death Valley, even though it was midsummer. They greatly misjudged their limited water supply, and finding no springs, ultimately had to kill their pack animals to drink their blood. Daunet and a friend left to seek help, and luckily found some Indians who provided water. They returned to rescue the two survivors (of the five) left in camp. His friends later moved on, but Daunet thought the white salt crusts they had crossed might have been borates. He persuaded three others to join in prospecting the area, and they located a 109-hectare (270-acre) claim.

Ulexite was common on the playa, along with halite, thenardite, and trona (Bailey, 1902). The men were fortunate in digging a well that intercepted a freshwater artesian strata at the edge of the playa, and they had plant equipment hauled in. It included a cylindrical leaching tank 1.5 m (5 ft) in diameter, 6.6 m (22 ft) long, and 0.9 m (3 ft) deep, and twelve 3800-liter (1000-gal) 16-gauge galvanized iron circular crystallizing vats that were wider at the bottom than the top. They purchased soda ash, and hired 50 men to harvest cotton-ball ulexite and gather mesquite to be burned under the leach tank. When the operation started, however, they found there was not sufficient cooling in the August evenings to crystallize borax, but by using evaporative cooling they finally recovered 34 metric tons (37 tons) of an impure product. It sold for only 8¢/lb at the distant Calico station, and by the time they produced 118 metric tons (130 tons) of higher-quality borax (it sold for 10¢/lb), their finances were exhausted and the plant (and playa) was never operated again (Hanks, 1883).

4.6.3.2 Harmony Borax Works

The establishment of this small operation also had an unusual origin. Aaron Winters was a poor homesteader living with his frail wife in a cave and one-room stone house with a tule-reed roof in nearby Ash Meadows. They lived much like the area's Piute Indians, eating mesquite beans, lizards, and chuck-walla's when their other rations ran short. They had a fine spring running off the rock cliff next to the house, which formed a pool for a number of ducks, and water for a few chickens, a pig, and a large dog. They shared this water with wandering prospectors, and in 1880 one mentioned the simple test for borax. Winters immediately sent for some alcohol and acid, and when it arrived he and his wife traveled to Death Valley and gathered cotton-ball and surface crusts from the Furnace Creek playa. When they ground the samples and added acid and alcohol, all of them burned with a green flame, the sure test for borax. Winters filed for mining claims and later sold them to the William T. Coleman & Company, wholesale merchants, for $20,000, a fortune for him and his wife. In 1882 he and others also filed on several playas in the Amargosa Valley, which again were sold to Coleman.

Coleman organized the Harmony Borax Mining Company, and in 1883 built a plant (Fig. 4.5) northwest of Furnace Creek, 35 km north of the Eagle operation. A wooden frame building was constructed for the 30–40 Chinese and Shoshone Indian workers (who earned $1.25/day for a 70-hr workweek), and an adobe structure 6.1-m long and 3-m wide (20 × 10 ft) was built for the plant's 5.2-m-long (9-cm-tubes) boiler. Water was piped in from springs several miles across the valley and stored in tanks on a hill above the plant. Process water was obtained from wells at the edge of the playa. Surface crusts

Figure 4.5 The Harmony Borax Works in Death Valley, and their 20-mule team. (From Ver Planck, 1956; used with permission of the Department of Conservation, Division of Mines and Geology.)

Figure 4.6 Collecting borax at Death Valley. (From Robottom, 1893.)

and 0.1- to 30-cm (0.25- to 12-in.) cotton-ball ulexite were harvested, hauled to the plant in wagons (Fig. 4.6), and dumped with soda ash into boiling water in one of three steam-heated tanks. The slurry was stirred, and when the density reached the desired level, the mesquite fire under the boiler was put out, the steam turned off, and the slurry allowed to settle. When clear, it was withdrawn and sent to some of the 57 6800-liter crystallizing tanks in which it cooled for up to 10 days. Borax crystallized on the vessel walls and many suspended iron rods. The end liquor was then drained and partially recycled, with the rest discarded. The borax was knocked off the walls and rods, removed from the vats, washed, drained, dried, and shoveled into burlap bags for shipment. The ulexite residue was re-leached before being discarded. The plant closed for 2–3 months during the summer because of the heat, and the product was shipped by 20-mule-team wagons the 266 km to Mojave (Fig. 4.7). Supplies were brought back on the return trip. This operation also was unprofitable, and in 1889 the claims and plant were sold in bankruptcy to the Pacific Coast Borax Company, who shut them down (Chalfant, 1936; Travis and Cocks, 1984; Woo, 1995).

Figure 4.7 The 20-mule team used to haul borax from Death Valley to Mojave, 1883–1890. (From Matthews, 1954.)

4.6.3.3 Resting Springs (Amargosa Valley)

The Coleman operations were handled by a central organization, the Amargosa Mining Company, which controlled about 800 hectares (2000 acres) of Amargosa and Death Valley's best borate playas. In 1884 this operation established the Meridian Borax Company to produce borax from ~100-hectare claims in the Amargosa Valley, 80 km (50 mi) to the southeast of Harmony. The playa was 29 km (18 mi) long, 5–6 km (3–4 mi) wide and had summer temperatures of only 43°C (110°F) compared with Death Valley's >54°C (130°F). The borates were most concentrated in the southern end of the playa, 9.6 km from Resting Springs, a well-known stopping place with its water, shade and fruit trees, alfalfa fields, and gardens. There was some borax (and trona) in the ore, and the ulexite was of fairly high grade, varying from "pinhead size to balls 10–12 in. (25–30 cm) in diameter," which could be gathered easily in "chunks and balls." It was loaded into wagons, and hauled to the comparatively small plant where it was reacted with soda ash, settled, then cooled in eight 10,600-liter (2800-gal) crystallizing tanks with many suspended wires and rods. Its "vats and works were of a primitive design," but had the advantage of not needing to close in the summer. It operated from 1882 to 1887, with some production in 1889 when it was sold to the Pacific Coast Borax Company and closed (Bailey, 1902; Chalfant, 1936; Hanks, 1883).

4.6.3.4 Other Death Valley Playa Deposits

The Greenland Salt and Borax Company also briefly produced borax on a small scale from one of the Death Valley playas. The Saratoga playa is typical of many others in the general Death Valley area that contained borates but were too small or had ore too low in grade to be worked commercially. It occupied the flats around Saratoga Springs at the base of the Funeral Range in the southern end of Death Valley. The crusts were 12- to 91-cm (1- to 3-ft) thick and contained 7–40% borax, 10–60% Na_2SO_4, 0–5% Na_2CO_3, 8–25% NaCl, and 10–50% insolubles, with traces of I and MgO. The richest deposits were in old 0.19-m^2 (2-ft^2) to 1-hectare (several-acre) "pools or basins," and along shallow river channels. The Saratoga springs issued with considerable volume from lava rocks, and its warm water (with some boron) formed a 0.4-hectare (1-acre) lake. Among the other borate-containing playas were Salt Springs, 100 hectares (several hundred acres), on the south fork of the Amargosa River, Coyote Holes, Owl Springs, Resting Springs, Tecopah, Bennett's Wells, Upper and Lower Canon Beds, Confidence, and crusts on the slopes of Round Mountain and Monte Blanco. The Pilot Beds were at the southern end of the Slate Range, southeast of Searles Lake (see Fig. 2.13; see Table 4.8; Bailey, 1902).

4.6.3.5 Twenty-Mule Teams

In the early days of the borax industry in the United States essentially all of the product was hauled from the source to the nearest railroad station in mule-drawn wagons (see Figs. 4.5, 4.7, and 4.8). Each unit consisted of 3 or 4 wagons, usually two for the ore, one for water, and the other (if there was a fourth) carrying food for the driver, his assistant, and the mules, or in the later days bringing oil for the plant's boiler. The back haul always brought food, supplies, and water because all of the operations were in comparatively remote areas, and many did not have a drinking water supply. The wagon train was drawn by 14–24 mules (sometimes the lead or last pair were horses), depending on the size of the wagons and the difficulty of the haul.

An eastern writer visiting Death Valley was intrigued by the mule trains and gave them a great deal of publicity. Later, the Pacific Coast Borax Company used them as the theme for a popular television series with Ronald Reagan. The trip from Death Valley to Mojave was 261 km (162 mi) each way. The haul was first made by contract with a local farmer (Charles Bennett from Bennett Wells), but then the Coleman Company decided that it could manage the costly trip more efficiently. The company superintendent, J. S.

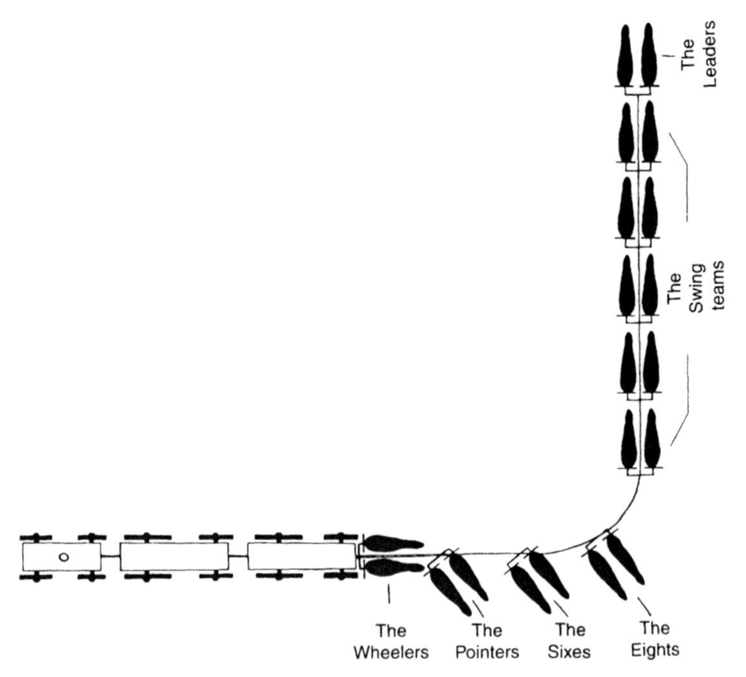

Figure 4.8 Schematic tactics of the mules in a 20-mule team in going around a curve. (From Matthews, 1954.)

Perry, followed the route of the early immigrants and improved the roadway by making a solid crossing over the playas, filling low areas and cutting down pinnacles. There were only three springs on the route, and because the teams could travel only 26–29 km/day (16–18 mi/day), many "dry camps" had to be made. There were 10 stations along the way for the 10-day trip, and special wagons with 4500-liter (1200-gal) tanks were built to supply water for the dry camps. Ten very sturdy wagons were then built, each 4.9 m (16 ft) long, 1.2 m (4 ft) wide, and 1.8 m (6 ft) high. The rear wheels were 2.1 m (7 ft) in diameter and the front wheels 1.5 m (5 ft), both having 2.5-cm-thick, 20-cm- (8-in.-)-wide steel "tires." The hubs were 46 cm (18 in.) in diameter, and the total wagon weighed 3.5 metric tons (7800 lb). It could carry 20.9 metric tons (46,000 lb) of borax or supplies.

Each wagon had a tapered 20-cm (8-in.) by 13-cm (5-in.) straight-grained oak tongue to which (on the front wagon) was attached the 1.3-cm ($\frac{1}{2}$-in.) steel chain leading to the mules. They pulled, two abreast, against a bar fastened to the chain and stretched out 37 m (120 ft) ahead of the driver. Each train consisted of two borax, one provision, and one 5700-liter (1200-gal) water-tank wagon, and there were two extra mules to fill in for casualties of heat [up to 65.6°C (150°F)], rattlesnake bites, and injury. The crew was a "skinner" (driver) and a "swamper" (helper) who became very skilled in handling the mules, and followed a strict time schedule. The driver had a whip with a 1.8-m (6-ft) handle and a 6.7-m (22-ft) lash, as well as a "box of rocks" to throw at mules beyond the whip's reach. The skinner used a 6-mm ($\frac{1}{4}$-in.) rope around the neck of the lead mule and his voice to pass on instructions: A pull meant a right turn, a jerk or slap meant a turn to the left. The mules were selected for strength and intelligence, and during the 5-year operation there were no breakdowns or lost mules (Chalfant, 1936).

Sections of the 20-mule teams were classified according to their specific work, and mules were selected and trained accordingly. The "captain" ("nigh leader") of the team, the smartest and most willing animal, was positioned first and to the left. The "pointers," "sixes," and "eights," used to get the team around a curve, were trained to jump over the chain either to the right or left (depending on the direction of the turn) at the command of the driver, then to pull hard at various angles to the direction taken by the wagons, and to step along "sideways." This prevented the wagons from going over a cliff or into a bank on a turn (see Fig. 4.8). The last two mules, the "wheelers," were usually the strongest and largest mules. The swing teams (10 mules) did not require as much training because they were just workers. However, even these mules had to be taught their names and the meaning of commands. The driver rode the first mule on the left, and from this position operated the front wagon brake (Matthews, 1954). The 20-mule teams stopped hauling borate out of the Death Valley in 1907 when the Tonopah and Tidewater Railroad connected to Death Valley Junction, and a narrow-gauge railroad

Figure 4.9 A narrow-gauge ore train in Death Valley. (From Matthews, 1954.)

linked the Lila C (and later Ryan) mine to the main line (Fig. 4.9). Other areas used their mule teams for a short period longer, but competition with colemanite ore and motor vehicles soon phased them out.

4.6.4 Rhodes Marsh, Nevada

This small playa (see Fig. 2.8) provides a classic example of how different portions of a dry lake's drainage basin deposits fairly pure salts in segregated areas (Fig. 4.10). Salt, ulexite, borax, glauber salt, thenardite, and trona all occurred in separate, limited, but mineable quantities in this playa, and all but trona were selectively recovered. Salt was the first mineral mined, with operations beginning in 1862 by Rhodes and Wasson. It was used as a reagent for the local gold and silver mines, being hauled to the mining camps (Virginia City, etc.) by camels. However, the camels frightened and stampeded the

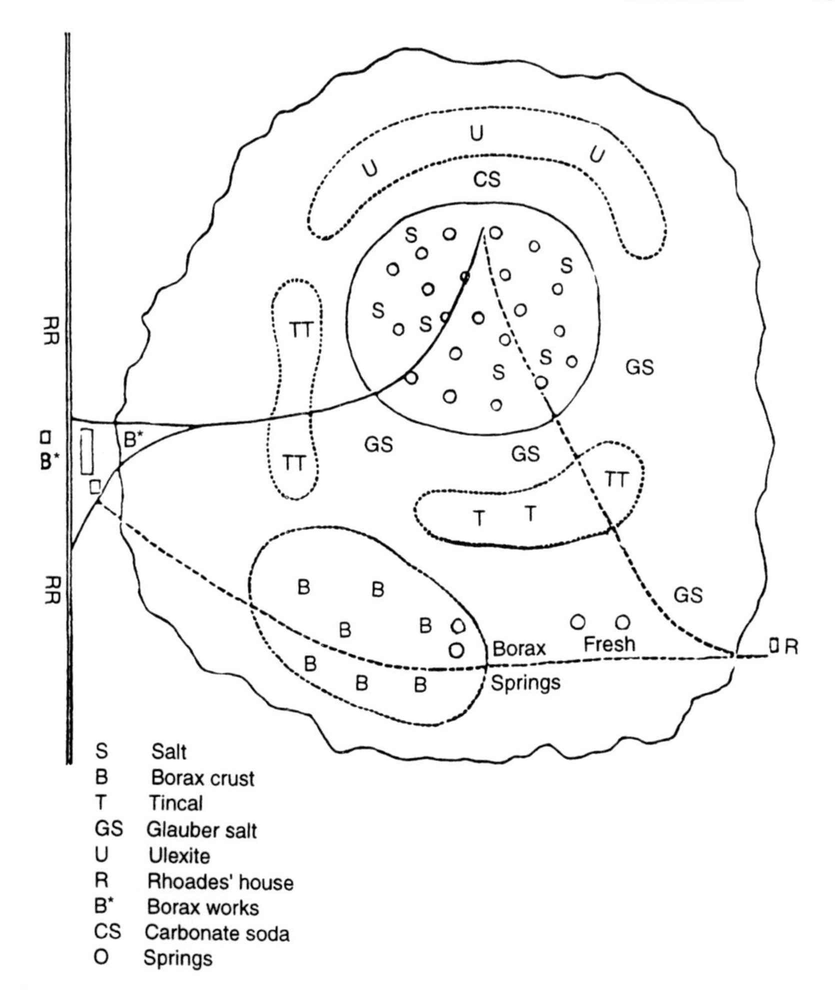

Figure 4.10 LeConte's version of the mineral distribution on Rhodes Marsh. (From Hanks, 1883; used with permission of the California Department of Conservation, Division of Mines and Geology.)

towns' horses, so a law was passed to limit their delivery hours. In 1874 a plant was built to produce 1 ton/day of refined borax, and from 1882 to 1900 the Nevada Salt and Borax Company marketed both salt and borax. In 1930–1933 glauber salt and thenardite were mined.

Rhodes Marsh is 4 km (2.5 mi) long and 2.4 km (1.5 mi) wide (4–5 km in diameter; Hanks, 1883), having an area of 8 km^2 and a drainage basin of 520 km^2. The Carson and Colorado (Tonopah and Goldfield) Railroad ran along its western edge, and a spur track was laid to the ulexite-rich area. The marsh is on the western side, in the lowest part of the 19-km (diameter) Soda Springs Valley at an altitude of 1340 m (4400 ft). The playa was generally dry, but could flood briefly to ~5 cm (a few inches) depth, and the water table was never more than 0–1.5 m (0–5 ft) below the surface. The surface was smooth and hard for a period after the rainy season, then rose ~5 cm (several inches) as it dried to form a loose, porous, pulverulent NaCl-containing crust. A ulexite zone was along the north and west margins (north, northwest, and northeast; Hanks, 1883; a 600-acre belt from central-west, north, and northeast). About 30 cm (1 ft) under the brownish salty crust, ulexite nodules were disseminated through 1.8–2.4 m (6–8 ft) of a wet, porous clay. They often occurred in thick beds that were "acres (hectares) in extent." The 0.6- to 10-cm (0.25- to 4-in.) nodules had a wavy white fibrous texture with a silky or pearly luster. Ulexite also was found in some other areas of the playa, including deposits under the massive salt bed.

The "richest borax zone on the marsh" covered 190 hectares (480 acres) and was 2.4 km (1.5 mi) long, adjacent to, and west, southwest, and southeast (southwest; Hanks, 1883) of the central salt mass. It contained 0.9–1.2 m (3–4 ft) of a stiff gray mud full of 0.3- to 5-cm- (0.125- to 2-in.-) long, <2.5 cm- (1-in.-) diameter borax crystals. Other descriptions were 1.3- to 2.5-cm (0.5- to 1-in.) crystals in a near-surface blue clay matrix south of the central salt mass. Borax also occurred in an 800-hectare (2000-acre) area east, south, and west of the crystalline borax (in the southern shore area) as a 2.5- to 7.6-cm- (1- to 3-in.-) thick 50% borax crust [Table 4.9; a 5- to 15-cm (2- to 6-in.) crust of borax, trona, and clay, Hanks, 1883], and brine (see Table 4.8). The total

Table 4.9

Surface Borax Analyses at Rhodes Marsh, wt% (Promoter's Samples)[a]

	Borax area			Other areas	
	SE area, 16 ha (40 ac), 2.5–12.7 cm (1–5 in.) in depth	N of 1st sample, 40 ha (100 ac), 5–20 cm (2–8 in.) in depth	S-central area, more than 0.9–1.2 m (3–4 ft) of tincal	ENE, ulexite area	Center, massive salt area
Borax	40.06	57.20	36.5	—	—
Ulexite	1.16	5.80	—	15.05	10.0
Na_2SO_4	16.00	10.70	16.5	20.10	8.7
NaCl	8.07	9.00	15.0	15.90	—[b]
Na_2CO_3	5.00	—	6.8	8.07	19.6
Insolubles, etc.	29.71	17.30	25.2	40.88[1]	61.7[b,c]

[a] Hanks, 1883.
[b] Included with the ulexite.
[c] Some NaCl included with insolubles, etc.

potential borate area was claimed to be 1620 hectares (4000 acres), of which 955 hectares (2360 acres) were the most valuable.

The lowest portion of the marsh was its 2.6-km^2 (1-mi^2) center, with its slightly undulating, thick hard crust of salt, borax, and ulexite. Under the crust was 2.4–3.6 m (8–12 ft) of pure, porous halite, and in the long 1.5-m- (5-ft-) deep, 3.6-m- (12-ft-) wide mining trenches, a clear, sparkling brine soon collected. It evaporated rapidly [i.e., 15–20 cm (6–8 in.)/10 days in the summer], and crystallized into a pure salt, which was washed, dried, ground, and sold as table salt. Trona was found as a surface crust 5–7.6 cm (2–3 in.) thick on the west (southeast; Hanks, 1883) side of the playa. Thenardite (Na_2SO_4) was found at depths of 0.6–1.8 m (2–6 ft) in lenses 0.9–1.8 m (3–6 ft) thick (~3 million tons) surrounding the salt but principally to the north and east. The thenardite was sometimes under the borax crystal zone. Glauber salt ($Na_2SO_4 \cdot 10H_2O$) occurred to the northeast immediately adjacent to the central salt mass at ~1 m (several feet) below the surface. It was often under the thenardite, but at least one 4.6-m- (15-ft-) thick near-surface lens was reported. The latter two salts were somewhat different from the others, being more massive and pure, and resembling ice.

The surface borax crust was by far the easiest and cheapest borate to gather and process, so it was the first ore mined, and the fairly pure surface ulexite was processed next. Both borates were scraped into piles to drain and be taken to the plant. Later, most of the ore was <5-cm (few-inches) nodules recovered from trenches a few inches to occasionally 1.8–2.4 m (6–8 ft) below the surface. The nodules were hand picked from the mud by Indian and Chinese laborers digging the trenches, then piled, washed to eliminate mud (this could be done in the plant), and air-dried. The nodular ulexite was next crushed and dried in a rotary oil-burning dryer, bagged, and sold as a finished product. The surface borax salts, and later ulexite–borax (with soda ash added) were boiled in high-boron spring water in four large tanks 2.1 m (7 ft) in diameter and 2.4 m (8 ft) deep. After settling, the clear liquor was cooled in 24 small galvanized tanks. The borax crystallized 5 cm (2 in.) thick on the walls or galvanized sheets hanging in the tanks. The liquor in time was siphoned off and discarded. The crystals were removed, washed, drained, dried, and then bagged as crude borax for shipment to the nearest railroad, initially 210 km (130 mi) away, but later adjacent to the plant.

The residual mud from the first leach was mixed with freshwater and soda ash, releached, settled, clarified, and allowed to cool and crystallize. The mud was discarded and the end liquor recycled. A recrystallized borax product could also be made if desired. The plant had a warehouse, engine and boiler room, waste tanks, salt mill, and later a railroad station. The Nevada Salt and Borax Company produced more than 1 ton of borax a day, or 50 tons of ulexite plus borax a month, and harvested salt with a staff of 15–20. As the marsh's production tapered off, the combined borate output was 700 tons in

1890 and 550 tons in 1892. Mill water came from shallow artesian wells (see Table 4.8), and in one 137-m (450-ft) well (drilled later), there were three zones of artesian freshwater strata. It could yield a combined strata flow of 12.6 liters/sec (200 gal/min; Anon., 1883; Hanks, 1883; Papke, 1976; Vanderberg, 1937; Williams, 1883; Yale, 1892).

4.6.5 Minor Playas

4.6.5.1 Dixie Marsh, Nevada

This playa covers 124 km^2 (48 mi^2) at an altitude of 1025 m (3363 ft). Salt forms a 0.3- to 1.5-m (1- to 5-ft) crust in the central 23-km^2 (9-mi^2) area, and was hoed into piles and shipped from 1861 to 1868 without further processing (it contained 96.5% NaCl, ~2% Na_2SO_4, 1% Na_2CO_3, and 0.12% insolubles). Brine from a 0- to 30-m (0- to 98-ft) depth was analyzed as containing 26.9% NaCl, 4.3% Na_2SO_4, 3.1% Na_2CO_3, and 0.2–0.8% KCl. No details have been reported on the lake's borate content or mining operation other than that "10 cars of borax were produced from the northern end of the marsh ... in the early 1870s" (p. 10, Papke, 1976) and hauled to the railroad at Lovelocks. The borate was said to have been cotton-ball ulexite, despite the preceding analyses of a high-carbonate salt and brine (Hance, 1914; Papke, 1976).

4.6.5.2 Eagle (or Hot Spring) Marsh, Nevada

The American Borax Company tried to produce borates from the north end of this playa from 1871 to 1872. A small plant was built, but even though it was located near the Central Pacific Railroad, the operation was not successful. The borates occurred as a 0- to 2.5-cm (0- to 1-in.) low-grade, limited crust (Yale, 1892). The two hot springs, however, were quite unique, with "beautifully clear water, boiling hot. No person had ever been able to find a bottom to these springs" (Robottom, 1893).

4.6.5.3 Mud Lakes, Nevada

"Extensive beds of borax" were reported in these playas, but were never commercialized. They are in northwestern Nevada, isolated by deserts, and farther than 160 km (100 mi) from a railroad (Williams, 1883).

4.6.5.4 Sand Springs Marsh (Salt Wells Area), Nevada

William Troop first identified cotton-ball ulexite in Nevada at the Sand Springs Marsh in 1870. This 32-km^2 (12.5-mi^2), 1210-m- (3960-ft-) elevation marsh is at the southeastern end of the 18-km- (11-mi-) long, 91-km^2 (35-mi^2) Salt Wells Area playa. In 1870 the American Borax Company quickly built a plant capable of producing 1 ton of borax a day in six 7600-liter (2000-gal) leach tanks. A second company also built a smaller plant the same year, but because of low-grade ore (average 10% equivalent borax; occasionally up to

30%), neither were successful and closed within 3 years. American Borax produced only 23.5 tons of borax and mined 160 hectares (400 acres) of ulexite crusts and cotton balls. It was claimed that the crust renewed itself within 2 years after harvesting. From 1863 to the present, the playa has been a commercial salt producer from a deposit 21 m (70 ft) thick over a 2.6-km^2 (1-mi^2) area. Product was delivered by camels in the early days. There were also high-sodium carbonate and sodium sulfate areas (Hance, 1914; Papke, 1976; Yale, 1892).

4.7 MISCELLANEOUS SOURCES

Gale (1917) noted that the first reports on nodular ulexite, colemanite, and boracite were from the Gobi Desert in China. Low-grade borate reserves of 74,200 metric tons have been estimated for the Leh district of Jammu and Kashmir. Other occurrences have been noted in the Surendra district of Gujarat and Juipur, and the Nagaur district of Rajasthan (Lyday, 1994). Orris (1995) stated that small amounts of ulexite have been harvested from the Iranian playas at Ashin, Deh-e-Shotoran, and Tenkar. Each playa has a halite–clay surface. Grabau (1920) stated that ulexite (or bechilite?) could be observed being deposited at the Peruvian hot springs of Bonos del Toro in the Cordilleras of Coquimbo.

References

Alonso, R. N. (1986). *Occurrences, stratigraphic position and genesis of the borate deposits of the Puna Region of Argentina.* Doctoral dissertation, Univ. Nacional de Salta, Dept. of Nat. Sci., Salta, Argentina.

Alonso, R. N. (1988). Los boratos de salares en la Argentina. *Asociacion Agentine de Geologos Economistas, Buenos Aires* **6**(6), 11–22.

Alonso, R. N. (1992). *Geologia de la Monte Verde Mine (Colemanita, Inyoita), Salta, Argentina,* pp. 215–225, 4th Nat. Cong. Econ. Geol., 1st Latinamer. Cong. Econ. Geol., Cordoba, Argentina.

Alonso, R. N., and Gutierrez, R. (1984). Zonacion de ulexita in Los Salares de la Puna Argentina. *Assoc. Geol. Argentina, Revista* **39**(1–2), 52–57.

Alonso, R. N., Jordan, T. E., Tabbutt, K., and Vandervoort, D. S. (1991). Giant evaporate belts of the Neogene Central Andes. *Geology* **19,** 401–404.

Alonso, R. N., and Menegatti, N. (1990). La formacion blanca Lila (Pleistoceno) y sus depositos de boratos. *Deaino Primer Congreso Geologico Argentino, San Juan* **1,** 295–298.

Alonso, R. N., and Viramonte, J. G. (1990). Borate deposits in the Andes. In *Stratabound Ore Deposits in the Andes* (L. Fontbote, G. C. Amstutz, M. Cordozo, E. Cedillo, and J. Frutos, eds.), pp. 721–732, Springer-Verlag, Berlin.

Alonso, R. N., and Viramonte, J. G. (1993, October 10–15). *La Cuestion Genetica de los Boratos de la Puna,* Vol. 5, pp. 187–194, 12th Congreso Geologico Argentino, 2nd Congreso de Exploracionde Hydrocarburos, Mendoza.

Anon. (1883). Borax. *U.S. Geol. Survey, Mineral Resources for the U.S. in 1882,* pp. 570–571.

Anon. (1909). The borate fields of Chile. *Chem. Trade J.* **45,** 380–382 (Oct. 23).

Anon. (1992a). Quiborax $10m Expansion. *Industrial Minerals,* No. 296, p. 18.

Anon. (1992b, January–February). Bolivia. *Phosphate and Potassium,* p. 11.

Ayers, W. O. (1882). Borax in America. *Pop. Sci. Monthly* **21,** 359–360.

Bailey, G. E. (1902). Borates. *The Saline Deposits of California, California State Mining Bureau, Bull. 24,* Pt. 2, pp. 33–90.

Battaglia, R. R., and Alonso, R. N. (1992, September). *Geologia y Mineria de Ulexita en el Grupo Minero Maggie Salar Centenario, Salta,* pp. 241–251, 4th Congreso Nacional del Geol. Econ., ler Cong. Latinamer. de Geol. Econ., Cordoba, Argentina.

Bowser, C. J., and Dickson, F. W. (1966). Chemical zonation of the borates of Kramer, California. *Second Symp. on Salt* **1,** 122–132.

Buttgenbach, H. (1901). Gisements de borate des Salinas Grandes de le Republique Argentine. *Anales Societe Geologique* **28,** 99–116.

Chalfant, W. A. (1936). *Death Valley, The Facts,* 3rd ed., pp. 120–127, Stanford University Press, Palo Alto, California.

Chamberlin, R. J. (1912). The physical setting of the Chilean borate deposits. *J. Geol.* **20,** 763–768.

Crozier, R. D. (1981, September). Chilean nitrate mining. *Mining Mag.,* pp. 160–173.

Dennis, W. B. (1902, April 26). A borax mine in southern Oregon. *Eng. Mining J.* **73,** 581.

Dupont, F. M. (1910, December). The borax industry. *J. Ind. & Eng. Chem.* **2,** 500–503.

Ericksen, G. E. (1993). Upper tertiary and quaternary continental saline deposits in the central Andean region. Geol. Assoc. Can., Spec. Paper 40. *Min. Deposit Modeling* (R. V. Kirkham, W. D. Sinclair, R. I. Thorpe, and J. M. Duke, eds.), pp. 89–102.

Ericksen, G. E., Vine, J. D., and Ballon, R. A. (1978). Chemical composition and distribution of lithium-rich brines in Salar de Uyuni and nearby salars in southwestern Bolivia. *Energy* **3,** 355–363.

Fonseca, E. (1874, February). Salitrera del toro i boratera de Maricunga (Copiapo). *Anales de la Universidad de Chile* **1,** 153–161.

Gale, H. S. (1914). Potash tests at Columbus Marsh, Nevada. *U.S. Geol. Surv. Bull. 540,* pp. 422–427.

Gale, H. S. (1917). Potash. *U.S. Bur. Mines Yearbook,* Pt. 2, pp. 397–481.

Garrett, D. E. (1985). Chemistry and origin of the Chilean nitrate deposits. *Sixth Int. Symp. on Salt* **1,** 285–301.

Grabau, A. W. (1920). *Geology of the Non-Metallic Mineral Deposits,* Vol. 1, pp. 337–339. McGraw-Hill, New York.

Hance, J. H. (1914). Potash in western saline deposits. *U.S. Geol. Surv. Bull. 540,* pp. 457–473.

Hanks, H. G. (1883). Report on the borax deposits of California and Nevada. *Calif. State Min. Bur. Third Ann. Rept.,* Pt. 2, pp. 1–102.

Helvaci, C., and Alonso, R. N. (1994). An occurrence of primary inyoite at Lagunita Playa, Northern Argentina. *Proc. 29th Int'l. Geol. Congr.,* Pt. A, pp. 229–308.

Hicks, W. B. (1916). The composition of muds from Columbus Marsh, Nevada. *U.S. Geol. Surv. Prof. Paper 95,* pp. 1–11.

Hicks, W. B. (1920). Potash. *U.S. Geol. Surv. Bull. 540,* pp. 407–417.

Jochamowitz, A. (1907, August 24). The borax deposit of Salinas, Near Arequipa. Peru. *Mining J.* **82,** 247.

Kistler, R. B., and Helvaci, C. (1992, 1994). Boron and borates. In *Industrial Minerals and Rocks* (D. D. Carr, ed.), pp. 171–186, Soc. Min. Met. Explor. Inc., Littleton, Colorado.

Kistler, R. B., and Smith, W. C. (1983). Boron and borates. In *Industrial Minerals and Rocks* (S. J. Lefond, ed.), 5th ed., pp. 533–560, Soc. Min. Eng., AIME.

Kunasz, I. A. (1980). Lithium in brines. *Fifth Salt Symp.* **1,** 115–117.

Leiser, T. (1996). *An Overview of Selected South American Borate Producers,* pp. 31–35. 12th Ind. Minerals Internat. Cong.

Libbey, F. W. (1960, October). Boron in Alvord Valley, Harney County, Oregon. *Oregon Dept. Geol. and Min. Ind.* **22**(10), 97–105.

Llambias, H. (1963). Sombre inclusiones halladas en cristales inderita, borax. *Asoc. Geol. Argent. Rev.* **18**(3–4), 129–138.

Luizaga, M. J. (1978). Evolucion de aguas dulces a salmueras en presencia boro y litio para la boratera de Rio Grande (Salar de Uyuni). *Revista Boliviano de Quimica* **1,** 89–100.

Lyday, P. A. (1992a). *Borate Mining and Manufacturing Expansion in Chile.* U.S. Dept. of Interior, Bureau of Mines.

Lyday, P. A. (1992b). History of boron production and processing. *Ind. Min.* (*London*) **303,** 19, 21, 23–25, 31, 33, 34, 37.

Lyday, P. A. (1977, 1984, 1986, 1990, 1991, 1994, 1995). *Boron.* Annual Report, U.S. Bur. Mines, Washington, D.C.

Matthews, C. W. (1954, September–October). *Conquerors of Death Valley: The 20 Mule Teams,* pp. 4–5, Deco Trefoil.

Michels, D. E. (1991, October). A heat/mass/solute balance model for Borax Lake, Oregon. *Geothermal Res. Council Trans.* **15,** 265–272.

Muessig, S. (1958a). First known occurrence of inyoite in a playa, at Laguna Salinas, Peru. *Am. Min.* **43,** 1144–1147.

Muessig, S. (1958b). Turilari, a borax crystal playa deposit in Argentina. *Geol. Soc. of America, Bull. 69,* pp. 1696–1697. [Abstract].

Muessig. S. (1966). Recent South American borate deposits. *Second Symps. on Salt* **1,** 151–159.

Nicolli, H. B. (1981). Geochemical composition of waters and brines of evaporitic basins of the Puna. *Anal. Acad. Nac. Cs. Fis. Nat. Buenos Aires* **33,** 171–190.

Nicolli, H. B., Suriano, J. M., Kimsa, J. F., and Brodtkorb, A. (1980). *Geochemical Characteristics of Brines in Evaporitic Basins, Argentine Puna.* 26th Int. Geol. Congress, Paris, S. 10 (Geochemistry) 1.01.0094 [Abstract].

Norman, J. C., and Santini, K. N. (1985). An overview of occurrences and origin of South American borate deposits with a description of the deposit at Laguna Salinas, Peru. In *Borates: Economic Geology and Production* (J. M. Barker and S. J. Lefond, eds.), pp. 53–69, Soc. of Mining Eng., AIMMPE, New York.

Orris, G. J. (1995). Borate deposits. *U.S. Geol. Survey Open-File Report 95-842.*

Papke, K. G. (1976). Evaporites and brines in Nevada playas. *Nevada Bur. Mines, Geology, Bull. 87,* pp. 18–24.

Reichert, F. (1909, June 12), The borate deposits of the Atacama Desert and Argentine Republic. *Mining J.* **85,** 730.

Rettig, S. L., Jones, B. F., and Risacher, F. (1980). Geochemical evolution of brines in the Salar de Uyuni, Bolivia. *Chem. Geol.* **30,** 57–79.

Risacher, F., and Eugster, H. P. (1979, April). Holocene pisoliths and encrustations associated with spring-fed surface pools, Pastos Grandes, Bolivia. *Sedimentology* **26**(2), 253–270.

Robottom, A. (1893). The tincal trade, pp. 30–33; Boracic acid, pp. 84–88; Crude borate of soda, pp. 143–146, 173–175; Boracite, pp. 159–163. *Travels in Search of New Trade Products,* Jarrold & Sons, London.

Salas, R. O. (1972). *Brief Report on a Geological Study of Salar de Surire, Arica,* Inst. Geol. Investigations, Chile.

Shaffer, L. L., and Baxter, R. P. (1975). Oregon borax: Twenty Mule Team Rose Valley History. *Oregon Hist. Q.* **73**(3), 228–244.

Travis, N. J., and Cocks, E. J. (1984). *The Tincal Trail-A History of Borax,* pp. 24–26, Harrap, London.

Van Denburgh, A. S., and Glancy, P. A. (1970). Water resources appraisal of Columbus Salt Marsh–Soda Springs Valley Area, Mineral and Esmeralda Counties, Nevada. *Nev. Dept. Conserv. and Natl. Resources, Water Resources Reconn. Ser. Report. 52,* pp. 2, 7, 34–37.

Vanderburg, W. O. (1937). Reconnaissance of mining districts in Mineral County, Nevada. *U.S. Bur. Mines Inf. Circ. 6941,* pp. 64–66, 77–78.

Vandervoort, D. S., Jordan, T. E., Zeitler, P. K., and Alonso, R. N. (1995). Chronology of internal drainage development and uplift, southern Puna plateau, Argentina Central Andes. *Geology* **23**(2), 145–148.

Ver Planck, W. E. (1956). History of borax production in the United States. *Calif. J. Mines Geol.* **52**(3), 273–291.

Vila, T. (1990). Salar deposits in Northern Chile. In *Stratabound Ore Deposits in the Andes,* (L. Fontbote, G. C. Amstutz, M. Cordozo, E. Cedillo, and J. Frutos, eds.), pp. 703–720, Springer-Verlag, Berlin.

Viramonte, J., Alonso, R. N., Gutierrez, R., and Arganaraz, R. (1984, September). Genesis del Litio en Salares de la Puna Argentina. *9th Cong. Geol. Argentino, Buenos Aires* **3,** 471–481.

Williams, A. (1883). Borax. *U.S. Geol. Survey, Mineral Resources for the U.S. in 1882,* pp. 568–569.

Woo, L. (1995). In harmony with history. *Borax Pioneer,* No. 5, pp. 16–18.

Yale, C. G. (1892). Borax. *U.S. Geol. Survey Mineral Resources for the U.S. for 1889 and 1890,* pp. 494–506.

Yang, S., and Zheng, X. (1985). The components of the saline lakes in Xizang, and an approach to their origin. *Chin. J. Oceanol. Limnol.* **3**(2), 251–264.

Chapter 5 Lake or Brine Deposits

5.1 ARGENTINA

5.1.1 Salar de Hombre Muerto

Many lakes, springs, geysers, or other brines contain commercial quantities of borates, but in 1996 only Searles Lake, California among these had a very large borate operation. Several others have (or had) smaller production, and additional small operations can be anticipated in the future because of their generally low production costs. The Salar de Hombre Muerto is such a deposit, which may produce boric acid as a by-product of its lithium recovery operation. It is a large brine-filled playa (with a perennial lake) in the Andes, southwest of Salta at an elevation of 4000 m (Fig. 5.1). It has a somewhat distorted rectangular shape, with several large peninsulas intruding into it (e.g., Lonja Negra, with its Tincalayu borax deposit), and the 400-km^2 Farallon Catal island in its center (Fig. 5.2).

The northern, southern, and eastern sides of the salar contain near-surface ulexite, and much of its surface has either a clean white, or dirty salt crust (Fig. 5.3). There is massive salt (>50 m thick) under the northwestern corner of the playa. The perennial Los Patos River enters from the southeast, forms a shallow lake (the Catal Lagoon; see Fig. 5.2), and supplies brine to the remainder of the salar, allowing it always to be <1 m (usually <20 cm) from the surface. Much of the eastern and some of the southern areas of the salar flood during the rainy season, but the western side has little drainage and only a few small springs. The area's average air temperature is 7°C, its minimum −6°C (June–July, coldest −32°C), and its maximum 13°C (December–January). The average rainfall is 60–80 mm/year (December–March). Near the Los Patos River entrance the Catal Lagoon contains abundant cyanophytas algae, a food source for large numbers of pink flamingoes. Wild burros and domesticated llamas graze on the bunch grass near the salar (Alonso, 1986).

Nicolli, Suriano, Mendez and Peral (1982) analyzed the brine in more than 100 0.2- to 1-m (mostly 0.7- to 0.9-m) holes (Fig. 5.4) in the Salar, and one 15-m hole at several depths. The latter had almost the same analyses for all of the 0.5-m intervals, but packers were not used to isolate the samples. The amount of clay (-5-μ particles) in the holes was 2–11%, and most of the

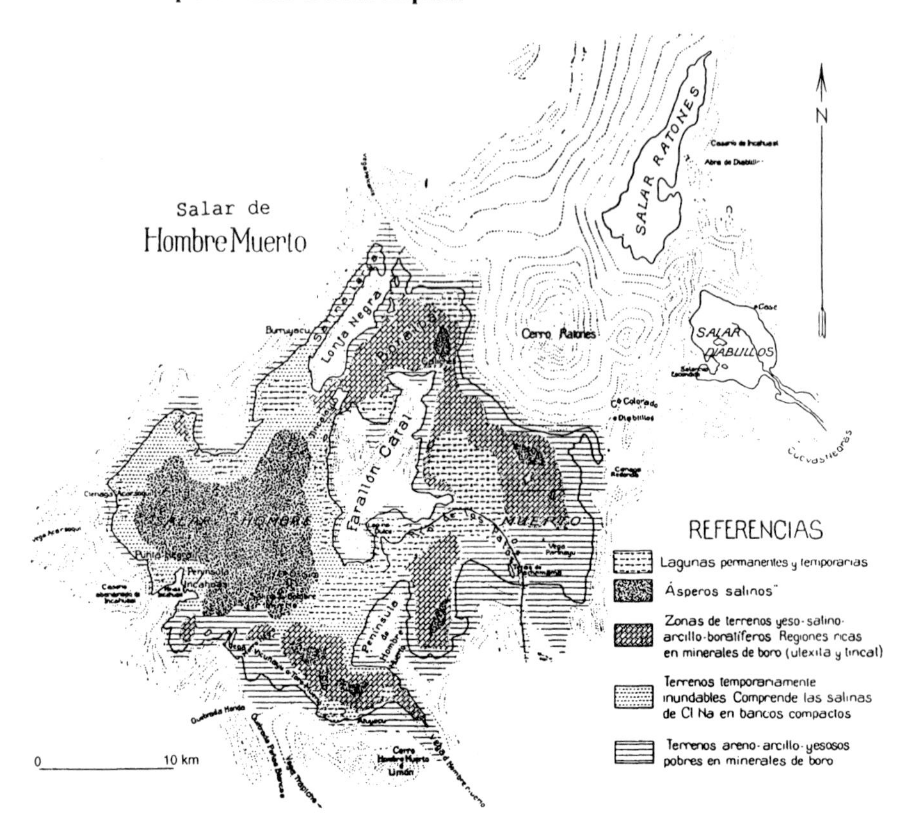

Figure 5.1 Borate location and surface structure of the Salar de Hombre Muerto. (From Catalano, 1964.)

sediments were quite porous, >170 mesh (88 μ) silt or salt. This allows free flow of brine to the entire area (565 km^2) of the salar, where both surface and subsurface (by capillary flow) evaporation can occur. Assuming an average 15-m brine depth with a 15% porosity, the brine contains 1.1 million tons of B_2O_3, 0.8 million tons of Li, and 80 million tons of K. Igarzabal and Poppi (1980) studied the brine concentrations in the the Catal Lagoon from near the entrance of the Los Patos River to the Farallon Catal (Table 5.1). The Na, Cl, Li, and B_2O_3 did not precipitate with evaporative concentration up to 20% total salts, but Ca, Mg, SO_4, and some of the potassium did. At salt saturation (~25% TDS, total dissolved solids), the NaCl, and later the other ions precipitated.

Catalano (1964) drilled 1439 0.2- to 1.2-m holes into the salar to determine its ulexite content. There was an average 7.6 cm of ulexite surface crusts in 10.8% of the salar's area, and an additional 21.5% of the area had an average

Figure 5.2 Catal Lagoon and the east side of the Salar de Hombre Muerto. The Farallon Catal Island is to the upper left.

Figure 5.3 Typical old, high-clay salt surface on the Salar de Hombre Muerto.

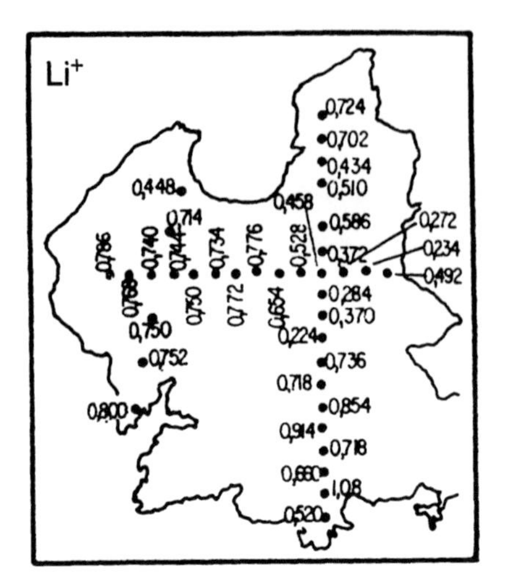
Li+
0,724
0,702
0,434
0,510
0,448
0,458
0,586
0,272
0,372
0,234
0,492
0,786
0,740
0,714
0,744
0,734
0,776
0,528
0,768
0,750
0,772
0,654
0,284
0,370
0,750
0,224
0,752
0,736
0,718
0,800
0,854
0,914
0,718
0,680
1,08
0,520

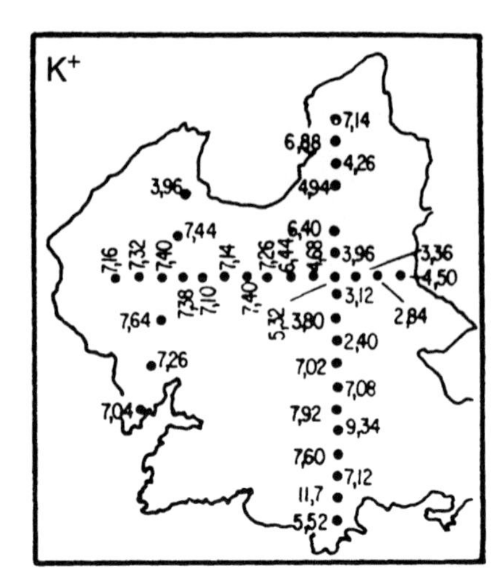
K+
7,14
6,88
4,26
3,96
4,94
7,44
6,40
7,16
7,32
7,40
7,14
7,26
6,44
4,68
3,96
3,36
4,50
7,38
7,10
7,40
3,12
7,64
5,32
3,80
2,84
2,40
7,26
7,02
7,08
7,04
7,92
9,34
7,60
7,12
11,7
5,52

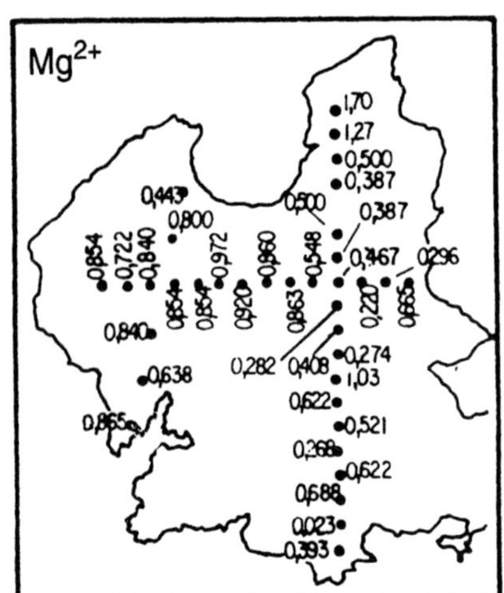
Mg2+
1,70
1,27
0,500
0,387
0,443
0,500
0,387
0,800
0,854
0,722
0,840
0,972
0,960
0,548
0,467
0,296
0,854
0,854
0,920
0,863
0,220
0,665
0,840
0,282
0,408
0,274
0,638
1,03
0,622
0,865
0,521
0,268
0,622
0,688
0,023
0,393

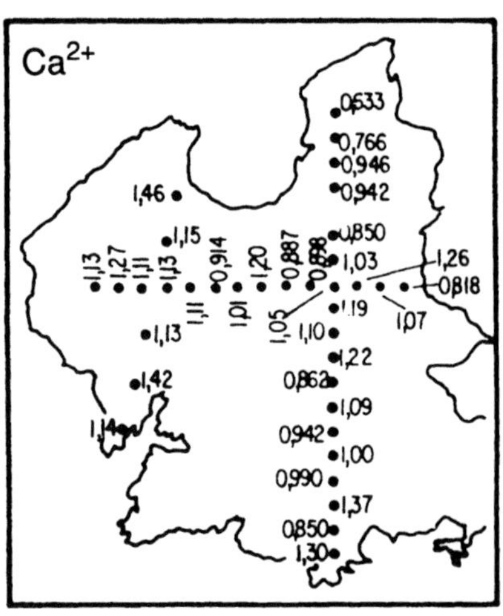
Ca2+
0,633
0,766
0,946
0,942
1,46
1,15
0,850
1,13
1,27
1,11
1,13
0,914
1,20
0,887
0,898
1,03
1,26
0,818
1,11
1,01
1,05
1,19
1,07
1,13
1,10
1,22
1,42
0,862
1,09
1,14
0,942
1,00
0,990
1,37
0,850
1,30

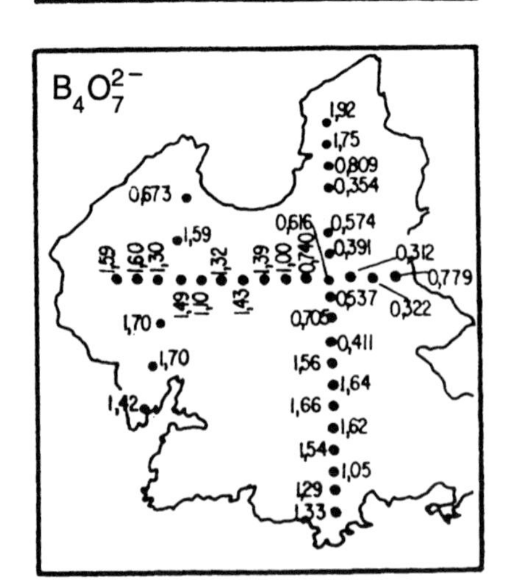
B4O7 2−
1,92
1,75
0,809
0,354
0,673
0,616
0,574
1,59
0,391
0,312
1,59
1,60
1,30
1,32
1,39
1,00
0,740
0,779
1,49
1,10
1,43
0,537
0,322
1,70
0,705
0,411
1,70
1,56
1,64
1,42
1,66
1,62
1,54
1,05
1,29
1,33

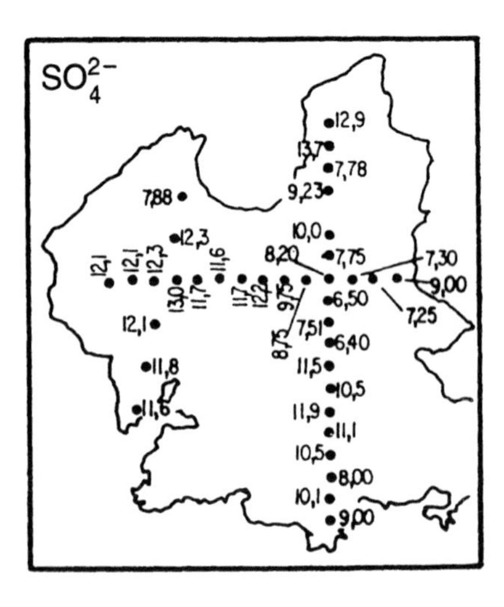
SO4 2−
12,9
13,7
7,78
7,88
9,23
10,0
12,3
12,1
12,1
12,3
11,6
8,20
7,75
7,30
13,0
11,7
11,7
12,2
9,75
6,50
9,00
7,25
12,1
8,75
7,51
6,40
11,8
11,5
10,5
11,6
11,9
11,1
10,5
8,00
10,1
9,00

of 12.7 cm of cotton ball ulexite, starting at a depth of 80 cm. The Salar's near-surface sediments were usually salt or sandy, porous, and fairly permeable. Many holes from which no ulexite was reported had flooded (and perhaps caved) at a shallow depth, and ulexite could be deeper than this survey's drilling depth, making the 7 million metric tons B_2O_3 estimate of ulexite conservative. The principal borate areas are shown in Fig. 5.1, and a later drilling program in an area where lava had intruded discovered some ulexite at a 10-m depth.

5.2 CHILE

5.2.1 Salar de Atacama

The Salar de Atacama in northern Chile is about 100 km east of Antofgasta in the intermediate zone of the coastal uplift (Fig. 5.5). It occupies 13,000 km^2, bounded by the Andes Mountains on the north and east, and smaller ranges on the other sides (Fig. 5.6). To the west is a ridge of Tertiary rock salt and gypsum, the Cordillera de la Sal, and on the east the rocks are largely volcanic. A recent volcanic cone (Cerro Miniques) rises to 5200 m in the south, and the perennial San Pedro River flows into the northern edge of the basin, forming partially flooded, shallow delta areas. A short-lived seasonal lake can form in the northeast corner. The high-lithium and -boron geothermal brines of the El Tatio geyser also flow into the basin (see Table 5.6).

The average surface elevation is 2300 m, and the playa floor covers 3,000 km^2. More than 40% of this area is covered with salt crusts, which form beautiful clear white-to-pink fresh crystalline polygonal (often hexagonal) patterns 1.5–2 m wide for the occasionally flooded areas. The ridges rise 5–80 cm, forming a magnificent mosaic, even while making travel very difficult. A complex brine (Table 5.2) is found within 0.6 m of the surface over most of the salar (Ide, Vergara-Edwards, and Pavlovic-Zuvic, 1983), and it extends to depths of >40 m in the areas tested. The salt typically has a porosity of 30, 20, 15, and 5% at depths of 0–0.5, 0.5–2, 2–25, and >25 m, respectively (another estimate: average porosity of 18% for 25 m and 8% for 25–40 m). Brine wells can pump at high rates (i.e., >31.5 l/sec, or >500 gal/min) with a small drawn down. Analyses over a 1860 km^2 area had a K concentration of 15–44 g/l, or >120 million tons of KCl, and 80 million tons of K_2SO_4. Lithium averaged 1500 ppm, or 4.5 million tons of Li (perhaps 45% of the total world

Figure 5.4 Sample locations and brine analyses of various ions at the Sarlar de Hombre Muerto (g/l). (From Nicolli, Suriano, Mendez, and Peral, 1982.)

Table 5.1

Brine Analyses at the Salar de Hombre Muerto (ppm or wt%)

	Na	K	Ca	Mg	Li, (ppm)	Cl	SO_4	B_2O_3 (B) (ppm)	Total solids	Density, (g/cc)	pH	Conductivity, (μmho/cm)
Initial brine in the Catal Lagoon (ppm)[a]	15	76	210	72	2.1	900	1100	25 (7.77)	3500	1.001	7.5	—
Catal Lagoon brine at NaCl saturation (wt%)[a]	9.45	0.55	0.02	0.16	930	15.8	1.06	1400 (435)	28	1.22	7.2	—
Average brine in the top 1 m of sediments (wt%)[b]	10.1	0.519	0.088	0.054	521[c]	16.0	0.846	750 (233)	27.8	1.204	6.9	1.74
Range (wt%)	9.9–10.3	0.24–0.97	0.068–0.121	0.018–0.141	190–900	15.8–16.8	0.53–1.14	260–1590 (87–535)	27.2–29.4	1.199–1.212	6.5–7.2	1.68–1.80

[a] Igarzabal and Poppi, 1980; 1000× B/TDS, ~2.
[b] Nicolli, Suriano, Mendez, and Peral, 1982, including the more dilute brine near the Catal Lagoon.
[c] As well as 29 ppm Rb and 33 ppm Cs.

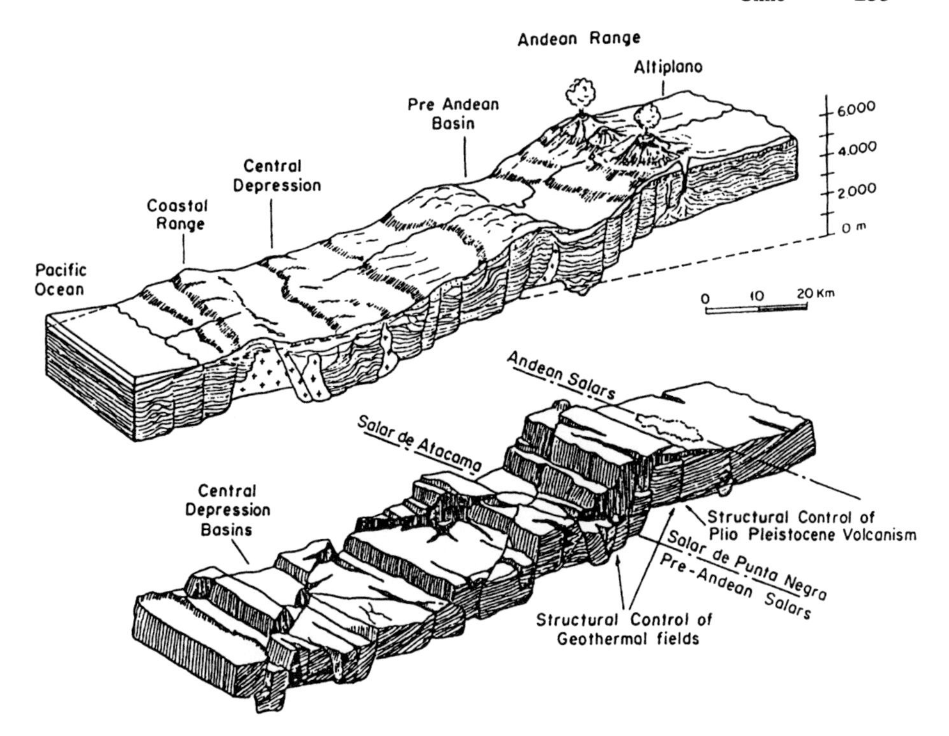

Figure 5.5 Structural relationship of the Andean Salars. (From Vila, 1990; reprinted by permission of Springer-Verlag Gmbh & Co. KG.)

reserves; Anon., 1984, 1985a, 1985b; Ferro, 1991). Ten million tons of B_2O_3 were indicated.

An example of the brine evaporation sequence is given in Table 5.2. In the southern area, typically 60% of the potash can be recovered in the sylvinite ponds, and an equal percentage of the sulfate (primarily as kainite, $KLiSO_4$, and $Li_2SO_4 \cdot H_2O$) in the sulfate ponds. The quantity of salts and the specific minerals vary with the season, but processing the average salts from each pond is quite simple. The sylvinite floats well at −6 mesh with high yields and purity. The lithium and borates dissolve in the flotation brine (or the schoenite end liquor) and may be returned to the solar ponds. The sulfate salts revert to schoenite in the K_2SO_4 end liquor, which can then be converted to potassium sulfate. By further solar evaporation of the schoenite end liquor all of the lithium and boron can be recovered easily by precipitation, salting out, extraction, or other means (Garrett and Laborde, 1983; Pavlovic, Parada, and Vergara, 1983). In 1996 Soquimich, the area's large nitrate company, announced plans to make 18,000 tons/year of boric acid as a by-product from its KCl and lithium operations (Anon., 1992, 1997; Donoso and Theune, 1990).

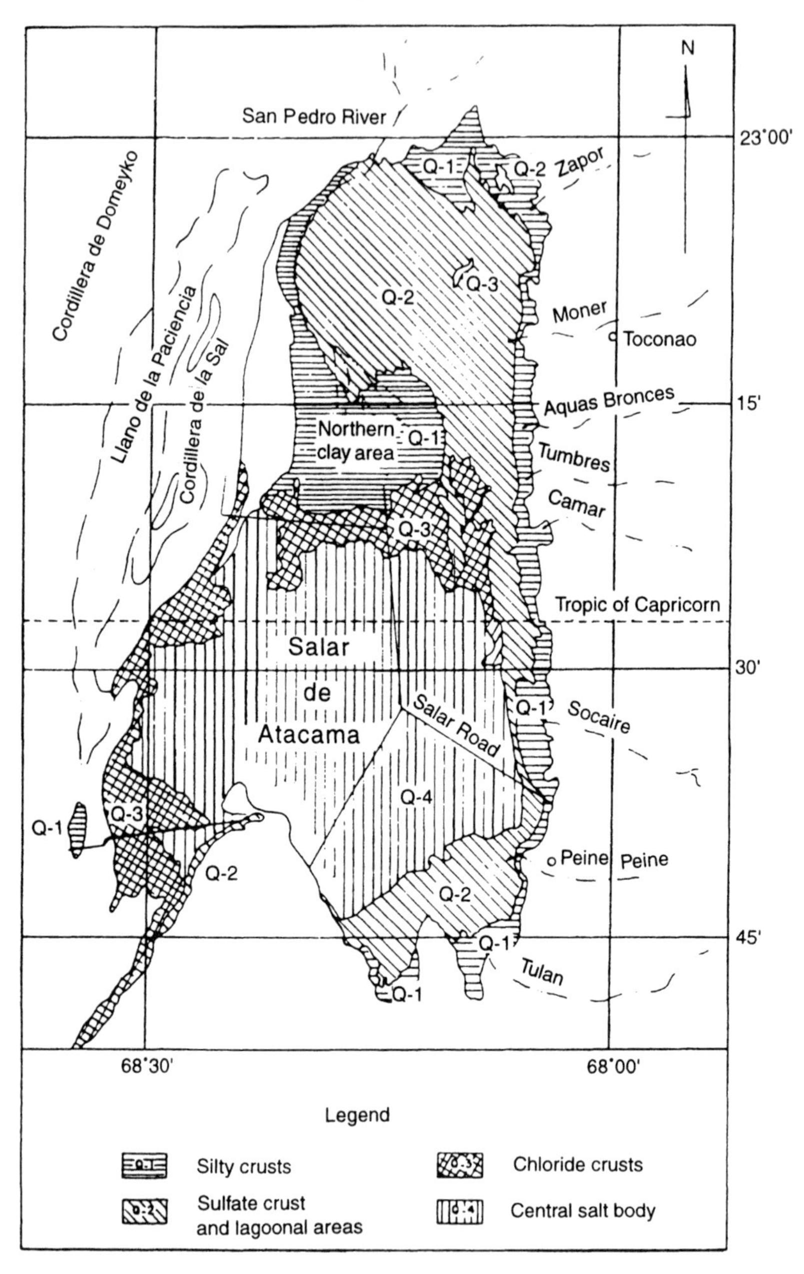

Figure 5.6 Geological map of Salar de Atacama. (From Ide, Vergara-Edwards, and Pavlovic-Zuvic, 1983; reprinted by permission of the Salt Institute.)

Table 5.2
Typical Brine Analysis in the Salar de Atacama Experimental Solar Ponds (g/l)[a]

	Original brine[b]	To sylvinite pond	To sulfate pond	From carnallite pond
Cl	192.0[c]	205	195	292
SO_4	23.3	45	88	23
H_3BO_3 (B)	4.4 (0.77)	9.2 (1.61)	18 (3.15)	50 (8.74)
Na	93.2[c]	72.0	40	4.0
Mg	12.3[c]	23.7	46	92
K	22.0[c]	46.8	37	4.0
Li	1.96[c]	3.66	7.07	8.9
H_2O	873	856	860	867
Density	1.227	1.258	1.284	1.323
1000 B/TDS	6.06	11.2	22.5	65.7

[a] Garrett, 1995.
[b] Also containing, as wt%: Ca 0.03, NO_3 0.012, CO_3 0.003, I trace.
[c] Vergara and Parada, 1983.

5.3 CHINA

5.3.1 Xiao Qaidam Lake

China reportedly has an equivalent borax production of 30,000 tons/year from the lakes in the Qinghai-Xizang (Tibet) plateau (including 2000 tons/year from Qinghai Lake; Lyday, 1987, 1992b). At least 57 lakes in China contain considerable amounts of boron, and many others some-to-moderate quantities. Typical of those with the most boron is Xiao Qaidam in the northern section of the Qaidam Basin (Fig. 5.7), which covers 40 km^2. At a depth of 0.26 m it has primarily pinnoite ($MgB_2O_4{\cdot}3H_2O$) and ulexite ($NaCaB_5O_9{\cdot}8H_2O$) in its southern border's subsurface muds, and ulexite and other borates in its surface crusts (Sun, 1990a; Sun, Ma, and Shan, 1991). The lake has boron reserves of 300,000 tons of B_2O_3 as solids, and 200,000 tons of B_2O_3 in its brine (analyses are listed in Table 5.3).

The upper 2.5–3.5 m of sediments in the lake are porous, and contain some gypsum and mirabilite. The next 1–2 m contain ulexite mixed with gypsum and lentiform mirabilite, and under it are 1–2 m of gray-to-black mud with gypsum crystals. Beneath that are 2–2.5 m of lamina and lentiform ulexite in black-to-gray mud with some aragonite and quartz sand cemented by calcite, and finally 1–2 m of a granular (sandy) carbonate layer with pinnoite. Halite and glauber salt have crystallized in the mud of the north, east, and west

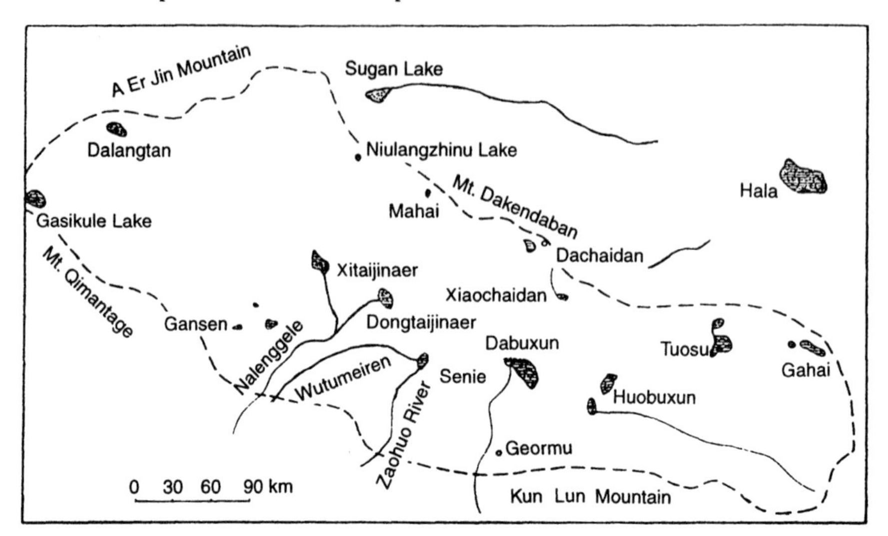

Figure 5.7 Distribution of saline lakes in Qaidam Basin, Qinghai. (From Qian and Xuan, 1983; reprinted by permission of the Salt Institute.)

edges of the lake, and glauber salt and gypsum are uniformly distributed through most of the upper sediments. Gypsum, some fine sand layers, gravel and clay beds are under the borate zones. The ulexite lenses are typically <30 m long and 2 m wide, they can be inclined or even vertical, and often have brine-feeder veins.

The pinnoite occurs in three forms: (a) *spherical* nodules separate or co-crystallized with aragonite, (b) *cumular massive,* tightly packed small crystals in quartz and feldspar sands, as small lenses or cement for the mass, and (c) *massive* beds, lenses, or veins with aragonite (occasionally dolomite; Sun, 1990b). Qain and Xuan (1983) also noted that the borates often appear to have roots, which are large or small fracture zones under the deposit, and filled with borates and brine. Age-dating tests with ^{14}C in the carbonates indicate that the near-surface muds' age is 10,000–11,700 years, ulexite 7000 years, pinnoite 4300 years, and sandstone above the ulexite 5500 years. Perhaps medium-strength borate brine first seeped through the older near-surface muds and met the slowly rising, relatively high-Ca groundwater, crystallizing ulexite in the cracks and parting plains as lentiform or vein-like deposits (instead of in the more common nodules). Much later, as a more concentrated $MgCl_2$ brine was formed by continued evaporation, it sank to a lower permeability zone to crystallize pinnoite.

The $\delta^{11}B$ values for the lake's brine and borates (see Tables 6.5 and 6.6) show a 1–1.2% preferential crystallization of ^{10}B as pinnoite (implying that it crystallized from only part of the borate-depositing brine). The ulexite and brine's $^{11}B/^{10}B$ ratios were about the same, indicating nearly complete boron crystallization. Sun (1990a) also let a sample of interstitial brine stay sealed at 16–26°C for 3 years, crystallizing 8% ulexite, 64% halite, and 28% aragonite. This implies that some ulexite could form even without a calcium supply from below. In other experiments, Sun and Li (1993) cooled different brines for 2 days–3 years, generally crystallizing inderite, but occasionally borax or a new form of boracite or inderite, along with natron, glauber salt, epsomite, and halite. More dilute brines in a cold area for 4 months–6 years also crystallized inderite. With a different brine at ambient temperatures for 3–9 years, ulexite, inderite, and kurnakovite were formed. The authors concluded that several different borate minerals can form from similar strong or dilute brines, when cooled or at ambient temperatures, when the residence times are long.

5.3.2 Da Qaidam (Dachaidan)

Da Qaidam Lake occupies the center in a line of three major borate lakes (Xiao Qaidam to the southeast and Mahi to the northwest) near the north shore of the Qaidam Basin, and just south of a rift zone on the Qillian mountains (see Fig. 5.7). Da Qaidam's brine has a small carbonate content (see Table 5.3), and the lake has ulexite and pinnoite lenses like those at Xiao Qaidam. However, in Da Qaidam there is more thermal and cool artesian spring activity, with more surface and subsurface borate mineralization on the eastern shore of the playa. Some of the springs form unique sediments in 2- to 200-m^2 zones, with frequent mounds typically composed of 0.5–1 cm of hungchaoite ($MgB_4O_7 \cdot 9H_2O$). Above it are 0.4–0.8 cm of mcallisterite ($Mg_2B_{12}O_{20} \cdot 15H_2O$), and below it is a gypsum–hydroboracite ($CaMgB_6O_{11} \cdot 6H_2O$) bed. These mounds appear to be formed by the reaction of lake brine with the more dilute high-Ca spring water during capillary evaporation. Ulexite also forms on the entire surface in this part of the lake, and kurnakovite ($Mg_2B_6O_{11} \cdot 15H_2O$), inderite ($Mg_2B_6O_{11} \cdot 15H_2O$), carboborite ($Ca_2Mg(CO_3)B_2O_4 \cdot 8H_2O$), and borax have been observed in small quantities (Qain and Xuan, 1983). Sun and Li (1993) noted that borax and glauber salt are the predominant salts crystallizing in the winter.

5.3.3 Nei Mongol Plateau

Many other lakes and playas in China contain lesser concentrations of boron, but major amounts in total, such as the soda ash lakes and playas of the Nei Mongol Plateau of Inner Mongolia (Fig. 5.8). The lakes generally are surrounded by sand dunes in dry and windy desert and semidesert country at fairly high altitudes with cold winters. The annual evaporation is 7 to 20 times the rain and snowfall, so the 2- to 100-km^2 playas are dry for most of the year, and may contain 5000–500,000 tons of B_2O_3. A typical brine analysis

Table 5.3

Analyses of Several Low-Carbonate Borate Lakes in China (wt% except as noted)

	Na	K	Mg	Ca	Li, (ppm)	Cl	SO_4	HCO_3	CO_3	B_2O_3	Br, (ppm)	Total salts	pH	Density (est.)	1000B/TDS
Da Qaidam															
Intercryst.[a]	5.63	0.44	2.02	0.02	310	13.42	3.41	0.06	0.02	0.20	58	25.05	—	1.234	2.48
Surface[a]	7.77	0.36	1.17	0.03	182	14.16	2.04	0.21	—	0.26	80	25.68	—	1.240	3.15
Surface[b]	5.66	0.89	3.69	0.32	—	14.52	3.78	—	—	0.69	—	27.87	7.11	1.254	7.69
Concen. pool[b]	tr.	0.11	8.92	0	—	22.22	2.79	—	—	0.22	—	36.90	5.04	1.344	1.85
Hot springs, (ppm)[b]	312	—	29	20	—	480	164	—	—	53	—	1,060	8.62	1.001	15.5
Spring, high carb. (ppm)[b]	630	—	77	51	—	644	341	315	176	179	—	2,410	8.34	1.002	23.1
Spring, low carb. (ppm)[b]	1,010	1,010	94	199	—	1,670	1,030	273	79	282	—	4,130	8.45	1.004	21.2
Lake springs, (ppm)[a]	1,400	48	822	707	—	1,320	4,620	384	684	2,180	—	12,160	—	1.011	55.7

Kiao Qaidam															
Intercryst.[a]	3.65	0.16	0.43	0.05	—	6.04	1.48	—	—	0.15	—	11.95	—	1.126	3.90
Mine pit[b]	3.75	0.16	0.44	0.05	—	6.20	1.52	—	—	0.15	—	12.28	8.30	1.111	3.80
Surface[a]	5.43	0.13	0.39	0.08	38	12.14	3.57	0.004	—	0.19	16.6	21.76	—	1.203	2.71
Surface[b]	7.99	0.15	0.36	0.08	—	10.4	4.38	—	—	0.11	—	23.53	7.85	1.186	1.45
River Tatalin (ppm)[b]	1,300	100	400	900	—	2,100	2,100	2,200	—	100	—	9,100	—	1.008	3.41
Zhacang Caka															
Concentrated[b]	2.60	2.26	4.84	0	—	7.45	2.35	—	—	0.193	—	14.23	5.5	1.133	4.21
Mahai															
Intercryst.[a]	8.08	0.16	0.96	0.07	51	10.84	2.33	—	—	434 ppm	—	22.38	—	1.208	6.03
Quinghai Lake[b,c]	3.93	0.16	0.79	0.01	0.84	5.79	2.35	0.68	0.52	15 ppm	1.5	14.23	—	1.133	0.33

[a] Qian and Xuan, 1983.
[b] Sun and Li, 1993; Sun, 1990a.
[c] Also (ppm): Si 0.93, P 0.50, Al 0.26, Cr 0.12, Ni 0.092, Fe 0.067, U 0.042, Sr 0.04, Ba 0.02, Cu 0.016, Mn 0.016, Ti 0.01, I 0.004, Zn 0.0021.

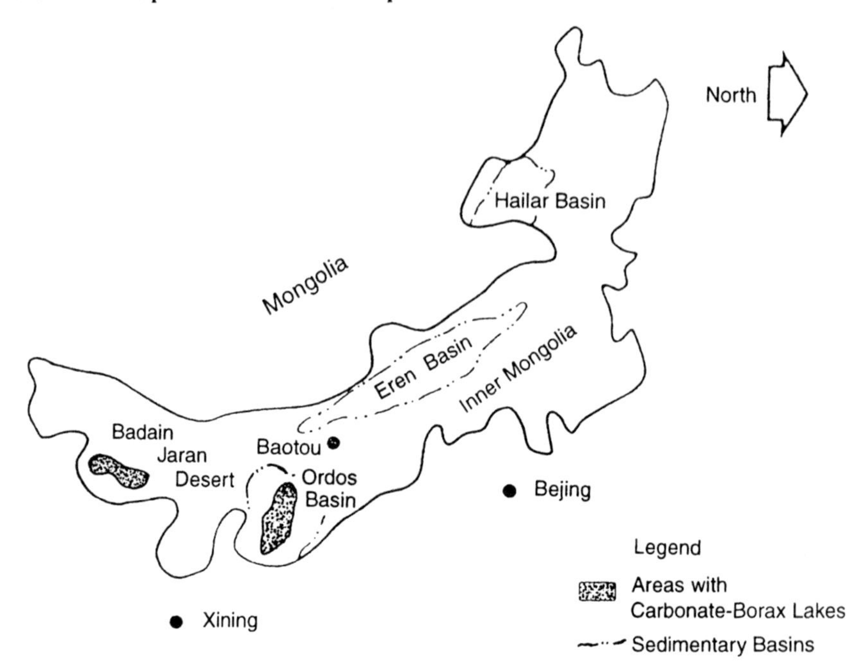

Figure 5.8 Map of the Nei Mongol Plateau, China. (From Sun, 1987.)

is noted in Table 5.3 (Quinghai). Two fair-sized natural soda ash plants at Chaganor and Ordus Lakes produced some borax as a by-product in 1996 (Zhang, 1997).

5.4 RUSSIA

5.4.1 Inder Lake and Other Sources

Some borax production has been reported from Inder Lake, Kazakhstan, a large salt-saturated lake north of the Caspian Sea and east of the Ural River. It receives drainage from the borate-containing Inder salt dome and has formed a porous, 36-m deep brine-filled salt mass. The brine contains potassium chloride, bromides, and 0.23% B_2O_3 (Kistler and Smith, 1983). Boric acid production has also been reported from the evaporation of underground waters with 0.045–0.202% boron in the Crimea and the Taman and Manayshlak peninsulas. It is likewise a by-product from salt cake and magnesium chloride operations on the Gulf of Kara Bogas on the east shore of the Caspian Sea (0.25% boric acid in the feed brine; Lyday, 1992a). The Sivash Lagoon also has been mentioned as a source of brine that produces several products, including borax (Lyday, 1991).

5.5 UNITED STATES

5.5.1 Hachinhama (Little Borax) Lake, California

When the processing plant at Borax Lake flooded in 1868, the equipment was moved 6.4 km (4 mi) to the <8-hectare (20-acre) Lake Hachinhama (Little Borax Lake). It was oval shaped, >400 m in diameter, 1 m deep and on the southwestern shore of Clear Lake in a shallow crater formed by a recent phreatic (caused by steam, mud, or gas; not lava) explosion. It had been fed by a boron-containing hot spring, but because it did not concentrate sufficiently in the summer, it had no borax crystals in its soft, plastic muds. Its brine had a higher B/TDS (born to total dissolved solids) than Borax Lake (see Table 2.5), and contained 75.4% Na_2CO_3, 8.3% NaCl, and 16.3% $Na_2B_4O_7$. The lake waters were first concentrated in solar ponds, and the evaporation completed in the plant. The resulting mixed salts were then leached with cold lake water, dissolving the NaCl and other salts. This left crude borax to be dissolved and recrystallized in lead-lined tanks, which resulted in a high purity product. The solar ponds were lined with brick and asphalt, and crusts from the lake were also harvested several times in the summer (they reformed in 1–3 weeks) and dissolved in the entering pond brine. The clarified hot slurry from the plant evaporator was cooled in 4000 small (2–3 gal) milk cans because it was thought that the borax supersaturation would be less in small vessels. With each winter's rains the waters slowly regained their salt content, and the process could be repeated. The plant had a modest production from 1869 to 1872, which supplied most of the U.S. borax demand.

However, even though the operation claimed an 80% borax recovery, it was an expensive process. Consequently, from 1872 to 1873 when ulexite became available from the Nevada playas, limited quantities were boiled with the concentrated solar pond brine, allowing the sodium carbonate to react with the ulexite. However, the reaction was very slow (the ulexite was not finely ground), requiring an excess of ulexite and prolonged boiling. The milky-looking $CaCO_3$–insolubles–ulexite slurry was then clarified and cooled to yield an impure product that still had to be leached and recystallized. The operation closed in 1873, and the lake has now been flushed clean (Ayers, 1882; Simoons, 1954; Ver Planck, 1956; Vonsen and Hanna, 1936).

5.5.2 Owens Lake, California

Owens Lake is the first in the chain of lakes on the eastern slopes of the Sierra Nevada mountains that overflowed during the recent Ice Ages, carrying most of their soluble salts to Searles Lake (see Fig. 2.19). The salts remaining in Owens Lake are about 3500 years old, the same age as the youngest salts in Searles Lake. The valley, 160 km long and 24 km wide, was formed by block-faulting of the Sierra Nevada Mountains to the west and the Inyo Range to the east. Present-day steam fumaroles at Hot Creek and Coso Hot Springs (where mercury has been produced) still flow into the Owens River, and were the lakes' primary source of boron, potassium, carbonates, and other geothermal salts.

The Owens Valley's elevation is 1081 m, and until 1917 the lake was a moderately deep, comparatively dilute alkaline body (Table 5.4) with an area of about 259 km^2 (100 mi^2). It supported steamboat traffic to bring ore from mines on its eastern side to the railroad at its southwestern corner. However, Los Angeles purchased the rights for the water feeding the Owens River, and when an Aqueduct was opened in 1917, the entire flow of the Owens River was diverted. The lake began to shrink in size, and by 1922 it was almost totally dry. It now covers 87.5 km^2, and in the summer contains a slushy mixture of 35 vol% brine and 65 vol% solids (70.4 wt%), averaging 0.8 m (<3 m) in thickness. Trona is the dominant salt, and there is also borax and some burkeite, mirabilite, natron, and thermonatrite.

In the winters considerable natron, glauber salt, and borax crystallize, but they "melt" in their own water of hydration and/or dissolve in the hot summer months. Trona (with NaCl, and burkeite at times) crystallizes during the

Table 5.4

Brine and Surface Solids Analyses of Owens Lake (wt%)[a]

	1912 analyses		Recent analyses		
			Brine		
	Brine	Total (million tons)	1985	1966[b]	Surface salts
Na_2CO_3	3.5	53.6	8.9	5.3	41.5
$NaHCO_3$	0.7	10.1	0.2	—	25.0
NaCl	4.4	64.5	18.0	16.6	2.0
Na_2SO_4	1.6	24.3	4.5	5.6	12.4
KCl	0.5	6.7	—	1.7	9.1
$Na_2B_4O_7$	0.3	4.1	—	0.56	9.4
Water	89.0	—	68.4	70.2	—
1000B/TDS	5.5	5.0	—	3.9	18.9
Density	1.0997	—	—	—	—

[a] Alderman, 1983.
[b] Jones, 1966; also TDS 31.0, SiO_2 653 ppm, B 1200 ppm, pH 10.3.

summer. The brine contains considerable dissolved organic matter ("humates"), and in the muds there is some sulfate-reducing bacterial action. There were several soda ash plants on the lake from 1915 to 1958, and some produced borax as a by-product. Each plant used the carbonation process on solar pond or naturally evaporated brine. After the sodium bicarbonate was removed lake brine was added to the end liquor (to convert the acidified pentaborate to borax), and the mixture cooled. This quite simply and selectively produced a reasonable yield of borax.

5.5.3 Searles Lake, California

The Searles Lake deposit has two massive brine (see Table 8.2) and mixed salt beds formed over the past 40,000 years. Their mineral source was a group of still-flowing hot springs (as noted above) in the Owens Valley area (Garrett and Carpenter, 1959). This 66-million-ton B_2O_3 deposit is unique among the world's large borax occurrences in that it contains both a high-boron brine and a mixture of borax with many other salts. It was the final basin that received waters from the Owens River (and its hot springs), flushed by Ice Age storms 190–270 km to the north (see Fig. 2.19). When filled 180 m (600 ft) above its present level, its stratified fresh water overflowed into the Panamint Valley, and then into Death Valley (Smith and Pratt, 1957).

During the first 7900 years that the hot spring water accumulated, only mud was deposited, but in the next 8500 years complex soluble salt beds, separated by comparatively thin mud seams, were crystallized to form the 12-m "Lower Structure," or Lower Salt body. Water next accumulated for another 13,500 years, laying down 4 m of "Parting Mud." Then followed 7000 years when the 15-m "Upper Structure" or Upper Salt was crystallized, which ended 3500 years ago.

The Upper and Lower Salts have a 35% void volume, which was originally filled by the lake's end liquor. This brine has been the raw material for the Lake's processing plants, and pumped since 1916. The original "mother liquor" has now been depleted, and a weaker equilibrium brine has taken its place, formed by the leaching of the Lake's salts (Smith, 1979).

It appears that the initial salts crystallized in the lake by both winter cooling and summer evaporation, and they compacted in a normal manner to a fairly low porosity as they were capped by halite and/or clay. Over the intervening years these salts reached the mean average ground temperature of the area (20–22°C) and slowly "melted" to redissolve the glauber salt (the NaCl-Na_2SO_4-$Na_2SO_4 \cdot 10H_2O$-H_2O transition temperature is 18°C), then natron and some of the borax. This transformation resulted in a high void volume as the other crystals provided a strong noncollapsing mass for these shallow beds. It also allowed convection currents to mix the solution, dissolving and recrystallizing the original salts into the more complex true equilibrium minerals such as burkeite, glasserite, hanksite, and many sparingly soluble carbonate, borate, and potassium compounds. This slow conversion is indicated by their often large and perfect crystal size and shape (e.g., frequent hanksite crystals 10–15 cm in diameter). Many of the minerals in the lake have been found only in a few other locations.

The Lower Salt (of middle Wisconsin age) is a series of alternating layers of mud and crystals about 1 m thick. Trona is the dominant mineral in beds 1–3 and 6–7; trona and burkeite in bed 4; and halite, trona, and burkeite in bed 5. Lesser amounts of northupite, thenardite, hanksite, and borax occur in all zones, and still smaller quantities of nahcolite, sulfohalite, and tychite are found locally. Most of the saline beds contain thin layers of mud, and some of these include a little gaylussite or pirssonite. The mud is composed of silt to clay-sized carbonates, silicates, and organics.

The 3.7- to 4.3-m (12- to 14-ft) Parting Mud separates the Uppėr and Lower Salts and contains crystals of gaylussite, pirssonite, and a little borax, along with its dark-green to black silicates and organic debris. In its upper one-third to two-thirds, thin laminae of white aragonite are numerous. Its outer limits (1000 km^2) are well beyond the salt areas, and there are scattered outcroppings around the basin.

The 114 km^2 Upper Salt is 15 m thick near its center and lens shaped. Its most abundant minerals are halite, trona, and hanksite, with smaller quantities of borax, burkeite, pirssonite, thenardite, glaserite, sulfohalite, and mud. Compared with the Lower Salt, it contains more halite and hanksite; about the same amount of borax; less thenardite, trona, and burkeite; and no significant mud layers. The brine is higher in K and Cl, and lower in CO_3 and B_4O_7 (Haines, 1959; Smith, 1979; Smith and Pratt, 1957).

5.5.4 Soda Lakes, Nevada (Big and Little Soda Lake, or Ragtown)

One of the earliest attempts at borax production in the United States was at Little Soda Lake (see Table 2.8). At an altitude of 911 m (2989 ft), this small, deep (45- to 52-m) lake had an area of 16 hectares (40 acres) and a diameter of 460 m (1500 ft). It was adjacent to the larger Big Soda Lake (Table 5.5), and its brine contained a small amount of borax (primarily NaCl, Na_2CO_3, etc.). The two lakes were separated by a ridge 366 m wide, and lay in an old volcanic crater with three sides rising abruptly 61 m (200 ft). Their brine source was alkaline groundwater penetrating the porous crater from the Carson River and Carson Sink. In 1867 a San Francisco financial group had brine pumped from Little Borax Lake into wooden troughs that carried it to ponds in a nearby small playa. They hoped to harvest borax, but after a short trial, the yield was small, and borax was only a minor component of the deposited salts. No further attempts were made to recover borax from these lakes, even as a by-product when soda ash production was initiated in 1868 (Garrett, 1992).

5.6 MISCELLANEOUS BRINE SOURCES

There are literally thousands of borate-containing lakes, brines, or waters throughout the world that might be considered sources of borax if it were a by-product or recovered by inexpensive methods. Random examples of such brines are listed in Table 5.5. Many of the brines are high in carbonate and/or bicarbonate, since boron and carbonates are often associated. Essentially all carbonate lakes contain at least some boron, but as seen in those of the West African Rift Valley (i.e., lakes Natron and Magadi), the amount may not be very large. An example of a noncarbonate brine is Lake Monte Rotondo near the Larderello fumaroles. This 7.3-hectare (18-acre) lake contained one part of boric acid to five hundred parts of water. For a brief period boric acid was recovered by evaporation, producing 64 tons in 1854 and 142 tons in 1855 (Hanks, 1883).

5.7 GEOTHERMAL BRINES

5.7.1 Salton Sea Geothermal Brine and Cerro Prieto, Mexico

The Salton Sea geothermal system extends from the Gulf of California to the Salton Sea, its hottest brines centered southeast of the Salton Sea and much farther south at Cerro Prieto, Mexico. The basin is an extension of the Gulf of California rift–fault system, having a 50-km^2 northern area and a somewhat smaller southern zone. In the north, 300°C brine has been found at 900 m, and 360°C brine at 2100 m (White, 1968). At Cerro Prieto, the power-production wells at 250–344°C are 780–1450 m deep (Dominguez and Vital, 1976). The Salton Sea trough is filled with 4300 m of recent deltaic sediments from the Colorado River as it empties into the Gulf of California, and the basement rocks are at 6,100 m. Various major faults of the San Andreas system extend from the north into the trough. The northern reservoir contains a concentrated brine with a very complex mineral content (Table 5.6), which is estimated to include more than 2 million tons of boron. At Cerro Prieto the brines are much more dilute (about one tenth the northern concentration), and yet they have the same temperature, probably because (as various isotopic studies have indicated) much of the brine has a local meteoric origin. White (1968) suggested that rainwater falling on the Chocolate Mountains several kilometers to the east seeped through fault zones to depths of 3000 m, became heated, mixed with some magmatic (geothermal) brine, and dissolved a small potash deposit. Thermal currents caused this mixture to spread through the basin, leaching rocks to obtain its present mineral content. Its variable dilution and mineralization depended on the groundwater and rocks that it encountered. The deuterium content of the geothermal brine, thermal springs in the

Table 5.5

Average Analyses of Miscellaneous U.S. Boron-Containing Brines (ppm)

	Paradox Basin, CO[a,b]	Mono Lake, CA[c,d]	Saline Valley, NE Corner CA[e]	Goose Lake, OR[f]	Deep Springs Lake, CA[g]	Summer Lake, OR[f]	Surprise Valley, Mid. Lake, CA[d]	Big Soda Lake, NV[c]	Green River Black Water, WY[b,h]
Na	25,200	21,600	67,600	697	99,700	4,010	2,740	8,610	83,600
K	26,700	1,160	3,760	33	17,550	191	8.2	39	4,070
Mg	30,900	36.3	26.6	4.2	1.1	0.4	11.7	195	22
Ca	43,500	4.4	43.4	14	3.5	2.2	14.8	7.9	18
Al	(66)	—	—	—	—	—	—	0.89	—
As	(20)	—	—	—	—	—	—	—	—
Mn	(260)	—	—	—	—	—	—	—	—
NH_4	(1,090)	0.9	—	—	—	—	—	—	—
Cl	201,100	14,300	93,200	127	110,500	2,090	2,480	7,570	60,600
CO_3	(800)	11,100	1,190	69	18,950	2,490	375	1,360	25,500
HCO_3	(882)	5,230	1,660	543	9,000	2,170	1,140	1,250	13,200
SO_4	(227)	7,740	18,800	61	47,350	714	530	6,220	1,780
B	1,690	350	315	8.0	408	34	30.3	51	40
SiO_2	(10)	14	15.5	68	3.5	139	35	3.3	236

Br	1,960	35	263	1	—	7.6	—	18	—
F	(25)	44	—	0.8	77	5.4	—	7.9	41
NO_3	—	16	—	3.8	—	—	—	2.2	84
I	(264)	6	—	—	—	—	—	1.1	—
PO_4	(1,000)	—	—	7.2	—	20	—	12	—
NO_2	—	0.4	—	—	—	—	—	—	
TDS	359,000	61,400	186,100	1,331	300,000	11,870	7,090	24,200	102,400
Density	—	—	—	—	—	1.010	—	1.017	1.076
pH	(6.2)	9.65	8.57	8.9	9.55	9.6	9.13	9.6	10.1
1,000B/TDS	4.7	5.7	1.7	6.0	1.4	2.9	4.3	2.1	2.3

[a] Mayhew and Heylumn, 1966.
[b] Also Fe 1,380, Sr 1,300, Li 110, Rb 95, Zn 50, Cs 16, Cu 8, Pb 6.
[c] Whitehead and Feth, 1961; for Big Soda Lake: Sr 0.5, Fe 0.10.
[d] Jones, 1966; also Fe 0.6.
[e] Hardie, 1968.
[f] Phillips and Van Denburgh, 1971.
[g] Jones, 1965.
[h] Phillips, 1976.
( ) = Limited number of analyses.

Table 5.6

Typical Analyses of Several Boron-Containing Geothermal Brines (ppm)

		Cerro Prieto				Steam Boat Springs,				
	Salton Sea[a]	From Power Plant[b]	From Salt Pond[c]	El Tatio Springs, Chile[d]	Tejon Rock Pile Fault, CA[e]	CO[f]	NV[g]	Yellowstone, Norris, WY[g]	Typical Volcanic Springs[g]	Lake Magadi Springs, Kenya[h]
Na	50,000–70,000	8,700	70,000	4,460	94.3	682	653	439	815	11,410
K	13,000–34,200	1,700	36,000	523	—	103	71	74	101	204
Mg	700–5,700	—	—	—	0.7	0.3	0.8	0.2	—	—
Ca	22,600–39,000	400	9,400	15.4	1.8	6.8	5.0	5.8	—	—
Fe	1,200–3,700	0.94	0.4	—	—	0.3	—	—	—	—
Mn	1,000–2,000	0.64	1.8	—	—	—	—	—	—	—
Sr	540–2,000	15.7	—	—	—	—	1.0	—	—	—
Zn	500–700	0.2	1.7	—	—	—	—	—	—	—
HN_4	650	—	—	—	—	—	<1	0.1	—	—
Te	520	—	4.0	—	—	—	—	—	—	—
As	312	1.20	6.5	—	—	2.7	2.7	3.1	—	—
Li	219–319	16	393	46	—	7.1	7.6	8.4	9.4	—
Ba	200	9.73	17.0	—	—	—	—	—	—	—
Pb	90–210	—	6.7	—	—	—	—	—	—	—
Rb	25–100	9.4	—	6.6	—	—	—	—	—	—
Cs	24	3.5	—	15.5	—	—	—	—	—	—
Cu	6–20	0.12	0.6	—	—	—	—	—	—	—
Ag	1–2	—	0.8	—	—	—	—	—	—	—

Sb	—	—	3.3	—	—	0.5	0.4	0.1	—	—
U	—	—	31.0	—	—	—	—	—	—	—
Cl	142,000–209,000	15,610	159,000	8,050	54.7	952	865	744	1,255	5,430
CO_3	—	—	—	—	0	—	0	0	17	4,180
HCO_3	—	—	—	45	136	246	305	27	177	12,750
SO_4	—	—	—	32	24.0	125	100	28	53	171
B	400–500	12	—	179	7	67.5	49	11.5	36	8.0
Br	200	20.5	—	—	—	—	0.2	0.1	—	24.9
Si	40	—	2.1	102	—	145	137	247	—	40.5
$S^{=}$	15–30	—	—	—	—	6.9	4.7	0	—	—
F	—	—	—	—	—	—	1.8	4.9	7.2	155
I	—	—	—	—	—	—	0.1	<0.1	—	—
PO_4	—	—	—	—	—	0.8	—	—	—	5
TDS	—	—	—	—	168.2	2,500	2,360	1,890	2,850	28,130
Density	1.18–1.26	—	1.250	—	—	—	—	—	—	1.025
pH	4.6–5.5	—	—	7.4	—	—	7.9	7.45	8.1	9.0
Temp.,°C	—	—	—	85	—	—	89.2	84	93	70
1,000B/TDS	1.34–1.84	0.45	—	13.3	41.6	27.0	20.8	6.1	12.6	0.3

[a] Various sources.
[b] Mercado, 1976; Galinzoga, 1981.
[c] Vasquez, 1981.
[d] Giggenbach, 1978.
[e] Schmidt, 1969.
[f] Bailey, 1902.
[g] White, 1957.
[h] Jones, 1966; Eugster, 1970.

area, and springs at the base of the Chocolate Mountains is quite similar for all, and far different from that of the ocean, Salton Sea, or Colorado River. The brine also has more ^{18}O, which is proportional to its temperature, and typical of very hot waters reacting with rocks much higher in ^{18}O. On the basis of isotopic and deuterium data it appears that <25% magmatic (or geothermal) water mixed with the meteoric water (White, 1968). The ratios of NaCl and KCl to Br, and of KCl to Rb and Cs are similar to that of a potash deposit, although no evidence of such a deposit has been found. The large amount of $CaCl_2$ in the geothermal brine and the small concentration of $MgCl_2$ appear to have resulted from a dolomitization reaction (Craig, 1966; Hardie, 1990). Finally, the brine's ^{34}S content is similar to that assumed to be in magma fluid, as is the wide variety and high concentration of many of the minor elements. There have been several experimental programs to recover the boron, potash, and other minerals from this brine, but none had been commercialized by 1996.

5.7.2 Miscellaneous Hot Springs

As with lake brines and other waters there are a large number of boron-containing geothermal springs in the world, with a few random examples listed in Table 5.6. The boron content of these springs is usually quite small, and long periods of evaporation and "fractional crystallization" would be required for them to form a borax deposit. However, the hot springs that formed the Searles Lake deposit began with only 7-20 ppmB, and it took only 40,000 years of desert evaporation to form this very large deposit. Also, the temperature of the springs need not be especially hot for them to have a high boron content. Some are above the boiling point, but many others are in only the 30–40°C range. It is interesting to note that some of the high-boron springs have a very low 1000 B/TDS (total dissolved solids) ratio, and thus it would be somewhat improbable that they could form a pure borate deposit. Mud volcanoes are a special form of geothermal spring, with their brines also varying from cool to very hot. They may accompany more active fumarolic geysers such as at Yellowstone National Park, but it is also fairly common for them to exist alone. Some of their brines have a high boron content, such as 52 ppm in the Barbados mud volcanoes, and a $\delta^{11}B$ of +28.0 (You, Spivack, Gieskes, Martin and Davisson, 1996). Various of the Russian mud volcanoes have been used as commercial borate sources.

References

Alderman, S. S. (1983). Geology of the Owens Lake evaporite deposit. *Sixth Int. Symp. on Salt* **1,** 75–83.

Allison, I. S., and Mason, R. S. (1947). *Sodium Salts of Lake County, Oregon.* Oregon Dept. Geol. and Mineral Ind., GMI Short Paper No. 17.

Alonso, R. N. (1986). *Occurrences, Stratigraphic Position and Genesis of the Borate Deposits of the Puna Region of Argentina.* Doctoral dissertation, Univ. Nacional de Salta, Dept. Nat. Sci., Salta, Argentina, 196 p.

Anon. (1984). New lithium frontier in Chile. *Foote Prints* **47**(1), 2–14.

Anon. (1985a, September). Lithium sales to reach $18 million in 1985. *Chile Econ. Report,* p. 41.

Anon. (1985b, April). Lithium. *Chile Econ. Report,* p. 7.

Anon. (1992, November–December). Salar de Atacama project edges closer to reality. *Phos. & Potas.,* No. 182, p. 12.

Anon. (1997, January 1–8). Minsal begins shipments. *Chem. Week,* **159**(1), 35.

Ayers, W. O. (1882, July). Borax in America. *Popular Sci. Monthly* **21,** pp. 350–361.

Bailey, G. E. (1902). The saline deposits of California: Borax. *Calif. Div. Mines, Mineral Resources, Bull. 24,* p. 347.

Bencini, A., and Duchi, V. (1988). Boron distribution in thermal springs from Tuscany and Northern Latium. *Rend. Soc. Ital. Mineral. Petrog.* **43,** 927–933.

Catalano, L. R. (1964). *Estudio Geologico-Economico del Salar de Hombre Muerto.* Minist. de Economia, Buenos Aires, No. 4.

Chatard, J. M. (1890). Natural soda ash: Its occurrences and utilization. *U.S. Geol. Surv. Bull. 60,* pp. 27–101.

Chen, Y. (1986). Hydrochemistry and evolution of interstitial brine in the Zabuye Saline Lake of Tibet. *Bull. of the Institute of Mineral Deposits, Chinese Academy of Geological Sciences,* No. 2, pp. 176–184.

Cohenour, R. E. (1966). Great Salt Lake, Utah, and its environment. *Second Symp. on Salt, Northern Ohio Geol. Soc.* **1,** 201–214.

Craig, H. (1966). Isotopic composition and origin of the Red Sea and Salton Sea geothermal brines. *Science* **154,** 1544–1548.

Craig, H. (1969). Geochemistry and origin of Red Sea brines. In *Hot Brines and Recent Heavy Metal Deposits in the Red Sea* (E. J. Degens and D. A. Ross, eds.), pp 208–242. Springer-Verlag, New York.

Dominguez, B., and Vital, F. (1976). *Repair and Control of Geothermal Wells at Cerro Prieto, Baja California, Mexico,* pp. 1484–1499. Comision Federal de Elec.

Donoso, R., and Theune, C. (1990). Nonmetallic mining in Chile. *Ind. Mineral.* (*London*), No. 277, pp. 63–71.

Ermakov, V. A., and Grebennikov, N. P. (1977). Structural characteristics of bischofite deposits in salt-containing strata of the Lower Volga River region. *Probl. Solenakopleniya* 2, 40–45.

Eugster, H. P. (1970). *Chemistry and Origin of the Brines of Lake Magadi, Kenya,* pp. 213–235. Mineral Soc. Am. Spec. Paper No. 3.

Ferro, R. (1991, November 6). Chemicals in Chile. *Chem. Week,* p. 24.

Flint, R. F., and Gale, W. A. (1958). Stratigraphy and radiocarbon dates at Searles Lake, California. *Am. J. Sci.* **256**(10), 689–714.

Friedman, I., Smith, G. I., and Matsuao, S. (1982). Economic implications of the deuterium anomally in the brine and salts in Searles Lake. *Econ. Geol.* **77**(3) 694–702.

Galinzoga, J. A. (1981). *Planta de Cloro de Potasio, Bid Package.* Fertilizantes Mexicanos.

Garrett, D. E. (1992). *Natural Soda Ash,* Van Nostrand and Reinhold, New York.

Garrett, D. E. (1995). *Potash: Deposits, Processing, Properties and Uses.* Chapman & Hall, London.

Garrett, D. E., and Carpenter, L. G. (1959). Tungsten in Searles Lake. *Mining Eng.* **11**(3) 301–303.

Garrett, D. E., and Laborde, M. (1983). Salting out process for lithium recovery. *Sixth Int. Symp. on Salt* **2,** 421–430.

Giggenbach, W. F. (1978). The isotopic composition of water from the El Tatio Geothermal Field, Northern Chile. *Geochim. et Cosmochim. Acta* **42,** 979–988.

Haines, D. V. (1959). Core logs from Searles Lake, San Bernardino County, California. *U.S. Geol. Surv. Bull. 1045-E,* pp. 139–317.

Hanks, H. G. (1883). *Report on Borate Deposits of California and Nevada.* Calif. State Mining Bureau, 3rd Ann. Rept., Pt. 2, pp. 6–111.

Hardie, L. A. (1968). The origin of recent nonmarine evaporite deposits in Saline Valley, Inyo County, California. *Geochim. et Cosmochim. Acta* **32,** 1279–1301.

Hardie, L. A. (1990). The roles of rifting and hydrothermal calcium chloride brines in the rigin of potash evaporites: An hypothesis. *Am. J. Sci.* **290**(1), 43–106.

Ide, F., Vergara-Edwards, L., and Pavlovic-Zuvic, P. (1983). Solar pond design for the production of potassium salts from the Salar de Atacama brines. *Sixth Int. Symp. on Salt* **2,** 367–375.

Igarzabal, A. P., and Poppi, R. F. (1980). El Salar de Hombre Muerto. *Acta Geol. Lilloana* **15**(2), 103–117.

Jones, B. F. (1965). The hydrology and mineralogy of Deep Springs Lake. *U.S. Geol. Surv. Prof. Paper 520A,* 52A, 56P.

Jones, B. F. (1966). Geochemical evolution of closed basin water in the Western Great Basin. *Second Symp. on Salt, Northern Ohio Geol. Soc.* **1,** 181–200.

Jones, B. F., Rettig, S. A., and Eugster, H. P. (1977). Hydrochemistry of the Lake Magadi Basin, Kenya. *Geochim. et Cosmochim. Acta* **41,** 53–72.

Kistler, R. B., and Smith, W. C. (1983). Boron and borates. In *Industrial Minerals and Rocks* (S. J. Lefond, ed.), 5th ed., pp. 533–560, Soc. Min. Eng., AIME, New York.

Lyday, P. A. (1992a). History of boron production and processing. *Ind. Min.* (*London*), **303,** 19, 21, 23–25, 31, 33, 34, 37.

Lyday, P. A. (1987, 1991, 1992b). *Boron.* Annual Report, U.S. Bur. Mines, Washington, D.C.

Mayhew, E. J., and Heylmun, E. B. (1966). Complex salts and brines of the Paradox Basin. *Second Symp. on Salt, No. Ohio Geol. Soc.* **1,** 221–235.

Mercado, S. G. (1976). *Cerro Prieto Geothermoelectric Project,* pp. 1394–1398. Commission Federal de Electricidad.

Mueller. G. (1969). Occurrence of mirabilite and ulexite at Ihn. Saarlouis District. *Nues. Jahrb. Mineral., Abs.* **110**(2), 188–198.

Nicolli, H. B., Suriano, J. M., Mendez, V., and Peral, M. A. (1982). Salmueras ricas en metales alcalinos del Salar del Hombre Muerto. *Quinto Cong. Latinamer de Geol., Argentina* **3,** 187–204.

Pavlovic, P., Parada, N., and Vergara, L. (1983). Recovery of potassium chloride, potassium sulfate and boric acid from the Salar de Atacama brine. *Sixth Int. Symp. on Salt* **2,** 374–394.

Phillips, J. E. (1976). *A Survey of Potential Methods for Resource Recovery from Black Water of the Upper Green River Basin.* Master's thesis, Univ. of Wyoming, Laramie, WY.

Phillips, K. N., and Van Denburgh, A. S. (1971). Hydrology and geochemistry of Abert, Summer,

Goose Lake, and other closed basin lakes in south central Oregon. *U.S. Geol. Survey Prof. Paper 502B.*

Qian, Z., and Xuan, Z. (1983). Borate minerals in salt lake deposits at Chaidamu Basin, China. *Sixth Int. Symp. on Salt, The Salt Institute* **1,** 185–192.

Ryan, J. E. (1951, May). Industrial salts: Production at Searles Lake. *Mining Eng.* **190,** 447–452.

Schmidt, K. D. (1969). *The Distribution of Boron in the Ground Water of the Arvin–Caliente Creek Area, Kern County, California.* Master's thesis, Univ. of Arizona, Phoenix, AZ.

Simoons, F. J. (1954). Borax. *Nineteenth Century Mines and Mineral Spring Reports of Lake County, Calif. J. Mines, Geol.* **50,** 300–303.

Smith, G. I. (1979). Subsurface stratigraphy and geochemistry of Late Quaternary evaporites, Searles Lake, California. *U.S. Geol. Survey Prof. Paper 1043.*

Smith, G. I., Friedman, I., and McLaughlin, R. J. (1987). Mineral, chemical and isotopic evidence of salt solution and crystallization processes in Owens Lake, California 1969–1971. *Geochim. et Cosmochim. Acta* **51**(4), 811–827.

Smith, G. I., and Pratt, W. P. (1957). Core logs from Owens, China, Searles and Panamint Basins, California. *U.S. Geol. Surv. Bull. 1045-A,* pp. 1–62.

Stuiver, M. (1964). Carbon isotopic distribution and correlated chronology of Searles Lake sediments. *Am. J. Sci.* **262**(3), 377–392; 1979, *U.S. Geol. Survey Prof. Paper 1043,* pp. 68–75.

Sun, D. (1987). Soda lakes and origin of their trona deposits on the Nei Mongol Plateau of China. *Chinese J. Oceanol. Limnol.* **5**(4) 351–362.

Sun, D. (1990a). Origin of borate deposits in Xiao Qaidam Lake, Qaidam Basin, China. *China Earth Sci.* **1**(3), 253–266.

Sun, D. (1990b). *Soda Lakes in Tibet.* Personal communication.

Sun, D., and Li B. (1993). Origins of borates in the saline lakes of China. *Seventh Int. Symp. on Salt* **1,** 177–194.

Sun, D., Ma Y., and Shan L. (1991, June). A preminiary investigation of ulexite obtained from natural brine. *Chinese Sci. Bull.* **36**(11), 922–927.

Vasquez, F. M. (1981, March 23). *Planta Piloto Cloruro de Potasio.* Fertilizantes Mexicanos, Dept. de Experimentacion.

Vegara, L., and Parada, N. (1983). Study of the phase chemistry of the Salar de Atacama Brine. *Sixth Int. Symp. on Salt* **2,** 345–366.

Ver Planck, W. E. (1956). History of borax production in the United States. *Calif. J. Mines, Geol.* **52**(3), 273–291.

Vila, T. (1990). Salar deposits in northern Chile. In *Stratabound Ore Deposits in the Andes* (L. Fontbote, G. C. Amstutz, M. Cardozo, E. Cedillo, J. Frutos, eds.), pp. 703–720, Springer-Verlag, Berlin.

Vonsen, M., and Hanna, G. D. (1936). Borax Lake, California. *Calif. J. Mines, Geol.* **32**(1), 99–108.

White, D. E. (1957). Thermal waters of volcanic origin: Magmatic, connate and metamorphic waters. *Bull. Geol. Soc. Am.* **68,** 1637–1682.

White, D. E. (1968). Environments of generation of some base metal ore deposits. *Economic Geology* **63**(4), 301–322.

Whitehead, H. C., and Feth, J. H. (1961). Recent chemical analyses of waters from several closed-basin lakes and their tributaries in the western United States. *Geol. Soc. Am. Bull.* **72**(9), 1421–1425.

You, C. F., Spivack, A. J., Gieskes, J. M., Martin, J. B., Davisson, M. L. (1996). Boron contents and isotopic compositions in pore waters. *Marine Geol.* **129**(3–4), 351–361,

Zheng, C, 1997. "The Natural Soda Deposits of China," 1st. Int. Soda Ash Conf., Rock Springs, WY, June 10–12, 11p.

Zheng, M. (1989). *Saline Lakes on the Qinghai-Xizang* (*Tibet*) *Plateau,* Beijing Science and Technical Press, Beijing, 431p. (in Chinese).

Zheng, X., Tong, Y., and Xiang, J. (1988). *Salt Lakes in Xizang (Tibet),* Chinese Science Press, Beijing, China, pp. 108, 123, 138 (in Chinese).

Chapter 6 Marine Borate Occurrences: Isotropic Distribution

6.1 MARINE BORATE OCCURRENCES

6.1.1 General

Seawater contains about 4.6 ppmB, and some boron is adsorbed or co-crystallized with each of the sediments during seawater's evaporation cycle. On the basis of typical analyses (Table 6.1), a boron material balance for each period of marine deposition has been prepared (Table 6.2). The brine concentration ratios were taken from small-scale evaporation experiments (Garrett, 1995), and the amount crystallized compared with boron is based on standard seawater analyses. Zherebtsova and Volkova (1966) evaporated Black Sea water (seawater diluted about half: 4.0 ppmB) to produce an end liquor with 355 ppmB and a 22.1% boron loss. Valeyev, Ozol, and Tikhvinskiy (1973) stated that 200–400 ppmB is a normal range for solar-evaporated seawater end liquor, and Vengosh Starinsky, Kolodny, Chivas and Raab (1992) obtained 400 ppmB. The latter's salts had slightly less boron than actual deposits, but their final $\delta^{11}B$ was similar that found in Table 6.2 without clay adsorption.

Valeyev *et al.* (1973) found that borates could not precipitate from seawater under dynamic, halite or potash crystallizing conditions until the concentration was 2500–4000 ppmB because of extreme supersaturation. They speculated that boron-supplemented seawater, however, could precipitate boracite during the final period of salt and potash-crystallization if residence times were long.

Many reports have been made on borates in marine formations (Table 6.3), and on the basis of their $\delta^{11}B$ values, at least part of the boron in these minerals originated from the seawater. However, in the early evaporation stages boron should be well below any borate's solubility (Table 6.4). Additional boron could enter the seawater from terrestrial or geothermal sources, or originally deposited boron could have been redissolved (as when anhydrite converts to gypsum), concentrated, and reprecipitated. In the former case, with stronger brines trapped in the void space, the borates should have been widely dispersed, very small, and perhaps only recognizable in insoluble residues. In a few deposits it appears that large amounts of boron mixed much later with the end liquor and entered through fractures or permeable zones (boracite in Thailand, various borates in the Zeichstein and Inder deposits,

Table 6.1

Examples of the Boron Content of Various Sediments in Marine Deposits (ppm B)[a]

	Zeichstein cycle				Russia		Oklahoma,[b] Others
	Z1	Z2	Z3	Z4	Byelorussia[c]	Donets Basin[d] (no potash)	Gypsum,[e] no halite or potash
Shale	300	670–2000	1600	500	10	—	160–290 (avg. 220)
Calcite	10	—	—	—	—	—	39–58[f]
Dolomite, average	20	70	—	—	—	—	—
Anhydrite	2–20	150–500	24–90	3–80	16[d]	12–25	19–215
Average[g]	12	200	55	40	—	8[h]	72, 43[h]
Halite, average	5	50	3	10	25	13	—
Hartsalz	—	10–2000	—	—	1200–1400[i]	—	—
Average	—	600	—	—	—	—	—
Carnallite	—	500–2000	—	—	60–70	—	—
Average	—	118	—	—	—	—	—
Polyhalite, average	—	800	200	—	—	—	—

[a] Harder, 1959.
[b] Ham, Mankin, and Schleicher, 1961.
[c] Sonnenfeld, 1984.
[d] Zaritskiy, 1965.
[e] Marine deposits with no halite or potash.
[f] Gaillardt and Allegre, 1995 (aragonite in corals).
[g] Average of all four Zeichstein zones, 70[b].
[h] In gypsum, not anhydrite.
[i] In sylvinite, not hartsalz.

Table 6.2

An Estimated Material Balance for Boron during the Solar Evaporation of Seawater

Precipitating-settling stage	Cumulative concentration ratios	Amount compared to B[a]	B in precipitate (ppm)[b]	% of original B removed	Final brine	
					B Concentration (ppm)	$\delta^{11}B$[c]
Original sea water	1	—	133[d]	0	4.6	40.0
Clay, silt[e]	1	267.6	500	13.38	3.99	61.8
Iron (Fe_2O_3)	2	1	300	0.03	7.97	61.9
Calcite (Dolomite)	3.79	24.8	10	0.248	15.1	62.3
Gypsum (Anhydrite)	8.15	267.6	20	0.535	32.2	63.3
Salt (pure)	23.68	5300	10	5.30	87.7	74.1
Salt-epsomite	53.20	483	20	0.996	195	76.3
Potash salts	~77	695	33	2.294	274	81.6
Maximum evaporation (start of bischofite precipitation)	~100	—	—	—	355	—
Total, final	—	—	—	22.8%	—	81.6

[a] Data from Garrett, 1995. All numbers are at the end of the primary precipitating period for each mineral.

[b] Data from Table 6.1.

[c] Assume that 2% more ^{10}B than in the original ratio of $^{11}B/^{10}B$ (4.20722) is lost at each precipitation stage. The $^{11}B/^{10}B$ standard is the 1989 U.S. Bureau of Standards 951, 4.0454.

[d] The average value of boron in the marine sediments if all the original B in the seawater (alone) precipitated with all of the sediments.

[e] Assume that the total amount of clay or silt in the entire evaporation cycle is present initially. Also, assume that its quantity equals that of the gypsum.

Table 6.3

Location and Host Rocks of Marine Borates

Location / Mineral	Formula
Asia Minor (Gale, 1917)	
Priceite (in gypsum)	$Ca_4B_{10}O_{19}\cdot 7H_2O$
Bosnia, Hercegovina (Bermanec, Armbruster, Tibljas, Sturman and Kniewald *et al.,* 1994)	
Tuzlaite (in halite)	$NaCaB_5O_9\cdot 4H_2O$
Canada (Newfoundland, Nova Scotia) (Roulston and Waugh, 1981; Papezik and Fong, 1975)	
Boracite (in halite)	$Mg_3B_7O_{13}Cl$
Colemanite (in halite)	$Ca_2B_6O_{11}\cdot 5H_2O$
Congolite (Hawthorne, Burns, and Grice, 1996)	$(Fe, Mg, Mn)_3B_7O_{13}Cl$
Danburite (in gypsum, halite, and anhydrite)	$CaSi_2B_2O_8$
Ginorite (in gypsum; How, 1857)	$Ca_2B_{14}O_{25}\cdot 8H_2O$
Hilgardite-1A and -4M (in halite)	$Ca_2B_5O_9Cl\cdot H_2O$
Howlite (in gypsum and halite)	$Ca_2SiHB_5O_{12}\cdot 2H_2O$
Hydroboracite (in halite)	$CaMgB_6O_{11}\cdot 6H_2O$
Inyoite (in gypsum)	$Ca_2B_6O_{11}\cdot 13H_2O$
Penobosquisite (Hawthorne *et al.*, 1996)	$Ca_2FeB_9O_{16}Cl\cdot 7H_2O$
Priceite (in halite)	$Ca_4B_{10}O_{19}7H_2O$
Pringleite (Hawthorne *et al.*, 1996)	$Ca_2B_{26}O_{46}Cl_4\cdot 25H_2O$
Ruitenbergite (Hawthorne *et al.*, 1996)	$Ca_2B_{26}O_{46}Cl_4\cdot 25H_2O$
Strontioginorite (Hawthorne *et al.*, 1996)	$Sr_2B_{14}O_{23}\cdot 8H_2O$
Szaiblyite (in halite)	$Mg_2B_2O_5\cdot H_2O$
Trembathite (Hawthorne *et al.*, 1996)	$(Fe, Mg)_3B_7O_{13}Cl$
Tyretskite (Hawthorne *et al.*, 1996)	$Ca_4B_{10}O_{19}\cdot 3H_2O$
Ulexite (in halite and gypsum; How, 1857)	$NaCaB_5O_9\cdot 8H_2O$
Veatchite (in anhydrite, or halite, dolomite, sylvinite veins)	$Sr_4B_{22}O_{37}\cdot 7H_2O$
Volkovskite (in anhydrite, or halite, dolomite, sylvinite veins)	$(Ca, Sr)_2B_{14}O_{23}\cdot 8H_2O$
England (Guppy, 1944; Milne, Saunders, and Weids, 1977; Stewart, Chalmers, and Phillips, 1954)	
Borocite (in halite)	$MgB_7O_{13}Cl$
Ericaite (in anhydrite)	$FeB_7O_{13}Cl$
Veatchite (in polyhalite, anhydrite, halite–anhydrite, dolomite-shales)	$Sr_4B_{22}O_{37}\cdot 7H_2O$
Germany (Kuhn, 1968; Braitsch, 1971)	
Ascharite (all Z1 and Z2 marine sediments)	$Mg_2B_2O_5\cdot H_2O$
Boracite (in all marine sediments of Z2, and some in Z1)	$Mg_3B_7O_{13}Cl$
Danburite (in anhydrite, Z2)	$Ca_2B_2Si_2O_8$
Datolite (in halite near basalt, Z1)	$CaHBSiO_5$
Ericaite (Kuhn and Schaacke, 1955)	$Fe_3B_7O_{13}Cl$
Fabianite (in halite, Z1)	$Ca_2B_6O_{11}\cdot H_2O$
Heidornite (in anhydrite, Z1)	$Na_2Ca_3B_5O_9(SO_4)_2Cl\cdot H_2O$
Howlite (in anhydrite–halite contact zone, Z1)	$Ca_2HSiB_5O_{12}\cdot 2H_2O$
Hydroboracite (in anhydrite; Siemroth, 1992)	$CaMgB_6O_{11}\cdot 6H_2O$
Kaliborite (in kainite cap rock, Z2)	$K_2Mg_4B_{24}O_{41}\cdot 19H_2O$
Luneburgite (in gypsum, Z2)	$Mg_3(PO_4)_2B_2O_3\cdot 8H_2O$
Pinnoite (in kainite cap rock, Z2)	$MgB_2O_4\cdot 3H_2O$
Probertite (in anhydrite and halite; Fischbeck, 1983; Siemroth, 1992)	$NaCaB_5O_9\cdot 5H_2O$

Table 6.3 (*continued*)

P-Veatchite (in halite; Z2)	$SrB_6O_{10}\cdot 2H_2O$
Strontioginorite (in halite, Z2)	$(Sr, Ca)\ _2B_{14}O_{23}\cdot 8H_2O$
Strontiohilgardite (in hartsalz, halite, Z2)	$(Sr, Ca)\ _2B_5C_9Cl\cdot H_2O$
Sulfoborite (in carnallite, Z2)	$Mg_3SO_4B_2O_5\cdot 5H_2O$
Tourmaline (in Z2 carnallite, Z3 halite, sylvinite)	$NaR_3B_3Al_6Si_6O_{27}\ (OH,F)_4$
Ulexite (in anhydrite; Siemroth, 1992)	$NaCaB_5O_9\cdot 8H_2O$
Russia	
Inder Region (Braitsch, 1971, Godlevsky, 1937)	
β Ascharite (in gypsum, clay, and halite)	$MgHBO_3$
Boracite (in all formations)	$Mg_3B_7O_{13}Cl$
Colemanite (in gypsum and red clay)	$Ca_2B_6O_{11}\cdot 5H_2O$
Hilgardite (in all formations)	$Ca_2B_5O_9Cl\cdot H_2O$
Hydroboracite (in gypsum, brown and red clay, and halite)	$CaMgB_6O_{11}\cdot 6H_2O$
Inderborite (in gypsum)	$CaMgB_6O_{11}\cdot 11H_2O$
Inderite (in gypsum, clay and halite)	$Mg_2B_6O_{11}\cdot 15H_2O$
Inyoite (in gray clay)	$Ca_2B_6O_{11}\cdot 13H_2O$
Kaliborate (in halite, kieserite, and sylvinite)	$K_2Mg_4B_{24}O_{41}\cdot 19H_2O$
Kurgantaite (in dolomite and gypsum)	$(Sr, Ca)\ _2B_4O_8\cdot H_2O$
Kurnakovite (in dolomite and gypsum)	$Mg_2B_6O_{11}\cdot 15H_2O$
Lunebergite (in dolomite and gypsum)	$Mg_3\ (PO_4)\ _2B_2O_3\cdot 8H_2O$
Preobrazhenskiite (in halite, epsomite, and sylvinite)	$Mg_6B_{22}O_{39}\cdot 9H_2O$
Priceite (in gypsum and red clay)	$Ca_4B_{10}O_{19}\cdot 7H_2O$
Sulphoborite	$Mg_3SO_4B_2O_5\cdot 5H_2O$
Ulexite (in red and sandy clay)	$NaCaB_5O_9\cdot 8H_2O$
Caspian Area, Satimolinsk (Valeyev, Ozol, Tikuinskiy, 1973)	
Boracite (with polyhalite and carbonate–clay)	$Mg_3B_7O_{13}Cl$
Ginorite (with anhydrite and carbonate–clay)	$Ca_2B_{14}O_{23}\cdot 8H_2O$
Hilgardite (with anhydrite and carbonate–clay)	$Ca_2B_5O_9Cl\cdot H_2O$
Hydroboracite (with polyhalite and anhydrite)	$CaMgB_6O_{11}\cdot 6H_2O$
Kaliborite (with polyhalite and carbonate–clay)	$K_2Mg_4B_{24}O_{41}\cdot 19H_2O$
Pinnoite (with polyhalite)	$MgB_2O_4\cdot 3H_2O$
Preobrazhenskite (with polyhalite)	$Mg_6B_{22}O_{39}\cdot 9H_2O$
Satimolite (with polyhalite, anhydrite, and carbonate–clay	$KNa_2Al_4B_6O_{15}Cl_3\cdot 13H_2O$
Sulphoborite (with polyhalite and carbonate–clay)	$Mg_3SO_4B_2O_5\cdot 5H_2O$
Veatchite (with anhydrite)	$Sr_4B_{22}O_{37}\cdot 7H_2O$
Caspian Area, Chelkar (Valeyev *et al.*, 1973)	
Aksaite (with halite, anhydrite and bischofite)	$MgB_6O_{10}\cdot 5H_2O$
Ascharite (with sylvinite)	$MgHBO_3$
Boracite (with halite, anhydrite, sylvinite, and carnallite)	$Mg_3B_7O_{13}Cl$
Ginorite (with halite and anhydrite)	$Ca_2B_{14}O_{23}\cdot 8H_2O$
Halurgite (with sylvinite, carnallite, and bischofite)	$MgB_4O_7\cdot 3H_2O$
Hilgardite (with halite, anhydrite, and bischofite)	$Ca_2B_5O_9Cl\cdot H_2O$
Hydroboracite (with halite and anhydrite)	$CaMgB_6O_{11}\cdot 6H_2O$
Kaliborite (with sylvinite)	$K_2Mg_4B_{24}O_{41}\cdot 19H_2O$
Pinnoite (with sylvinite, carnallite, and bischofite)	$MgB_2O_4\cdot 3H_2O$
Preobrazhenskite (with carnallite and bischofite)	$Mg_6B_{22}O_{39}\cdot 9H_2O$
Strontiohilgardite (with halite and anhydrite)	$(Ca, Sr)\ _2B_5O_9Cl\cdot H_2O$
Sulfoborite (with sylvinite, carnallite, and bischofite)	$Mg_3SO_4B_2O_5\cdot 5H_2O$

continues

Table 6.3 *(continued)*

Donets Basin (in gypsum; Zaritskiy, 1965)	
Kernite	$Na_2B_4O_7 \cdot 4H_2O$
Hydroboracite	$CaMgB_6O_{11} \cdot 6H_2O$
Inderite	$Mg_2B_6O_{11} \cdot 15H_2O$
Inyoite	$Ca_2B_6O_{11} \cdot 13H_2O$
Probertite	$NaCaB_5O_9 \cdot 5H_2O$
Ulexite	$NaCaB_5O_9 \cdot 8H_2O$
Siberian Platform, Angara Series, Irkutsk, Siberia (in dolomite–anhydrite; Yarzhemskii, 1969)	
Danburite	$CaB_2Si_2O_8$
Strontiohilgardite	$(Ca, Sr)\ _2B_5O_9Cl \cdot H_2O$
General	
Aldzhanite (Kondrateva, 1964)	$(Ca, Mg)_2\ BO_3Cl \cdot nH_2O$
Boracite (in sylvinite; Christ, 1966)	$Mg_3B_7O_{13}Cl$
Chambersite (in sylvinite; Yarzhemskii, 1969)	$Mn_3B_7O_{13}Cl$
Chelkarite (Avrova, Bocharov, Khalturina and Yunosova, 1968)	$CaMgB_2O_4Cl_2 \cdot 7H_2O$
Danburite (in dolomite and anhydrite; Yarzhemskii, 1969)	$CaB_2Si_2O_8$
Ericaite (in sylvinite; Yarzhemskii, 1969)	$Fe_3B_7O_{13}Cl$
Fluoborite (Diarov, 1966)	$Mg_3\ (OH, F)\ _3BO_3$
Ginorite (in dolomite; Yarzhemskii, 1969)	$Ca_2B_{14}O_{25} \cdot 8H_2O$
Halurgite (in halite and sylvinite; Yarzhemskii, 1968)	$Mg\ B_4O_7 \cdot 3H_2O$
Ivanovite (Nefedov, 1953)	$Ca_xB_2O_3Cl_y$
Kurnakovite (in sylvinite; Borchert and Muir, 1964)	$Mg_2B_6O_{11} \cdot 15H_2O$
Lunebergite (in mixed sulfate salts; Yarzhemskii, 1969)	$Mg_3\ (PO_4)\ _2B_2O_3 \cdot 8H_2O$
Metaborite (Avrova *et al.*, 1968)	HBO_2
Preobrashenskiite (in sylvinite; Borchert and Muir, 1964)	$Mg_6B_{22}O_{39} \cdot 9H_2O$
Satimolite (Bocharov, Khalturina, Avrova and Shipovalov, 1969)	$KNa_2Al_4Cl_3B_6O_{15} \cdot 13H_2O$
Searlesite (in halite and volcanic tuff; Yarzhemskii, 1969)	$NaBSi_2O_6 \cdot H_2O$
Strontioborite (in halite; Yarzhemskii, 1969)	$SrB_8O_{13} \cdot 2H_2O$
Strontiohilgardite (in dolomite and anhydrite; Yarzhemskii, 1969)	$(Sr, Ca)\ _2B_5O_9Cl \cdot H_2O$
Tyretskite) (Ivanov and Yarzhemskii, 1954)	$Ca_4B_{10}O_{19} \cdot 3H_2O$
Volkovskite (in halite; Christ, 1966)	$(Ca, Sr)\ _2B_{14}O_{23} \cdot 8H_2O$
United States	
Alabama (in halite; Simmons, 1988)	
Boracite	$Mg_3B_7O_{13}Cl$
Danburite (in adjacent shale)	$CaB_2Si_2O_8$
Hilgardite	$Ca_2B_5O_9Cl \cdot H_2O$
Szaibelyite	$Mg_2B_2O_5 \cdot H_2O$
Volkovskite	$(Sr, Ca)\ _2B_{14}O_{23} \cdot 8H_2O$
Louisiana (in halite; Hurlbut, 1938; Taylor, 1937)	
Boracite	$Mg_3B_7O_{13}Cl$
Danburite	$CaB_2Si_2O_8$
Hilgardite	$Ca_2B_5O_9Cl \cdot H_2O$
Parahilgardite	$Ca_2B_6O_9Cl \cdot H_2O$

Table 6.3 (*continued*)

Mississippi (in halite; Gann, Dockery, and Marble, 1987; Sundeen, 1990)	
Boracite	$Mg_3B_7O_{13}Cl$
Danburite	$CaB_2Si_2O_8$
Hilgardite	$Ca_2B_5O_9Cl \cdot H_2O$
New Mexico (in sylvinite; Schaller and Henderson, 1932)	
Luneburgite	$Mg_3\ (PO_4)\ _2B_2O_3 \cdot 8H_2O$
Oklahoma (in gypsum; Ham, Mankin, and Schleicher, 1961)	
Priceite	$Ca_4B_{10}O_{19} \cdot 7H_2O$
Probertite	$NaCaB_5O_9 \cdot 5H_2O$
Ulexite	$NaCaB_5O_9 \cdot 8H_2O$
Texas (in halite; Honea and Beck, 1962)	
Chambersite	$Mn_3B_7O_{13}Cl$
Utah (in anhydrite, or dolomite or sylvinite veins in halite; Raup, 1972)	
Braitschite	$(Ca, Na_2)\ _6RE_2B_{24}O_{45} \cdot 6H_2O$

and searlesite in Russia). In many of the occurrences, due to the large number and variety of borates present, it appears that there was considerable postdepositional change.

6.1.2 Asia Minor, Bosnia, and Hercegovina

Nodules of priceite ($Ca_4B_{10}O_{19} \cdot 7H_2O$) have been found embedded in sedimentary deposits of oceanic gypsum in Asia Minor (Gale, 1921). A new mineral, tuzlaite ($NaCaB_5O_9 \cdot 4H_2O$), was reported in the Tuzla salt mine in Bosnia, occurring "as nearly monomineralic veinlets" in dolomitic marl beds in halite (that may have been altered). Intergrowths with halite as well as tuzlaite covered with recrystallized halite were also present (Bermanec, Armbruster, Tibljas, Sturman and Kniewald, 1994).

6.1.3 Canada

6.1.3.1 Newfoundland and Nova Scotia

Borate minerals have been observed frequently in the Carboniferous-period gypsum and anhydrite beds of Canada's Atlantic Provinces (Fig. 6.1). How (1857) first reported ulexite ($NaCaB_5O_9 \cdot 8H_2O$) and smaller quantities of ginorite ($Ca_2B_{14}O_{23} \cdot 8H_2O$) in a Nova Scotia gypsum mine. Later, he also (1861) found howlite ($Ca_4Si_2B_{10}O_{23} \cdot 5H_2O$) "in nodules sometimes as large as a man's head" in another mine. Much later, others reported similar 5- to 15-cm nodules in various gypsum quarries, as well as ulexite and danburite ($CaSi_2B_2O_8$) in the "shore outcrops." Inyoite, ($Ca_2B_6O_{11} \cdot 13H_2O$) "as beautiful colorless crystals," has been noted also.

Table 6.4
Solubility of Various Borate Minerals (ppm B)

Mineral	Formula	Solubility in Water 25°C	Solubility in Water 100°C	Solubility in Conc. brine
Tourmaline[a]	Na (Mg, Fe) $_3Al_6Si_6O_{27}$ $(OH, F)_4$	<0.1	0.2	—
Howlite[a]	$Ca_4Si_2B_{10}O_{23}\cdot5H_2O$	34	104	—
Boracite[b]	$Mg_3B_7O_{13}Cl$	—	—	155
Kurnakovite[c]	$Mg_2B_6O_{11}\cdot15H_2O$	—	—	160
Bakerite[a]	$Ca_8B_{10}Si_6O_{35}\cdot5H_2O$	112	176	—
Szaibelyite	$Mg_2B_2O_5\cdot H_2O$	25[d], 167[e]	300[e]	330[b]
Colemanite	$Ca_2B_6O_{11}\cdot5H_2O$	42[f], 87[g], 128[h]	385[a]	—
		144[i], 169[a], 284[e]	600[e]	—
		Ave. 142	Ave. 493	—
Preobrazhenskite[e]	$Mg_6B_{22}O_{39}\cdot9H_2O$	300	500	—
Hexahydroborite[j,k]	$CaB_2O_4\cdot6H_2O$	323	—	—
Priceite[e]	$Ca_4B_{10}O_{19}\cdot7H_2O$	346	692	—
Hydroboracite	$CaMgB_6O_{11}\cdot6H_2O$	330[e], 714[d]	691[e]	360[b], 700[c]
Inderite	$Mg_2B_6O_{11}\cdot15H_2O$	346[d]	—	310[c], 345[b]
Inyoite	$Ca_2B_6O_{11}\cdot13H_2O$	327[e], 652[d]	643[e]	245[b]
Inyoite-ulexite[l]	$Ca_2B_6O_{11}\cdot13H_2O$– $NaCaB_5O_9\cdot8H_2O$	503	—	—
Ulexite	$NaCaB_5O_9\cdot8H_2O$	618[h] 627[e]	1250[e]	340[c], 630[c]
Ulexite	$NaCaB_5O_9\cdot8H_2O$	640[d], Ave. 628	—	670[c], Ave. 547
Pinnoite[e]	$MgB_2O_4\cdot3H_2O$	858	1390	—
Ulexite-meyerhofferite[l]	$NaCaB_5O_9\cdot8H_2O$– $Ca_2B_6O_{11}\cdot7H_2O$	900	—	—
Ulexite–inyoite–hexahydroborite[k]	—	1150	—	—
Kaliborite	$K_2Mg_4B_{24}O_{41}\cdot19H_2O$	1620[d], 2150[e]	4475[e]	—

[a] Graham, 1957.
[b] Kuhn, 1968; carnallite decomposition end liquor (similar to seawater end liquor).
[c] Sun, 1990; concentrated Tibetan brine (high $MgCl_2$); samples stored for 4–9 years.
[d] Yarzhemskii, 1968.
[e] Karazhanov, 1963; Spiryagina, 1953.
[f] Yarar, 1985.
[g] Gulensoy and Kocakerim, 1978; 1-hr contact; initial pH 5.5, final 9.08.
[h] Celik and Bulut, 1996.
[i] Gale, 1917.
[j] Nikolaev and Chelishcheva, 1940; one month storage.
[k] Kurnakova, and Nikolaev, 1948.
[l] Inan, Dunham, and Esson, 1973. Two weeks storage. Ca solubility for inyoite 43 ppm, meyerhofferite 180 ppm.

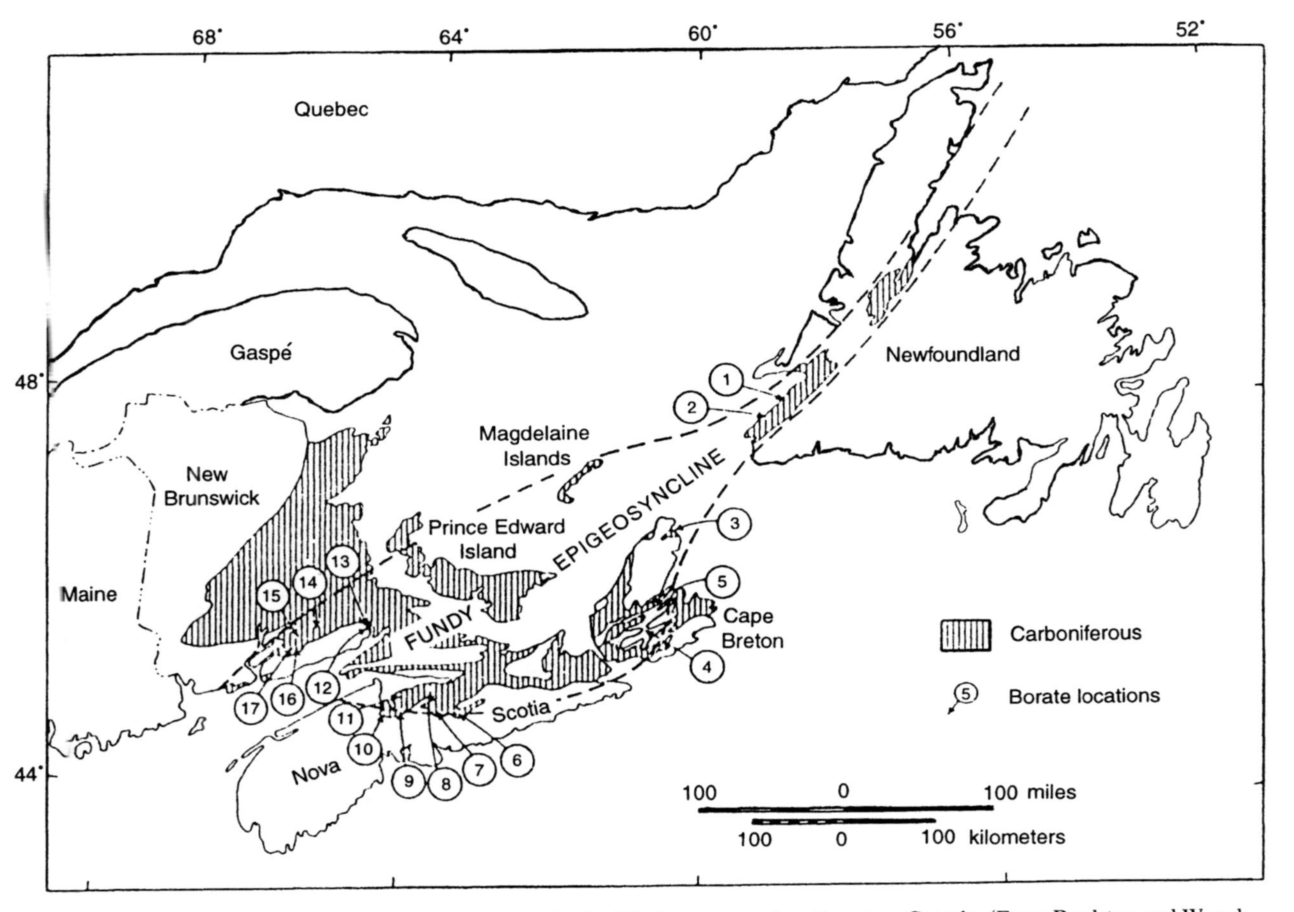

Figure 6.1 Location of the marine borate occurrences in the Windsor group of southeastern Canada. (From Roulston and Waugh, 1981; reprinted courtesy of Canadian Mineralogist and the authors.)

In Newfoundland Papezik and Fong (1975) also reported ulexite and howlite in Lower Carboniferous gypsum beds. The ulexite was the most abundant, occurring as irregular 1- to 3-cm patches of white fibrous crystals or as "cotton balls" with fine-felted, somewhat crumbly fibers that were less resistant to weathering than howlite. The howlite forms "composite framboidal 0.5- to 2-cm nodules consisting of thin platy <0.1-mm crystals." In one mine the howlite nodules appeared regularly at about $11/m^2$ ($1/ft^2$) of exposed anhydrite, and resembled small cauliflowers with a vitreous to pearly luster. The area has no volcanic rocks or signs of thermal springs, but in Nova Scotia the beds are locally contorted or steeply dipping. They often are accompanied by thin veins of purple fluorite, which like the borates may have a geothermal origin (Ham, Mankin, and Schleicher, 1961).

6.1.3.2 New Brunswick

Potash drill cores in the Penobsquis and Salt Springs evaporite deposits of southern New Brunswick contain borates in two separate strata of the Carboniferous Windsor Group (Mississippian) formations. The borates are found principally in halite (mainly above potash) of the Fundy Epigeosyncline that contains these potash basins (Fig. 6.2). Boracite ($Mg_3B_7O_{13}Cl$) is the most common of the borates, even though howlite ($Ca_4Si_2B_{10}O_{23}\cdot5H_2O$) is dominant at Penobsquis. Boracite occurs as colorless to pale green or blue 0.8- to 2-mm cubes or 2- to 4-cm aggregates, frequently as laminations near high-clay halite. At Salt Springs the boracite occurs with one or more of the following minerals: hydroboracite ($CaMgB_6O_{11}\cdot6H_2O$), hilgardite ($Ca_2B_5O_9Cl\cdot H_2O$), szaibelyite ($Mg_2B_2O_8\cdot H_2O$), priceite ($Ca_4B_{10}O_{19}\cdot7H_2O$), ulexite, colemanite and, danburite ($CaB_2Si_2O_8$). At Penobsquis the howlite usually is found with hydroboracite and volkovskite [$(Ca,Sr)_2B_{14}O_{23}\cdot8H_2O$], or with one or more of the following: hilgardite, danburite, szaibelyite, colemanite, and veatchite ($Sr_4B_{22}O_{37}\cdot7H_2O$). It occurs as patches and bands of white <1.5-cm (often 0.5- to 1-cm) nodules in high-clay anhydrite–claystone laminae in halite beds.

Hydroboracite occurs as clear to light brown or greenish <1-cm prismatic crystals, often forming <2-cm rosettes or "swallow tails" resembling gypsum. The crystals are evenly distributed in the salt, and not concentrated in high-clay laminae. Volkovskite is abundant at Penobsquis and forms the largest borate nodules (up to 15 cm), of which many are light orange, uneven granular aggregates. They usually are found in the high-clay halite with howlite and to a lesser extent with hydroboracite. Hilgardite occurs as clusters of brownish, triangle-shaped <2-cm (usually 0.5- to 5-mm) crystals, often in parallel multiple growths with other borates. Szaibelyite occurs as fine radiating acicular white <9-mm crystals forming small aggregates, usually near hilgardite and boracite. Priceite is found in the middle and lower halite–anhydrite interfaces at Salt Springs as chalky white <1-mm nodules resembling danburite, and

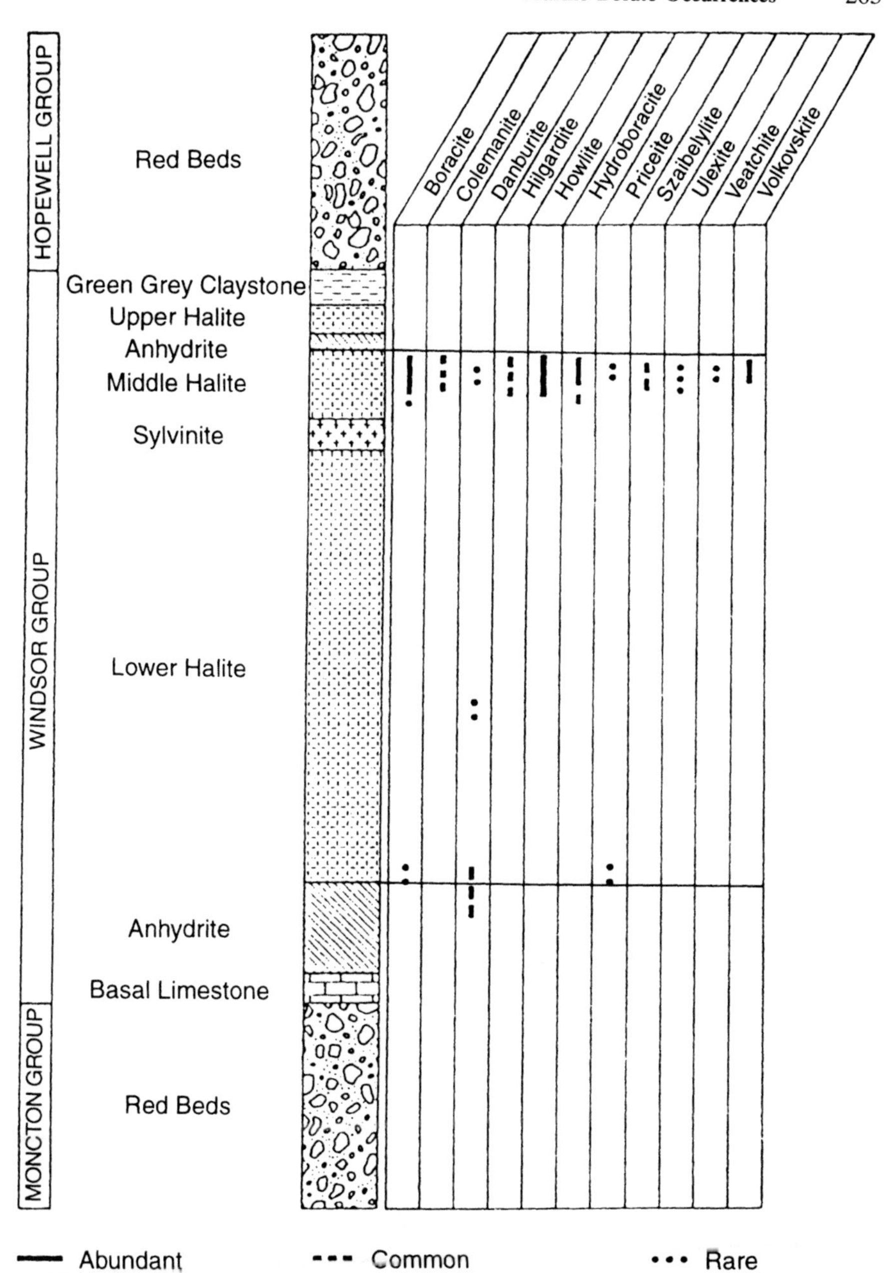

Figure 6.2 Generalized stratigraphy of the New Brunswick borate occurrences. (From Roulston and Waugh, 1981; reprinted courtesy of Canadian Mineralogist and the authors.)

often is associated with boracite and hydroboracite. Ulexite is found at Salt Springs only as fine, fibrous small white aggregates. Danburite forms amorphous to microcrystalline 1- to 8-mm nodules. It is common at Penobsquis and rare at Salt Springs. Veatchite is rare, occurring only at Penobsquis as brownish to red platy crystalline aggregates resembling volkovskite (Roulston and Waugh, 1981).

In later studies of core insolubles from two potash mines, many other borates have been discovered (the total in 1996 was 21, and the search continued). The new minerals were congolite $(Fe,Mg,Mn)_3B_7O_{13}Cl$, ginorite $(Ca_2B_{14}O_{23}{\cdot}8H_2O)$, hilgardite-1A and -4M $(Ca_2B_5O_9Cl{\cdot}H_2O)$, inyoite $(Ca_2B_6O_{11}{\cdot}13H_2O)$, penobosquisite $(Ca_2FeB_9O_{16}Cl{\cdot}7H_2O)$, pringleite and ruitenbergite (both $(Ca_2B_{26}O_{46}Cl_4{\cdot}25H_2O)$, strontioginorite $(Sr_2B_{14}O_{23}{\cdot}8H_2O)$, trembathite $([Fe,Mg]_3B_7O_{13}Cl)$, and tyretskite $(Ca_4B_{10}O_{19}{\cdot}3H_2O$; Burns, Grice, and Hawthorne, 1995; Grice, Gault, and Van Velthuizen, 1996; Hawthorne, Burns, and Grice, 1996). Some of the borates have more than 100 atoms (not counting hydrogen) and very complex polymer structures, implying very long crystallization times. The salt domes in which these minerals formed are highly faulted and folded, and on the basis of the borates present, appear to have been contacted by a mixture of geothermal brine, seawater end-liquor, and dolomitization brines (just as the German Zeichstein).

6.1.4 England

In the Permian Z3 Zeichstein potash formation in Yorkshire, England, veatchite $(Sr_4B_{22}O_{37}{\cdot}7H_2O)$ occurs as colorless, transparent, 3 mm × 0.5 mm needles or acicular and platy 2 mm × 0.5 mm crystals. They were found in 1330- to 1390-m-deep cuttings in polyhalite, anhydrite, halite–anhydrite, or dolomite shales, with one polyhalite–kieserite rock containing 0.9% Sr and 200 ppmB. Boracite $(Mg_3B_7O_{13}Cl)$ also was found at 1250–1270 m in halite's insoluble residues from an interbedded halite–anhydrite zone with some dolomite and quartz. The <1-cm clusters were colorless-to-grayish white and translucent, with a dull vitreous luster, averaging about 0.25% of their section of the cores. Small quantities of celestite $(SrSO_4)$ were also present (Guppy, 1944; Stewart, Chalmers, and Phillips, 1954). A bed of nodular ericaite $(Fe_3B_7O_{13}Cl)$ also was found in the base of an anhydritic shale layer, crosscut by sylvinite veins, overlaying sylvinite in the Z3 Boulby potash mine. It contained small amounts of Mn. The 1- to 15-cm × 0.5–7.5 nodules were ellipsoidal and tightly packed in their bed. Some were fairly pure, whereas others contained impurities of sylvinite, magnesite, and minor talc and hematite. The individual crystals were either large (<3 mm), blue-green, euhedral, and pseudocubic with well-developed faces, or brown-purple, anhedral, and smaller (<0.5 mm). In clusters of nodules the outer surfaces had larger crystals (with some twinning), and the center contained anhedral crystals (Milne, Saunders, and Woods, 1977).

6.1.5 Germany

The extensive Upper Permian Zeichstein evaporates of Germany contain four cycles of complete deposition from shale through potash, numbered Z1–Z4. Their most common borate mineral is boracite ($Mg_3B_7O_{13}Cl$), which almost certainly formed in a strong $MgCl_2$ brine, such as seawater end liquor, and at a high temperature (its cubic crystals are formed only at >265°C). Larger clusters of boracite ("stassfurite") are found primarily in highly faulted salt dome formations (Stassfurt Z2) that could be reached easily by an intruding mixture of end liquor and geothermal brine. Their crystals are tetrahedrons and cubes that form rounded 1–700 kg aggregates in the form of fine-radiating, double-refracting, orthorhombic lamellae. Boracite also is found in the Z1 (Werra), with its basalt intrusions, and the stassfurite nodules make up about 1/17 of the deposit's total boron content. There are several substituted boracite minerals with Fe^{++}, Mn, and other elements.

The next most common borate is ascharite ($Mg_2B_2O_5 \cdot H_2O$), which is widely distributed in the Z1 and Z2 halite and hartsalz fissures, and occasionally in the carnallite, gypsum, and cap rock of Z2 salt domes. It is especially common in Z2 langbienite (a high-temperature potash double salt)-containing hartsalz in the Hanover region where ascharite nodules with <1 m^3 volume (containing 23% carnallite, 4% bischofite, and 2.3% kainite inclusions) have been found. Higher boron concentrations in the precipitating brine are said to favor ascharite crystallization over boracite (Kuhn, 1968). Other high-temperature borates found in Germany are danburite ($CaB_2Si_2O_8$), lower Z1 anhydrite; datolite ($Ca_2B_2Si_2O_9 \cdot H_2O$), Z1 halite near and in the basalt; fabianite ($Ca_2B_6O_{11} \cdot H_2O$), Z1 halite; heidornite ($Ca_3Na_2[SO_4]_2B_5O_9Cl \cdot H_2O$), Z1 anhydrite and its fractures, often with glauberite; and the rare strontiohilgardite ($[Sr,Ca]_2B_5O_9Cl \cdot H_2O$), Z2 hartsalz. Tourmaline ($Na[Mg,Fe]_3Al_6[[OH,F]_4 B_3Si_6O_{27}$), is quite common, especially in one zone of black carnallite, two Z2 carnallite mines, and Z3 halite and sylvinite. Kuhn (1968) noted that some basement rocks under the Z2 formation contain "extensive felsic porphyrite," or "acidic volcanism" in the underlying Rotliegendes sandstone. These features, in addition to the ferroan and manganoan varieties of boracite, and the relatively large amounts of Sr, F, Mn, Cu, Pb, Ba, and Zn in the halite and potash all tend to indicate considerable geothermal intrusion into the Zeichstein. It also appears that there were considerable secondary changes, given the wide variety of lower-temperature or hydrated borates.

Most of the German borates are very small and disseminated, indicating that some geothermal water entered the basin while the seawater was evaporating or after the salts had redissolved, and remained in the void spaces to crystallize later. In the center of the Zeichstein basin boracite and ascharite are more concentrated in the Z2 potash by as much as 200 ppm over that at the edges. For instance, the hartsalz averages about 600 ppmB in the (central) Hanover area, and 400 ppm at the (outer) Hildesia mine (Kuhn, 1968).

Several mines that leached their potash ore have recovered boron as a saleable product, particularly in the area that was formerly East Germany (Braitsch, 1971). It was recovered from the insoluble residues and yielded about 4200 tons/year of borates in 1974–1976 (Borchert and Muir, 1964; Lyday, 1976).

6.1.6 Russia

6.1.6.1 Inder Borate Deposit

The Inder uplift is a 20–30 m high, 250-km^2 borate-containing plateau 150 km north of the Caspian Sea and east of the Ural River on a broad plain in Kazakhstan. The borates were discovered in it in 1934. It was formed by the rising pressure of a salt dome, had a 50-m-thick gypsum–anhydrite interbedded-clay outcropping caprock, and is one of the largest salt domes in the world. The halite, with some sylvinite in its upper section, is 100 m thick and is part of the Lower Permian period Kungur Formation. The caprock may be classified into at least four series, with the borates occurring primarily in the second and fourth (from the top) beds, with each having a high clay content. The gypsum's large crystal size and evidence of many faults, fractures, and water channels suggest that the anhydrite has been largely rehydrated since it was raised to a position readily assessable by surface water.

The dominate borate mineral in the deposit is hydroboracite ($CaMg B_6O_{11}\cdot6H_2O$), but β ascharite ($Mg_2B_2O_5\cdot H_2O$) may well have been the parent mineral. Priceite ($Ca_4B_{10}O_{19}\cdot7H_2O$), colemanite ($Ca_2B_6O_{11}\cdot5H_2O$), inyoite, ulexite, inderite, and boracite also are present, and all but the last two have had commercial value. In 1937 there were 56 known areas of borate deposition, but only a few were being worked. The majority of the borates (mainly hydroboracite) were concentrated in the northeastern quadrant in No. 2 bed's gray clay–gypsum layers, which were above the water table. The hydroboracite occurred as lenses or tabular blocks in the clay–gypsum, or as nodules in predominantly clay beds. The nodules had a felty texture of fine, needle-like crystals, which were coarser at the edges of the beds. Thin interbeds in gypsum contained parallel-fibrous hydroboracite, sometimes banded with 1-mm to 5-cm-spherulite aggregates.

The principal industrial mineral was β ascharite, which was most concentrated in the southern part of bed 2 in a narrow northeast trending band, but in lesser amounts throughout the area, with some found even in beds 3 and 4. It sometimes appears that ascharite was derived from hydroboracite (or colemanite and inyoite) because there were ascharite pseudomorphs of these minerals in the frequent alteration zones containing mixtures of all four minerals. However, ascharite had to be formed by a high-Mg, hot (deeply buried) brine, so only the reverse reaction should be possible in the present high-calcium, moderate-temperature environment. Ascharite occurred predominantly in clay matrixes as lenses or bedded deposits. Its crystals could have a felty and parallel fibrous texture, be chalk-like in powdery massive structures, or be coarse needles in larger radial or spherulite nodules. Ascharite sometimes occurred as priceite–ascharite or as ulexite–inyoite–ascharite mixtures.

Lower-grade deposits, such as in bed 4's red clay in the weathered or leached southern part of the uplift, contained a variety of other borates, often occurring as tabular blocks, crevice fillings, and other irregular accumulations. These deposits averaged about 2% B_2O_3, and rarely exceeded 7–8%. In crevices gypsum often co-crystallized with hydroboracite, and in blocks it was finely stratified with parallel fibrous crystals. Small amounts of inyoite were found mainly in fissures or as nodules with ulexite and inderite in clay, whereas β ascharite could form small incrustations on hydroboracite, and each mineral showed some signs of having replaced hydroboracite.

In the gray clays of the northeastern section, ulexite was common and sometimes occurred as cotton balls or as older much more dense nodules. Inyoite could form as a fissure filling or as nodules in the clay, and in some areas hydroboracite had been leached to calcite, forming beautiful pseudomorphs. Priceite occasionally was the dominant borate, forming snow-white nodules up to several meters in size, and composed of micron-size crystals. They generally were with colemanite or inyoite, and considerably compacted the surrounding clay. Also found in the southern section in a narrow band with a northwest trend along fault lines, priceite was the second most valuable mineral in that area. In general, most of the richer deposits were near the periphery of the uplift in well-drained gypsum. Above the gypsum beds was a sandy strata that in the northwest contained numerous <5-cm ulexite nodules. They were soft and white with long silky needles, elongated downward as if supplied by a borate solution from below, or as if they were near an evaporating surface. They occasionally formed low-grade commercial deposits. In the halite or sylvinite, borates such as kaliborite ($K_2Mg_4B_{24}O_{41} \cdot 19H_2O$) occasionally were found in small amounts (i.e., <2% B_2O_3) in clay layers or fracture zones. Many other borates in trace amounts also have been recognized in the deposit (Godlevsky, 1937).

Lyday (1992) briefly described the processing of Inder borate ores, which typically had an analysis of 18.9–21.2% B_2O_3, 4.2–6.4% SO_2, 4.3–5.8% CO_2, 14.1–21.1% MgO, 13.2–16.8% CaO, 0.9–1.7% Al_2O_3, 0.8–1.6% Fe_2O_3, and 18.9–22.2% H_2O. The ore was first ground to −2 mm(9 mesh) and repulped at 60°C in part of the prior filtrate and wash water. Sulfuric acid (10–15% less than stoichiometric) was added to form pentaborate, and the slurry was agitated at 95°C for 1 hr. The mixture was then settled and the clarified brine was removed and filtered. Next, the rest of the sulfuric acid was added, and the solution was cooled and seeded to crystallize boric acid. The residual mud was washed, filtered, and discarded, and the wash water was recycled. A large bleed stream was taken to prevent an excess of magnesium and other soluble salts from building up in the system, before the remaining end liquor was recycled. The bleed stream was evaporated, then cooled to crystallize epsom salts ($MgSO_4 \cdot 7H_2O$), and the remaining liquor was recycled. An 88% boric acid recovery was claimed of a fairly pure product. Matterson (1980) noted that low-grade ore with 1.5–25% B_2O_3 in halite could be deslimed and leached to 44% B_2O_3, while ores high in clay and gypsum could be beneficiated to 18–34%B_2O_3.

The initial reserves of "massive" and >28% B_2O_3 ore were 400,000 tons, but by 1994 there was only 2–5% B_2O_3 disseminated ore left. It was still being produced, but the Kazakhstan government was seeking companies for further development. The estimated reserves were 33.7 million metric tons of 2.15%B_2O_3 hydroboracite and boracite(?) (along with some potash) in four areas with beds 1–10 m thick and <2 km long. The government also sought developers for the Satimola deposit with its 43 million metric tons of 9.15%B_2O_3 nodules and veins of hydroboracite and ulexite in the gypsum–clay cap of another salt dome (Lyday 1994, 1995).

6.1.6.2 Upper Kama, Byelorussia, Carpathia

In the Upper Kama (Solikamsk), potash deposit kurnakovite ($Mg_2B_6O_{11} \cdot 15H_2O$) and preobrashenskiite ($Mg_6B_{22}O_{39} \cdot 9H_2O$) have been reported (Borchert and Muir, 1964), and volkovskite [$(Ca, Sr)_2B_{14}O_{23} \cdot 8H_2O$]) was found in halite drill core insoluble residues. The volkovskite consisted of colorless vitreous luster 1.5 × 0.5 × 0.05-mm plates accompanied by hilgardite, boracite, anhydrite, and sylvite (Christ, 1966). In the other two Russian potash mining basins borates are almost totally unknown. The two exceptions are very small amounts of ericaite ($Fe_3B_7O_{13}Cl$) and chambersite ($Mn_3B_7O_{13}Cl$) in Byelorussia, and lunebergite ($Mg_3[PO_4]_2B_2O_3 \cdot 8H_2O$) in Carpathia.

6.1.6.3 Kungurian Sequence (Upper Kama to the Caspian Region)

A number of borates have been noted in the scattered, usually complex potash deposits extending from the Urals to the Caspian area. Most appear to be in the halite or potash strata, with a strong correlation between the borates and clay in the deposit. All also are found within 5 m of the vertical extension of basement rock fracture lines (Valeyev *et al.*, 1973). When borates were present in the Satimolinsk salt mass, the insolubles averaged 12.9% compared with 3.9% of the surrounding nonborate salts. In the Chelkar marine structure the corresponding numbers were 19.1% and 7.1%, respectively, indicating that considerable terrestrial water had entered the basin (to transport insolubles) during the potash-depositing period. Much of the drainage area contains ultrabasic rocks (diabases, serpeninites, etc.) with 100–200 ppmB, and sandstones with <170 ppmB. The runoff water thus had a high boron content, but because of the strong adsorption on clays, only those areas low in illites and hydromicaceous clays contained borates.

Deep wells in the region have encountered geothermal waters of the NaCl type containing <500 ppmB, but no F, Li, or metals. Deeply penetrating groundwater may have been heated and circulated through the marine formation, thus leaching some of its boron content. When later cooled and in anhydrite (converting it to gypsum), it liberated more boron and allowed the borates to concentrate into nodules, lenses, and larger crystals (Valeyev *et al.,* 1973). In an earlier report on the Kunguru Salt Dome in the pre-Caspian Depression, sizeable deposits of searlesite ($NaBSi_2O_6{\cdot}H_2O$) were noted. The searlesite is banded with layers of halite containing volcanic ash and thenardite (Yarzhemskii, 1969).

6.1.6.4 Donets Basin

The Lower Permian Artemovka Formation in the Bakhmuta Depression of the Donets Basin also has been found to contain a number of borates. They occur only on the east side of the depression where it outcrops. There halite has been dissolved and the overlying anhydrite partly hydrated into gypsum. The borates occur in several gypsum mines in small, almost spherical, fairly uniformly dispersed pockets at the junction of a 1-m anhydrite bed that is still being hydrated. Some are filled with tabular gypsum, thenardite, and occasionally borates. Less frequently, there are horizontal blind fissures healed with tabular or columnar gypsum and borates. No borates are in the dense gypsum and anhydrite.

Ulexite ($NaCaB_5O_9{\cdot}8H_2O$) is the most common borate, usually with some Mg, K, Cu, and V (and trace amounts of Ba, Sr, Al, Ti, Si, and Mn) replacing Na and Ca. Microscopic amounts of thenardite, celestite, calcite, or dolomite are found with it. Probertite ($NaCaB_5O_9{\cdot}5H_2O$) is much less common, occurring as round or oval 1- to 2-mm nodules of dense, wavy crystals with a vitreous silky luster. Its hardness is 3–3.5, its general appearance resembles ulexite, and small amounts of kernite ($Na_2B_4O_7{\cdot}4H_2O$) are found with it. In two drill holes there were mixtures of hydroboracite ($CaMgB_6O_{11}{\cdot}6H_2O$), inyoite ($Ca_2B_6O_{11}{\cdot}13H_2O$), and inderite ($Mg_2B_6O_{11}{\cdot}15H_2O$). The hydroboracite occurred as single crystals or as nodules of radial to fibrous aggregates with a vitreous luster. The inyoite consisted of small transparent granular aggregates, also with a vitreous luster, generally not with hydroboracite, and occasionally found alone. Inderite could be detected only by x-ray diffraction, and all of the borates contained trace amounts of Al, Ba, Cu, Fe, Mn, Na, Si, Sr, Ti, and V (Zaritskiy, 1965).

6.1.6.5 Irkutsk Area, Siberia

Both strontiohilgardite [$(Sr,Ca)_2B_5O_9Cl{\cdot}H_2O$] and danburite ($CaB_2Si_2O_8$) have been found in the Upper Angara Member of the Lower Cambrian marine dolomite–anhydrite rocks (interstratified with halite) in the Ilga Trough in eastern Siberia. On some of the strontiohilgardite aggregates there is a 1-mm layer of acicular–fibrous ginorite ($Ca_2B_{14}O_{23}{\cdot}8H_2O$), with a chalcedony-like structure (typically 58% strontiohilgardite and 25% ginorite). Locally, the ginorite can form discrete chains of 0.1- to 0.2-mm nodules. Most of the strontiohilgardite is white or yellowish and <0.01 mm in size, but some is coarsely crystalline with a thick, radiating fibrous appearance. Inside the occasional small (0.2 mm) strontiohilgardite nodules there is some dolomite, halite, silicates, or other solids. This mixture typically contains 0.1- to 0.15-mm crystals that are 87% dolomite, 3% strontiohilgardite, 4% anhydrite, and 6% halite, with some chalcedony–quartz. In fissures and dolomite–anhydrite contact zones the strontiohilgardite layer can be 10–15 mm thick.

In the high-anhydrite sections of the dolomite–anhydrite rock [with thin layers (20 bands/17 mm) of anhydrite and anhydrite–dolomite], oolitic inclusions of danburite are common. A typical analysis might be 52% anhydrite, 19% dolomite, 25% danburite, and 1% each of celestite, silica, and R_2O_3. There is more danburite higher in the formation. The very thin (0.001- to 0.005-mm) danburite layers often connect to danburite nodules. The nodules are 0.05–8 mm across, 0.5–3 mm being the most common, and all have a microcrystalline structure. Frequently, they are spherically shaped and like beads strung on the danburite laminae layer, sometimes hanging down like a raindrop and sometimes pointing upward. There is intense silicification (opal, chalcedony, quartz) in some dolomite structures, forming karst cavities and caverns. The danburite occurs more randomly and locally, whereas the strontiohilgardite is found fairly uniformly throughout the entire formation (Yarzhemskii, 1969).

6.1.7 Thailand

The massive Khorat Plateau terrestrial (not marine, but similar; Garrett, 1995) salt–carnallite–tachyhydrite–sylvinite deposit contains a large amount of boracite ($Mg_3B_7O_{13}Cl$) in its sylvinite and carnallite. In three drill holes a few thin intervals of sylvinite near the top of the deposit contained up to 8.5% boracite (with no other insolubles), whereas the insolubles lower in the hole averaged 1.5% boracite in anhydrite. In other cores of the carnallite section boracite comprised about 85% of the carnallite's insolubles, and in one hole averaged 3.7% of a 1.86-m sylvinite zone. The <3 mm boracite consisted of either pale yellow tetrahedrons disseminated through the other salts, or irregular white 0.1- to 10-mm nodules (Hite, 1986; Hite and Japakasetr, 1979).

6.1.8 Turkey

Another terrestrial (but similar to marine) borate occurrence was near Bandirma, first commercially operated in the late 1800s (perhaps about 1865). The product was initially called boracite, but almost certainly it was priceite, which was later mined. Gypsum had been quarried for centuries from a deposit inland from Bandirma on the Marmara Sea, and used as a polishing agent for marble from a nearby quarry. The workers frequently removed large nodules of a white, compact, chalk-like impurity. Some of these nodules were set aside, whereas others were carved into figures or sculptures, which were then given or sold locally. Eventually, a nodule was analyzed and found to contain boron, which resulted in the organization of a company to mine the nodules. It prospered sufficiently to warrant the establishment of a special refinery near Paris for converting the priceite to boric acid. After a number of years of successful operation, the company was sold to Borax Consolidated of London, and it was closed (Robottom, 1893).

6.1.9 United States

6.1.9.1 Alabama, Louann Salt

The Upper Louann Salt and its overlying Norphlet Shale of a salt dome in southwestern Alabama at a depth of 3719–3727 m contains five different borates. They occur in fractures or can be found in the salt's 0.5% insoluble residue (mainly anhydrite and talc, with some quartz, magnesite, phlogopite, illite, and K feldspar). Boracite ($Mg_3B_7O_{13}Cl$) is the dominant borate, occurring as bluish-green cubic 0.1- to 1-mm crystals, with a hardness of 7 and several multicrystalline structures. Hilgardite-1Tc ($Ca_2B_5O_9Cl{\cdot}H_2O$), the next most common borate, is found as aggregates of clear wedge-shaped crystals, and as white to grayish-white spherical microcrystalline <2cm agglomerates. Szaibelyite ($Mg_2B_2O_5{\cdot}H_2O$) is found in acicular masses and radiating sprays, with 0.2 mm $\times$ 0.005 m thin blades and even smaller needles. A few crystals of volkovskite [$(Ca,Sr)_2B_{14}O_{23}{\cdot}8H_2O$; analyzed as $KCa_4B_{22}O_{37}Cl{\cdot}9H_2O$] were found as small rounded plates, and danburite ($CaB_2Si_2O_8$) was in the overlying halite–shale junction and cracks in the shale within 1 m of the halite. It occurred with halite as white opaque <4-cm nodules. The shale fractures had been clearly distorted by the growth of the nodules (Simmons, 1988). The boron content of the halite and shale was considerably enriched compared with that normally found in similar Gulf Coast evaporites. Swihart, Moore, and Callis (1986) found that the boracite and hilgardite had $\delta^{11}B$ values of +28.1 and +29.8, respectively, indicating that they were deposited from a mixture of seawater end liquor and geothermal brine at great depth. Leaching of silica from shale would have allowed danburite (as well as the quartz and talc) to form as the brine left the halite.

6.1.9.2 Louisiana, Choctaw Salt Dome

A small amount of borates has been found in the insoluble residue that fell to the floor during the solution mining of the Choctaw salt dome. Hilgardite ($Ca_2B_5O_9Cl{\cdot}H_2O$ I) and parahilgardite ($Ca_2B_5O_9Cl{\cdot}H_2O$ II), twin right- and left-hand crystals, were identified in the 2–4% insolubles (mainly anhydrite, but also, in descending order of abundance: dolomite, clay, shale, magnesite, hilgardite, parahilgardite, pyrite, calcite, boracite, danburite, quartz, sulfur, limonite, hematite, marcasite, hauerite, and gypsum). The hilgardite occurred as well-formed monoclinic 0.5- to 15-mm crystals that resembled small arrow-heads with one right angle. Parahilgardite, always found as 0.3- to 8-mm twins attached to the negative end of the α axis of a hilgardite crystal, was usually smaller but proportional to the attached hilgardite. The surfaces of the twins were more etched than that of the hilgardite and appeared to be of the pedial class (i.e., lacking all elements of symmetry; Hurlbut, 1938; Hurlbut and Taylor, 1937).

6.1.9.3 Mississippi

Unidentified borate crystals with a formula of $Mg_{10}B_{23}O_{45}Cl_2$ (and a description suggesting boracite, $Mg_3B_7O_{13}Cl$) were found in two wells at the fluid-penetrated upper boundary of the Louann salt formation. The samples were from cuttings at a 5200- to 5303-m depth in a poorly consolidated sandstone strata under the Smackover carbonates. The 1-mm borate was in the form of clear pseudoisometric cubes or tetrahedrons with a hardness of 6.2–6.8, with a density of 2.88, and in a matrix suggestive that it grew in its present location. X-ray diffraction analysis gave results similar to those of the boracite group, but not the same as those for any known borate. The molecular structure also may have contained 0.01 moles of Na, Al, Fe, and Ba and 0.08 moles of Ca and Si, with a major element formula of $Mg_3B_7O_{13}(Cl_{0.6}Si_{0.1}[SO_4]_{0.3})$ (Gann, Dockery, and Marble, 1987). Later Sundeen (1990) examined the cuttings from 39 wells drilled to the Louann salt in an area about halfway between the Gann *et al.* (1987), and Simmons (1980, 1988) in Alabama. Nine 3660- to 4570-m deep wells contained small amounts of boracite ($Mg_3B_7O_{13}Cl$), and one had a single white nodule of hilgardite ($Ca_2B_5O_9Cl{\cdot}H_2O$). The 0.5- to 2-m euhedral boracite crystals were clear, pale yellow, or pale bluish-green, having a hardness of 7, and generally cubic with some tetrahedrons and combinations. About 16% of the boracite contained enough iron to be classified as ericaite, with some Al, Ca, and Mn. A study of the insolubles in halite from 20 Gulf Coast locations also has been made by Taylor (1937): 17 in Louisiana, 2 in Texas, and 1 in Mississippi. Only the latter sample contained borates: <3-mm colorless danburite ($CaB_2Si_2O_8$) and hilgardite, and 0.1- to 0.5-mm cubic boracite, along with several other minerals not of marine origin.

6.1.9.4 New Mexico

In the Permian Carlsbad potash deposit, in halite or clay in the sylvinite, lueneburgite ($Mg_3[PO_4]_2B_2O_3 \cdot 8H_2O$) has been identified in many localities (Schaller and Henderson, 1932). Aquifer brines in the salt below the potash also contain a large excess of boron (1130 ppmB) over that of seawater end liquor, as well as high Sr, Mn, Fe, Al, and F, indicating the presence of a geothermal brine (Abitz, Myers, Drez, and Deal, 1990).

6.1.9.5 Oklahoma

Probertite ($NaCaB_5O_9 \cdot 5H_2O$) as well as small amounts of ulexite ($NaCaB_5O_9 \cdot 8H_2O$) and priceite ($Ca_4B_{10}O_{19} \cdot 7H_2O$) have been found as small nodules in drill cores and gypsum quarries in three districts separated by about 110 km in the Blaine (Middle Permian) and Cloud Chief (Late Permian) of west central Oklahoma. The gypsum beds are 152–183 m apart and nearly horizontal. They have no major faults, and there is no evidence of nearby volcanic or geothermal activity. The borates occur in <10% of the gypsum, and concentrate along bedding plains near beds or lenses of anhydrite. There seldom is more than 0.16% B_2O_3/0.3 m thick section, and often some celestite and thenardite, as well as a small amount of halite in the anhydrite (Ham *et al.* 1961).

In one area of the Cloud Chief formation probertite with some ulexite pseudomorphs of probertite were found as small nodules in a halo 11–22 m above, below, and around a 4.8 × 9.7 km lens of anhydrite. Elsewhere it was within 1.5 m above massive anhydrite or 46 cm below an anhydrite lens (with a few 7.6- to 12.5-cm nodules of ulexite and smaller nodules of priceite). In the Blaine formation, only small 1- to 15-mm (average 4-mm) nodules of probertite have been found, which have a porcelaneous and cauliflower appearance of tightly compacted 0.1- to 1-mm crystals. The crystals are slightly curved (or loosely coherent silky aggregates of randomly oriented tufts) with a 10/1 length-to-diameter ratio. The probertite is found only in the high-purity (96–98%) Nescatunga gypsum, and its amount and location vary widely. The ulexite also occurs as white, cauliflowerlike nodules, or as encrustations on selenite. The 2.5- to 12.7-mm (most are 6.4- to 12.7-mm) nodules are arranged as bundles of fine, silky, curved crystals and may be somewhat flattened along the gypsum bedding plain. The relatively rare priceite is found as small nodules of chalk-like appearance, and usually with selenite.

Several laboratory experiments were conducted to determine the nature of the boron in anhydrite (average 72 ppmB. Cloud Chief 87 ppm, Blaine 53 ppm). There appears to be no solid-phase borate as inclusions, and leaching left no borates in the residue. In the case of one −8 micron or one -8μ sample with many liquid inclusions, only 17–37% of the original 133–173 ppmB could be washed away after fine grinding. Also, when anhydrite was precipitated from 1–6% H_3BO_3 solutions, it contained 28–46 ppmB, both tests implying that much of the boron is in the anhydrite's lattice network. The near-shore anhydrite had more boron (<215 ppmB) than the center (>19 ppm), similar to the German Z2 (shoreward anhydrite reached <500 ppmB; Ham *et al.* 1961).

6.1.9.6 Texas

Chambersite ($Mn_3B_7O_{13}Cl$) has been found in insoluble sediments from a gas storage cavern in the Barber's Hill salt dome. The 1-mm to 1.2-cm crystals were subhedral to euhedral tetrahedrons, colorless to deep purple (from oxidation of the manganese), with a hardness of 7 and a density of 3.49. The cavern is 707–829 m deep, and the insolubles were predominantly anhydrite (Honea and Beck, 1962).

6.1.9.7 Utah, Paradox Basin

Braitschite ($[Ca,Na_2]_6RE_2B_{24}O_{45} \cdot 6H_2O$) has been found in the Cane Creek (Moab) potash deposit, which is in the Paradox Member of the Hermosa Formation of the Late Carboniferous (Middle Pennsylvanian) period in southeastern Utah. It occurs as roughly spherical white to pinkish red 0.05- to 8-mm nodules containing needle-like crystal aggregates, with anhydrite, halite, and some hematite (average, 65% braitschite). Some nodules have combined to form aggregates. They are distributed through a 15-cm (up to 1-m) anhydrite bed (containing variable amounts of dolomite and halite, and some pyrite, quartz and mica) overlying sylvinite. Braitschite also was found in the same position at 910- to 1000-m depth in drill holes 1.7 km south and 3.3 km north of the mine shaft. Nodules occasionally are in an upper dolomite bed, the lower sylvinite, or thin halite veins that extend about 2 m into the anhydrite. Smaller nodules may not contain hematite, and larger ones can contain chalcopyrite. The distribution of rare earth elements in the braitschite is similar to that in rocks in the adjacent drainage basin, and quite different from that in seawater. Also, the quantity of rare earths is far greater than seawater (0.00034 ppm) could have provided (based on the amount of NaCl). It thus appears that the rare earths and boron entered the basin from geothermal springs at the end of the No. 5 potash cycle (Raup, 1972; Raup and Gude, 1968).

6.2 ISOTOPE RATIO, $^{11}B/^{10}B$; $\delta^{11}B$

6.2.1 General

In nature boron occurs in only two isotopic forms, ^{10}B and ^{11}B. The relative amount of these isotopes in boron-containing minerals and waters varies considerably, as is seen in Table 6.5. A $\delta^{11}B$ value is commonly used to measure this variability as follows:

$$\delta^{11}B = 1{,}000 \left\{ \frac{^{11}B/^{10}B\ (\text{mineral})}{^{11}B/^{10}B\ (\text{reference})} - 1 \right\}$$

This magnifies the basic values of interest, the percentage of ^{10}B or ^{11}B, as well as their inaccuracy, by as much as 13-fold. Its disadvantage is that the author must establish a reference number, then let the reader know what it is. Over the years, every author listed in Table 6.5 on $\delta^{11}B$ values used a different reference number, so some of them have been recalculated using the U.S. Bureau of Standard's 1989 reference value for boric acid, NBS 951 (Anon., 1991). Boric acid contains 80.18% ^{11}B and 19.82% ^{10}B (the material analyzed was U.S. Borax Inc.'s commercial boric acid, produced from kernite in their Boron, California mine). This gives a reference $^{11}B/^{10}B$ ratio of 4.0454, with the isotope analysis having a stated ±0.02% (^{10}B = 19.82% ± 0.02; ^{11}B = 80.18% ± 0.02) accuracy. The uncertainty (±0.02) can change the $\delta^{11}B$ value by 1.3 units, so the numbers in the table have been rounded off to one decimal point. Most of the authors' $\delta^{11}B$ values appear to be fairly consistent, but with some scattering. The Shima (1963) numbers, for instance, appear to be reasonable, except +6.3 for seawater (average of others is +40.0) and +49.1 for borax (average of others is −2.6). The out-of-line numbers were not shown in the table. All of the $^{11}B/^{10}B$ values of Thode, MacNamara, Lossing and Collins (1948) were very high, but when their Boron, California value of 4.322 for $^{11}B/^{10}B$ was used as the reference number, their $\delta^{11}B$ data agreed well with the others. For Gregoire (1987), a correction of +8.8 $\delta^{11}B$ to compensate for his 32.5 seawater brought his other numbers in line. More typically, Vengosh *et al.* (1992), like many other authors, appeared to have a consistent error (their seawater was 25.7), so their reference number and $\delta^{11}B$ values were not corrected.

6.2.2 Isotope Distribution

Boron is a highly unusual element in that the ratio of its isotopes varies comparatively widely. The $\delta^{11}B$ values of different boron sources have been recorded from +56.6 to −49.3. This indicates an 8.84% range in the ^{10}B content, and causes the molecular weight of boron from its different sources to vary from 10.807 to 10.824 (assuming atomic weights of ^{10}B = 10.016122, and ^{11}B = 11.012805). The value of 10.811 has arbitrarily been established as the official molecular weight, although the reference ore has a value of 10.815. To have a molecular weight of 10.811, the reference atomic ratio should have been 3.9383 (79.75% ^{11}B, 20.25% ^{10}B) or a $\delta^{11}B$ of −26.5 (based on the present reference number). This value is very far from the actual isotopic average

Table 6.5
Reported $\delta^{11}B$ Values for Boron-Containing Substances

	Average $\delta^{11}B$	Reported $\delta^{11}B$ Values; location
A. Marine		
1. Seawater	40.0	44.0[a], 42.7[b], 41.1[c], 40.0[d,e], 40.1[f], 39.6[g], 39.5[h,i,j], 39.1[k,l], 39.0[m,n], 38.4[o]
Brine in		
Marine sediments[n]	40.0	25–49 (10.6 ppmB)
Red Sea[p]	33.0	30–39 (8 ppmB)
Mediterranean	46.0	54.2 (5.2 ppmB)[q], 37.7 (5.3 ppmB)[m]
2. Borates		
Ulexite[r]	32.5	Inder
Hilgardite[r]	29.8	Alabama
Howlite[r]	24.9	Nova Scotia
Inyoite	23.5	23.6[s], 23.1[r] New Brunswick; 23.6[s], 23.1[t] Germany; 24.7[aaa], 22.6[r] Inder
Boracite	21.8	28.1[r] Alabama; 23.1[t], 19.2[r], 16.7[r] Germany
Average marine borates	26.4	
3. Others		
Andydrite[u]	31.5	
Oil field water[k]	27.7	(123 ppmB), U.S. Gulf Coast
Ocean vents	26.6	33.4 (450 ppmB) East Pacific Rise[v]; 32.9 (5 ppmB)[j,w] 31.4 (5.5 ppmB)[p] Midocean Ridge; 30.0 (8 ppmB) Red Sea[b]; 26.4 (10.5 ppmB)[v,x,w], 20.0 (9.5 ppmB)[p] Six other systems; 33.5 (48 ppmB, 314°C)[i] 21.5 (17 ppmB), 315°C)[j], 19.2 (17.5 ppmB)[p] Guaymas Basin; 26.1 (9 ppmB)[p], 20.0[h], 19.2 (18 ppmB)[w] Marianas Back Arc
Carnalite	25.3	36.0 (39 ppmB) marine[m]; 14.6 inland, China[y]
Corals[d]	25.1	24.9 (53 ppmB) modern; 25.2 (46 ppmB) ancient
Biogenic carbonates	24.0	39.0 Conodont fossil[u]; 23.3 (32 ppmB)[z]; 22.1 (41 ppmB)[d]; 18.6 (14 ppmB) sediments[z]; 17.1 (20 ppmB)[o]
Aragonite[e]	23.1	(45 ppmB)

Dolomite[e]	22.7	(54 ppmB)
Calcite[e]	22.0	(37 ppmB)
Ocean foraminifera	21.6	24.0 glacial[aa]; 21.7 Holocene[aa]; 19.2 (20 ppmB)[z]
Ocean vent crust[bb]	21.5	Guaymas Basin
Gypsum	13.4	13.6[u]; 13.2[m]
Clay, shale, slate	12.2	15.3[u], 14.5 (155 ppmB)[cc] Japan; 13.0[c], 10.9[b] New Hampshire; 10.0 Russia[dd]; 9.5 (55 ppmB) Illite[dd]
Sepentinized peridotites	11.1	12.0 (65 ppmB)[ee]; 10.9 (64 ppmB)[i]; 10.5 (66 ppmB)[ff]
Halite	10.5	16.5 (9.1 ppmB) marine[m]; 4.4 inland, China[gg]
Ocean floor	8.9	15.3[u], 14.5[cc] sediments; 2.8 (1–8 ppmB) crust[ff]
Smectite[hh]	5.8	2–9
Basalt, altered ocean vent	4.8	6.4 (10 ppmB)[ff]; 4.8 (28 ppmB)[ee]; 4.7 (39 ppmB)[h]; 4.5 (25 ppmB)[j]; 3.5[i]
Tourmaline, metapelite[ii]	4.4	
Biogenic silica ooze, recent[hh]	3.5	−4 to 11 (62 ppmB)
Tourmaline in marble[ii]	1.8	−1 to 4
B. Other High $\delta^{11}B$ Sources		
Dead Sea[n]	56.6	(47 ppmB)
Dead Sea hot springs[n]	52.9	(24 ppmB)
Australian salt lakes[o,y]	39.7	30–57 (3 ppmB)
Dead Sea, brackish, fresh springs[n]	37.3	(0.6 ppmB)
Mud volcanoes	35.6	43.1 (26 ppmB) Taiwan[w]; 28.0 (5 ppmB) Barbados[f]
Australian springs, rivers[o]	31.8	(0.3 ppmB)
Sborgite[q]	31.1	Larderello power plant scale
Inland fumarole water	28.3	33.0 (129 ppmB) Rumsey Hills[f]; 30.2 (454 ppmB) Iceland[g]; 25.7 (33 ppmB) Larderello[q]; 24.2 (94 ppmB, 305°C) Salton Sea[f]
Kimberlite, Meisnechite[jj]	17.6	Russia
Groundwater	13.7	28.8 (0.1 ppmB) Israel[kk]; 16.2 (0.2 ppmB) Ariz., Tex., Wisc.[ll]; 15.7 (4.3 ppmB) Larderello[q]; 15.2[a]; −7.4 (0.9 ppmB) Australia[o]
Sassolite	13.6	21.1[t], 20.1[q] Larderello; 9.8[a,cc], 3.2[mm], −3.0[mm] Japan

continues

Table 6.5 (*continued*)

	Average $\delta^{11}B$	Reported $\delta^{11}B$ Values; location
Gabro	10.9	14.7 South Africa[2]; 7.1[mm]
Sewage effluent[kk]	9.7	5.3–12.9 (0.6 ppmB) Israel
Metapelite marble[c]	8.2	
Boron compounds[cc]	7.6	(B, B_2O_3, boron carbide)
Sodium perborate	3.1	U.S. Borax
C. Skarns		
Kotoite[a]	18.2	Korea
Suanite[nn]	11.0	China
Ludwigite[nn]	10.2	China
Tourmaline	9.5	9.5 China[oo]; 9.5 Finland[a]
Szaibelyite (ascherite)	7.6	10.0 China[oo]; 5.7[aaa], 4.6 B.C. Canada[s]
Datolite	3.5	4.4[a] Turkey; 3.1[cc], 2.9[a] New Caldonia[g]
Skarn average	10.0	
D. Hot spring borates (borate deposits)		
Inderite[a]	5.8	Argentina
Hot springs water	5.7	14.7[pp], 10.7 (9 ppmB, 59°C)[qq], 10.4 (2 ppmB)[rr], 10.3 (65 ppmB)[ss], 10.0 (39 ppmB, 294°C)[l], 5.5 (3[bb]–15[l] ppmB), 5.1 (20 ppmB, 436°C)[mm] Japan; 5.5[ff]; 3.7 (7 ppmB)[tt] China; 0.9 Tibet[t]; −1.8 (238 ppmB) Iceland[g]; −6.3 (11 ppmB, 88°C) Yellowstone[nn]
Playa lake brine	5.6	12.5 Dabuxan[y]; 8.1 (620 ppmB) interstitial, 5.0 (598 ppmB) surface DaQuidam[tt]; −8.5 Xiao Qaidam[gg] China; 11.1 Searles Lake[pp]
Playa sediments	−0.3	5.1 (6.3 ppmB) Australia[o]; −5.6 (6880 ppmB; clay, halite, gypsum) China
Priceite (pandermite)	−0.4	5.8[a], −6.5[t] Turkey
Borax	−2.6	(11.1[t], 9.0[uu], 2.4[pp], −1.2[e]) 5.3ave. Searles Lake; (1.9[uu], 0[vv,t], −0.1[r], −1.1[ww]) 0.1 ave. Boron; (−2.6[j], −4.5[a]) −3.6 ave. Argentina; −3.7[xx] Turkey; (0.9[t], −8.8[cc]) −4.0 ave. Tibet; (−8.8[cc], −9.5[j], −11.5[a]) −9.9ave. Kashmir
Kernite	−3.3	0.0 NBS951; 0.1[uu] Boron; −7.0[xx] Turkey

Hydroboracite	−4.2	+4.4[a] Argentina; −8.5[ll] China; −12.0[xx] Turkey
Tunellite[xx]	−4.3	Turkey
Ulexite	−4.5	(6.8[a], −1.7[uu], −5.5[ww], −7.2[xx]) −1.9 ave. Turkey; (−4.7[vv], −3.6[a], −3.6 ave.[uu]) −4.0 ave. Boron; −4.2 Columbus Marsh[yy]; (+5.8[a], −10.7[cc], −12.0[t]) −5.6 ave. Argentina; (−8.2[ll], −13.2[gg]) −10.7 ave. China; −15.2 San Bernardino[r,gg]; +7.3[j], −1.7[uu] Misc.
Hamburgite[s,g]	−4.8	Madagascar
Tourmaline	−5.4	−1.0[cc], −6.5[u]; −8.8[h] Sulfide deposits
Inyoite	−8.3	(−2.1[a], −5.5[ii], −9.0[xx]) −5.5 ave. Turkey; −11.0[r] Boron
Colemanite	−8.6	−0.1[a] Argentina (−5.3[uu], −7.3[u], −12.5[s], −13.4[a]) −9.6 ave. Death Valley; (−8.6[vv], −11.4[aaa], −12.9[b], −13.2[uu]) −11.6 ave. Boron; (−8.8[uu], −11.4[aaa], −12.5[s], −13.0[xx], −13.0[a]) −13.2[qq], ave. Turkey
Veatchite-A[xx]	−9.0	Turkey
Meyerhofferite	−9.4	−8.5[r] Death Valley; −10.2[xx] Turkey
Mcallisterite[ll]	−9.7	China
Probertite[xx]	−12.5	Turkey
Average	−5.4	Primary hot spring borates
Secondary hot spring borates		
Howlite	−20.9	−20.6[s], −21.4[r] San Bernardino; −20.8[xx] Turkey
Pinnoite[gg]	−21.8	China
Teruggite[xx]	−25.0	Turkey
Priceite[xx]	−17.2	Turkey
Average	−21.2	Secondary hot spring borates
Total	−8.7	All hot spring borates
E. Miscellaneous sources		
Rainwater[ll]	16.7	China
Alder wood[t]	12.3	
River sediments[x]	4.6	6.4 (80 ppmB) suspended; 2.7 (134 ppmB) bulk
Carbonate rocks[z]	4.4	2–8 (7 ppmB) Israel
Granite[c]	3.2	5.3; 1

continues

Table 6.5 (*continued*)

	Average $\delta^d B$	Reported $\delta^d B$ Values; location
Feldspar, pyroxene[bb]	−0.5	(17 ppmB)
Kimberlite[a]	−1.1	(4 ppmB)
Basalt	−1.1	+3.6 Russia[jj]; +1.3[mm], −3.4[ee] (1 ppmB), −3.5[a] Hawaii; −0.8[cc], Mohole; −2.9 (0.5 ppmB)[h]; −3.0 (0.4 ppmB)[j]
Serpentine[u]	−2.6	−11 to −0.1
Beryl[a]	−3.3	(7 ppmB)
Altered hyaloclastite[u]	−3.7	
Coal fly ash[ll]	−3.9	+15.8 to −7.9 (10 ppmB), U.S.
Midocean basalt	−4.2	−3.0[ff], −3.4[zz] (0.9 ppmB), −5.9 (1.1 ppmB)[ff], −3.1[i] (0.4 ppmB)
All terrestrial rocks	−4.2	+4.8[cc], −4.1[b], −7.0[zz], −10.5[ee]
Rhyolite	−7.5	−7.5[c]; −7.5 (5 ppmB) Yellowstone[nn]
Aragonite[tt]	−8.5	China
Tourmaline in pegmatites, granite	−9.2	−6.5[u], −7.1[ww], −8.5[ii], −9.5, −11.0 Finland[a], −12.7[ww]
Meteorites, general	−9.4	+17.5 (0.5 ppmB) Stony-iron[a], +4.6 (0.6 ppmB) Canyon Diablo[a]; −2.0 (0.1 ppmB) Orgueil[ff,zz]; −5.5 (23 ppmB) Tektites[a]; −3.0[zz], −4.0 (0.6 ppmB)[a], −12.0 (14 ppmB)[ff], −19.2[cc], −21.1 Achondrite[cc]; −49.3[cc] Chondritic
Upper mantle[ff]	−10.2	(0.02–0.2 ppmB)
Continental crust[ff]	−10.5	−8 to −12 (10 ppmB)
Various ores[u]	−12.2	−7.3 Mn, Fe oxide; −11.0 Sn silicate; −18.3 Zn, Pb sulfide

[a] Agyei and McMullen, 1968.
[b] Schwarcz, Agyle and McMullen, 1969.
[bb] Musashi, Nomura, Okamota, Ossaka, Oi and Kakihana, 1988.
[cc] Shima, 1963.

[c] Barth, 1993.
[d] Gaillardet and Allegre, 1995.
[e] Hemming and Hanson, 1992.
[f] You, Spivack, Gieskes, Martin and Davisson, 1996.
[g] Aggarwal, Palmer and Ragnarsdothr, 1992.
[h] Palmer and Slack, 1989.
[i] Spivack and Edmond, 1987.
[j] Spivack and Edmond, 1986.
[k] Land and Macpherson, 1992.
[l] Nomura Kanzaki, Ozawa, Okamoto, and Kakihana, 1982.
[m] Vengosh, Starinsky, Kolodny, Chivas and Raab, 1992.
[n] Vengosh, Chivas, McCulloch, Starinsky, and Kolodny 1991a.
[o] Vengosh, Kolodny, Starinsky, Chivas and McCulloch, 1991b.
[p] Palmer, 1991a.
[q] Klotzli, 1992.
[r] Swihart, Moore and Callis, 1986.
[s] Finley, Eberle and Roddin, 1962.
[t] Thode, MacNamara, Lassing and Collins, 1948.
[u] Gegoire, 1987.
[v] Palmer, 1991b.
[w] You 1994.
[x] Spivack, Palmer and Edmond, 1987.
[y] Vengosh, Chivas and McCulloch, 1989.
[z] Vengosh, Starinsky, Kolodny, and Chivas, 1991c.
[aa] Sanyal, Hemming, Hanson and Broecker, 1995.
[dd] Shergina and Kaminskaya, 1967.
[ee] Chaussidon and Jambon, 1994.
[ff] Chaussidon, 1995.
[gg] Sun, 1990.
[hh] Ishikawa and Nakamura, 1993.
[ii] Swihart and Moore, 1989.
[jj] Cherepanov, 1967.
[kk] Vengosh, Heumann, Juraske and Kasha, 1994.
[ll] Davidson and Bassett, 1993.
[mm] Kanzaki, Yoshida, Nomura, Kakihana and Ozawa, 1979.
[nn] Palmer and Sturchio, 1990.
[oo] Peng and Palmer, 1995.
[pp] McMullen, Cragg and Thode, 1961.
[qq] Oi and Kahihana, 1993.
[rr] Musashi, Oi, Ossaka and Kakihana, 1988.
[ss] Kakihana, Ossaka, Oi, Musgshi and Okamoto, 1987.
[tt] Xiao, Sun, Wang, Qi, and Jin, 1992.
[uu] Oi, Nomura, Musashi, Ossaka, Okamoto and Kakihan, 1989.
[vv] Swihart, McBay, Smith and Siefke, 1996.
[ww] Smith and Yardley, 1996.
[xx] Palmer and Helvaci, 1995, 1996.
[yy] Swihart, McBay, Smith, and Carpenter, 1993.
[zz] Chaussidon and Robert, 1995.
[aaa] Finley and Leuang, 1961.

$\delta^{11}B$ of about -5. Seawater has a comparatively high (positive) $\delta^{11}B$ number, as does the boron in rocks or minerals from marine sources, many fumaroles, skarns, and residues from the results of considerable ^{10}B fractionation (during adsorption, volitization, crystallization, etc.).

As shown in Table 6.6, there is a slight preferential loss of B^{10} by any of its reactions or physical changes because of its smaller size than ^{11}B, generally about 2% for each phase change. For instance, with clay *adsorption* experiments Schwarcz, Agyei, and McMullen (1969) made a brine with an initial $\delta^{11}B$ value of -0.7, and after adsorption on illite, its $\delta^{11}B$ was $+18.6$. This $\delta^{11}B$ increase of $+19.3$ indicates a 1.5% preferential ^{10}B adsorption. In another test, Shergina and Kaminskaya (1967) contacted clay in an adsorption column with a pure water solution containing 600 ppmB. The ^{10}B selective adsorption was initially 4%, then 0.9%, and finally 0.5%. The solution's $\delta^{11}B$ increased 24.6 units in the initial contact, and the clay's $\delta^{11}B$ decreased 35.8 units. The clay's boron content increased from 55 to 1020 ppmB. Several other studies on clay adsorption are also listed in Table 6.6.

An interesting example of the large shift in $\delta^{11}B$ caused by preferential ^{10}B *adsorption* on clay and organic matter when the initial water has a very *low boron content* was shown by a number of Australian saline lakes, mainly from the interior of the country. Their brine's $\delta^{11}B$ was 30–56, averaging 39.7, and their boron content was 0.2–18.7 ppm, averaging 3 ppm. When seawater's most characteristic elements, Br and Mg were considered, none of the brines showed any resemblance to seawater. Preferential ^{10}B adsorption could give such high residual $\delta^{11}B$ brine values, even though the spring and river waters sampled also had high $\delta^{11}B$ readings. The mud from the lakes had a moderately high boron content, indicating a 3.5% selective ^{10}B adsorption. Experimental error must be considered with these unusual results, even though their granite and shale numbers were in general agreement with others. Similar high values also were found by the same author for the Dead Sea and its adjacent springs. Because very few low-B $\delta^{11}B$ values are available from other saline lakes, springs, or rivers, these high numbers may be more common than expected (Vengosh, Chivas, McCulloch, Starinsky and Kolodny, 1991a, Vengosh, Starinsky, Kolodny and Chivas, 1991c).

An interesting aspect of seawater boron chemistry has been reported in a study of boron *co-crystallization* in the aragonite skeleton material of corals by Gaillardet and Allegre (1985). They found a $\delta^{11}B$ range in modern and ancient corals of 23.3–27.3 (24.9 average), and a boron content range of 39–58 ppm (49.5 average). A 1.3% preferential co-crystallization of ^{10}B was indicated, which is slightly lower than the $\Delta^{10}B$ removal found in other marine carbonate (Hemming and Hanson, 1992) and biogenic carbonate (several authors) studies.

A quite different study on the isotopic distribution of boron during *reaction and crystallization* has been made on tourmalines from massive sulfide deposits of Cu, Pb, Zn, and Ag (Palmer and Slack, 1989). These authors examined tourmaline [$Na(Mg,Fe)_3Al_6B_3Si_6O_{27}(OH,F)_4$] from 60 metallic sulfide deposits and found $\delta^{11}B$ values of -22.8 to $+18.3$. The average was -8.9, and Mg was dominant in the tourmalines' structure (dravite), with Fe^{++} (schorl) being the next most common. Trace element as well as oxygen and hydrogen isotope analyses indicated that the tourmaline (and the massive metallic sulfide deposits) originated from hot ocean floor vent fluids. Most vents have a pH of ~6, allowing the αAl_2O_3 structure of the contacted rocks to be leached, thus creating a solution that would crystallize tourmaline. If present-day deep ocean vents are similar to those that formed these deposits, there would be an average ^{10}B enrichment in the tourmaline of about 3%. In a laboratory experiment making synthetic tourmaline, however, the ^{10}B enrichment was only 0.2–0.7% (Palmer, London, Morgan, and Babb, 1992).

According to several authors, deep ocean vents have an enriched boron content, with $\delta^{11}B$ values of $+19$ to $+37$. On the basis of corresponding Mg analyses, none of the vent's brines would appear to have a large amount of entrained seawater (a maximum of 2–3%). The vents also appear to be similar in temperature, boron content, and $\delta^{11}B$ values to some of the active terrestrial fumaroles in Japan and Italy. Ocean crusts near to thermal vents have $\delta^{11}B$ values of +16.5–23.2, whereas fresh basalt near the vents contained only 0.39–0.46 ppmB, with a $\delta^{11}B$ of -3.0 (indicating that its boron content had been leached). Chaussidon and Jambon (1994) found 0.91 ppmB and an average $\delta^{11}B$ of -3.4 in 40 similar samples, which are typical $\delta^{11}B$ values for other basalts. The relative ease with which seawater's boron later reacts with the basalt (even at its 2°C temperature) is indicated by the increase in its boron content with aging to 8.9–69 ppm and by the increase of the $\delta^{11}B$ value to $+4.5$. Even further reaction by seawater to form serpentinized peridotites increased the B content to 51–81 ppm with a $\delta^{11}B$ value of $+10.9$.

The ^{10}B fractionation during the *cooling–crystallization* of molten meteorites also indicates preferential $\delta^{10}B$ fractionation values in the 2–3% range. When meteoritic chondrules crystallize, there develops a wide range of $\delta^{11}B$ values in individual crystals (+40 to -50). The average $\delta^{11}B$ of the meteorite would have become fixed once vaporization of the boron stopped as it originally formed, and the mass cooled and crystallized to eliminate all but perhaps some surface reaction with "galactic cosmic rays" (low energy cosmic rays reacting with H or He to give a $^{11}B/^{10}B$ of about 4.5, with the product of very energetic protons and α-particles to produce a $^{11}B/^{10}B$ of about 2.5). As the molten meteorite crystallized, the crystals' ^{10}B would be somewhat preferentially increased, but the boron content (as well as that of other trace elements such as Be) would be slightly lower due to the tendency of the crystallization processes for purity of the major species. Thus the first crystals would have had the lowest $\delta^{11}B$ and boron analyses, and the residual melt's $\delta^{11}B$ would have increased. The final liquid to solidify would have the highest $\delta^{11}B$ and

Table 6.6

Examples of the Preferential Adsorption, Crystallization or Volitization of ^{10}B

Process					Amount of ^{10}B Excess (%)		
					Individual	Group	Average
A. Adsorption, on clay, mud or shale from brine							
1. Australian lake mud in lake brine[a]							
a. Kielambete					3.79		
b. Cadibarrowirracanna					3.51		
c. Eyre					3.34		
d. Gnotuk					3.29		
e. Acraman					2.24		
Average						3.48	
2. Oceanic sediments in sea water[b,g]						2.67	
3. Guaymas Basin sediments in ocean vent brine[c]						2.11	
4. Marine shale in sea water, Japan[d]						4.00	
5. Deep ocean shale (clay) in sea water							
a. Water desorbable[e]					3.58		
b. Nonwater desorbable[e]					1.99		
Average[e]						2.76	
c. Pelagic clay[d]						3.53	
6. Brief laboratory experiments with mud or clay							
a. Hydromica–montmorillonite clay in a column, contacted with synthetic brine[f]							
	0–10 cm	70–89 cm	166–176 cm	Ave.			
1) Water desorbable.	6.15	2.80	0.0	2.98			
2) Non water desorb.	0.72	0.47	0.59	0.59			
Total	4.17	0.94	0.49	—		1.87	
b. Mississippi river and delta sediments, batch contact with a synthetic brine[c]							
1) Water desorbable					3.59		
2) Nonwater desorbable					2.28		
Total						2.94	
c. Illinois illite, batch contact, synthetic brine[f]						1.52	
7. Average of adorption on clay in brines							2.76

B. Carbonates or silicates crystallized from seawater			
1. Ocean ooze[d]		3.04	
2. Biogenic silica[d]		3.13	
3. Carbonate sediments			
a. Carbonate sediments[h]		1.91	
b. Biogenic carbonates[d]		2.45	
c. Biogenic carbonates[d]		1.37	
d. Carbonates[i]		1.33	
e. Corals[j]; modern 1.19; old 1.16		1.18	
4. Average of carbonates and silicates in sea water			2.06
C. Marine salts crystallized from sea water			
1. Gypsum and/or halite			
a. Halite, gypsum; Australia[k]		0.12	
b. Gypsum[b]		2.09	
c. Laboratory experiment, gypsum[l]		2.40	
d. Laboratory experiment, halite[b]		2.52	
e. Brine change, halite to carnallite; Chinese playa[m,w]		0.84	
2. Halite–epsomite, laboratory experiment[l]		2.50	
3. Potash salts, laboratory experiment[l]		1.01	
4. Average of marine salts			1.64
D. Tourmaline crystallization			
1. Crystallized in ocean vents[n]		2.99	
2. Crystallized in granite: fine vs coarse crystals[x]		0.32	
3. Laboratory experiments:[y]			
50 Mpa, 350°C	0.68		
50 Mpa, 450°C	0.47		
100 and 200 Mpa, 750°C	0.21	0.45	
4. Average tourmaline			1.25
E. Borate deposits, only crystallized partially to completion			
1. Borax, Lower Structure, Searles Lake[m]		0.61	

continues

Table 6.6 (*continued*)

Process	Amount of ^{10}B Excess (%)		
	Individual	Group	Average
2. Colemanite:			
a. Kirka, Turkey[o]		0.75	
b. Boron, USA[e]		0.69	
3. Pinnoite, ulexite; Da Qaidam Lake, China[p]		1.14	
4. Hydroboracite; Da Qaidam Lake, China[q]		1.17	
5. Mcallisterite; Da Qaidam Lake, China[q]		1.26	
6. Gypsum from Larderello, Italy groundwater[r]		1.38	
7. Sassolite:			
a. Larderello fumarole briner[r]		0.38	
b. Kurome, Japan[s]		0.70	
c. Japanese average[s]		0.14	
8. Sporgite precipitated in Larderello cooling towers[r]		0.29	
9. Average of borate deposits			0.77
F. Fumarole volatilization of borate waters[*]			
1. Larderello:			
a. cooling tower evaporation of groundwater[r]		1.51	
b. Groundwater to fumarole condensate[r]		0.73	
Average Larderello brine			1.12
2. Cosco geothermal brine: laboratory sample flashed[t]		−0.11*	
3. Gysers[f]		0.05*	
4. Laboratory brine flashed, 440°C, 376 bars[u,z]		0.19*	
G. Anion exchange resin: borate absorbed from synthetic brine			
1. Brine through a small column:[s]			
a. First mMB/cc resin absorbed	3.6–5.3		
b. Second mMB/cc resin absorbed	3.4–0.5		
c. Third mMB/cc resin absorbed	0–1.0		
Average (to total borate saturation)			2.3
2. Acidified brine through boron-sective resin, average[v]		8.0	

H. Meteorites: initial crystallization compared to the average $\delta^{11}B$[aa]		
1. Semarkona	2.88	
2. Allende	2.92	
3. Hedjaz	4.06	
4. Average		3.29
I. Highly acid (pH2) geothermal brine–rock (feldspar) leaching		−0.14

* Leeman, Vocke, and McKibben (1992) stated that the ^{11}B (and not the expected ^{10}B) volatilized in excess, but showed no data to substantiate this claim. Actually, from their very small graphs, in the only measurable point, ^{10}B did volatilize in excess, and any excess of one form over the other was negligible on the other points. In the data of other authors, the ^{10}B did volatilize in excess.

[a] Vengosh, Starinsky, Kolodny and Chivas, 1991c.
[b] Gregorie, 1987.
[c] Spivack and Edmond, 1987.
[d] Ishikawa and Nakamure, 1993.
[e] You, Spivack, Gieskes, Martin and Davisson, 1996.
[f] Schergina and Kaminskaya, 1967.
[g] Schwarcz, Agyie, and McMullen, 1969.
[h] Vengosh, Kolodny, Starinsky, Chivas and McCulloch, 1991b.
[i] Hemming and Hanson, 1992.
[j] Gaillardet and Allegre, 1995.
[k] Kakihana and Kotaka, 1977.
[l] Vengosh, Starinsky, Kolodny, Chivas and Raab, (1992).
[m] McMullen, Cragg and Thode, 1961.
[n] Palmer and Slack, 1989.
[o] Palmer and Helvaci, 1992.
[p] Sun, 1990.
[q] Xiao, Sun, Wang, Qi and Jin, 1992.
[r] Klotzli, 1992.
[s] Kanzaki, Yoshida, Nomura, Kakihana and Ozawa, 1979.
[t] Leeman, Vocke and McKibben, 1992.
[u] Spivack, Berndt and Seyfried, 1990.
[v] Oi, Nomura, Musashi, Ossaka, Okamoto and Kakihana, 1982.
[w] Vengosh, Chivas, and McCulloch, 1989.
[x] Smith and Yardley, 1996.
[y] Palmer, London, Morgan, and Babb, 1992
[z] Leeman, Vocke, and McKibben, 1990.
[aa] Chaussidon and Robert, 1995.

B content. When individual crystals (chondrules) were analyzed, this is exactly what was found, as well as a good correlation of the B with Be (both similarly formed minor elements; Chaussidon and Robert, 1995).

The *vaporization* of boron also experiences a strong preferential ^{10}B removal. Data on the Larderello geothermal brines indicate that the ^{10}B concentrates in the vapor (see Table 6.5), as would be expected from its smaller-sized molecule giving it greater volatility. This property also is used commercially to prepare pure ^{10}B by the fractionation of volatile boron compounds. The separation tendency is so great that only 40-tray columns are required to produce >95% ^{10}B. Somewhat surprisingly, in laboratory tests Leeman, Vocke, and McKibben (1992) indicated a small enrichment (0.11%) of ^{10}B in the residue (and not the vapor) when Salton Sea geothermal brine ($\delta^{11}B$ −1.9) was flashed (residue −3.3). However, the vapor was not condensed to check this result, nor was pH, TDS, or temperature of the solution noted, making it almost certainly in error.

An equally anomalous but inconclusive result has been reported on the *leaching* of boron from rocks. Musashi, Oi, Ossaka, and Kakihana (1991) examined the unaltered base rock and the leached throat of a very acidic (pH 2) hot spring. The throat rock had been altered from 60 to 93% silica, but its boron content changed only from 17 to 14 ppmB (which is totally inconsistent with other boron leaching observations where the boron leached rapidly and fairly completely). Their data inferred a 0.14% ^{10}B excess in the unleached boron in the rock, but probably indicates only that the analyses were in error or there was no boron leaching. The spring's $\delta^{11}B$ was 10.0, which is fairly normal for geothermal brine but not for rock, further indicating little or no boron leaching during the period of their test.

The *absorption* of boron in ion exchange resins is a final example of the greater activity, and thus preferential removal of ^{10}B. There have been many studies of this phenomena, and it has been used to prepare solutions of ^{10}B more than 98% pure. For instance, Oi, Takeda, Hosoe, and Kakihona (1992) used both Diaion $CRBO_2$ and Amberlite IRA 743, 20- to 50-mesh, boron-specific resins in the Cl or SO_4 form. They passed a dilute acidic borate solution through resin-filled columns, and in one experiment the initial solution through the column had removed a 16% excess of the ^{10}B, the next portion 12%, then 5%, 1.5%, and finally no selective removal. On elution of the boron from the column, the average solution had about an 8% ^{10}B enrichment. This process could then be repeated to reach the desired ^{10}B purity.

With hot spring deposits that had *almost complete boron deposition,* the ^{10}B fractionation would be averaged, and thus appear to be much less. For instance, Sun (1990) found only a 1.1% ^{10}B preference in the crystallization of pinnoite from the brine in Chinese playa muds. Xiao, Sun, Wang, Qi, and Jin (1992) found a similar ^{10}B preferential crystallization of 1.3% for mcallisterite and 1.2% for ulexite [Sun (1990) reported 0.44%]. McMullen,

Cragg, and Thode (1961) found a 0.6% preferential ^{10}B fractionation for borax crystallizing from Searles Lake's brine. In each case much of the boron had been crystallized from the incoming brine when their sample was crystallized, thus making the $\delta^{11}B$ of the borate more similar to that of the brine.

Secondary borates represent a special case of $\delta^{11}B$ values. Since every borate transformed into another mineral must first be dissolved and then recrystallized, all of the boron in the first mineral usually is not totally reprecipitated, and some escapes in the leach solution. This again results in preferential ^{10}B crystallization and further lowers the $\delta^{11}B$ value. Thus, the primary borate's $\delta^{11}B$ is assumed by the secondary mineral, but it is usually further reduced, often causing the secondary mineral to have a very low value. Examples of this are presented in the article by Palmer and Helvaci (1997), such as the Sultancayiri deposit in which it appears that colemanite was primary and had a $\delta^{11}B$ of -11.1. Both priceite and howlite are almost certainly secondary, and have values of −20.2 and −24.5, respectively. Examples in the article also show the influence of specific geothermal spring's $\delta^{11}B$ within a similar group of deposits. With colemanite in the Turkish deposits, the $\delta^{11}B$ at Bigadic was −10.2 (2 samples), Sultancayiri −11.1 (1 sample), Kirka −12.8 (8 samples), Emet −15.1 (12 samples), and Kestelek −15.8 (1 sample). Within a hydration series at a given deposit (Bigadic), the $\delta^{11}B$ of inyoite was −9.0 (2 samples), meyerhofferite −10.2 (2 samples), and colemanite −10.8 (18 samples). These values all are roughly the same, indicating an initial precipitation from only part of the borate solution. At the same deposit, the $\delta^{11}B$ of ulexite was −6.8 (6 samples), implying its later crystallization and a (cumulative) larger fraction of the boron deposited. Interestingly, it appeared that tunellite crystallized late in the sequence, and was primary (at Kirka its $\delta^{11}B$ was −4.4; borax was −3.8), as also was veatchite-A (at Emet −9.0; colemanite was −15.1). Teruggite, however, is indicated to be a secondary mineral with a $\delta^{11}B$ of −25.0.

It appears that the high value for seawater's $\delta^{11}B$ has remained in dynamic equilibrium throughout history. Presumably, it is the result of the slight selectivity for ^{10}B by adsorption (i.e., on clays), reaction (i.e., tourmalines), or co-crystallization (i.e., marine carbonates, gypsum, halite, and potash). This increased the seawater's ^{11}B content, as did the waters from deep ocean vents and some terrestrial fumaroles. The leaching of various high-$\delta^{11}B$-content rocks also helped. Apparently these mechanisms for increasing the $\delta^{11}B$ of seawater were just sufficient to balance the input from terrestrial water with its much lower $\delta^{11}B$ value.

It appears that fumaroles and geothermal waters randomly have high or low $\delta^{11}B$ values, indicating a variable magma $\delta^{11}B$ content and/or that their boron has been at least partly derived from the leaching of crustal rocks. These brines, along with some ^{10}B selectivity for those deposits in which the borate did not totally crystallize or there was adsorption or reaction, give rise to the very wide range of $\delta^{11}B$ values manifested by terrestrial borates in nature. However, terrestrial borates average −6.4, which is quite similar to the suggested average value for all other terrestrial rocks, which is −4.2.

In the late 1980s and the 1990s a very large number of articles were published on boron isotope distribution. Boron isotopes have been suggested as tracers for groundwater contamination from ash leachate (Davidson and Bassett, 1992), sewage pollution (Vengosh, Heumann, Juraske and Kosher, 1994), and many other applications. They also have been used as indicators for an endless number of geologic processes, some as previously discussed, and as a factor in determining the source material in borate deposits (Malinko, Lisitsyn, and Sumin, 1987). Land and Macpherson (1992) suggested that boron isotopes helped to determine the source of U.S. Gulf Coast oil field and other formation waters. Factors of paleosalinity, fluid processes in forming ore deposits, and the history of marine sediments, mantle rocks, meteorites, and so forth have all been examined on the basis of differing boron isotopic ratios.

6.2.3 Factors Determining $\delta^{11}B$

A "theory" has been proposed to explain the different $\delta^{11}B$ values of various borate minerals. It was initially stated by Kakihana and Kotaka (1977). Then on the basis of this theory Oi, Nomura, Musashi, Ossaka, Okamota and Kakihana (1989) presented simplifying assumptions for calculating $\delta^{11}B$ values. The authors assumed that the $\delta^{11}B$ of sodium borates > the $\delta^{11}B$ of sodium-calcium borates > the $\delta^{11}B$ of calcium borates. They further assumed that the distribution of two isotopes between two dissolved ionic (or nonionic) species depends on the isotopic reduced partition function ratio (RPFR) for the species. The RPFR values were estimated by first guessing the vibrational frequencies for the vaporized hydrides of ^{10}B and ^{11}B, and then estimating how these values might relate to ^{10}B- and ^{11}B-containing ions in solution. The authors then assumed that ^{10}B preferentially exists in the $B(OH)_4^-$ ion, and that dilute borate solutions contain only two ionic components: (a) the triangular, B-centered $B(OH)_3$ cluster, and (b) the tetragonal, B-centered $B(OH)_4^-$ group. At 25°C, $B(OH)_3$ was estimated to have an RPFR value of 1.2008, and $B(OH)_4^-$ 1.1780 (Table 6.7).

On this basis, if a solution of borax with a $^{11}B/^{10}B$ ratio of 4.048, or a $\delta^{11}B$ of +0.64, were acidified to contain primarily $B(OH)_3$ (i.e., boric acid), the $^{11}B/^{10}B$ ratio of any $B(OH)_4^-$ ion that was present would be $4.048 \times 1.1780/1.2008 = 3.971$. (It would have a $\delta^{11}B$ of −18.4, implying that there was an ionic shifting for some reason of the ^{10}B by 1.55%.) If a small amount of a calcium borate such as colemanite or inyoite crystallized from this boric acid solution, because their atomic configurations contain one $B(OH)_3$ group and two $B(OH)_4^-$ groups, the solution would now be assumed to contain the same ionic ratio, and not to have changed its $^{11}B/^{10}B$ ratio during the crystallization. Under these conditions, inyoite crystallizing from a $\delta^{11}B$ +0.6 boron solution would have a $^{11}B/^{10}B$ ratio of $4.048^{1/3} \times 3.971^{2/3} = 3.997$, or a $\delta^{11}B$ of -12.0. This value is somewhat close to the average from springs (-8.3), but not the average from marine deposits ($+23.2$). A pH correction for borate solutions was next added to the procedure by replotting the pH-ionic composition estimates of Ingri (1963; see Fig. 1.4) as if only $B(OH)_3$ and $B(OH)_4^-$ were present (by breaking the more complex ions into these components; Fig. 6.3). The data in Fig. 6.3 can then be compared with the pH correction estimates in Table 6.7B to predict a wide range of $\delta^{11}B$ values for various compounds.

A critical review of the RPFR method indicates that the concept of different isotopes having slightly different physical and chemical reactivities and properties is certainly valid. With boron there is a 10% difference in the weights of its isotopes, so differences in reactivity should be quite apparent. With small gaseous elements this difference can be related to their vibrational frequencies, and even though no such data is available for boron, a rough extrapolation might be made for boron compared with lighter elements. This was done, and the estimate was a 1.94% greater reactivity for ^{10}B than for ^{11}B, a value that agrees quite well with experimental data (see Table 6.6).

However, beyond this point the RPFR method has serious flaws. It assumes that ^{10}B is preferentially in the tetragonal [$B(OH)_4^-$] form, which has no basis in theory or fact. Actually, the excess ^{10}B reactivity is the same for both acidic (Δ form) and basic (T form) solutions, and as seen in the Japanese hot springs, there is no dependence of $\delta^{11}B$ on the water's pH. The procedure then selects as a reference value a near-actual $^{11}B/^{10}B$ for borax, but immediately assumes that this value is the same as the average for sassolite (or the $B[OH]_3$ group), which is far from the case. From this value, the procedure then estimates the $\delta^{11}B$ of other minerals solely on the basis of their component Δ/T ratio. Again, there is no validity to this parameter having any influence on ^{10}B's excess reactivity, or being a measure of it. Furthermore, it implies that all minerals with the same Δ/T ratio have the same $\delta^{11}B$, such as the 2T,1Δ of inderite, hydroboracite, inyoite, meyerhofferite, colemanite, fabinite, and many others. However, as seen in Table 6.5 they vary widely. The method next predicts a 226% decrease in the excess reactivity of ^{10}B over the temperature range of -73 to 227°C, which appears to be far too great. Finally, the method attempts to predict the effect of pH on the $\delta^{11}B$ by assuming that all the complex borate ions act as if only their component Δ and T groups were present, and that the ^{10}B reactivity then varies with these simple groups. Again, there is no basis in theory or fact to assume a different excess ^{10}B reactivity with either the Δ/T ratio or pH. However, even if the T and Δ groups were important, the complex borate ions in solution would have retained essentially no residual properties of these component Δ or T structures. Even more importantly, the borate-precipitating solution has no independent means of changing its pH. It is fixed by the solution's concentration of HCO_3, CO_3, $MgCl_2$, $CaSO_4$, SiO_2, and borate ions. In summary, it would appear that there is no $\Delta^{10}B$

Table 6.7

Reduced Partition Function Ratios[a]

A. RPFRs of $B(OH)_3$, $B(OH)_4^-$ and polyborate anions at several temperatures						
Temperature (°K)	200.0	273.1	298.1	323.1	400.0	500.0
$B(OH)_3$	1.3796	1.2315	1.2008	1.1758	1.1225	1.0827
$B(OH)_4^-$	1.3449	1.2066	1.1780	1.1549	1.1061	1.0705
$(BO_3 + BO_4)/2$	1.3622	1.2191	1.1894	1.1654	1.1143	1.0766
$(2BO_3 + 3BO_4)/5$	1.3588	1.2166	1.1871	1.1633	1.1127	1.0754
$(BO_3 + 2BO_4)/3$	1.3565	1.2149	1.1856	1.1619	1.1116	1.0746

B. The effect of pH on the $^{11}B/^{10}B$ ratio (and $\delta''B$), as estimated by Figure 6.3

Mole fraction of $B(OH)_3 = \Delta$	1.0	0.9	0.8	0.7	0.6	0.5	0.4	0.3	0.2	0.1	0.0
Mole fraction of $B(OH)_4^- = T$	0.0	0.1	0.2	0.3	0.4	0.5	0.6	0.7	0.8	0.9	1.0
$B(OH)_3$ (Δ)	4.048 (+1.1)	4.056 (+3.8)	4.064 (+5.1)	4.071 (+6.8)	4.079 (+8.8)	4.087 (+10.7)	4.095 (+12.7)	4.103 (+14.7)	4.111 (+16.7)	4.118 (+18.4)	4.126 (+20.4)
$B(OH)_4^-$ (T)	3.971 (−18.0)	3.976 (−16.7)	3.986 (−14.2)	3.994 (−12.3)	4.002 (−10.3)	4.009 (−8.6)	4.017 (−6.6)	4.025 (−4.6)	4.033 (−2.6)	4.040 (−0.9)	4.048 (+1.1)
Na borate (2Δ + 2T)	4.009 (−8.6)	4.017 (−6.6)	4.025 (−4.6)	4.033 (−2.6)	4.040 (−0.9)	4.048 (+1.1)	4.056 (+3.1)	4.064 (+5.1)	4.071 (+6.8)	4.079 (+8.8)	4.087 (+10.7)
Na/Ca borate (2Δ + 3T)	4.002 (−10.3)	4.009 (−8.6)	4.017 (−6.6)	4.025 (−4.6)	4.033 (−2.6)	4.040 (−0.9)	4.048 (−1.1)	4.056 (+3.1)	4.064 (+5.1)	4.071 (+6.8)	4.079 (+8.8)
Ca borate (Δ + 2T)	3.997 (−11.5)	4.004 (−9.8)	4.012 (−7.8)	4.020 (−5.8)	4.027 (−4.1)	4.035 (−2.1)	4.043 (−0.2)	4.051 (+1.8)	4.058 (+3.6)	4.066 (+5.5)	4.074 (+7.5)

Note. The $^{11}B/^{10}B$ value of original solution is assumed to be 4.048 ($\delta''B$ = +1.1 assuming a reference ratio of 4.0436). In each row, the numbers are the $^{11}B/^{10}B$ values and the numbers in the parentheses are $\delta^{11}B$ values.

[a] Oi, Nomura, Musashi, Ossaka, Okamoto and Kakihana, 1989.

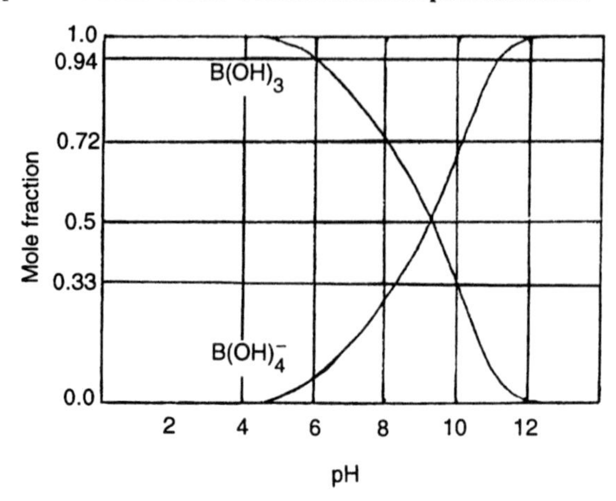

Figure 6.3 Hypothetical plot of pH versus a borate solution's $B(OH)_3$ and $B(OH)_4^-$ content (as estimated by breaking the ions indicated in Figure 1.6 into their component groups). (From Oi, Nomura, Musashi, Ossaka, Okamoto and Kakihana, 1989; reprinted by permission of Elsevier Science Ltd, The Boulevard, Langford Lane, Kidlington OX5 1GB.)

reactivity dependence on the borate minerals' molecular structure or the solution's pH, and thus the RPFR method has little value.

What Table 6.5 does clearly reveal, however, is that the borate minerals' $\delta^{11}B$ is highly dependent on three factors: (a) the $\delta^{11}B$ of the solution from which the borate mineral crystallized, (b) the fraction of the boron initially present that crystallized, and (c) whether the mineral was primary (to follow the solution's $\delta^{11}B$) or secondary (to assume the initial mineral's $\delta^{11}B$ as modified by the secondary solution's $\delta^{11}B$ and the amount crystallized). This is illustrated by the average of +26.4 for marine borate minerals (with seawater's $\delta^{11}B$ of +40.0), +24.1 for Larderello's minerals, +10.0 for skarn minerals (with an assumed +19 to +33 magma solution, as in ocean vents), and −5.4 $\delta^{11}B$ for geothermal spring borates (with the present-day spring water's average of +5.7 $\delta^{11}B$).

The first two factors are shown in Fig. 6.4 for three typical geothermal spring $\delta^{11}B$ values. For the sequential crystallization of colemanite, ulexite, and borax it indicates typical colemanite $\delta^{11}B$ values of −6 to −20, with 0–80% of the initial boron being crystallized. Then ulexite has $\delta^{11}B$ values of −2 to −10 (the average for colemanite and ulexite together), and 20–80% (cumulative) of the boron crystallized. Finally, the borax crystallized with a total deposit average of +10 to −10 $\delta^{11}B$, and 60–90% of the borate crystallized (the total of all three minerals). The uncrystallized portion of the boron can be assumed to be that portion that escaped (by seepage or overflow) with the other ions in the brine (that did not crystallize).

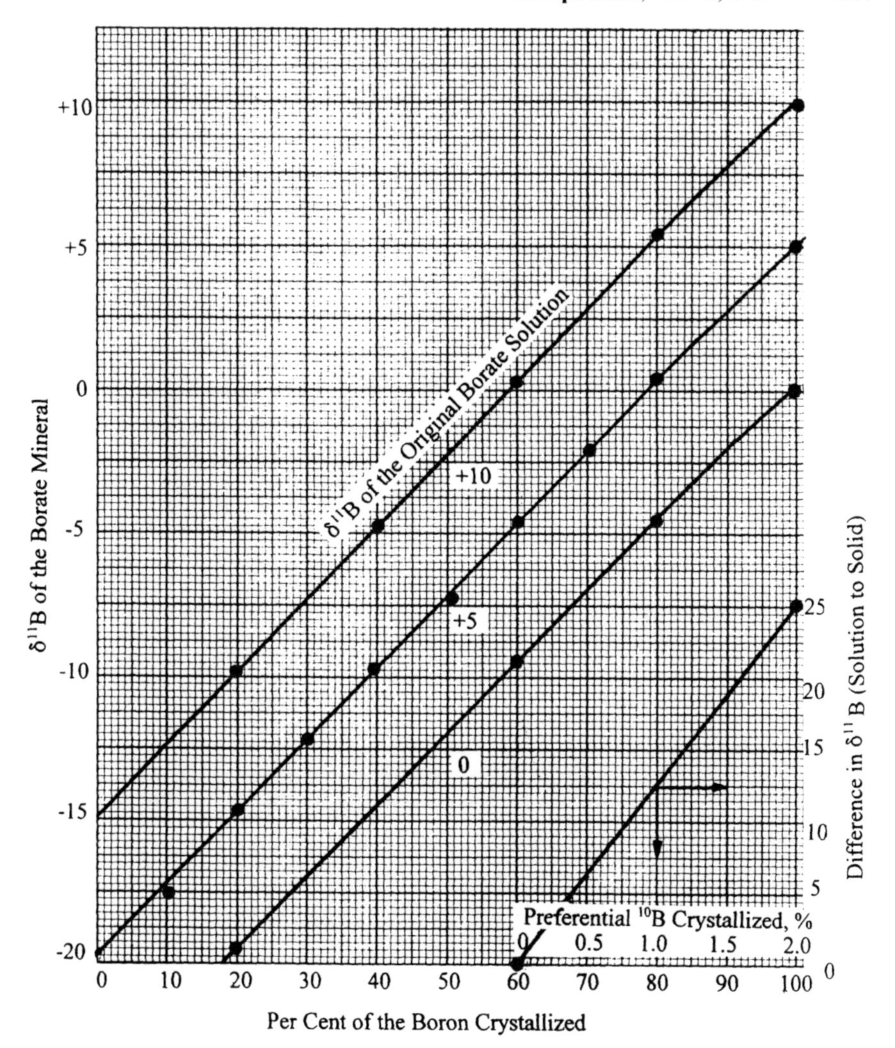

Figure 6.4 Change in $\delta^{11}B$ with the amount of boron crystallized.

References

Abitz, R., Myers, P., Drez, P., and Deal, D. (1990). Salado formation brines recovered from the waste isolation pilot plant (WIPP) repository. *Proc. Symp. Waste Mgt., Waste Mgt. 90* **2,** 381–392.

Aggarwal, J. K., Palmer, M. R., and Ragnarsdottir, K. V. (1992). Boron isotopic composition of Icelandic hydrothermal systems. In *Water–Rock Interaction* (Y. K. Kharaka and A. S. Maest, eds.), pp. 893–895, Balkema, Rotterdam.

Agyei, E. K., and McMullen, C. C. (1968, August). A study of the isotopic abundance of boron from various sources. *Can. J. Earth Sci.* **5**(4, Pt. 1), 921–927.

Anon. (1991). Isotopic composition of the elements, 1989. *Pure & Appl. Chem.* **63**(7), 991–1002. (National Bur. Stds. No. 951)

Avrova, N. P., Bocharov, V. M., Khalturina, N. P., Yunosova, Z. R. (1964). Metaborite. Geol. Razved. Mestorozhd. Tverd. Polez. Iskop., pp. 169–173.

Barth, S. (1993). Boron isotopeic variations in nature: A synthesis. *Geol. Rundsch.* **82,** 640–651.

Bermanec, V., Armbruster, T., Tibljas, D., Sturman, D., and Kniewald, G. (1994). Tuzlaite, a new mineral with a pentaborate sheet structure from the Tuzla Salt Mine, Bosnia and Hercegovina. *Am. Min.* **79**(5–6), 562–569.

Bocharov, V. M., Khalturina, N. P., Avrova, N. P., and Shipovalor, Y. V. (1969). *Satimolite.* Tr. Mineral. Muz. Akad. Nauk SSSR, No. 19, pp. 121–125.

Borchert, H., and Muir, R. O. (1964). *Salt Deposits: The Origin, Metamorphism and Deformation of Evaporites.* D. Van Nostrand, London.

Braitsch, O. (1971). *Salt Deposits: Their Origin and Composition.* Springer-Verlag, Berlin.

Burns, P. C., Grice, J. D., and Hawthorne, F. C. (1995). Borate minerals: Polyhedral clusters and fundamental building blocks. *Can. Mineralogist* **33,** 1131–1151.

Celik, M. S., and Bulut, R. (1996). Mechanism of selective flotation of sodium–calcium borates with anionic and cationic collectors. *Sep. Sci. Tech.* **31**(13), 1817–1829.

Chaussidon, M. (1995). Isotope geochemistry of boron in mantle rocks, tektites and meteorites. *Isotopic Geochemistry,* pp. 455–472, C. R. Acad. Sci., Paris, Vol. 321, No. 2a.

Chaussidon, M., and Jambon, A. (1994). Boron content and isotopic composition of oceanic basalts. *Earth and Planetary Sci. Lett.* **121**(314), 277–291.

Chaussidon, M., and Robert, F. (1995, March 23). Nucleosynthesis of ^{11}B-rich boron in the presolar cloud recorded in meteoritic chondrules. *Nature* **374,** 337–339.

Cherepanov, V. A. (1967, January–February). Boron in the kimberlite and meimechite of Siberia. *Dokl. Acad. Sci., U.S.S.R.* (*Earth Sci. Sect.*) **172,** 187–190.

Christ, C. L. (1966, September–October). A new hydrous calcium–strontium borate, volkovskite. *Am. Min.* **51**(9–10), 1550. [Abstract]

Davidson, G. R., and Bassett, R. L. (1993). Application of boron isotopes for identifying contaminants such as fly ash. *Environ. Sci. Technol.* **27,** 172–176.

Diarov, M. (1966). Fluoborite. *Gohkimiya,* No. 5, pp. 594–599.

Finley, H. O., Eberle, A. R., and Rodden, C. J. (1962). Isotopic boron composition of certain boron minerals. *Geochim. et Cosmochim. Acta* **26,** 911–914.

Finley, H. O., and Leuang, E. E. (1961). *Mass Spectrometric Determination of Boron Isotope Abundance in the Near-Normal Range,* U.S. AEC Report NBL-170.

Fischbeck, R. (1983). Probertite in the Gorleben Salt Massif. *Kali und Steinalz* **8**(11), 384–386.

Gaillardet, J., and Allegre, C. J. (1995). Boron isotopic compositions of corals; Sea water or diagenesis record. *Earth, Planatary Sci. Lett.* **136**(3–4), 665–676.

Gale, H. S. (1917). Potash. *U.S. Bur. Mines Yearbook,* Part 2, pp. 397–481.

Gann, D. E., Dockery, D. J., and Marble, J. C. (1987, September). A new borate mineral from the upper Louann formation in the northern Gulf. *Mississippi Geol.* **8**(1), 1–11.

Garrett, D. E. (1995). *Potash: Deposits, Processing, Properties and Uses.* Chapman & Hall, London.

Godlevsky, M. (1937). Mineralogical investigation of the Inder borate deposits. *Memoires de la Societe Russe de Mineralogie* **66**(2), 345–368.

Goldschmidt, V. M. (1954). Group 3, Boron. *Geochemistry,* pp. 280–291, Clarendon Press, Oxford.

Graham, E. R. (1957). The weathering of some boron-bearing materials. *Soil Sci. Soc. Am. Proceed.* **21,** 505–507.

Gregoire, D. C. (1987). Determination of boron isotope ratios in geological materials by inductively coupled plasma mass spectrometry. *Anal. Chem.* **59**(20), 2479–2484.

Grice, J. D., Burns, P. C., and Hawthorne, F. C. (1994). Determination of the megastructures of the borate polymorphs pringleite and ruitenbergite. *Can. Mineralogist* **32,** 1–14.

Grice, J. D., Gault, R. A., and Van Velthuizen, J. (1996). Penobsquisite: A new borate mineral with a complex framework structure. *Can. Mineralogist* **34,** 657–665.

Gulensoy, W. B., and Kocakerim, M. M. (1978). Solubility of colemanite in CO_2-containing water, and geological formation of this mineral. *Bull. Mineral Res. Explor. Inst.,* Foreign ed., No. 90, pp. 1–19, Turkey.

Guppy, E. M. (1944). Boracite from a boring at Aislaby, Yorkshire. *Min. Mag.* **27**(187), 51–53.

Ham, W. E., Mankin, C. J., and Schleicher, J. A. (1961). Borate minerals in Permian gypsum of west-central Oklahoma. *Oklahoma Geol. Surv. Bull. 92.*

Harder, H. (1959). *Bor in Meeressalzen, Ch. 7, pp. 167–175, Beitrag zur Geochemie des Bors, Teil II, Gottingen Akad., Wiss. II, Math. Physickal K1, Nachr.*

Hawthorne, F. C., Burns, P. C., and Grice, J. D. (1996). The crystal chemistry of boron. In *Boron Mineralogy, Petrology and Geochemistry* (Grew, E. E., Anovitz, L. M., eds.), Vol. 33, Ch. 2, Min. Soc. Am., MSA Reviews in Mineralogy.

Hemming, N. G., and Hanson, G. N. (1992). Boron isotopic composition and concentration in modern marine carbonates. *Geochim. et Cosmochim. Acta* **56,** 537–543.

Hite, R. J. (1986). Potash deposits of the Khorat Plateau, Thailand. *Fert. Min in Asia, Pacific, U.N. Econ. Soc. Com. Asia, Pacific* **1,** 51–73.

Hite, R. J., and Japakasetr, T. (1979). Potash deposits of the Khorat Plateau, Thailand and Laos. *Econ. Geol.* **74,** 448–458.

Honea, R. M., and Beck, J. R. (1962, January–February). Chambersite, a new mineral. *Am. Min.* **47**(1–2), 665–671.

How, H. H. (1857). On the occurrence of natroborocalcite with glauber salt in gypsum of Nova Scotia. *Am. J. Sci.* **24,** 230–235.

How, H. H. (1861). Natroboracalcite and another borate occurring in the gypsum of Nova Scotia. *Am. J. Sci.* **32,** 9–13.

Hurlbut, C. S. (1938). Parahilgardite, a new triclinic–pedial mineral. *Am. Min.* **23,** 765–771.

Hurlbut, C. S., and Taylor, R. E. (1937). Hilgardite, a new mineral species from Choctaw Salt Dome, Louisiana. *Am. Min.* **22,** 1052–1057.

Inan, K.A., Dunham, A. C., and Esson, J. (1973). *Mineralogy, Chemistry and Origin of the Kirka Borate Deposit, Eskishehir Province, Turkey,* pp. B114–B124, Trans. Inst. Min. Metal., Nottingham.

Ishikawa, T., and Nakamura, E. (1993). Boron isotope systematics of marine sediments. *Earth, Planetary Sci. Lett.* **117,** 567–580.

Ivanov, A. A., and Yarzhemskii, Y. Y. (1954). Tryetskite. *Trudy Vses. Nauch. Issled. Inst. Galurgii* **29,** 210–214.

Kakihana, H., and Kotaka, M. (1977). Equilibrium constants for boron isotope exchange reactions. *Bull. Res. Lab. Nucl. Reactors* **2,** 1–12.

Kakihana, H., Kotaka, M., Satoh, S., Nomura, M., and Okamoto, M. (1977). Fundamental studies on the ion exchange separation of boron isotopes. *Bull. Chem. Soc. Japan* **50**(1), 158–163.

Kakihana, H., Ossaka, T., Oi, T., Musashi, M., and Okamoto, M. (1987). Boron isotopic ratios of some hot spring waters in the Kusatsu–Shirane Area, Japan. *Geochem. J.* **21**(3), 133–137.

Kanzaki, T., Yoshida, N., Nomura, M., Kakihana, H., and Ozawa, T. (1979, November). Boron isotopic composition of fumarolic condensates and sassolites from Satsuma Iwo-jima, Japan. *Geochim. et Cosmochim. Acta* **43**(11), 1859–1863.

Karazhanov, N. A. (1963). The kinetics of dissolution of borates and other minerals. *J. Appl. Chem. USSR* **36,** 2560–2567.

Kitano, Y., Okumura, M., and Idagoki, M. (1978). Coprecipitation of borate–boron with calcium carbonate. *Geochem. J.* **12,** 183–189.

Klotzli, U. S. (1992). Negative thermal ionization mass spectrometry: A new approach to boron isotopic geochemistry. *Chem. Geol.* (*Isotope Geosci. Sect.*) **101,** 111–122.

Kondrateva, V. V. (1964). Aldzhanite. *Akad. Nauk SSSR* **4,** 10–18.

Kuhn, R. (1968). Geochemistry of the German potash deposits. In *Saline Deposits* (R. B. Mattox, ed.), pp. 460–467, Geol. Soc. Am., Spec. Paper 88.

Kuhn, R., and Schaake, I. (1955). Vorkommen und Analyse der Boracit und Ericait Krystalle aus dem Salzhorst Wathingen-Hanigsen. *Kali und Steinsalz* **11,** 33–42.

Kurnakova, A. G., and Nikolaev, A. V. (1948). The solubility isotherm of the system Na_2O-CaO-B_2O_3-H_2O at 25°C. *Akad. Nauk USSR Izv. Otd. Khim. Nauk* **1,** 377–382.

Land, L. S., and Macpherson, G. L. (1992). Origin of saline formation waters, Cenozoic section, Gulf of Mexico sedimentary basin. *Am. Assoc. Petrol. Geol. Bull.* **76**(9), 1344–1362.

Leeman, W. P., Vocke, R. D., and McKibben M. A. (1990). Boron isotopic studies in geothermal fluids. *Abs. Eos* **71**(1–2), 1686–1687.

Leeman, W. P., Vocke, R. D., and McKibben, M. A. (1992). Boron isotopic fractionation between coexisting vapor and liquid in natural geothermal systems. In *Water–Rock Interaction* (Y. K. Kharaka and A. S. Maest, eds.), pp. 1007–1010, Balkema, Rotterdam.

Lyday, P. A. (1992). History of boron production and processing. *Ind. Min.* (*London*) **303,** 19, 21, 23–25, 31, 33–34, 37.

Malinko, S. V., Lisitsyn, A. E., and Sumin, L. V. (1987). Boron isotopes in minerals as indicators of the source of ore-forming matter. *Sov. Geol.* **3,** 89–97.

Matterson, K. J. (1980). Borate ore discovery, mining and benefication. In *Inorganic And Theoretical Chemistry,* pp. 153–169, Longman, New York.

McMullen, C. C., Cragg, C. B., and Thode, H. G. (1961). Absolute ratio of $^{11}B/^{10}B$ in Searles Lake brines. *Geochim. et Cosmochim. Acta* **23,** 147–149.

Milne, J. K., Saunders, M. J., and Woods, P. J. (1977). Iron boracite from the English Zechstein. *Mineralogical Mag.* **41**(319), 404–405.

Musashi, M., Oi, T., Ossaka, T., and Kakihana, H. (1991). Natural boron isotope fractionation between hot spring water and rocks in direct contact. *Isotepenpraxis* **27**(4), 163–166.

Musashi, M., Nomura, M., Okamota, M., Ossaka, T., Oi, T., and Kakihana, H. (1988). Regional variation in the boron isotopic composition of hot spring waters from central Japan. *Geochem. J.* **22**(5), 205–214.

Nefedov, E. I. (1953). Ivanovite. *Zap. Vses. Mineral. Obshch.* **82,** 311–317.

Nikolayev, A. V., and Chelishcheva, A. G. (1940). The 25°C isotherms of the systems MgO- and CaO-B_2O_3-H_2O. *Comp. Rend.* (*Doklady*) *Akad. Sci. USSR* **28**(2), 127–130.

Nomura, M., Kanzaki, T., Ozawa, T., Okamoto, M., and Kakihana, H. (1982). Boron isotopic composition of fumarolic condensates from some volcanoes in Japanese Island arcs. *Geochim. et Chosmochim. Acta* **46,** 2403–2406.

Oi, T., and Kakihana, H. (1993). Influence of sea water on boron isotopic compositions of hot spring waters in Japan. *Seventh Symp. on Salt* **1,** 159–163.

Oi, T., Nomura, M., Musashi, M., Ossaka, T., Okamoto, M., and Kakihana, H. (1989). Boron isotopic composition of some boron minerals. *Geochim. et Cosmochim. Acta* **53,** 3189–3195.

Oi, T., Takeda, T., Hosoe, M., and Kakihana, H. (1991). Basic data of ion exchange separation of boron isotopes. In *New Dev. Ion Exch. Proc., Int. Conf. Ion Exch.* (Abe, M., Kataoka, T., Suzuki, T. eds.), pp. 293–298, Kodansha, Tokyo.

Orti, F., Helvaci, C., Rosell, L., Gundogan, I. 1997. Sulphate–borate relations in an evaporitic lacustrine environment: The Sultancayir gysum member (Miocene, western Anatolia). *Sedimentology* in press, 26 p.

Palmer, M. R. (1991). Boron isotope systematics of hydrothermal fluids and tourmaline: A synthesis. *Chem. Geol.* (*Isotope Geosci. Sect.*) **94,** 111–121.

Palmer, M. R., and Helvaci, C. (1995). The boron isotope geochemistry of the Kirka borate deposit, western Turkey. *Geochim. et Cosmochim. Acta* **59**(17), 3599–3605.

Palmer, M. R., and Helvaci, C. (1997). The boron isotope geochemistry of the Neogene borate deposits of western Turkey. *Geochemica Cosmochimica Acta* V. 61, No. 15, pp. 3161–3169.

Palmer, M. R., and Slack, J. F. (1989). Boron isotopic composition of tourmaline from massive sulfide deposits and tourmalinites. *Contrib. Mineral. Petrol.* **103,** 434–451.

Palmer, M. R., and Sturchio, N. C. (1990). The boron isotope systematics of the Yellowstone National Park (Wyoming) hydrothermal system: A reconnaisiance. *Geochim. et Cosmochim. Acta* **54,** 2811–2815.

Palmer, M. R., London, D., Morgan, G. B., Babb, H. A. (1992). Experimental determination of fractionation of $^{11}B/^{10}B$ between tourmaline and aqueous vapor. *Chem. Geol.* (*Isotope Geosci. Sect.*) **101,** 123–129.

Papezik, V. S., and Fong, C. C. (1975). Howlite and ulexite from the carboniferous gypsum and anhydrite beds in western Newfoundland. *Can. Mineralogist* **13,** 370–376.

Peng, Q. M., and Palmer, M. R. (1995). The Palaeoproterozoic boron deposits in eastern Liaoning, China: A metamorphosed evaporite. *Precambrian Res.* **72,** 185–197.

Raup, O. B. (1972). Origin of rare-earths in a marine evaporite mineral. In *Geology of Saline Deposits* (G. Richter-Bernburg, ed.), Publ. UNESCO, Paris pp. 103–109.

Raup, O. B., and Gude, A. J. (1968, July–August). Braitschite, a new hydrous calcium rare earth borate from the Paradox Basin, Grand County, Utah. *Am. Min.* **53,** 1081–1095.

Robottom, A. (1893). The tincal Trade, pp. 30–33; Boracic acid, pp. 84–88; Crude borate of soda, pp. 143–146. *Travels in Search of New Trade Products,* Jarrold & Sons, London.

Roulston, B. V., and Waugh, D. C. (1981). A borate mineral assemblage from the Penobsquis and Salt Springs Evaporite Deposits of Southern New Brunswick. *Can. Mineralogist* **19,** 291–301.

Sanyal, A., Hemming, N. G., Hanson, G. N., Broeker, W. S. (1995, January). Evidence for a higher pH in the glacial ocean from boron isotopes in foraminifera. *Nature* (*London*) **373**(6511), 234–236.

Schwarcz, H. P., Agyei, E. K., and McMullen, C. C. (1969, April). Boron isotopic fractionation during clay adsorption from sea water. *Earth and Planetary Sci. Lett.* **6**(1–5), 1–5.

Semmons, W. B., and Berger, M. K. (1980). A borate mineral assemblage in the Louann Salt accompanied by boron metasomatism of the Norphlet Shale, Clarke County, Alabama. *Abs. SW Sect. Geol. Soc. Am.* **12**(4), 208–209.

Sergina, Y. P., and Kaminskaya, A. B. (1967). Experimental simulation of the natural separation of boron isotopes. *Geochemistry Int.* **4**(5), 991–996.

Shaller, W. T., and Henderson, E. P. (1938). Mineralogy of drill cores from the potash field of New Mexico and Texas. *U.S. Geol. Survey Bull. 883.*

Shima, M. (1963, July). Geochemical study of boron isotopes. *Geochim. et Chosmochim. Acta* **27**(7) 911–913; *Tokyo Inst. Phys. Chem. Res. Reports* **39,** 207–210.

Siemroth, J. (1992). The minerals of the anhydrite quarry at Niedersachswerfen near Nordhausen in the Harz. *Mineralien Mag. Lapis* **17**(1), 52–56.

Simmons, W. B. (1988, October 19–21). Boron mineralization in the Louann Salt and Norphlet Shale in Clark County, Alabama. *Trans. Gulf Coast Assoc. Geol. Soc.* **38,** 553–560.

Smith, M. P., and Yardley, B. W. (1996). The boron isotopic composition of tourmaline as a guide to fluid processes in the southwestern England ore fields. *Geochim. et Cosmochim. Acta* **60**(8), 1415–1427.

Sonnenfeld, P. (1984). *Brines and Evaporites,* pp. 220–222, 469–472, Academic Press, New Jersey.

Spiryagina, A. I. (1953). *Determination of the Solubility of Borates.* Tr. VNII-Galurgii, No. 27.

Spivack, A. J., Berndt, M. E., and Seyfried, W. E. (1990). Boron isotope fractionation during supercritical phase separation. *Geochim. et Cosmochim. Acta* **54,** 2337–2339.

Spivack, A. J., and Edmond, J. M. (1986). Determination of boron isotope ratios by thermal ionization mass spectrometry of the dicesium metaborate cation. *Anal. Chem.* **58,** 31–35.

Spivack, A. J., and Edmond, J. M. (1987). Boron isotope exchange between seawater and the ocean crust. *Geochim. et Cosmochim. Acta* **51,** 1033–1043.

Spivack, A. J., Palmer, M. R., and Edmond, J. M. (1987). The sedimentary cycle of the boron isotopes. *Geochim. et Cosmochim. Acta* **51,** 1939–1949.

Stewart, F. H., Chalmers, R. A., and Phillips, R. (1954). Veatchite from the Permian evaporites of Yorkshire. *Mineralog. Mag.* **30**(225), 389–392.

Sun, D. (1990). Origin of borate deposits in Xiao Qaidam Lake, Qaidam Basin, China. *China Earth Sci.* **1**(3), 253–266.

Sundeen, D. A. (1990, September). *Determination of the Type and Distribution of Boron Minerals Associated with the Subsurface Evaporite Formations of East-Central Mississippi.* Bureau of Mines Open File Report 46–90.

Swihart, G. H., and Moore, B. P. (1989). A reconnaissance of the boron isotopic composition of tourmaline. *Geochim. et Cosmochim. Acta* **53,** 911–916.

Swihart, G. H., Moore, P. B., and Callis, E. L. (1986). Boron isotopic composition of marine and non-marine evaporite borates. *Geochim. et Cosmochim. Acta* **50,** 1297–1301.

Swihart, G. H., McBay, E. H., Smith, D. H., Carpenter, S. B. (1993). Boron isotopic study of the tertiary bedded borate deposits of Furnace Creek, California. *Geol. Soc. Amer., Abs.* **25**(6), 319.

Swihart, G. H., McBay, E. H., Smith, D. H., and Siefke, J. W. (1996). A boron isotopic study of a mineralogically zoned lacustrine borate deposit: The Kramer Deposit, California, U.S.A. *Chem. Geol. (Isotope Geosci. Sect.)* **127,** 241–250.

Taylor, R. E. (1937, October). Water-insoluble residues in rock salt of Louisiana (Gulf Coast) salt plugs. *Bull. Am. Assoc. Petrol. Geol.* **21**(10), 1268–1310.

Thode, H. J., MacNamara, J., Lossing, F. P., and Collins, C. B. (1948). Natural variation in the isotopic content of boron and its chemical atomic weight. *J. Am. Chem. Soc.* **70,** 3008–3011.

Valeyev, R. N., Ozol, A. A., and Tikhvinskiy, I. N. (1973). Genetic characteristics of the halide-sedimentational type of borate deposits. *Int. Geol. Rev.* **15**(2), 165–172.

Vengosh, A., Chivas, A. R., and McCulloch, M. T. (1989). Direct determination of boron and chlorine isotopic compositions in geological materials by negative thermal ionization mass spectrometry. *Chem. Geol. (Isotope Geosci. Sect.)* **79,** 333–343.

Vengosh, A., Chivas, A. R., McCulloch, M. T., Starinsky, A., and Kolodny, Y. (1991a). Boron isotope geochemistry of Australian salt lakes. *Geochim. et Chosmochim. Acta* **55,** 2591–2606.

Vengosh, A., Kolodny, Y., Starinsky, A., Chivas, A. R., and McCulloch, M. T., (1991b). Coprecipitation and isotopic fractionation of boron in modern biogenic carbonates. *Geochim. et Cosmochim. Acta* **55,** 2901–2910.

Vengosh, A., Starinsky, A., Kolodny, Y., and Chivas, A. R. (1991c). Boron isotope geochemistry as a tracer for the evolution of brines and associated hot springs from the Dead Sea, Israel. *Geochim. et Cosmochim. Acta* **55,** 1689–1695.

Vengosh, A., Starinsky, A., Kolodny, Y., Chivas, A. R., and Raab, M. (1992, September). Boron isotope variations during fractional evaporation of sea water: New constraints on the marine vs non-marine debate. *Geology* **20,** 799–802.

Vengosh, A., Heumann, K. G., Juraske, S., and Kasher, R. (1994). Boron isotope application for tracing sources of contamination in groundwater. *Environ. Sci. Technol.* **28,** 1968–1974.

Wardlaw, N. C., and Nicholls, G. D. (1972). *Cretaceous Evaporites of Brazil and West Africa and Their Bearing on the Theory of Continent Separation,* Vol. 6, pp. 43–55, Int. Geol. Cong., 24th Mtg., Montreal.

Yarar, B. (1985). The surface chemical mechanism of colemanite–calcite separation by flotation. In *Borax* (S. J. Lefond and J. M. Barker, eds.), pp. 219–233, AIME New York.

Yarzhemskii, Y. Y. (1968). Possibility of sedimentation of borates from eutonic brine in salinogenic basins of the marine type. *Int. Geol. Rev.* **10**(10), 1096–1102.

Yarzhemskii, Y. Y. (1969, September–October). *Formation Conditions of Strontium Hilgardite and Danburite in the Lower Cambrian Saliferous Deposits of Eastern Siberia,* 634–640, Lithology and Mineral Resources, No. 5.

You, C. F. (1994). *Boron, Lithium and Beryllium Isotope Geochemistry: Implications for Fluid Processes in Convergent Margins.* Doctoral dissertation, Univ. of Calif., San Diego, 237 p.

You, C. F., Spivack, A. J., Gieskes, J. M., Martin, J. B., Davisson, M. L. (1996). Boron contents and isotopic composition of pore waters. *Marine Geol.* **129,** 351–361.

Xiao, Y., Sun, D., Wang, Y., Qi, H., and Jin, L. (1992). Boron isotopic composition of brine, source water and sediments in the DaQaidam Lake, Qinghai, China. *Geochim. et Cosmochim. Acta* **56,** 1561–1568.

Zaritskiy, P. V. (1965). Boron mineralization in the Artemovka Formation in the Bakhmuta Depression of the Donents Basin. *Dokl. Acad. Sci. U.S.S.R.* (*Earth Sci. Sect.*) **149,** 157–159.

Zherebtsova, L. K., and Volkova, N. N. (1966). Experimental study of the behavior of trace elements in the process of natural solar evaporation of Black Sea water and Sasyk-Sivash brine. *Geochemistry Int.* **3**(4), 656–670.

Chapter 7 Mining

Borate mining in general follows the basic mining practices for other minerals, except that many of its operations are or have been comparatively small, and some of the deposits are (or were) close enough to the surface to allow open pit mining. When little data was available on some of the deposits in previous chapters, brief mining details were given in the geology sections. This chapter will review many of the other deposits.

7.1 OPEN PIT MINING

7.1.1 Argentina

7.1.1.1 Loma Blanca

S.R. Minerals (Barbados) Ltd. selectively mines beds of borax, ulexite, and inyoite–colemanite from the Loma Blanca deposit, with an estimated overburden ratio of <1.75/1 for the first 20 years of the deposit's life. Both overburden and ore zones usually can be mined by bulldozers or front-end loaders, then hauled in 35-metric-ton trucks to the waste dump or 0.5 km to the plant. Each ore type is placed on separate drying pads at the plant for 3–5 days before being processed (Solis, 1996).

7.1.1.2 Sijes

Ulex SA mined the Esperanza zone of the Sijes deposit in 1996 for both colemanite and hydroboracite. Most of the work at their Sol de Manana mine was performed by bulldozers (explosives were occasionally used on overburden), and front-end loaders filled the trucks. Work was curtailed for several months in the winter because of the cold and rain, which averaged 90 mm/year (Leiser, 1996).

7.1.1.3 Tincalayu

The initial open pit mine at Tincalayu was called the "Spiral Pit," with an overburden to ore ratio of 1.1/1. It was 100 × 75 m in size, had a 60° slope, and the ore averaged 30% B_2O_3. The pit was later expanded to 150 × 175 m, 60-m depth, with an overburden ratio of 3/1, and 26% ore. It was next decided

Figure 7.1 The Tincalayu open pit mine and ore storage pile area in 1995.

to blend >14% B_2O_3 ore with richer ore to form 18% B_2O_3. In 1981 the pit was expanded to 500 × 1000 m, and a 100 m depth. The area had numerous fault, fold, and displacement problems, and a weak sedimentary structure. A major fault bordered the deposit on the west, and after a number of cave-ins, a slope of 33° (requiring a 10/1 overburden ratio) was specified for that area. To the south there was a 10- to 12-m vertical offset in the borax from a southeastern fault.

More than 150 core samples had been drilled in the deposit [obtaining 16,500 m of 36.4-mm ($1\frac{1}{16}$ in.) core] on a 100-m grid, or 25–50 m in areas of discontinuity. Both >14% B_2O_3 ore and overburden thickness maps were prepared, and reserve figures calculated (assuming 1.87 g/cc for ore and 2.5 g/cc for <10% B_20_3 gangue).

Usually the borax was blasted in benches 5.5 m wide and 5 m deep in a pattern of 5–6 drill holes 4.5 m deep. The adjacent barren claystone was machine excavated with cuts 3 m deep for benches 6.6 m wide and 6 m high (Fig. 7.1). The bench slopes were 68° with a 1.5% gradient inward for drainage. Below the salar level of 4060 m, a 5-m-deep sump with a 50,000 l/hr (220 gal/min) pump ran 6–8 hr/day to remove seepage brine. Two 12-m-wide access roads with 6–8% slopes left the benches at the 4082 m level (later lowered to 4070 m), and one from 4070 m to the pit at 4035 m.

The overburden was removed by a 4-m^3 (5-yd^3) Caterpillar Model 983 front-end loader filling 12.5-ton Fiat end-dump trucks, and two 2-m^3 (3-yd^3)

Caterpillar Model 977 loaders filling 7-ton Ford dump trucks. About 7500 tons/day of overburden was hauled 600 m to a tailings pile in two 8-hour shifts. The borax was blasted (500 tons/day; 100 tons/hole) using a Caterpillar wagon drill, 50-mm shot holes on a 5-m grid, and 10 kg/hole of 79% Ammonite. Cuttings from the blast holes were analyzed to determine the ore's grade and to check the mining plan. A 1.3-m^3 Caterpillar Model 950 front-end loader then filled 7-ton Ford dump trucks. The ore was crushed by a 50–80 tons/hr Hazemag impactor to a −10- to 15-cm (4- to 6-in.) size before it went to the storage area. From there it went to a 110-ton silo, where it was reanalyzed if necessary and blended in the storage piles to 18% B_2O_3. It was then trucked (4000–8000 tons/month) to Salta. The mine also had maintenance and electric shops, a laboratory, truck scales, road maintenance, electric generation and distribution equipment, and staff quarters (Cornejo and Raskovsky, 1981).

7.1.2 Turkey

7.1.2.1. Bigadic, Emet, Kestelek

In 1996 there were open pit mines in the Bigadic area at Tulu and Yenikoy, and in the Emet area at Hisarcik and Espey (Fig. 7.2; Etibank, 1996). At

Figure 7.2 The colemanite open pit mine at Emet. (From Etibank, 1994; pictures courtesy of Etibank.)

Hisarcik the 14 m thick and 27% B_2O_3 ore dipped 5–7°, and both the ore and overburden were 20–100 m thick, with a stripping ratio of 1.73/1. A contiguous lower bed, 7 m thick and 16% B_2O_3, was not mined. At Kestelek the overburden was first blasted and removed by bulldozers and trucks, and the ore then mined by front-end loaders. There was a landslide in 1980, and 1,000,000 m^3 of overburden had to be removed (Albayrak and Protopapas, 1985).

7.1.2.2 Kirka

The initial open pit was in the northwestern corner of this deposit, 300 m from an old mine shaft where the ore grade was the highest. In 1974, however, a massive rock slide occurred, filling the mine and forcing the pit to be moved to the south near the deposit's center. Signs of a new slide on the extensions of the old one appeared in 1987 as the western slopes started moving toward the center. A tension crack 400 m long with a 1- to 1.5-m gap was formed. It was initially thought that the slide was caused by a clay layer between limestone beds having reduced cohesiveness when penetrated by water. However, it later was found that the main cause was mining in the toe of an ancient rock slide (Fig. 7.3). The old landslide was not stable when excavated at an 18° angle, but was stable at 16°. Also, blasting was done step-wise to put less stress on the slide area (Turk and Koca, 1990).

In 1984 the stripping ratio was 1.3/1, the ore had a density of 1.92 g/cc, 25.4% B_2O_3 (70.4% borax, 10% clay), the bed dipped fairly steeply from 2 to 150 m, and the overburden thickness averaged 70 m. The open pit covered 5.2 km^2; the benches were 8 m high and 25 m wide (later planned to be 14.5 m), the bench sloped 85° and the pit 20–22°. In 1990 the pit was 400 × 500 m in size, 60 m deep, and sloped 16–25°. The overburden could be blasted or ripped, loaded and smoothed with Komatsu 320-hp bulldozers, and hauled in 24 Soviet Belaz end-dump trucks to a dump 1 km east of the mine. Caterpillar bulldozers (175 hp) pushed overburden over the benches when necessary and

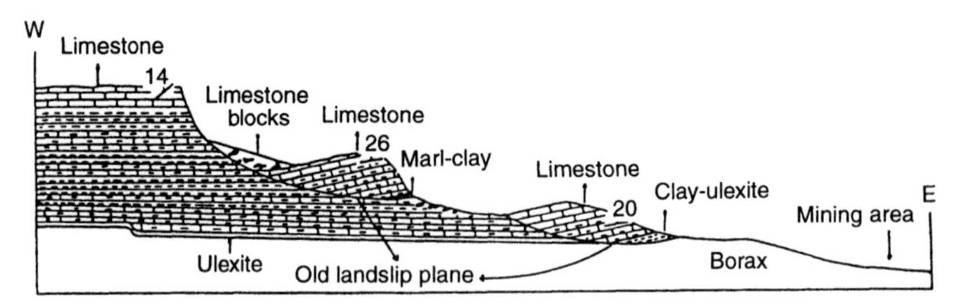

Figure 7.3 Cross-section of the old circular landslides in the Kirka borax mine. (From Turk and Koca, 1990; reprinted by permission of A. A. Balkema, Old Post Road, Brookfield, Vermont 05036, (802) 2763162.)

did general utility work. The ore was usually blasted, picked up by five 1.5- to 2.5-m^3 front-end loaders, and trucked the 13- to 14-min trip to the plant. Soft overburden could also be machine mined by three Russian 4.6-m^3 electric excavators. They could reach up 10.3 m, down 6.7 m, and out 7.8 m for dumping; travel 0.58 km/hr; load in 30–40 sec; and move 300 metric tons/hr. Harder areas were blasted each day with 160-mm blast holes and ANFO. The mine operated two 8-hour shifts 6 days/week (Albayrak and Protopas, 1985; Dickson, 1985; Turk and Koca, 1990).

7.1.3 United States

7.1.3.1 Kramer (Boron) Deposit

Construction began on the Kramer open pit (Fig. 7.4) in 1956, and operations started in 1957. By 1994 it was 1.7 km (1.1 mi) wide and 200 m (660 ft) deep [1.83 km long (east–west) and 1.52 km wide (north–south); Siefke, 1995], and nearly all of the former underground workings of the West Baker and Jenifer mines had been consumed by the pit (Siefke, 1995). The total U.S.

Figure 7.4 The boron open pit mine. (Photo courtesy of U.S. Borax Inc.)

Figure 7.5 Mining borax ore from the open pit at Boron, California. (From Anon., 1996b; photo courtesy of U.S. Borax Inc.)

Borax staff at that time was 1100 (Apodaca, 1994). In 1996 a new $2 million O&K RH120c hydraulic, 223-metric-ton, 15-m^3 (20-yd^3) diesel-powered (two 700-hp engines) shovel with a 23–27 metric tons bucket (Fig. 7.5) could load and dump in 30–40 sec, filling a 145-metric-ton truck in 5 or 6 passes. It could mine 1550 metric tons/hr and move anywhere in the mining area in an hour because it had no trailing electric cables. It replaced an older shovel purchased in 1965, and has an advantage over front-end loaders by using a slightly downward action compared with their in-and-upward movement. This reduces the load on the shovel and gives a better separation of the ore and overburden (Anon., 1996b).

In 1991 the overburden was stripped with three P&H 14.5-m^3 (19-yd^3) electric shovels [a fourth had a 11.5 m^3 (15-yd^3) bucket] filling 154- to 163-metric-ton (170- to 180-ton) trucks. The ore was loaded with three 10-m^3 (13-yd^3) Marathon LeTourneau L-800 front-end loaders in 1987; 11 and 12 m^3 (15 and 16 yd^3) units into 91-metric-ton (100-ton) trucks in 1991; and 17-yd^3 loaders and some 154-metric-ton trucks in 1995. Both the overburden and borax were drilled and blasted on a 5-day schedule, producing 36,000 metric tons/day of overburden and 8200–9100 metric tons/day of ore. Because the stripping ratio could increase by 50% as thinner ore beds to the east and deeper zones to the south were encountered, mining reverted to 7 days/week in 1994. It was always done at several levels to blend the ore, a very important plant need (Anon., 1987; Siefke, 1991, 1995). In 1987 the overburden and ore were blasted with ANFO in drill holes on a 5.5- to 7.3-m pattern to form benches 22.5m high. All of the haulage was by a fleet of Unit Rig diesel–electric Lectra-Haul trucks: six M-36s with a 154-metric-ton capacity, and twelve M-100s capable of hauling 82 metric tons. Because of the ore's low density (1.6 g/cc), the smaller trucks had tail gates to carry 82–86 metric tons instead of the usual 72.5 metric tons for unmodified trucks. The stripping ratio was 5/1, and 18,000,000 metric tons/year were mined (Anon., 1987).

The overlying and adjacent ulexite and colemanite zones of the deposit are mined and stockpiled separately for possible future use, whereas the overburden is sent to waste piles north and south of the pit. The ore is taken to a 1100-metric-ton/hr hammermill at the pit bottom to be crushed to −20 cm (8 in.; Siefke, 1991) [6 in.; Anon., 1987]. It is then conveyed 2 km (1.25 mi) up and out of the pit at an 18° angle and sent to a 60,000-ton surface storage pile (Anon., 1996a). The pit conveyor, powered by a 450-KW (600-hp) drive, had a 1.066 m (42 in.) wide single-ply, woven-steel rubber belt with good resistance to ripping and tearing. It traveled 3 m/sec (600 ft/min) and could carry up to 1090 metric tons/hr of ore. Kernite ore was handled separately and taken to a separate storage pile for the boric acid plant. When there was a surplus over the plant's needs the kernite was further crushed in a hammermill to −6.4 mm ($\frac{1}{4}$ in.) and sprayed with water in a pit to be hydrated to borax. When converted, it was periodically fed to the primary crusher from bottom withdrawal belts (Anon., 1987).

The slope stability of the pit is affected by the poorly consolidated near-surface sediments, water intrusion, faults, and great depth [it was scheduled to ultimately be 396 m (1300 ft) deep]. The west half of the north wall collapsed in 1985, primarily because of an intruding aquifer. Thus, before the benches were rebuilt several groundwater drainage wells were drilled north of the pit, and overburden was used to construct dykes for protection against surface runoff water and for tailing ponds. In the south, the former deep Western Borax mine acts as a drainage sump for that area. Between 1913 and 1995, 150 scattered surface wells, 225 underground core holes, and 675 mine-control holes have been drilled, many on a 61-m (200-ft) grid and with a 98% core recovery. Near faults or discontinuities the spacing often was 30.5 m (100 ft). From this data a mine plan computer model was made, which has been used for "a variety of geologic and mine engineering functions" (Siefke, 1991, 1995).

In the pit's early operation, the haulage roads had a 7% grade, and the first ore was encountered at 42 m (137 ft). The distance from the pit floor to the overburden dump was about 1.6 km (1 mi). The overburden benches were 15 m (50 ft) high, and 12–18 m (40–60 ft) wide. The borax benches were

9.1 m (30 ft) high and made by blasting 11.7-cm ($4\frac{5}{8}$-in.) holes 10.7 m (35 ft) deep on 4.9-m (16-ft) centers, up to 3.7 m (12 ft) from the outer bench wall. ANFO was made with 36 kg (80 lb) of ammonium nitrate and 2.8 l (3 qt) of fuel oil per hole. The primer was 10 cm (4 in.), 4.5-kg (10-lb) EP203 ignited by 50-grain Primacord and an electric blasting box. Electric auger drills were used for the blast holes. The broken ore was loaded by 3-yd^3 2300- to 4160-volt BE 54B electric shovels into 335-hp turbocharged 802 Kenworth end-dump trucks. These trucks had tubeless tires and Allison 30-speed Torquematic transmissions for hauling the ore 0.43 km (0.7 mi) to the 107 × 168-cm (42 × 66-in.) primary hammer mill, which was then near the top of the pit. The ore was reduced to −10 cm and sent by covered conveyer belt to be placed in horizontal layers by a mobile stacker on the plant's 54,000 metric tons, 223-m (730-ft) long, 34-m (112-ft) wide (at the base), 13.7-m (45-ft) high storage pile (Dayton, 1957).

7.1.3.2 Death Valley Region: Boraxo

Tenneco Mining, Inc. opened this mine in 1970 and closed it in 1977, having removed 9,100,000 metric tons of ore and overburden, with a stripping ratio of 2.6/1. About 100 employees worked in the mine. The pit was 671 m (2200 ft) long, 91 m (300 ft) wide in the west; 305 m (1000 ft) wide in the east; and 69 m (225 ft) deep [the design depth was 137 m (450 ft) with the same surface dimensions]. The benches were 12 m (40 ft) high, and the face slope was 38° in the north wall and 40° in the south wall. The deposit had been thoroughly drilled, showing that the surface zone of yellow-gray mudstone and shale was 3–46 m (10–150 ft) thick and could be ripped by D-9 Caterpillar bulldozers. It was taken to the dump in 14 and 18 m^3 (18 and 24 yd^3) push-loaded scrappers. The transition to nonoxidized green or blue-gray sediments and the colemanite ore averaged 6.1 m (20 ft), but could change quickly.

The lower rock on the south wall required blasting with 0.18–0.23 kg of powder per m^3 (0.5–0.65 lb/yd^3) of broken rock. Two rotary drills prepared the 20-cm (8-in.), 6.1-m (20-ft) deep, 3.4 × 6.7-m (11 × 22-ft) staggered-spacing shot holes to form the benches. The north wall was more difficult to blast, requiring 0.36 kg of powder/m^3 (0.8 lb/yd^3). The pit floor was developed with 7.6-m (25-ft)-deep shot holes on a 3.7 × 3.7-m (12 × 12-ft) spacing. Dry holes were loaded with ANFO, and wet holes with nitro-carbo-nitrate. The borates were also blasted in 6.1-m (20-ft) lifts, and the broken ore was loaded by 6-, 8-, and 12-m^3 (8-, 10-, and 15-yd^3) front-end loaders into 32- and 45-ton, rear-dump, off-highway trucks. Near the edges of the colemanite and ulexite–probertite zones the beds were thin and interlayered, requiring careful mining. Ore with $<10\%$ B_2O_3 was first sent to a screening plant to remove the lower-grade fines. About 230,000 metric tons of ore were left in the mine when it became too deep for an open pit operation (Evans, Taylor, and Rapp, 1976).

7.2 UNDERGROUND MINING

7.2.1 Turkey

7.2.1.1 Bigadic

Mining in the 12 or so small underground mines of this area in 1980 was done by a variety of methods (Table 7.1; Fig. 7.6). All of the mines entered the ore at outcrops, and then followed it down-dip. Both the floor and the roof presented problems, with the lower clay tending to rapidly swell and heave irregularly when exposed to air, greatly impeding or even closing the entries. Because the ore was stronger than the overlying clay, unmined ore was left in the ceiling to provide a more stable roof, but extensive timbering was still required [i.e., 20 cm (8 in.) timbers on a 1.5-m spacing, or 0.04 m^3 of wood per ton of ore]. The wood was both expensive and scarce, the entries were still unstable, the ore recovery was low, and the mining costs were high.

In the Begendikler mine blasting was not required, and only picks and shovels were used. The 1.5- to 1.75-m thick ore zone averaged 50–60% colemanite, and the rest was fairly soft clay. The adhering clay was brushed from the colemanite at the mine face, and both were separately taken in 50-kg wheelbarrows up the 13° slope to the surface. The entries were 2.2 m high and 2 m wide, required extensive timbering (0.14 m^3/ton of ore), and the floor heaved and swelled badly. When stoping was attempted the entire stope volume had to be timbered (Albayrak and Protopapas, 1985; Murdock, 1958).

Table 7.1
Mining Practice in 1985 at the Bigadic Borate Area[a]

Mine	Ore thickness (m)	No. of beds	Mining method	Ore capacity, (mt/yr)
Acep	4	1	Longwall	40,000
Arkagunevi	7.5	3	Longwall	49,000
Kireclik	4	3	Longwall	40,000
Kurtpinar	4	3	Open stope	3,000
Mezarbasi	5	4	Open stope	6,000
Ongunevi	7.5	3	Longwall, room and pillar	50,000
Simav	10	4	Longwall	67,000
Tulu	15	1	Open pit[b]	173,000
Yenikoy	6	2	Longwall (open pit later)	30,000
Total	—	—	—	458,000

[a] Albayrak and Protopapas, 1985.
[b] Since 1979.

Figure 7.6 Underground colemanite mining at Bigadic. (From Etibank, 1994; pictures courtesy of Etibank.)

The Gunevi mine was somewhat larger, the colemanite–ulexite ore of a higher grade, and the beds sloped at 7°. Blasting was required, but because the ore was 3.3 m thick, beams of it could be left in the ceiling. When the ore thinned, however, extensive timbering on a 1-m spacing was required. The nearby Kurt Pinari mine had similar ore, but its 3-m colemanite bed was free of clay partings, so there could be both floor and roof beam of unmined ore. The Acep mine was the largest in the area, and a 3-m ulexite bed with some colemanite at its base was mined. Rails (30 cm) with 1-m^3 cars were used, and for steep dips the ore was winched up the slope. For lesser dips hand labor moved the cars to the surface. Farther into the mine, standard timbering with 15- to 25-cm (6- to 10-in.) posts and caps had to be reinforced with timber pens, or walls or posts of quarried marl (Murdock, 1958).

7.2.1.2 Emet: Espey Mine

The colemanite beds at Espey varied from 5 to 10 m, and the overburden was 50–150 m thick. Because considerable ore was left in the roof, longwall mining was employed in 1985, operating at two faces. The face length was 80 m, the cutting height 2.5 m, and standard movable roof supports and a chain face conveyor were used. Conveyor belts then hauled the ore out of the mine. However, the recovery was still low, so the mine changed to an open pit operation (Albayrak and Protopapas, 1985). Brown and Jones (1971) noted that earlier the mine had nine parallel entries, each 200 m long. The colemanite roof beam was quite stable, and only a small amount of timbering was required. The 43% B_2O_3 ore was shoveled into 1-ton rail cars, hauled up the 6° slope, and shipped 290 km to the ports of Bandirma or Derince.

7.2.1.3 Kestelek

Conventional room and pillar mining was converted to a retreating longwall at this mine in 1981 (Albayrak and Protopapas, 1985). The mine had problems with an unstable roof and floor, and even with closely spaced timbering the floor heaved and the back fell. Water also was encountered, and pumping was necessary (Brown and Jones, 1971). Bekisoglu (1962) and Murdock (1958) noted that the 3- to 8-m thick beds had sudden dips and 0.5- to 1-m cavities (containing drusy surfaces of minute crystals). Some of the ore was low in grade (it averaged 39–42%), and the black shale foot wall contained lignite. A compressed air hoist hauled ore cars up rails with a 20° dip, and a diesel generator provided lights and powered an air compressor for the hoist and rock drills. Only occasionally was blasting required.

7.2.2 United States

7.2.2.1 Calico Region: Borate Mine

The initial inclined shaft at Borate (in 1888) was in the colemanite outcrop's highest-grade ore at High Point. It was at a 45° angle, followed the ore's dip, and eventually was 183 m (600 ft) deep. The mine's roof was very unstable, requiring close set 25 × 30-cm (10 × 12-in.) Douglas fir timbering, and the entry had to be replaced in 1 year because the caps had been compressed to 5 cm (2 in.). The floor also swelled, causing the narrow-gauge rail to "hump like a gentle roller-coaster." Gas engines (50 hp) hoisted the ore up the shaft, initially in a 50-gal whiskey barrel, after the ore had been broken and sorted. The high-grade colemanite was hauled 18 km (11 mi) to the rail line at Daggett, whereas the low-grade ores and fines went 10 km (6 mi) to a calcining plant at Marion. The route from the mine was tortuous and steep, but it was handled well by mule teams. However, a steam-powered traction engine, Old Dinah, was a complete failure. Later, a narrow-gauge rail line "with some sharp bends, phenomenal gradients, and a flimsy high trestle bridge" (over the very steep Mule Team Canyon) finally did replace the mules.

In 1899 a second shaft was sunk [also in time 183 m (600 ft) deep], boosting production to 20,000 metric tons/year, making it the first underground (and colemanite) mine in the United States, and the largest borate mine in the world at that time. Numerous drifts and stopes (they were worked upward from the drifts) were taken off the shafts, and about 250 men were employed. The mining was done by hand labor, including breaking slabs of various sizes into low- and high-grade ore. Only comparatively small amounts of fines were produced (Storms, 1893; Willey, 1906; Travis and Cocks, 1984).

7.2.2.2 Coastal Mountain Area: Lang

Two vertical shafts, each 107 m (350 ft) deep, were employed at Lang from 1908 to 1922. The main entries were on the 30-, 61-, and 91-m (100-, 200-, and 300-ft) levels, with mining along strikes of 244–305 m (800–1000 ft). One segment, 183 m (600 ft) long, was mined on all three levels with numerous auxiliary entries. Square set timbering [12 × 18 m (40–60 ft) wide, 21 m (70 ft) long, and 18 m (60 ft) high] was needed in some stopes, and shrinkage stopes were mined in the others. The latter were about 21 m (70 ft) long, 18 m (60 ft) high, and 6.1 m (20 ft) wide at their base above the ore chutes. Pumping was constantly required, and 80 miners were employed, producing 360 metric tons/day (400 tons/day). The mine yielded 91,000 metric tons (100,000 tons), and from 1908 to 1920 was the dominant borate operation in the United States (Gay and Hoffman, 1954).

7.2.2.3 Death Valley Region: Billie

The Billie colemanite, ulexite–probertite mine is in the Death Valley National Park, 24 km (15 mi) southeast of Furnace Creek, and no entry shaft, above ground tailings, subsidence, or plant site was allowed. The mining also had to be quite selective because the ulexite–probertite was to be sold "as is," with tight specifications, and the colemanite needed to be high grade to reduce shipping costs. The ore's unconfined compressive strengths were as follows: colemanite 3–17 Mpa (500–2500 psi, average 1200), ulexite 14–20 MPa (2000–3000 psi), probertite 41 Mpa (6000 psi), and calcareous shale 20–34 MPa (3000–5000 psi). Their hardness (on the Mohs scale) was: colemanite 4.5, and ulexite 2.5. The ore body was 46–390 m (150–1300 ft) deep, dipped 20–30° to the southeast, and 46–53 m (150–175 ft) thick. It extended 1100 m (3700 ft) along the dip, and averaged about 200 m (700 ft) in width [Blakely (1980) stated the dimensions as 1524 m (5000 ft) long, in places over 91 m (300 ft) thick, and 122–366 m (400–1200 ft) deep]. The ore's bedding was intact, but the colemanite (not probertite's interlocking radial crystals) was considerably fractured and somewhat recemented by clay or calcite. The rock temperatures were 29–46°C (85–115°F).

A cut-and-fill, tall narrow stoping method with continuous, single and twin-boom miners was planned, with two long parallel rooms mined together by the machines, and a temporary pillar with cross-cuts between them. They would be 7.6 m (25 ft) wide, the length of the ore body's width, and gradually ascend with a cut and fill operation. Both permanent and temporary pillars were 9.1 m (30 ft) wide, and there were haulage entries above and below the ore body at 232 and 341 m (760 and 1120 ft). Each cut in the stope was 3.7 m (12 ft) high, with the ore dropped to the haulage level, segregated, and stored for later removal. After a room had been completely mined, it was filled with waste rock and smoothed. The mining machine then climbed onto the fill and made the next cut. This process continued to the top of the ore. Then the equipment was disassembled and moved to a new stope pair (Garrett, 1985; Norman and Johnson, 1980).

Blakely (1980) noted that there were 5 years of planning and environmental approval and 2 years of developmental work before mining began in 1980. The 4.9-m (16-ft) diameter, concrete-lined, 366-m (1200-ft)-deep shaft for the mine was 610 m (2000 ft) to the west of the ore body, outside the Death Valley National Park. The haulage drifts were then made to the deposit, and 2.1-m (7-ft)-diameter man and material raises were drilled between the levels. There were also two 1.2-m (4-ft)-diameter ore chutes to the haulage level, and a 1.2-m backfill chute from the surface to the upper level. Entries and development headings were made by 59 metric-ton AEC SuperRock 330, 500-hp ripping-type boom miners, and the mining was done by 91 metric-ton, twin-boom milling-type Dosco TB600, 1000-hp machines. Continuous miners had not operated on ore with >10,000 psi compressive strength (probertite could be this high), but both machines had the capability after modifications. They were also capable of very selective mining, even on thin seams. The H configuration of the stopes and cross-cuts allowed four headings per machine, and the temporary pillars were removed after both pillars had been mined.

Ore haulage was by low profile 1.5-m (5-ft), 10-ton, 8.5-m (28-ft)-long, 2.7-m (9-ft)-wide, battery-operated coal-type PB136S machines. These machines also picked up fill from the drop tubes, then placed, smoothed and compacted it. Their mainframes and buckets had been altered to be dismantled and hoisted through the 2.1-m (7-ft) raises, and their scissor-type ore unloading system could be used as a bulldozer blade. On the ore haulage level the ore or waste rock was picked up from the drop chutes by a 3.8-m^3 (5-yd^3) load-haul-dump (LHD) unit and loaded into 20-ton diesel trucks for the 610-m (2000-ft) trip to the shaft. Waste rock was crushed on the surface to −5 cm (2 in.), and during every fill cycle dumped to a hopper on the 232-m level where 20-ton trucks hauled it to dump holes. The scoops then took it to the freshly mined room in 0.61-m (2-ft) lifts with watering and compaction. When done it was smoothed, and the cutting cycle resumed. The backfill was given a 7% moisture content and 85% of maximum compaction, thus attaining an unconfined compressive strength of about 750 psi.

The mine was ventilated by 250- and 400-hp fans, each exhausting air through its own vent on the 232-m level. Air entered the main shaft to the lowest level, flowed through the stopes and out the upper level. The face being mined received air from auxiliary 50-hp fans and flexible ducting. The airflow was complicated by the mine's high temperature and thermal gradient (Chandler, 1996). Entries were supported in areas of faulting with W6 × 20 steel arches on 1.2-m (4-ft) centers. Other areas used 1.5-m (5-ft) split sets with landing mats and wire mesh. In the stopes resin-grouted fiberglass roof bolts were used for temporary roof support (Blakely, 1980).

During the initial operation just described it developed that the entries had to be smaller and all of the overhead rock immediately bolted with long, high-strength bolts (and wire mesh installed to completely cover the roof) to prevent dangerous rock falls. Because the continuous miners could not cut through the roof bolts and wire mesh, a new mining plan had to be established. Also, the mining machine's waiting to fill the small loaders, or dump ore on the ground was slow and made the roadways hard to travel. Because the mining machines were trapped in a stope until finished, the maintenance was poor and time consuming when mining stopped. Finally, the disassembling of the machines for each new pair of stopes was difficult and lengthy.

The new mining procedure (Fig. 7.7) had the continuous miners cut sublev-

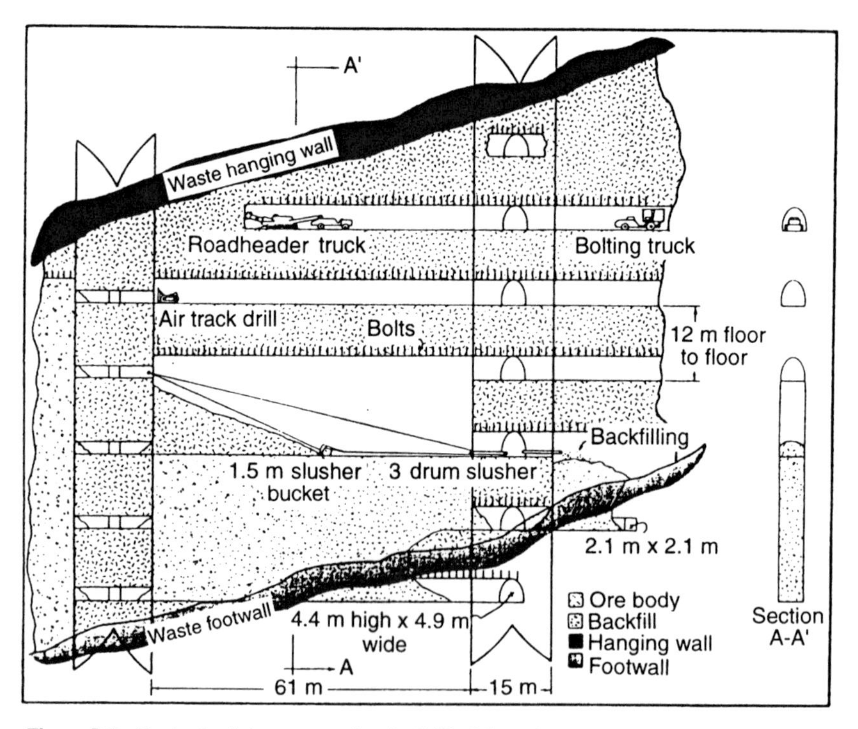

Figure 7.7 Revised mining system for the Billie Mine. (From Garrett, 1985; reproduced by permission of the Society for Mining, Metallurgy and Exploration, Inc.)

els, then blast the ore between them. As before, each room was backfilled after it was mined, but now slushers (pillar-mounted drag lines) removed the blasted ore. The mining machines fed ore directly into 12-metric-ton dump trucks, and spiral ramps connected the sublevels. The deposit was divided into 76-m (250-ft) stoping blocks with permanent 15-m (50-ft) pillars between the blocks. The pillars between the stope pairs were 24 m (80 ft) wide, and later they would be reduced to 9.1 m (30 ft). The sublevels were at 12-m (40-ft) intervals, and the mining machines cut an arched profile 4.4 m (14.5 ft) high and 4.9 m (16 ft) wide. The machines could not advance past where the operator was under the last row of roof bolts, or about 4.6 m (15 ft). Then the roof was bolted and the machine moved to another heading. Three to five headings were worked simultaneously.

The roof bolts were on 1.2-m (4-ft) centers and were 2.4 m (8 ft) long. Their holes were drilled in two 1.2-m cuts, with the second drill swiveled in the middle. The drilling platform allowed two rows of bolts to be drilled and inserted from one position of the truck. Resin bolts were used in the haulage and access drifts, but the stope bolts were galvanized. A 7.6-m (25-ft)-long, 2.4-m (8-ft)-wide roll of wire mesh was used in each position on the arched roof, with a 15-cm (6-in.) overlap on the adjacent section. Roof bolting took about twice as long as the mining period. When done, the ore between the sublevels was blasted with 6-cm (2.25-in.) holes on 1.5-m centers five rows at a time. After blasting, the broken ore was removed with a 1.5-m^3 bucket pulled by a 56-kw motor. A LHD unit then delivered it to the ore drop hole. When the room was empty, a remote-controlled 3.8-m^3 LHD unit finished cleaning it.

The room was next backfilled from the upper drift with a pneumatic system. Screened −25-mm (1-in.) gravel from a nearby open pit mine was fed to the bin of a rotary airlock feeder, and a 300-kw blower delivered it through 25-cm (10-in.) down or 20-cm (8-in.) horizontal pipes to the stope. Water was added 15 m (50 ft) from the end to reduce dust and aid in compaction. The stope was filled to the upper floor level to allow space for the next blasting–slushing cycle. Maintenance was done on the mining machines during the third shift, and only lubrication and broken bit changes were done between shifts. A Voest-Alpine AM-75, 40-metric-ton drum miner was used for the latter period of operation (Garrett, 1985). The American Borate Company shut down the 227,000-metric ton/year Billie mine in 1986 (because of its high costs, with a staff of 200), and the company was sold. It resumed small-scale production in 1991, employing 25 people (Norman, 1993). Figure 7.8 is a picture of its head frame and surface installations.

7.2.2.4 Death Valley Region: Lila C

This deposit was mined from 1907 to 1915 through four inclined shafts entering the outcrop and following the bed as it deepened. Each shaft had

Figure 7.8 The head frame for the Billie Mine.

drifts at different levels, with stopes being formed between them. The main level extended for 1067 m (3500 ft; Gale, 1912). Ore was sent by narrow-gauge rail to the new Tonopah & Tidewater railroad at Death Valley Junction, which had been built specifically for it and the metal mines to the north. When the Lila C closed, it was replaced by the series of mines near Ryan (Fig. 7.9).

7.2.2.5 Kramer (Boron) Deposit: Jenifer Mine

The last underground mine at Kramer (from 1952 to 1957) was the Jenifer Mine, considered to be highly automated for its time. It used boom-type mining machines and automatic ore lifting in the shafts. It produced 730–910 metric tons/day (800–1000 tons/day) from three borax beds, the top one at a depth of 61–152 m (200–500 ft). The total thickness of the three beds was 7.6–24.4 m (25–80 ft), and they were mined by a room and pillar-stoping method (see Fig. 7.10) with dual entries for better ventilation, safety, and movement of men and materials. All rooms and entries were the mining machines' 2.4-m (8-ft) height and 5.2-m (17-ft) width, and the pillars were 7 × 8.5 m (23 × 28 ft) [also quoted as 6.7–8.2 m (22 × 27 ft)], superimposed over each other in the three levels. The room heights were the bed thickness less >1.5 m (5 ft) left in the ceiling as a strong roof beam, and often 1.5–

Figure 7.9 The town of Ryan.

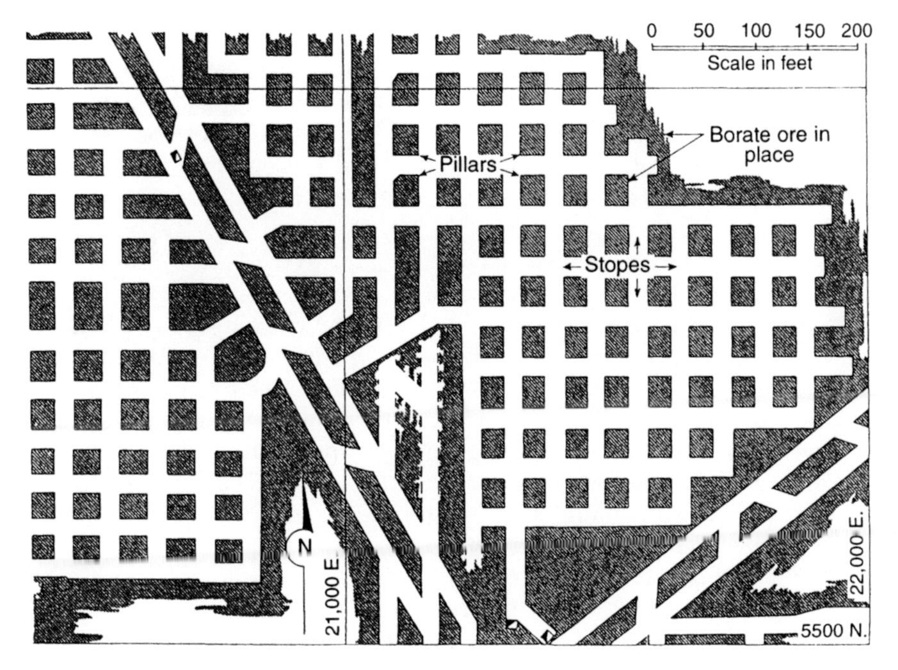

Figure 7.10 Typical mining plan for the Jenifer Mine. (From Obert and Long, 1962.)

2.4 m (5– 8 ft) left in the floor to prevent heaving and sinking of the pillars. This allowed an ore recovery of only 30–45%, not counting the three unmined low-grade borax beds and the contained or adjacent colemanite and ulexite. However, the mine was comparatively stable, needing little extra support (Obert and Long, 1962).

The machinery included 5 Joy JCM and 1 Joy IPM continuous miners, 12 Joy 10SC 6.4- to 7.3-metric-ton (7- to 8-ton) shuttle cars, 9 Jeffrey entry conveyors, and a mainline belt conveyor. The cutting head of the JCMs consisted of six rotating chains, each with 168 Carboloy or tungsten carbide-tipped bits, which could be resharpened 4–6 times/average life (36,000 metric tons of ore). The head was 76 cm (30 in.) wide and cut 46 cm (18 in.) at each position of the machine. The machines' climbing limit was <15%, and ore was picked up continuously from the floor as it was cut. It was conveyed to an attached shuttle car acting as a surge bin to allow more continuous operation. The car delivered the ore to a 4-wheel steering, dynamic-braking, dual-control shuttle car, which took it to the nearest entry belt, up to <183 m (600 ft) away. The loading and unloading time was about 1.5 min. A vibrating feeder fed the 76-cm (30-in.) entry belt, which delivered the ore to a 1.5 × 1.5 m (5 × 5 ft) drop hole, or directly to the mainline conveyor, which took it to a crusher. The miners averaged 3.6 hrs of actual cutting time/shift and an advance of 12.2 m (40 ft)/shift.

When the bed (plus beams) was thicker than 2.4 m (8 ft), the mining machines started at the top, then cut at progressively lower levels until the desired height was cut. The miners and shuttle cars were powered with 2300-V AC current, which was transformed to 440 V AC for the miners and rectified to 250 V DC for the shuttle cars. The 100-hp, high-tensile-strength mainline conveyor belt had troughing idlers and vulcanized splices, and was 351 m (1150 ft) long and 76 cm (30 in.) wide, carrying 320 metric tons/hr. It was in the ore for greater support of its 3 × 4.6 m (10 × 15 ft) passageway, and climbed from the lowest level at a 20% slope to the 113-m (370-ft) level. Here it discharged into small rail cars with a 350-hp hoist (a "winze") that took the ore to a hopper feeding an underground crushing station. The ore was screened, and the oversize pieces were crushed to 10 cm (4 in.) by a Pennsylvania 76 × 152-cm (30 × 60-in.) single-toothed roll crusher. The −10 cm ore fell into a 180-metric-ton bin that automatically filled the two 4.3-metric-ton counter-balanced skips in 4 sec for hoisting by 3.2-cm (1.25-in.) cables to the surface at a rate of 6.1 m (20 ft)/sec. They discharged into two storage bins with vibrating feeders for the plant or stockpile conveyors. The hoist was a double-drum, 500-hpNordberg unit, 132 cm (52 in.) wide and 2.1 m (7 ft) in diameter, operating semi-automatically to lift 360 metric tons/hr of ore. The men and materials (service) shaft, with its 150-hp hoist, was 107 m (350 ft) northwest of the ore shaft (Anon., 1958).

Obert and Long (1962) noted that roof bolting was required in a few areas with clay interbeds. The bolts were 1.9-cm (0.75-in.) steel and 1.2–1.8 m (4–6 ft) long. They were anchored in solid borax because the overlying shale would not hold. Wire mesh could be used between the bolts, and in a few permanent entries there were heavy timbers with caps, cribs, or steel lagging. The very slow plastic flow of borax was a factor in some roof failures, although most roofs and floors were stable for their lifetime (5 year). Mine dust was controlled by water trucks, but water adversely affected the magnetic separators, so it was not sprayed at the mining machines. Instead, each machine had a 20,000-cfm fan, and blowers in the entries directed 25,000 cfm of air to each working face. A 150,000-cfm fan at the base of the service shaft, and a 75,000-cfm fan at the West Baker shaft supplied air to the mine. A double-entry system with crossovers every 30.5 m (100 ft) supplied fresh air in the right entry and return air in the left. Brattice curtains were used to control the airflow, with plastic spray used to help seal the seams and joints. The mine temperature was 21°C (70°F). There were large underground shops where preventative maintenance was practiced and records for service and lubrication were kept on each machine. The continuous miners and shuttle cars were serviced every 16 hrs, and after 100,000 tons of ore had been cut or conveyed, they were totally disassembled, inspected, and rebuilt (Dayton, 1957).

7.2.2.6 Kramer Deposit: Baker Mine

The Baker mine was operated from 1927 to 1951 at a rate of 540–730 metric tons/day (600–800 tons/day). The combined thickness of the borax beds was 2.4–35 m (8–115 ft), averaging 22.9 m (75 ft), and the depth to the upper bed was 110–152 m (360–500 ft). The beds in about half of the mine were horizontal, and the others sloped from 17–24°. There was some kernite in the lower sections of the mine. Room and pillar open stoping was used (Fig. 7.11), with the stopes mined both downward and upward from the entry levels. The broken ore was removed by portable drag lines ("slushers"), with the kernite being handled separately. When borax was mined the entry, rooms, and pillar widths were 6.1 m (20 ft), whereas for kernite they were 6.7 m [22 ft; pillar width 5.5 m (18 ft)]. Pillar and room lengths varied up to 30.5 m (100 ft; Obert and Long, 1962). The mine had two 143-m (470-ft) single-compartment shafts and one 165-m (540-ft) three-compartment shaft, with loading levels at 110, 128 and 146 m (360, 420 and 480 ft). A 305-m (1000-ft)-long, 9° (16%) slope winze serviced the lower levels, carrying the ore to the shaft pockets. There it was dumped and hoisted to the surface in balanced 1.4-metric-ton skips. Horizontal drifts from the bottom of each bed were laid out on 36.6-m (120-ft) centers (with cross-cuts) from the base (foot wall) to intercept the sloping bed's top (hanging wall). Drag lines filled five 2.3-metric-ton automatic side-dump rail cars, and a battery-operated locomotive took

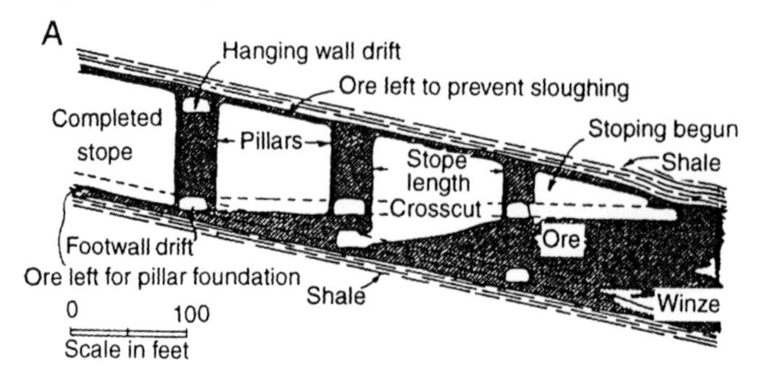

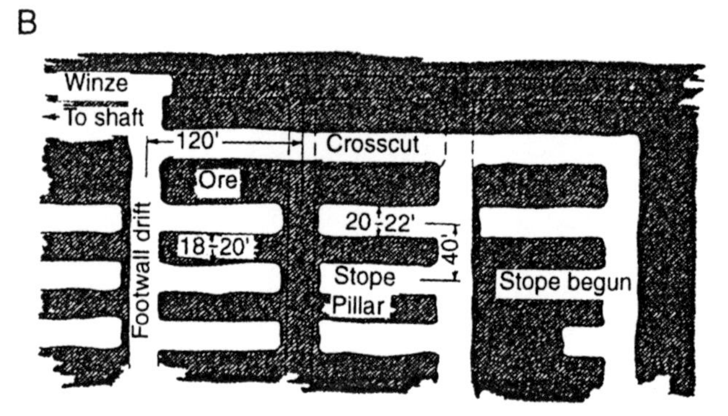

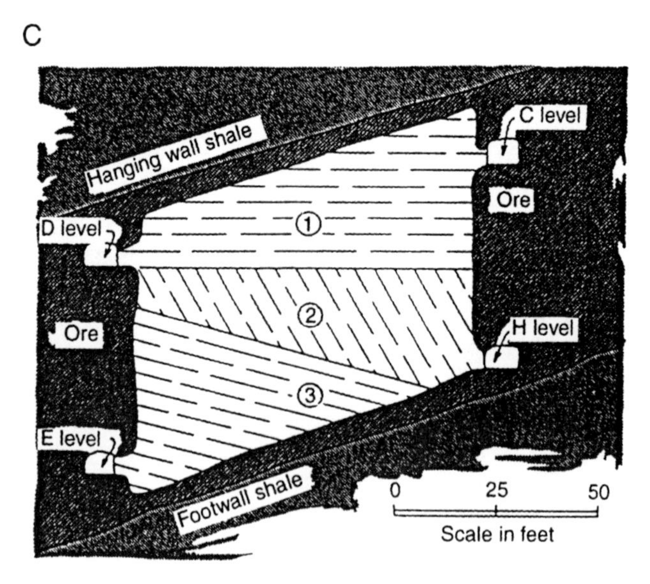

Figure 7.11 Typical mining pattern for the Baker Mine. (From Obert and Long, 1962.)

them to the winze. When a stope had been mined, it was filled with sand through 30-cm (12-in.) drill holes by a high-speed fan (Tucker, Sampson, and Oakeshott, 1949).

7.2.2.7 Kramer Deposit: West Baker (Suckow) Mine

This mine operated from 1929 to 1953 at a rate of 730–910 metric tons/day (800–1000 tons/day). It was southwest and adjacent to the Jenifer mine, and later merged with it. It also had three borax beds, with an average combined thickness of 15–40 m (50–130 ft). The depth to the upper bed was 36–107 m (150–350 ft), and the first shaft was 131 m (431 ft) deep. The second was 146 m (480 ft) deep with levels at 116 and 146 m. The ore was usually dry, but a small artesian spring developed in one of the rooms. A room and pillar-stoping procedure was used, with the pillars at the three levels placed over each other. All rooms, entries, and pillars were 6.1 m (20 ft) wide; 1.5 m (5 ft) of ore was left in the roof, and 3–4.3 m (10–14 ft) of ore was left in the lower bed's floor. The lower stopes were mined (and back filled) first, then the middle bed, and finally the upper bed. The ore was hoisted to the surface in 2.3-metric ton balanced skips, and the mined rooms were filled with sand. However, no effort was made to fill the rooms completely or compact the sand to provide roof support. On examination in 1957 none of the roofs had caved or the floors heaved in the unfilled rooms (Gale, 1946; Tucker *et al.*, 1949; Obert and Long, 1962).

7.2.2.8 Kramer Deposit: Western Borax Mine

This mine was operated from 1928 to 1933 on four levels connected by raises and winzes, with layouts as shown in Fig. 7.12. A stoping procedure was used following the quite variable high-grade kernite ore. This gave the upper levels (A and B) odd-shaped rooms and pillars, but in the lower levels the layout was far more uniform. The room height averaged 9.1 m (30 ft) on level A, 3–16.8 m (10–55 ft) on level B, and 6.1–12.2 m (20–40 ft) on levels C and D. The roof spans averaged 9.1 m, but were as much as 18–24 m (60–80 ft) on level A. Ore was left in the roof [1.8–2.4 m (6–8 ft)] and in the floor [3–3.7 m (10–12ft)]. After 24 years only one major roof fall had occurred (on level A), and it appeared to have been caused by water entry. On the lower level in 1957 a roof fall covered the entire floor with 0.9–1.8 m (3–6 ft) of 0.15–0.6 m (0.5–2 ft)-thick kernite slabs, and many of the pillars had spalled. The mine's 261-m (856-ft) No. 1 shaft was always dry; the 290-m (950-ft) No. 3 shaft continuously pumped 31 l/sec (0.8 gal/min) of water; and the No. 2 shaft was abandoned because of heavy water inflow at a depth of 60 m (197 ft). The compressive strength of the ore appeared to increase with depth (Obert and Long, 1962). In 1955 the Mudd interests deepened the No. 3 shaft to 344 m (1130 ft) and installed a new head frame and ore hoist. They dewa-

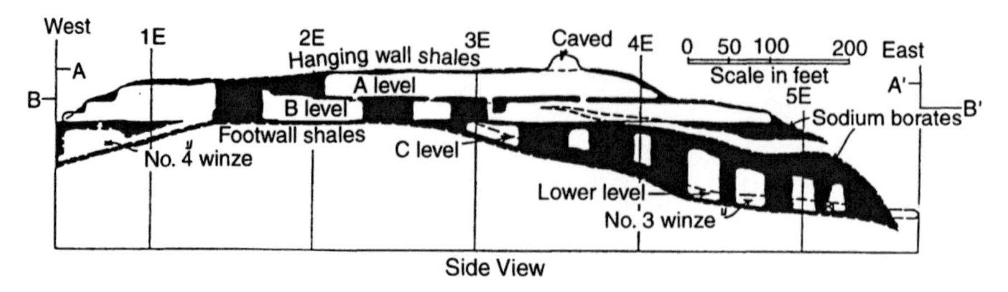

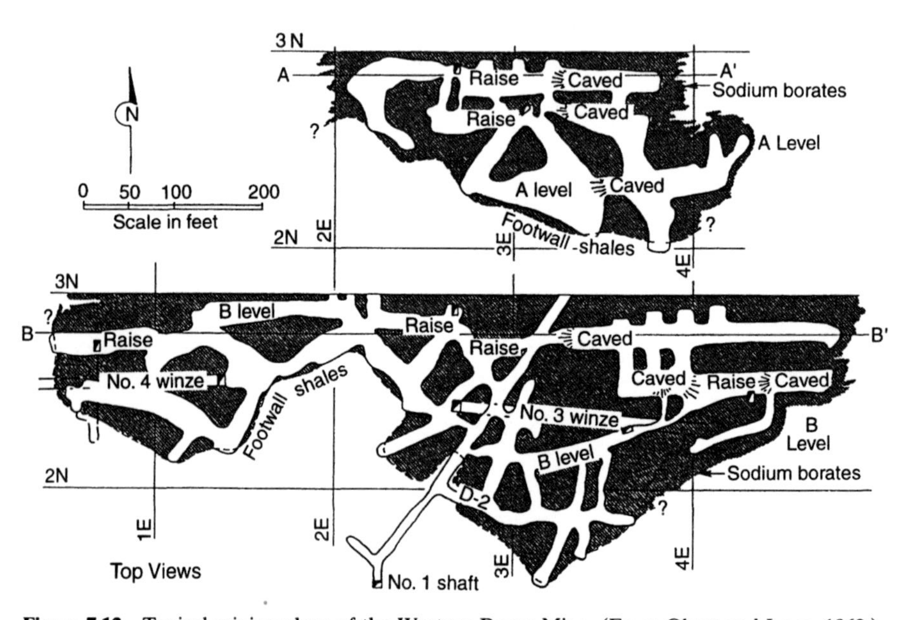

Figure 7.12 Typical mining plans of the Western Borax Mine. (From Obert and Long, 1962.)

tered the No. 2 shaft and prepared to mine from the 344-m level at a rate of 270 metric tons/day (300 tons/day; Anon., 1955). Little production was achieved.

7.2.2.9 Muddy Range, Nevada: Anniversary Mine at Calville Wash

The Anniversary mine operated from 1922 to 1927, producing 91,000 metric tons (100,000 tons) of ore. Its colemanite bed averaged 4.6 m (15 ft) in thickness and dipped 45–55°, so it was mined by overhead stoping. However, the dip was not steep enough, and the colemanite broke into tabular pieces which did not slide easily. Thus, when sufficiently steep, the stopes were kept filled with colemanite to provide a mining platform, but with less dip, as much ore as possible was retained in the stope, and small pillars were left at intervals. The ore was removed at the bottom by scrapers, and a 90% recovery was obtained. Timber supports were required only in the entries and ore chutes.

Figure 7.13 Ore cars from the haulage level of the anniversary mine. (From Castor, 1993; reprinted courtesy of the Nevada Bureau of Mines and Geology.)

The latter were spaced at 6.1-m (20-ft) intervals from the mining to the haulage level, reaching all of the ore. The ore was conveyed in small manually pushed rail cars (Fig. 7.13) to the bins of an electrically operated cable tramway, which took the ore 366 m (1200 ft) horizontally and 76 m (250 ft) vertically to the processing plant. The operation was semi-automatic, requiring only one operator to load, run the machinery, and empty the ore buckets (Young, 1924). The mine had two shafts and about 2100 m (7000 ft) of underground workings. As much as 160 metric tons/day (175 tons/day) were mined, and 360,000–730,000 metric tons of ore remained above the haulage level (the remaining total was 3,200,000 metric tons of 26% B_2O_3) when the operation was abandoned (Castor, 1993; Papke, 1976).

7.3 PLAYAS

7.3.1 Argentina

7.3.1.1 Centenario

This salar has both massive beds and nodules of ulexite, starting at a depth of 2–15 cm and extending to a layer of travertine or caliche. Machinery cannot operate on the salar because of its soft mud, so only hand labor is used in mining. The surface crust is first removed with 30 × 24 cm, square-nosed shovels and loaded into 50-kg wheelbarrows. This "overburden" is then hauled to a previously mined area as fill. The ulexite zone is next mined with pick and shovel, and the mud separated from the ulexite. The "ore" is then loaded into wheelbarrows and taken to a drying area. The individual slabs or nodules are mostly 10–25 cm in size, and each 50-kg wheelbarrow load is kept separate to facilitate production accounting, inventory, and man-power scheduling. Even during the winter the ulexite dries in 15–20 days when assisted by frequent turning (with pitchforks). When dry it is sent to the plant, using 4- to 5-ton dump trucks when possible.

At the plant, the ulexite is further broken (by picks) for later ease of bagging, and again laid out to dry. Some clay is removed and the moisture content reduced to 5–10% (the ore's bulk density is 0.7 g/cc). Each 3–5 m long, 1-m wide and 30- to 40-cm high drying area holds 30 wheelbarrow loads, and they are periodically turned with some clay removal. When dry (in 4–5 days in the summer) the ore is shoveled into 10-ton conical piles, from which workers fill 50-kg polypropylene bags (with coal shovels). The bags are manually sewn shut with 10-cm needles and 85 cm of polypropylene thread. A worker can fill 750–1000 bags/week or sew 750 bags/week. The bags (550, or 25–28 metric tons) are then loaded in trucks, and about 1500 metric tons/year is taken to Salta (Battaglia and Alonso, 1992).

7.4 SOLUTION MINING

Several experimental tests have been done on borate solution mining, such as a brief study of both single and dual connected wells on borax at the Kramer deposit in 1948 (Taylor, 1970). An 8.8-m (29-ft)-thick layer of 72% borax, 107 m (350 ft) deep, and with "a layer of competent shale above (the) ore" was solution mined, and the cavities later entered to inspect the results. No roof padding (with air or oil) was employed, and typical broad morning glory holes were formed, with extensive insolubles in their base (Fig. 7.14). The clay and shale had fallen as "sand and gravel," and some slabs of ore fell into the mud and were not leached. Hot water at 102–110°C (215–230°F) was injected into the 21°C (70°F) formation, and at one point a 15-l/min (4-gpm) flow produced a saturated, sparkling clear borax solution in a single well at 77°C (170°F). The exit brine temperature was steadily increasing, as was the allowable flow rate to produce a saturated solution. The experiment lasted for 38 days, and 230 metric tons (250t) of borax were leached. Many patents have been issued on borate solution mining, as indicated in the References.

At the Hector deposit in the Calico region (also called the Fort Cady deposit), hydrochloric acid has been injected into colemanite beds. The formation is saturated with a 20,000-ppm brine containing 10,000 ppmSO_4, 6000 ppmNa, and 1000 ppmK. Approximately 200 22.5-cm (8.875-in.) wells have

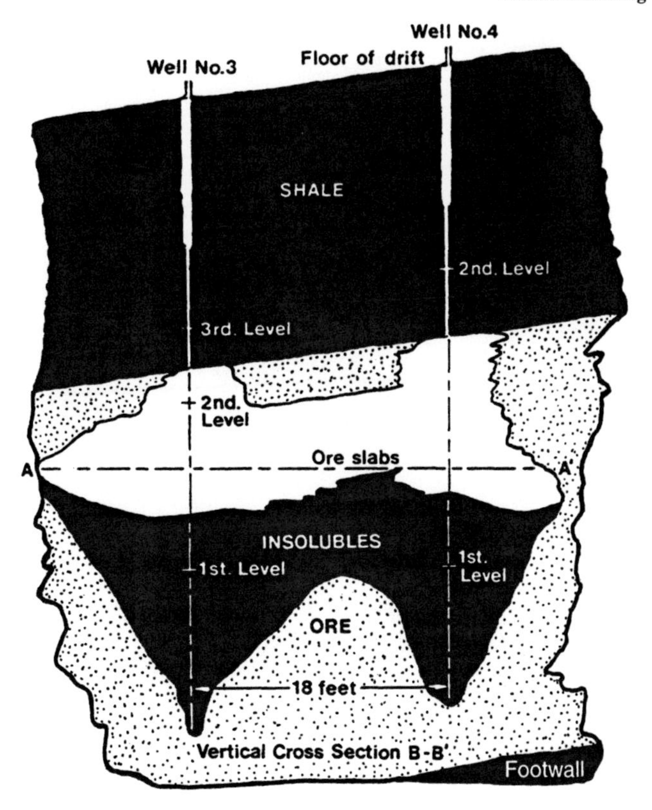

Figure 7.14 Connected single-well solution mining test at U.S. Borax. (From Taylor, 1970; reprinted by permission of the Northern Ohio Geological Society.)

been drilled 411 m (1350 ft) deep in 139 hectares (343 acres) of the 200-hectare (500-acre) deposit. A 17.8-cm (7-in.) OD fiberglass pipe was then placed in each hole and sealed with acid-resistent cement. Dilute (5%) hydrochloric acid was injected at the 365-m level (also reported as 244–366 m), and allowed to stand for 8 hours. The solution was then pumped from the wells with corrosion-resistant 7.5-hp, 0.57-l/sec (9-gal/min) submersible pumps, and solar evaporated to crystallize crude boric acid (410 metric tons of 95–98% H_3BO_3 by 1994). Alternately (and in 1996 the preferred process), lime was added to the leach liquor in steps to produce >43% B_2O_3, fine (81% −325 mesh), synthetic colemanite containing less than 1% SiO_2, 600 ppmCl, SO_4 and Fe, and 10 ppmAs. The end liquor was then reacted with sulfuric acid to precipitate gypsum, as well as regenerate hydrochloric acid to be reinjected with 6 l/sec (100 gal/min) of make-up water. A 6-month test in 1995 of a 3900 tons/year pilot plant led to the design of a 30,000 tons/year plant, which later was expanded to 90,000 tons/year (Hartman, 1994, 1996; Norman 1991, 1995).

References

Albayrak, F. A., Protopapas, T. E. (1985). Borate deposits of Turkey. In *Borates: Economic Gelogy and Production* (J. M. Barker and S. J. Lefond, eds.), pp 71–85, Soc. Min. Eng., AIMMPE, New York.

Anon. (1955, November). Development of new borax mine rushed in southern California. *Mining World* (Sect. 41-42) **17**(12) 57–58.

Anon. (1958, April). U.S. borax has integrated a complex industrial plant. *Eng. Mining J.* **159,** 101–105.

Anon. (1986, November 24). Boric acid is made in situ from calcium borate. *Chem. Eng.*, p. 12.

Anon. (1987). U.S. borax. *Mining Mag.* **157**(5), 455–457.

Anon. (1996a, July). The company behind the 20-mule team. *Compressed Air Mag.*, pp. 30–36.

Anon. (1996b). *Top Gear Mining,* Borax Pioneer, U.S. Borax Inc., No. 6, 24p.

Apodaca, P. (1994). Desert dig. *Los Angeles Times, Valley Business,* Sect. F, pp. 10–12.

Battaglia, R. R., and Alonso, R. N. (1992). *Geologia Mineria de Ulexita en el Grupo Minero Maggie, Salar Centenario, Salta,* pp. 241–245, 4th Congreso Nacional de Geologia Economica, les Congreso Latinoamericano de Geologia Economico, Cordoba, Argentina.

Bekisoglu, K. A. (1962, July). Boron deposits in Turkey. *Turk. Econ. Rev.* **3**(4), 12–34.

Blakley, J. W. (1980, April). American Borate's Billie Mine in Death Valley. *World Mining,* pp. 57–60.

Blumenberg, H. (1927, July 19). *Process of Mining Insoluble Boron Compounds.* U.S. Patents 1,636,455; 1,636,456.

Blumenberg, H. (1928). *Method of Mining Boron Compounds.* U.S. Patents 1,696,075 (December 18); 1,649,385 (November 15, 1927), each 3p.

Brown, W. W., and Jones, K. D. (1971). Borate deposits in Turkey. In *Geol. History of Turkey* (A. S. Cambell, ed.), pp. 483–492, Petrol Explor. Soc. Libya, Tripoli.

Castor, S. B. (1993). Borates in the Muddy Mountains, Clark County, Nevada. *Nevada Bur. Mines, Geol., Bull. 107,* 31 p.

Chandler, C. D. (1996, January–February). The Billie Mine, Death Valley, California. *The Mineralogical Record* **27,** 35–40.

Cornejo, R. J., and Raskovsky, M. A. (1981). Projecto minero y explotacion a cielo abierto del yacimiento Tincalayu. *Segundas Jornadas Argentinas de Ingenieria de Minas, San Juan* **1,** 232–242.

Dayton, S. H. (1957, June). $20,000,000 face lifting forges new era for Pacific Coast Borax. *Mining World* **19**(7), 36–45.

Dickson, J. (1985). Etibank at Kirka: From ore to derivatives. *Ind. Min.*, No. 210, pp. 65, 67.

Evans, J. R., and Anderson, J. P. (1976). *Colemanite Deposits near Kramer Junction, San Bernardino County, California,* Calif. Div. Mines, Geol., Spec. Publ. 50, 8p.

Evans, J. R., Taylor, G. C., and Rapp, J. S. (1976). Borate deposits (colemanite and ulexite–

probertite). *Mines and Mineral Deposits in Death Valley National Monument.* Calif. Div. Mines, Geol., Special Rpt. 125, 61p.

Gale, H. S. (1912). The Lila C borax mine at Ryan, California. *U.S. Geol. Survey, Mineral Resources, Borax,* pp. 861–866.

Gale, H. S. (1946). Geology of the Kramer Borate District, Kern County, California. *Calif. J. Mines, Geol.* **42**(4), 325–378.

Garrett, C. R. (1985). Development of a roadheader mining system at the American Borate Company. *Rapid Excavation and Tunneling Conference Proceedings,* Ch. 53, pp. 886–901, Soc. Min. Eng., Am. Inst. Min., Met. Petrol. Eng., New York.

Gay, J. E., and Hoffman, S. R. (1954). Borates. *J. Mines Geol.,* Resources of Los Angeles County, Calif. Mines and Mineral **50**(3–4), 506–508.

Hartman, G. J. (1994, 1996). Fort Cady in Situ Borate Mining Project, pp. 37–39, 1994 Soc. Min. Met. Explor., SME Ann. Mtg., and 1996, Proc. 12th Ind. Min. Int. Cong., Chicago, Albuquerque, Preprint 94-125.

Kistler, R. B., and Helvaci, C. (1994). Boron and borates. In *Industrial Minerals and Rocks* (D. D. Carr, ed.), 6th ed., pp. 171–186, Soc. Min. Metal, Expor., Littleton, Colorado.

Kistler, R. B., and Smith, W.C. (1983). Boron and borates. In *Industrial Minerals and Rocks* (S. J. Lefond, ed.), 5th ed., pp. 533–560, Soc. Min. Eng., AIME. New York.

Leiser, T. (1996). *An Overview of Selected South American Boron Producers,* pp. 31–35, 12th Ind. Min. Int. Cong.

Murdock, T. G. (1958, May). The boron industry in Turkey. *U.S. Bur. Mines, Mineral Trade Notes, Suppl. No. 53* **46**(5), 47 p.

Norman, J. C. (1991, 1993, 1995). Boron: A review of the year's activities. *Mining Eng.* **43**(7), 740–741; **45**(7) 718–720 (July, 1993); **47**(7) 660 (July, 1995).

Norman, J. C., and Johnson, F. C. (1980). The Billie Borate ore body, Death Valley, California. In *Geology and Mineral Wealth of the California Desert* (Fife and Brown, eds.), pp. 268–275, South Coast Geol. Soc., Santa Ana, California.

Obert, L., and Long, A. E. (1962). *Underground Borate Mining, Kern County, California.* U.S. Dept. Interior, Bur. Mines, Rept. of Invest. 6110, 67p.

Papke, K. G. (1976). Evaporites and brines in Nevada playas. *Nevada Bur. Mines, Geol., Bull. 87,* 35p.

Siefke, J. W. (1991, 1995). The boron open pit mine at the Kramer borate deposit. In *1991, The Diversity of Mineral and Energy Resources of Southern California* (M. A. McKibben, ed.), SEG Guidebook Series **12,** 4–15; 1995, Calif. Dept. Conserv., Div. Mines Geol., Spec. Publ. 110, pp. 181–188.

Solis, A. R. (1996), *Summary of Mining and Processing at Loma Blanca, Argentina,* Univ. Nac. Juju, Inst. Invest. Tech. Miner Ind., Juju, Argentina, 4p.

Storms, W. H. (1893). *The Calico Mining District,* pp. 345–348, Mineral Resources of San Bernardino County, Calif. State Mining Bur., 11th Ann. Rept.

Taylor, D. S. (1970). Experiments on solution mining of borax at Boron, California. *Third Symp. on Salt* **1,** 412–416.

Travis, N. J., and Cocks, E. J. (1984). *The Tincal Trail,* Harrap, London

Tucker, W. B., Sampson, R. J., and Oakeshott, G. B. (1949). Borates. Mineral Resources of Kern County. *Calif. J. Mines, Geol.* **45**(2) 241–244.

Turk, N., and Koca, M. Y. (1990). In *Slope Stability Problems of Kirka Borax Open Pit Mine (Turkey)* (D. G. Price, ed.), Vol. 3, pp 2305–2310, 6th Int. IAEG Congress, Balkema, Rotterdam.

Willey, D. A. (1906, October 6). Borax mining in California. *Eng. & Mining J.* **82,** 633–634.

Young, G. J. (1924). Mining borax in the Muddy Range. *Eng. & Mining J. Press* **117**(7).

Chapter 8 Processing

The processing of each of the commercial borate minerals is a comparatively simple procedure, but the desire for good purity, low costs, high recovery and efficiency has led to many interesting and sometimes complex variations on the operations. This technology is discussed in the following section, in addition to the brief reviews given for some of the deposits in the geology chapters.

8.1 BORAX AND KERNITE

8.1.1 Argentina

8.1.1.1 Loma Blanca

S.R. Minerals (Barbados) mines borax, ulexite, and inyoite from different beds in the Loma Blanca deposit, and delivers the ore to drying pads near the mine. After drying 3–5 days, each borate is separately run through a magnetic separator to remove some of the clay, producing ores with at least 26% B_2O_3. These are then trucked 180 km southeast to a 4-hectare processing plant at Palpada (near Juju). There the ore is further crushed, kiln dried, and again sent through magnetic separators, taking the B_2O_3 content to 38%. The products can also be calcined to 54–55% B_2O_3, and sized from 2 mm to 325 mesh. Shipments are made in 45- to 50-kg sacks or 900- to 1000-kg bulk bags (Solis, 1996).

8.1.1.2 Tincalayu

The borate processing capacity of Boroquimica SAMICAF in 1990 was rated at 37,700 metric tons/year at its Campo Quijano (Quyano) plant near Salta, 227 km (250 mi) northeast of the Tincalayu mine (Norman, 1990). The 16–18% B_2O_3 ore was first given a "cold washing" treatment to remove the soluble salts and some slimes, thus raising its purity (on a dry basis) to 28% B_2O_3. Next the ore was dissolved in hot mother liquor and water, the brine settled, and the clarified hot liquor sent to tanks to cool and crystallize. When completed, the brine was decanted and the crystals removed, centrifuged, washed, and dried to produce 99.9% borax or pentahydrate. Some anhydrous product was also claimed (Dublanc, Malca, and Leale, 1993).

8.1.2 Tibet

Before the 1900s, the processing of borax from Tibet was a closely guarded secret and difficult because the sheep caravan drivers who delivered the borax "ore" coated the crystals with any available fat, oil, or grease to reduce the water-dissolving loss during transport. The Indian merchants acting as wholesalers also often adulterated the borax, making each shipment quite variable. Dissolving and recrystallization was always the first processing step, but these crystals also had to be redissolved and their brine purified. The initial solution may have required treatment by activated carbon, clay, or equivalent adsorbant to remove the organics, as well as a precipitant for some impurities, and then a bleed stream removed. For the second crystallization, perhaps only adsorbants and filtration were required. There was also a need to produce large crystals as a sign of high quality. Poor refining yielded "small crystals with a yellow color imparted by the grease with which the tincal was covered." Because of its complexity, processing became somewhat of an art, with the Dutch operation in 1773 having its "secret step," such as clarifying the initial leach solution with "the aid of the white of an egg or its equivalent slaked lime and slate." A picture of borax processing in 1556 is shown in Fig. 8.1 (Travis and Cocks, 1984).

8.1.3 Turkey

8.1.3.1 Kirka

Borax shipments from the open pit mine at Kirka began in 1972. Mined ore, −40 cm in size, was first distributed onto a 3000-metric-ton stockpile, and starting in 1974 it was was then sent to a 200-metric-ton silo to feed a 400,000-metric-ton/year washing plant (Fig. 8.2; later expanded to 600,000 metric tons/year; Anac, 1988). In it the ore was first conveyed to a series of crushers working in closed circuit with screens, where it was initially reduced to −10 cm, then to −2.5 cm with hammer mills, and finally to −6 mm with roll mills. The −2.5-cm ore had been sent to 10,000-metric-ton surge bins to be metered into the roll mills, and the mills' discharge was screened at a 1-mm (6-mesh) size. The two fractions (+6 and −6 mesh) were repulped and sent to separate scrubbers (vigorously agitated tanks). Then the 1 to 6-mm fraction was pumped to spiral classifiers (Fig. 8.3), and its coarse fraction centrifuged and sent to product bins. The original −1-mm particles were repulped in the spiral classifier's undersize slurry. Then the mixture was cy-cloned and the overflow sent to a hydroclassifer. Both the cyclone and hydro-classifier's underflow streams (+65 mesh) were filtered. The filter cake was added to the coarser fraction if the concentrate was to be sold directly, the mixture containing about 32% B_2O_3 and 6–8% water. It was then dried and stored in silos for shipment to the port at Bandirma. Alternately, the 6- to

Figure 8.1 Borax refining in the sixteenth century. (A) Dissolver, (B) evaporator, (C) crystallizer, (D) copper rods, and (E) grinding. (From Travis and Cocks, 1984; picture courtesy of U.S. Borax Inc.)

65-mesh filter cake could be sent to the refinery for further processing. The overflow from the hydroclassifer was sent to a thickener, its underflow (the slimes) discarded, and the overflow brine was returned to the repulper. There was more than a 10% product loss in the washing operation (Table 8.1). In 1985 the concentrator plant operated three shifts, 7 days/week, with one of the Sunday shifts used for maintenance. There were 95 employees, and the maximum output was 2400 metric tons/day. Demircioghi (1978) reviewed the various potential borax ore beneficiation methods.

The derivatives (or purified product) plant in 1988 had a capacity of 160,000 metric tons/year of pentahydrate, 17,000 metric tons/year of borax, and 60,000 metric tons/year of anhydrous borax (Anac, 1988). The washed ore entered by conveyor belt at 50 metric tons/hr from either a stockpile or direct production. It was first dissolved in steam-heated tanks at 98°C to form a 26% $Na_2B_4O_7$ solution (saturated at 89°C). The resultant slurry was sent to countercurrent

Figure 8.2 The Kirka borax processing plant. (From Etibank, 1994; pictures courtesy of Etibank.)

thickeners where the slimes were settled, washed and discarded. Gur, Turkay, and Balutcu (1994) noted that 1200 ppm (of the dry slimes present) of hydrolyzed polyacrylamide (PAM) coagulant, 10% anionic, produced a clear brine with hot, saturated high-montmorillonite slimes (with a "blue color"), but did little for high-calcite or dolomite slimes. Nonionic polyethylenoxide (PEO) gave a clear overflow with dolomite slimes at 1200 ppm, and had a slightly poorer performance at 400 ppm. Cebi, Yersel, Poslu, Behar, Nesner and Langenbrick (1994) found that 55- to 61-wt% slurries could be obtained from centrifuged (solid bowl) slimes with the aid of a coagulant, but the centrate still contained 0.4–2.6% solids.

The clear brine from the hot thickeners was next filtered in eight pressure filters and sent to a 1000-metric-ton surge tank. It then went to either of the vacuum-cooled, growth-type crystallizer circuits. In the pentahydrate unit it was cooled to 66°C at a 0.23-atm vacuum. Excess fines were removed in cyclones and redissolved to allow the production of a coarse product that could be centrifuged to a 4–5% moisture content. Borax was produced in the same manner, but the exit temperature was 46°C. The centrifuged products were dried in oil-fired rotary dryers (except one for borax, which was a steam-

Figure 8.3 Spiral classifiers at the Kirka borax washing plant. (From Etibank, 1994; pictures courtesy of Etibank.)

tube unit). The dried products were screened and sent to storage for bulk or bagged shipment.

Part of the borax was dehydrated in a special furnace at 1100°C to produce anhydrous borax. The resultant melt was then cooled, solidified, crushed, and screened. Oversized material was recrushed and screened, whereas the undersized material was remelted. The processing plant also had its own 3.2-megawatt power plant, both because of its remote location and to achieve better thermal efficiency with its steam-power balance. The products were trucked either to the port at Bandirma, or 18 km to a rail connection at Degirmenozu with its truck dumps into 12 small silos, and its eight 10,000-metric-ton storage buildings. Automated vibrating feeders on conveyor belts could deliver product to all of the storage buildings, and reclaim conveyors took the products to 50-ton bins over the rail lines. Three cars could be loaded simultaneously, using hopper cars for the purified products and gondolas for the concentrates. The products were weighed and analyzed before shipment, and the dust from belts and bins was returned to the plant (Dickson, 1985).

Table 8.1

Examples of the Kirka Borax Ore and Concentrator Streams[a]

A. Chemical analyses (wt%)

Elements	Ore	Concentrate	Slimes[b]
B_2O_3	26.40	34.56	15.18
MgO	7.42	2.36	13.08
CaO	5.83	1.65	11.84
SiO_2	5.60	1.54	13.58
R_2O_3	0.12	0.12	0.80
H_2O, etc.	54.63	59.77	45.52
Minerals, dry basis (wt%)			
Borax	69.11	88.43	41.44
Dolomite	13.76	4.35	23.09
Clay	11.14	2.60	27.16
Calcite	3.80	1.63	7.91
Ulexite	2.19	2.99	0.40
Total	100.00	100.00	100.00

B. Concentrate screen analyses (wt%)

Mesh size, Tyler no.	Size (mm)	Wt% on screen
+3.5	+6	3.5
5	4–6	17.5
16	1.02–4	50.5
65	0.2–1.02	28.0
−65	−0.2	0.5

[a] Albayrak and Protopapas, 1985.

[b] In some samples a high percentage of montmorillonite clay was present (Gur, Turkay, and Buluteu, 1994).

The initial plant to refine Turkish borate ore was located at the shipping port of Bandirma (Fig. 8.4), and starting in 1975 produced about 55,000 metric tons/year of borax and 35,000 metric tons/year of boric acid (made from colemanite; Anac, 1988). In 1984, 20,000 metric tons/year of sodium perborate, and in 1986 an additional 100,000 metric tons/year of boric acid capacity were added. The handling capacity of the port for refined products and concentrates was 1.5 million metric tons/year in 1995 (Norman, 1995). All of the borate operations were owned and operated by Etibank, a state-owned company.

8.1.4 United States

8.1.4.1 Boron

A new processing plant was built in 1956 at the Kramer (boron) deposit (at the same time that it converted to open pit mining) to handle the anticipated

Figure 8.4 Views of the Bandirma borate processing and shipping facility. (From Etibank, 1995; pictures courtesy of Etibank.)

lower-grade ores (from 28 to 23% B_2O_3). More handling and washing capacity of insolubles was also needed, and an allowance was made for future expansion, causing the site to be enlarged to 32 hectares (0.32 km^2 or 80 acres). The high-intensity magnetic separators and some of the crude borax partial dehydration equipment (to produce pentahydrate by heating the ore to 110–121°C; Corkill, 1937) was then shut down, but the facilities to produce crude, anhydrous borax (called "Rasorite") were maintained.

In the new plant in 1958 the process started with the blending of different grades of ore, keeping the B_2O_3 content within a range of several percentage points. The −10-cm (−4-in.; also quoted as −8-in.) ore from the mine went to an automatically programmed stacker belt that discharged it in horizontal layers onto a 54,000-metric-ton, 223 × 34 × 13.7-m (730 × 112 × 45-ft) storage pile. A reclaiming tunnel 3 m (10 ft) in diameter, 244 m (800 ft) long, and 0.3 m (1 ft) deep was located under the pile, with 16 withdrawal chutes feeding a 76-cm (30-in.) conveyor belt (Dayton, 1957). However, it soon developed that outside (machine) reclaiming was necessary for all of the ore, and two overhead 1100-metric-ton/hr bucket–wheel reclaimers for each plant (borax and boric acid) were later installed (Anon., 1996b). After passing by a tramp iron magnet, the ore went to an impactor working in closed circuit with two screens to reduce its size to −1.9 cm (−3/4 in.). It was then sampled and conveyed to four 910-metric-ton (1000-ton) ore silos (Fig. 8.5). A fifth silo contained crude trona from Owens Lake to convert metaborate in the leach solution back to borax, to minimize calcite scaling, and to slightly react with the colemanite and ulexite in the ore.

The 20–25% B_2O_3 ore and a small amount of trona were removed from the silos (with some additional blending) and repulped with heated recycle brine. The slurry went in series flow through three (a fourth was maintained as a spare) 7570-liter (2000-gal) steam-jacketed, agitated dissolving tanks with internal steam coils, and having a residence time of 3–4 min. It then was passed over 1.9-cm (40-mesh; another reference said 60-mesh) Tyrock vibrating screens to remove the coarser undissolved ore, which was washed and discarded. The slurry next went to four countercurrent hot thickeners, where the slimes settled (originally aided by Separan coagulant), and were washed with the limited amount of water allowed by the plant's water balance. They were then discarded, still containing 3–4% B_2O_3. Later this mud was centrifuged (solid bowl), repulped in water (causing some difficulties because it tended to peptize to a colloidal state), combined with the screened large solids, and pumped to a tailings pond. Heat exchangers reheated the slurry as it was transferred between the 70-m- (230-ft-) diameter, covered, insulated, 2.4-m- (8-ft-) high (holding 13.2 million liters, or 3.5 million gal of slurry) thickeners. The clarified hot leach solution overflowed from the first (later also the second) of the thickeners, and was polish-filtered in six Sweetland 46.5-m^2 (500-ft^2) pressure leaf filters, except for that portion formerly made into a slightly

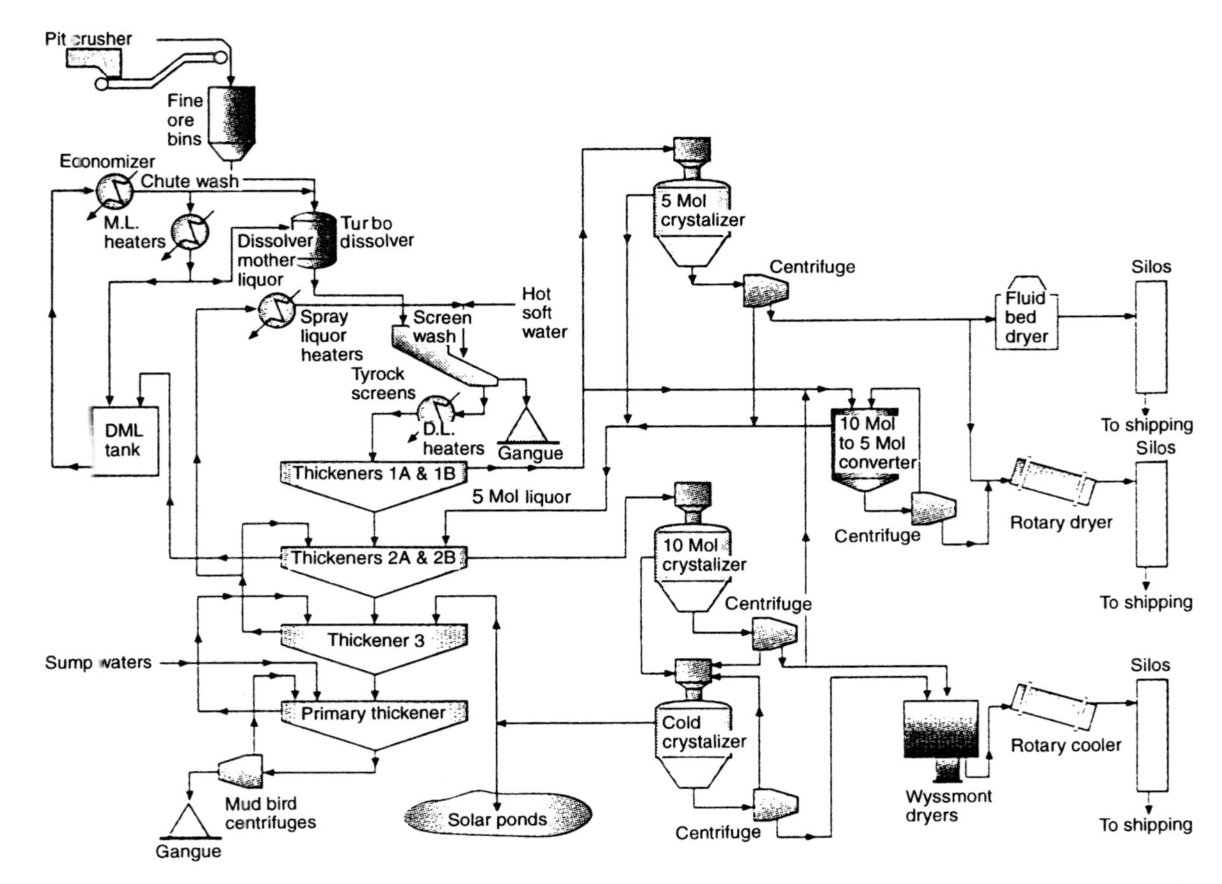

Figure 8.5 U.S. Borax's published flow sheet of its boron plant. (From Anon., 1996b; drawing courtesy of U.S. Borax Inc.)

impure product (classified as an ore, "Rasorite," to reduce European import duties).

The hot 99°C (210°F) brine next passed into one of three (a fourth was a spare) Struthers Wells, Oslo-type vacuum crystallizers, which made different products on a campaign basis. The brine was cooled to 37.8°C (100°F) when borax was being made, and 65.6°C (150°F) for pentahydrate (and formerly Rasorite). Several patents were issued (Taylor and Connell, 1953, 1956, 1957) on additives to reduce the borate supersaturation, to minimize equipment scaling, and to provide more controlled crystallization. The pentahydrate crystallizers had a 3123-l/min (825-gal/min) feed rate and 95,000-l/min (25,000-gal/min) internal circulation flow. Washouts were required on a cycle of about 4 weeks, and an organic phosphate ester defoamant was used. The product (50% +50-mesh borax; 10% +30-mesh pentahydrate) was centrifuged originally in 16 Sharples automatic basket centrifuges for the refined products, and three solid bowl centrifuges (operating at 800 rpm) for the crude pentahydrate. The later units were replaced in 1996 with vacuum belt filters, which provided much better product washing (three stages) and reduced crystal breakage and dust (from 12 to 5%). The impurities: SO_4, CO_3, Cl, and As were also reduced, as was product loading times. The moisture content did increase, however, along with caking problems, until the dryers were modified (Anon., 1996c). A dilute boric acid wash was used to reduce the product's caking tendency.

The borax was next dried in one fluid-bed, three oil-fired rotary, and seven steam-heated Wyssmont Turbo dryers. The Turbo dryers had multiple rotating trays, with rakes turning the product and advancing it from tray to tray. This provided very gentle heating to prevent dehydration of the products (except for the pentahydrate, which was purposely sold with 4.75 moles of water). Some of the borax was next partly dehydrated in four rotary kilns and then melted in four rectangular, side-fed, V-shaped-bottom, gas-fired reverbratory furnaces. The molten borax was solidified on 1.2-m- (4-ft-) diameter chilled rolls, further cooled on 0.91 × 18.3-m (3 × 60-ft) Carrier vibrating conveyors, broken with rotating-arm sheet breakers, crushed to −4 mesh in a hammer mill, and then cooled again on a 0.9 × 18-m vibrating conveyor. It was finally reduced by a hammermill to the desired particle size, and screened on Sweco screens. The fines were recycled and remelted.

The pentahydrate and anhydrous product storage was in 12 concrete silos, each 27.4 m (90 ft) tall and 10.7 m (35 ft) in diameter. Six had a combined capacity of 12,200 metric tons and were used initially for Rasorite. Three at 5300 metric tons were for the refined pentahydrate, and three at 6300 metric tons were formerly for anhydrous Rasorite. The refined borax was stored in an "angle-of-repose" warehouse because severe caking problems had been encountered when it was initially stored in the silos. When withdrawn the product was rescreened and then shipped by rail or trucks. Plant operation was monitored from a central control room, and there were administration, maintenance, warehouse, laboratory, boiler, and other buildings. Well water from a fossil aquifer went to a 38-million-liter (1-million-gal) storage tank, and then to a 36.6-m- (120-ft-) high, 3,800,000-liter (100,000-gal) pressure tank. Boiler water was treated by the lime–soda process (Anon., 1958; Garrett, 1959). Later, boric acid fusion furnaces were installed to produce some anhydrous B_2O_3 (Anon., 1965b).

Jensen and Schmitt (1985) noted that the plant had been "expanded and modified in numerous ways to increase capacity" since its original design. A new twin-boom automatic ore stacker with articulated arms (to reduce the drop height and thus dust) was installed to form separate ore stock piles for borax and kernite. Six clay-lined solar evaporation ponds were also built to recover part of the borax that was in the slimes and excess plant wash water. The largest was 0.49 km^2 (120 acres) and 12.2 m (40 ft) deep. An additional yield of 6% was claimed, bringing the stated total to 91%. (It actually may have been less than 70%, including the mine, ulexite, colemanite, and other borate losses.) In 1985 the borax plant worked 7 days/week, 3 shifts/day, and the boric acid plant as needed to meet the sales demands.

A 46-megawatt gas turbine cogeneration plant to generate 181 metric tons (400,000 lbs) of steam per hour at 173 psi and 192°C (378°F) was also built to replace five of the original six oil/gas-fired boilers. It had waste-heat boilers, a 3500-hp air blower, and evaporative air coolers to increase its efficiency. Its 50% excess capacity over the plant needs was sold to the local utility to provide a 5-year pay-out period. Some plant capacity was switched from borax to pentahydrate, and a 363-metric-tons/hr bulk-loading facility was installed to automatically fill and weigh rail cars and trucks. Automatic bagging, handling, and shipping costs were reduced, and dust control was improved. New instrumentation and automation, as well as computer assistance for plant maintenance, purchasing, and sales were also installed (Jensen and Schmitt, 1985).

In 1985 the Rasorite line of products ceased being made because the European import duties on refined products were dropped (Lyday, 1992). In 1976 twin air-supported product storage structures were erected (large plastic envelopes held to the ground by corrosion-resistant cables, with fans creating a small internal pressure). There were airlocks for the conveyor belts and personnel entry. The structures were 91 m (300 ft) long, and could store up to 91,000 metric tons of product (Anon., 1976). Later, two 18,000-metric-ton "domes" (hemispheric storage structures) were added (Anon., 1996b).

U.S. Borax's 91,000-metric-ton/year boric acid plant originally was at the Los Angeles harbor (Wilmington), 240 km (150 mi) away. However, shipping and mud disposal costs became prohibitive, so in 1980 a new 180,000-metric-ton/year, $80-million boric acid plant was built on a 0.16-km^2 (40-acre) site at Boron. It had 12 44.5-m (146-ft; also quoted at 100-ft; Anon., 1996a) silos, and featured "a continuous rotary drum reactor." It operated on kernite

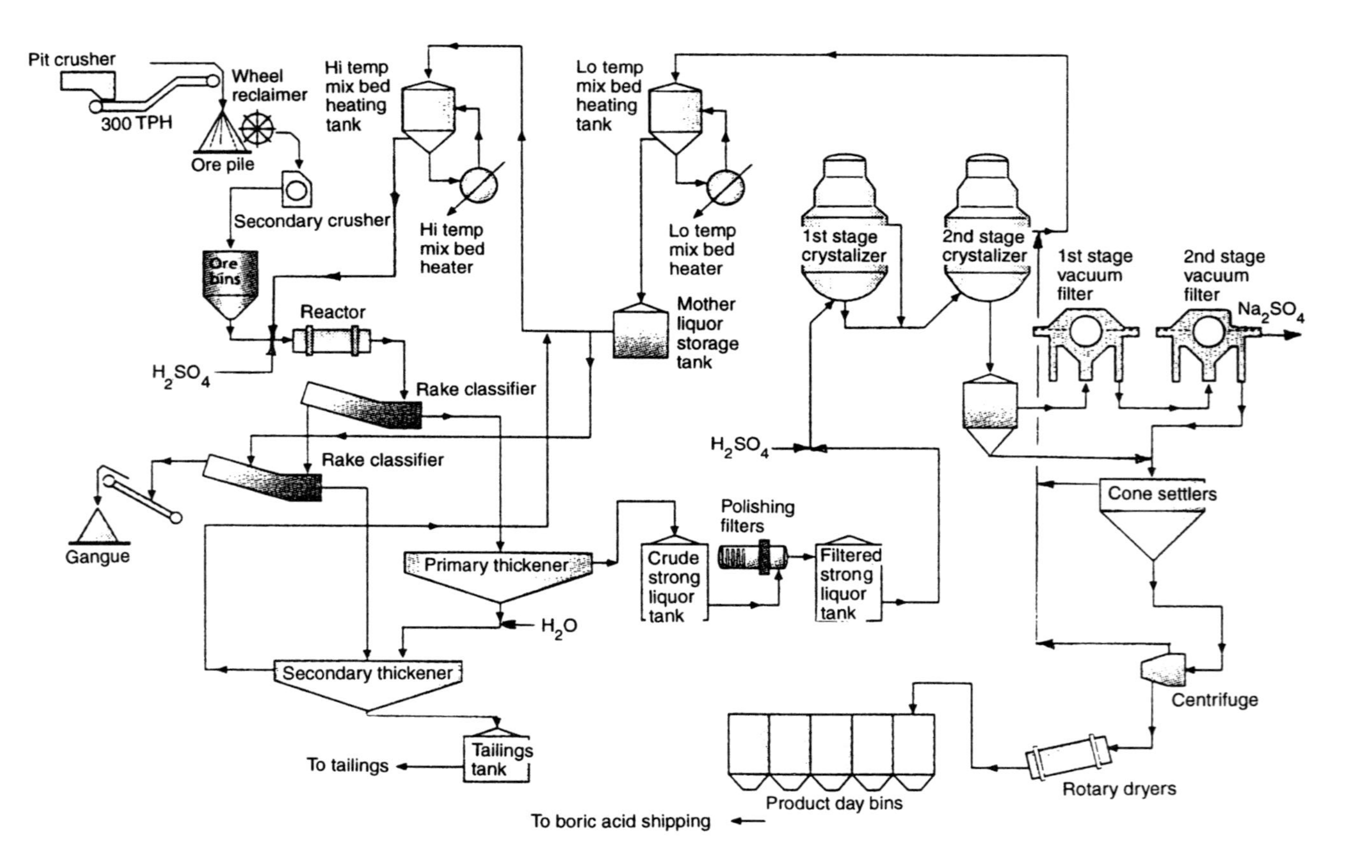

Figure 8.6 U.S. Borax's published flow sheet of its boric acid process. (Slightly modified by author; from Anon., 1996b; drawing courtesy of U.S. Borax Inc.)

ore, with its slimes and waste water being sent to a 0.49-km^2 (120-acre), clay-lined tailings pond. Sodium sulfate was made as a by-product (Anon., 1981).

The flow sheet for the process (Fig. 8.6) appears to follow the numerous U.S. Borax patents that suggest reacting the ore with sulfuric acid in a brine that immediately precipitates by-product sodium sulfate, removing it and then cooling the solution to crystallize boric acid. Kernite was employed because when used to make borax, it dissolves more slowly, requiring finer grinding and longer leaching times, or hydration in the mine. Also, when ground to a fine size for borax production kernite cleaves and forms fibers that tend to mat and clog the handling equipment. The kernite is added to 49°C (120°F) recycle liquor in the reactor, along with part of the sulfuric acid. The Na_2SO_4 concentration is kept just below its saturation point at that temperature. After the dissolving process, the coarse undissolved ore is then removed in rake classifiers (Anon., 1996b), the slimes are thickened, washed, and discarded, and the clarified brine is sent through polishing filters. The brine is then sent to the final sulfuric-acid reactor, heated to 99°C (210°F) and the sodium sulfate precipitated. The salt cake is thickened, filtered, washed and dried, and the remaining brine sent to a boric acid crystallizer. Here it is cooled, a boric acid slurry withdrawn, and the crystals thickened, centrifuged, washed and dried. After a bleed stream is removed to control the impurity level, the remaining solution is recycled to the kernite leach tanks. Sulfur dioxide gas could be introduced (i.e., 1.36 kg/5000 gal/100 ppmFe) to reduce the ferric iron to ferrous, thus minimizing Na_2SO_4 supersaturation and iron contamination. For higher-purity boric acid, the crude cake is redissolved, the solution purified (i.e., filtration, the precipitation of metallic impurities, carbon adsorption, etc.), and the boric acid recrystallized.

On the basis of U.S. Borax's dehydration patents (e.g. Corkill, 1937), a fusion furnace such as shown in Fig. 8.7 might be employed to produce anhydrous borax or B_2O_3 (Anon., 1965b). Partially dehydrated boric acid (as with borax) is fed into the rear of a reverbratory-type furnace to protect the refractory lining. As it slowly melts, it is withdrawn at the far end and solidified on water-cooled rolls. It is next ground to size and packaged in moisture-tight containers (U.S. Borax, 1962).

8.1.4.2 Processing

Many processing articles have been written on sodium borate or boric acid production (see References), but they do not significantly add to the descriptions of the major producers' operations given here. Also, few, if any, have been commercialized (which is the general criteria for review), because of space limitations so these articles are not further discussed in this chapter.

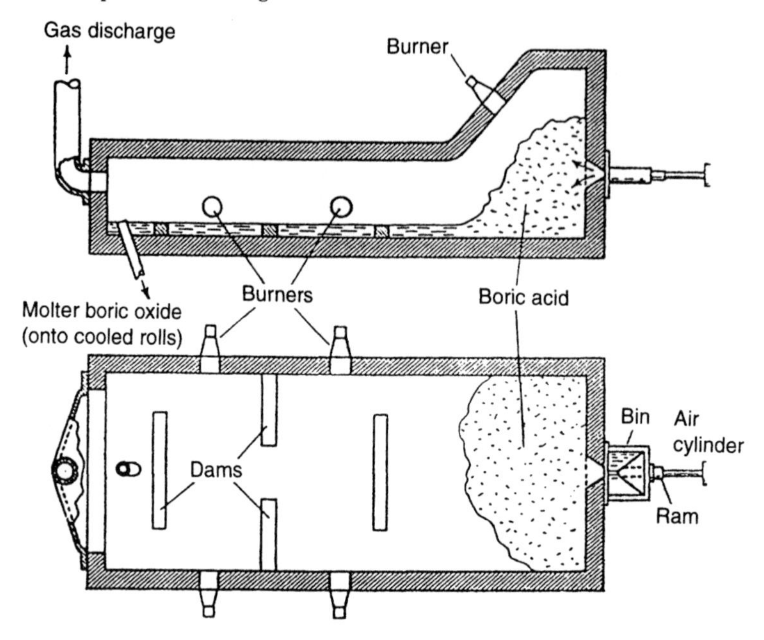

Figure 8.7 Side and top views of a borax–boric acid fusion furnace. (From U.S. Borax, 1962.)

8.1.5 Beneficiation

8.1.5.1 Magnetic Separation

Magnetic separation was used to upgrade the borax ore at the Boron plant until 1957. An early patent noted that the shale was slightly magnetic, but the borax and kernite were not. To be processed the ore was first dried and crushed to two size fractions: 1.6–3.2 mm (1/16–1/8in. or 6–10 mesh), and −1.6 mm by rolls and hammermills working in closed circuit with Tyler Hummer screens. The separation was conducted in four stages. First, tramp iron was removed. Then the ore was passed over a high-intensity induction-type Exolon magnetic roll and through a small gap between magnets. The shale was deflected slightly outward by the magnetic attraction, and a splitter blade under the falling ore divided the deflected from the free-falling portions. By repeating this separation twice, and then rerunning the product cuts, it was claimed that an 80% borax yield of a high-purity product could be obtained (Anon., 1956; Johnson, 1936). Downing (1995) claimed that borax and kernite ore were now capable of being upgraded from 60 to 90% purity by high-intensity magnetic separators, with only a 3–7% loss of their B_2O_3 content.

8.1.5.2 Borax and Boric Acid Flotation

Many articles and patents have been presented on the flotation separation of borax from other salts or gangue material. For example, Chemtob and

White (1972) and Garrett, Kallerud, and Chemtob (1975) noted that borax can be separated from other soluble salts by 75–500 g/ton (of the solids present) of 8–22C atom-sulfonated straight-chain fatty acids such as sulfonated oleic acid or tall oil. Some hydrocarbon sulfates such as sulfsucinates were also effective, as were a few secondary or tertiary alcohols when applied at 15–50 g/ton. Boric acid is a naturally floating material because air bubbles selectively adhere to its surfaces, something that does not happen to most other substances unless a special reagent has been added to the slurry. As a consequence, if any of the borax minerals were converted to boric acid, at least on their surface, they could be separated readily by flotation. The References contain other articles on both borax and boric acid flotation.

8.2 COLEMANITE

8.2.1 Argentina

Ulex SA's Sol de Manana mine at Sijes first hauls colemanite and hydroboracite 15 km to their El Paso facility to be hand sorted, crushed, washed, and dried to a 42% B_2O_3, −1-mm (16-mesh) product. It is shipped in bulk or containers to the railroad, and from there to customers via the ports of Antofagasta or Tocopilla, Chile, or Buenos Aires. It, along with ulexite from their salars, can also be sent to Salta to be further dried and passed through a Raymond mill to make a $+45\mu$ (325-mesh) product (Leiser, 1996).

8.2.2 Turkey

8.2.2.1 Bigadic, Emet, Espey, Kestelek

In the mid-to-late 1970s large washing plants were constructed at these facilities to produce higher-grade and more uniform colemanite and ulexite concentrates (Fig. 8.8). Products were made in different sizes, grades, and purities (see Table 10.9) to meet different market demands. The ore was first crushed to −30 cm, and then in closed-circuit stages with screens to −25 mm. It was next repulped to 60–75% solids, sent to attrition scrubbers, and classified (and the slimes removed) in a series of equipment. The −65-mesh slimes were discarded, and the concentrate was centrifuged, dried, screened, and shipped (Albayrak and Protopapas, 1985; Anac, 1988; Ozkan and Lyday, 1995).

8.2.3 United States

8.2.3.1 Calico Area: Borate Mine

The borate beneficiation plant was located at Marion, 6.4 km (4 mi) north of Daggett, and 19.2 km (12 mi) [9.6 km (6 mi); Yale, 1905] from the mine.

Figure 8.8 Views of the Bigadic (upper) and Emet (lower) colemanite and ulexite processing facilities. (From Etibank, 1995; pictures courtesy of Etibank.)

The ore was first shipped by mule teams, and later by a narrow-gauge rail line with spectacular curves and a high trestle (see page 315). In 1900, after 2 years of attempted start-up, a calcining plant began processing the −35% B_2O_3 ore and fines. It used a two-hearth, 70- to 90-metric-tons/day Holthoff-Wethey reverbratory furnace fired at 500°C (932°F) by No. 6 fuel oil. The ore entered at the top, and was mixed and advanced by rabble arms (countercurrent to the flue gas) to drop holes for the lower hearth. Here the process was repeated, and the calcine left at the bottom. It was then screened, the coarse fraction rejected, and the fines shipped to refineries in Alameda, California, or Bayonne, New Jersey. The fine product averaged about 45% B_2O_3, and it took 2–4 tons of ore to produce a ton of product. Any ulexite or priceite (pandermite; called "dry bone") in the ore was lost (sometimes 50% of the ore), because it did not decrepitate, and often fused into a greenish glass.

At the Alameda (on San Francisco bay) refinery, the colemanite was ground to a fine powder by a sequence that included a Blake crusher (reducing the uncalcined ore to nut size), a coffee-type grinder (taking all of the ore to a sand size), and finally buhr mills (making it fine enough to pass through silk screens). This powder was reacted with a hot soda ash (Na_2CO_3) solution for several hours in vigorously agitated tanks. The slurry was then settled, and the mud and by-product $CaCO_3$ sent to a filter press, washed, and discarded. The clarified brine and filtrate were pumped to tanks containing many suspended wires, where the brine cooled and crystallized, producing crystals with a slightly dark color. The end liquor was then removed, and part of it returned for another leach cycle. The borax was knocked off the wires and tank walls, drained, washed, and finally redissolved and recrystallized (Bailey, 1902; Ver Planck, 1956).

At the Bayonne refinery the high-grade ore was first crushed to 1.3–3.8 cm, and then it and the calcined ore were sent to a Griffin mill, which "reduced it to the fineness of flour," so it "could pass through No. 8 silk bolting cloth." It was then sent to holding bins from which it would later be weighed and sent to a 91-metric-ton, 11,000-liter (3000-gal) tank with internal steam coils, and partly filled with recycle brine and water. Soda ash (and some bicarbonate to prevent metaborate formation) was added, and the boiling slurry agitated for a lengthy period. When complete, the slurry was pumped to steam-heated settling tanks, and 1.4–1.8 kg (3–4 lb) of glue dissolved in "3 buckets of hot water" was added as a settling agent (without it the solution was "milky-looking"). After the solids had settled a clear brine was withdrawn. The residual mud was sent to a 50-psi filter press, washed, and discarded. The brine and filtrate went to mild steel crystallizers, 2 m (6.5 ft) deep, 1.8 m (6 ft) wide, and 6.1 m (20 ft) long. Two 6.7-m (22-ft) by 5-cm (2-in.) iron pipes were laid across the tanks, with 1.5-m (5-ft) long, 6.4-mm (0.25-in.) wires on a 20-cm (8-in.) spacing extending into the tanks. After cooling 6–10 days the liquor was withdrawn and part of it returned for a new leach cycle (the bleed stream was processed separately). The borax was first knocked off the wires, since it was pure enough to be sold directly. The crystals from the sides and bottom were then removed, washed, drained, redissolved, and recrystallized. The product crystals were washed, drained, and dried in a rotary dryer by a temperature-controlled hot air stream. It was then sent to a roll crusher and screened into three product sizes: refined crystals, granulated borax, and refined screenings. A fine product was occasionally made in a cyclone pulverizer (Hanks 1883; Baily, 1902; Yale, 1905; Dupont, 1910).

8.2.3.2 Calico Area: American Borax Company

The American Borax Company treated "borate mud" at Dagget by the patented Blumenberg (1918) sulfur dioxide process. It was claimed that with it the SO_2 dissolved less silica, iron, magnesium and other impurities, and the filtration of the residue was simpler, since most of the gypsum precipitated in the ponds. The 4–20% B_2O_3 (average 7%) ore was first broken in a "rock breaker," and then crushed to −6.4 mm (−0.25 in.) in a "Chili mill." It was next conveyed to a reactor, 3 m (10 ft) in diameter and 9.1 m (30 ft) tall, filled with water. Sulfur was burned nearby at 9 psi and distributed into the base of the tower. This agitated some of the ore into a "fluidized" slurry, and was continued until much of the SO_2 began to escape from the top.

The slurry was then settled, and a dark muddy-appearing liquor was decanted and sent to a series of shallow solar evaporation ponds made of concrete and coated with an asphalt sealant, and arranged so that the liquor flowed from one pond to another. The first ponds settled much of the solids, and gypsum began to precipitate (the soluble calcium sulfite and bisulfite were oxidized by air to calcium sulfate). When this reaction was complete, the brine was pumped by a windmill to the top of a natural-draft, open cooling tower. It was distributed over the staggered baffles placed about 30 cm (1 ft) apart, and it trickled back to the solar ponds. The flow rate in the ponds was regulated so that boric acid crystallized in the last ones. When full, the ponds were drained, the residual brine recycled, the mud removed (and discarded), and boric acid shoveled onto drainage areas between the ponds. It remained there until dry, and then was shoveled into cases and shipped to refineries as about an 85% H_3BO_3 product with impurities of $CaSO_4$, $MgSO_4$, water and insolubles. The plant ran 24 hrs/day with a staff of four at night, and eight during the day. The operation was never very profitable, so it closed when the Ryan mines opened and lowered the price of borax (Yale, 1905; Keys, 1910).

8.2.3.3 Calico Area: Palm Borate, Western Mineral Companies

These companies also leached borate mud: Palm with sulfur dioxide, and Western with sulfuric acid. Sulfur was burned at Palm under pressure and distributed into the bottom of a 76,000-liter (20,000-gal) agitated (by a slowly revolving rake) redwood tank. When excessive SO_2 came off the top the burner was shut off and the tank allowed to settle. Clarified brine was sent to solar ponds, and the mud filtered. Second-stage pond brine was pumped over a natural-draft cooling tower (Yale, 1905). Western was the largest of the four Calico companies processing "mud," but the company leached the ore with sulfuric acid in agitated tanks operating near the boiling point. Washing the heavy muds to recover more of the boric acid after the first decantation produced a very weak brine, which was concentrated in solar ponds. The pond brine could reach temperatures up to 60°C (140°F), and it was pumped over piles of brush to increase the evaporation. Crude boric acid crystallized in the last ponds at a rate of 27 metric tons/day (Bailey, 1902).

8.2.3.4 Coastal Mountain Area: Lang

Four oil-burning colemanite calciners were employed at Lang: two wedge furnaces (each producing 36 metric tons/day), and two rotary kilns (each producing 54 metric tons/day). Fine-calcined colemanite was blown from the calciners and recovered in cyclone dust collectors. The product was shipped by rail to the Lang siding on the Southern Pacific Railroad, and from there to several eastern refineries (Gay and Hoffman, 1954).

8.2.3.5 Death Valley Region: Boraxo

Tenneco's Boraxo mine in Death Valley shipped its 18–20% ore in the 1970s by trucks 50 km (31 mi) to a processing plant near Lathrop Wells, Nevada (13 km north of Death Valley Junction). The ore was dumped onto a 30-cm (12-in.) grizzly, the oversize manually reduced, and the undersize fed by belts to a hammermill that reduced it to −6.4 mm ($-\frac{1}{4}$ in.). It was then repulped to a 55% slurry at a rate of 45 metric tons/hr and sent to 1 × 1-m (40 × 40-in.) attrition scrubbers. The −65-mesh particles (about 33% of the ore) were removed in a 1.2 × 7.2-m (48-in. × 23.5-ft) spiral classifier, with sprays on the coarse solids to reduce fines carry-over. The classifier underflow slurry (the fines) was sent to a 25-cm (10-in.) cyclone, and its underflow solids were returned to the classifier. The cyclone's overflow was thickened, the solids sent to a tailings pond (from which 75% of the water was recycled), and the thickener overflow returned to the ore repulper. The coarse classifier solids (containing only about 3% −150-mesh particles and 15–18% water) went to a stockpile where further draining occurred. Their B_2O_3 content increased 3–5% in the washing step and there was a 15% colemanite loss. If ulexite was present the loss increased because of its friability. The plant's rated capacity was 45 metric tons/hr, but with dry ore it could be as much as 72 metric tons/hr, and with wet or high clay ore as low as 27 metric tons/hr. The plant had its own water wells.

A front-end loader took the washed ore to a hopper with a variable-speed apron feeder (also stated to be two slide-plates and a belt feeder). Belts then delivered it to a 2.1 × 12.8-m (7 × 42-ft) 10.9 metric tons/hr (12 tons/hr) rotary dryer, heated in counterflow by gas from the calciner, which entered at 343°C and left at 93°C. The coarser entrained solids from this flue gas were cycloned and added to the dryer discharge. The gas then went to a wet scrubber. The ore to the dryer averaged 9% moisture, 22% B_2O_3, and left at 177°C. It then entered (for about a 30-min residence time) the 2.4 × 12.8-m (8 × 42-ft) countercurrent-flow, direct contact stainless steel rotary calciner with lifters only at its front end. The flow rate of the gas in the dryer was 2.8- to 3.1-m/sec (9.2- to 10-ft/sec), and it left at 427°C. This entrained the decrepitated colemanite, which was recovered by six cyclones operating in parallel. The product was split into two size fractions (+ and −40 mesh), and cooled in air classifiers before being sent to silos. Part of the air from these coolers was used for combustion in the No. 2 fuel oil burners. Typical calciner recoveries were 70% (giving a 60% overall recovery), or a yield of 1 part product per 2.7 parts of ore. The plant operated with 37 people 24 hrs/day, 7 days/week, and shipped 47–49% B_2O_3 colemanite. The product specifications were >47.5% B_2O_3, <0.3% Fe_2O_3 and <0.9% SO_3, and since low-grade ore could not meet them, ore blending was practiced.

Later, equipment was added to crush the calciner tailings and screen them at 35 mesh. The undersize contained about 50% of the tailings B_2O_3 at a 30% B_2O_3 grade. A 10 to 35 mesh fraction was then processed on (air) density-classification tables, with a B_2O_3 yield of 70%, and 30% B_2O_3 product. The −35 mesh and this fraction combined with the calcine gave an overall yield of 71%.

The +40 mesh product was trucked to rail loading facilities in North Las Vegas, whereas the undersize was sent in 22.7 metric-ton (25-ton) pneumatic trucks to a modern, converted talc-grinding plant at Dunn Siding, 48 km (30 mi) west of Baker, California. It was unloaded into 109-metric-ton (120-ton) storage or blending silos, and then sent to a Raymond mill (with its built-in air classifier) to produce −70-mesh particles. From there it went to five 109-metric-ton product silos, and then was shipped in 75-metric-ton (85-ton) hopper cars. The plant had nine bag filters for dust control, and employed 25 people (Walters, 1975; Evans, Taylor and Rapp, 1976; Smith and Walters, 1980).

8.2.3.6. Death Valley Region: Billie

After the Boraxo mine had closed, ore from the Billie mine was also sent to the Lathrop Wells processing plant (Fig. 8.9). It was crushed and ball-milled to −32.5 mesh, and deslimed (at 70 mesh) to reach 37% B_2O_3. The 32.5–70 mesh fraction was then floated to produce a 42–45% B_2O_3 product (Lyday, 1996; Norman and Johnson, 1980).

8.2.3.7 Death Valley Region: Lila C, Ryan

From 1907 to 1916 the Lila C mine processed its lower-grade ore in a mill on the site. The ore was first reduced in size to −2.5 cm (−1 in.) by a roll

Figure 8.9 The Lathrop Wells (or Amargosa Valley) borate processing plant.

crusher, and sent through an indirectly heated, oil-fired rotary kiln with a flame temperature of 650°C (1200°F). The calcined ore was then screened, and the fine fraction shipped to the Bayonne refinery. The coarse material went to a tailings pile (Gale, 1912).

During 1916–1928 the dominant colemanite mines in Death Valley were at Ryan. Their "first-grade" ore went directly to refineries at Bayonne or Alameda, but the fines and lower-grade ore was sent to a calcining plant at Death Valley Junction. Ore from the Upper Biddy (Biddy McCarthy or Ryan), Lower Biddy, Grand View, Lizzie V Oakley, Played Out and Widow mines was first reduced to −5 cm (−2 in.) in a gyratory crusher, and then to −2.2 cm ($-\frac{7}{8}$ in.) in 61-cm (24-in.) roll crushers. Conveyor belts took the ore to two 73-metric-ton (80-ton) storage bins, which in turn fed six 1.8 × 15.2-m (6 × 50-ft) dual-tube, indirectly heated rotary calciners (Fig. 8.10). They were fired at 704°C (1300°F) by crude oil. The calcined ore was screened in double-deck units to +6.4-mm ($\frac{1}{4}$-in), 6.4-mm to 24-mesh, and −24 mesh fractions. The −24-mesh material was immediately sent to the product silos, whereas the other fractions (56% of the ore with a 1% B_2O_3 content) went to silos for further processing. Both oversize fractions of calcine were reground in roll crushers and rescreened at 24 mesh. Each fraction was repulped, and the +24 mesh calcine was sent to Hartz jigs, while the undersize went to Wilfrey tables (density separators). The light colemanite fractions were thickened in drag

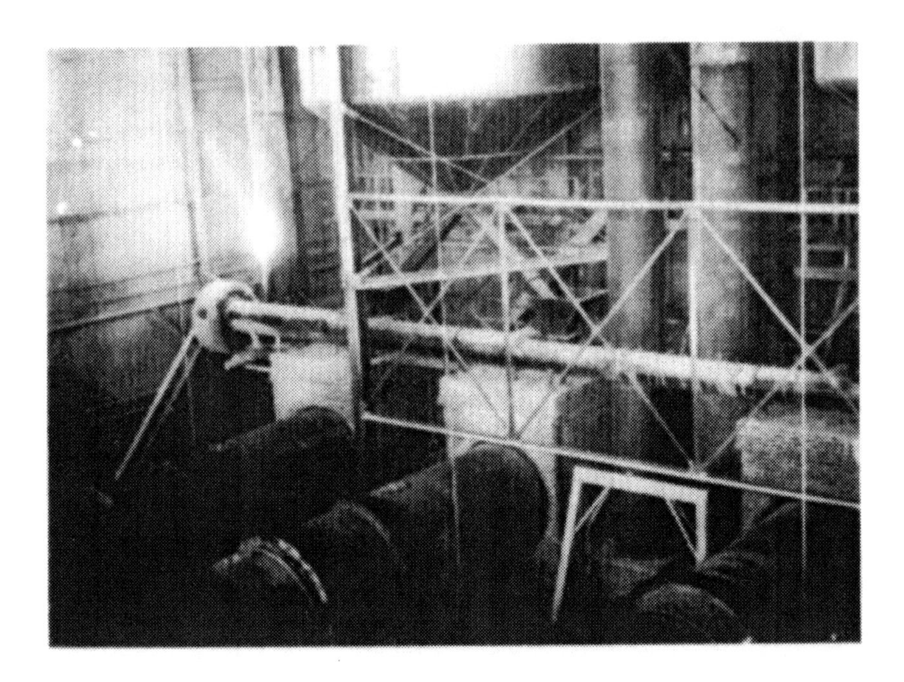

Figure 8.10 Roasting colemanite and bagging the concentrates at Death Valley Junction. (From Ver Planck, 1956; used with permission of the California Department of Conservation, Division of Mines and Geology.)

Figure 8.11 West End Chemical Company camp and mill at the Anniversary mine. Buildings from lower right to center: Club house, bunk house, boaring house, office; Upper right: processing plant. (From Castor, 1993; reprinted courtesy of the Nevada Bureau of Mines and Geology.)

(ladder) classifiers, filtered, and dried in table dryers. The combined calcine and low density streams were bagged for shipment, and contained 42–45% B_2O_3, with an overall B_2O_3 recovery of 85%. The plant capacity was 272 metric tons/day of ore; 85 men were employed; and power was supplied by a 500-hp diesel engine (Hamilton, 1921).

8.2.3.8 Muddy Range, Nevada: Callville Wash

The processing plant (Fig. 8.11) of the Anniversary Mine also employed the calcining process. An aerial tramway from the mine dumped ore into a bin at the plant, and from there it was withdrawn and crushed to −2.5 cm (1 in.). It was next fed into a 1.5 × 18.3-m (5 × 60-ft) indirectly heated, oil-fired rotary kiln (two kilns; Papke, 1976) whose flame temperature was kept at a minimum to avoid fusing the ore. The discharged calcine was screened, and the fine product hauled 43 km to the railway. The newly developed 4.6-metric-ton (5-ton) tractors or truck and trailers were used, and supplies brought back on the return trip. The production was 68 metric tons/day in 1923, and 50 people were employed in the mine, mill, and trucking operations. Electric power was provided by oil- (27 gravity-) fired engine–generator sets, and plant water from shallow 12.2-m (40-ft) wells at the site. Drinking water was hauled from the railroad (Young, 1924).

8.2.4 Processing

There have been many other processes developed to recover the boron values from colemanite, as indicated in the References and the following examples.

The chlorine, or Moore, process was used during the 1800s in England, with chlorine introduced into an agitated slurry of powdered colemanite at 70°C. When the chlorine began to escape the slurry was allowed to settle. The clear liquid was next decanted and sent to crystallizers to cool, and when withdrawn, was recycled until its calcium chloride and other salt content began to interfere with the borax crystallization.

In a similar process colemanite was treated with hydrochloric acid in an agitated, boiling slurry. When the reaction was completed the slurry was settled and the clear liquor sent to crystallizers for the fractional crystallization of $CaCl_2$ and H_3BO_3. In the Bigot process, 100 parts of fine colemanite were heated with 150 parts of ammonium sulfate in a closed vessel. After the reaction was complete further heating drove off the ammonia, and it was reacted with sulfuric acid to form ammonium sulfate to be recycled. The solids were then leached with water, and boric acid crystallized from the solution (Bailey, 1902). Winkler (1907) reported on a German process in which flue gas (CO_2) under pressure was forced into an agitated pressure vessel containing a colemanite–water slurry until the solution was saturated with $Ca(HCO_3)_2$. The pressure was then released, precipitating $CaCO_3$, and the cycle repeated until the colemanite was dissolved. The remaining slurry was settled, the liquor drained off, and boric acid crystallized. The process was stated to be less expensive, and the boric acid contained fewer impurities than with acid leaching.

8.2.5 Beneficiation

8.2.5.1 Calcining, Gravity Separation

When colemanite is to be *calcined,* the ore is usually crushed to −1.9 or −2.5 cm ($\frac{3}{4}$ or 1 in.; smaller if it is to be deslimed) and heated in a calciner to about 450–500°C (850–930°F; well below the normal mixed borate ore fusion point of 485–550°C). In the residence time of most calciners, 85–90% of the colemanite decrepitates to a predominantly −28 to −35 mesh size. If the flue gas velocity is greater than 2–5 m/sec (4–15 ft/sec), most of the colemanite is entrained and carried out to be recovered in dust collectors. If the kiln is indirectly fired, or if it has a low flue-gas velocity, the colemanite can be separated by vibrating screens. The hot flue gas from the dust collector or indirect fired kiln is often sent through a colemanite ore rotary dryer to remove its free moisture content (sometimes up to 16%), carry out some clay, and preheat the ore. In the dryer the flue gas often leaves at 90–120°C (194–265°F), and the ore is heated to >149°C (300°F). Both the calciner and dryer are operated in a counterflow manner to increase the ore's exit temperature, and to effect better efficiency (Miles, 1973; there are many articles on calcining in the References).

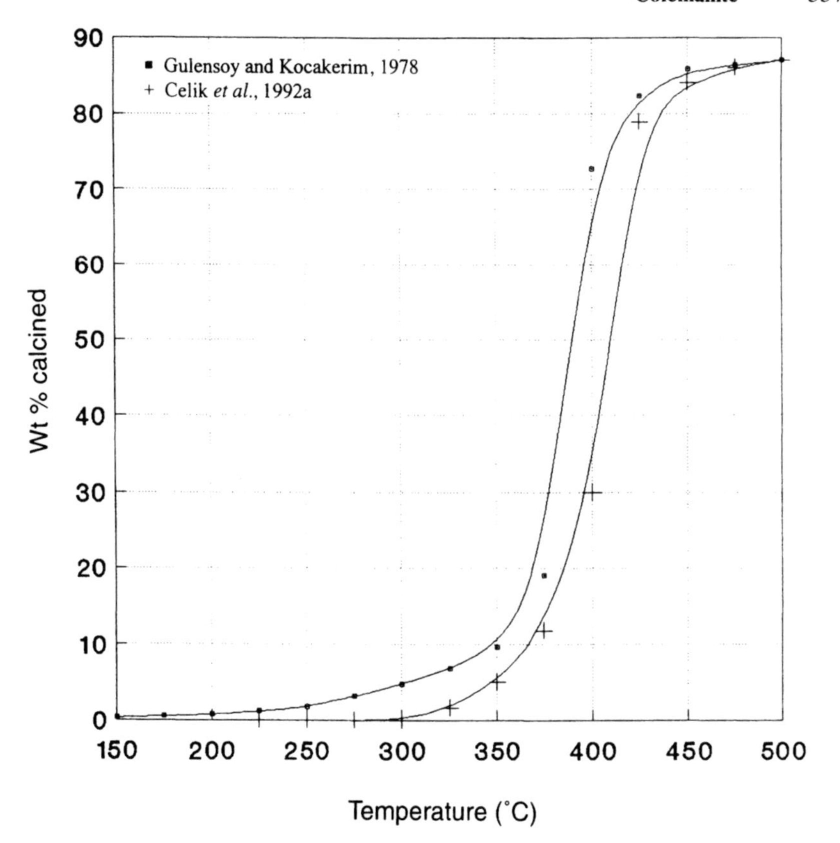

Figure 8.12 Colemanite calcination.

Small-scale laboratory calcining tests with about 15 min of heating have been made on hand-picked, high-purity Turkish (Bigadic) colemanite (Fig. 8.12). Regardless of the ore size (19–0.6 mm; 0.74 in.–10 mesh) or time (15 min–5 hr), at 500°C, 87–89% of the colemanite decrepitated to a −0.2-mm (65-mesh) size. With 10- to 16-mesh ore, the same results were obtained in 10 min. Ulexite did not decrepitate, but as much as 20% of the 10- to 16-mesh ulexite could pass through a −65-mesh screen merely because of its friability and ease of grinding (Atman and Baysal, 1973; Celik, Uzunoglu, and Arslan, 1992a).

Many studies have also been made on the *gravity separation* of the coarse residue from colemanite calcining (see References), such as that by Emrullahoglu, Kara, Tolun, and Celik (1993). They employed the rejects (43% of the total, 19.5% of the B_2O_3) from a pilot plant test calcining 0.2- to 25-mm (1-in. to 65-mesh), 30% B_2O_3 Turkish (Bigadic) colemanite ore. On screening, the +1-mm (+16-mesh) fraction contained 11.6% B_2O_3, and was 60% of the sample's weight. It was ground to −16 mesh and screened to two fractions: 1–0.5 mm (16–32 mesh) and −0.5 mm. They were then separately repulped and run over a laboratory diagonal concentrating table. The (light) concentrate plus middling analyses for the 1- to 0.5-mm and −0.5-mm fractions were 45.5wt%, 51.9% B_2O_3 and 42.2wt%, 51.0% B_2O_3, respectively. Microscopic studies indicated a 1-mm liberation size, and the densities were: calcined colemanite 1.85, and the gangue 2.72 g/cc.

Heavy media separation has also been attempted on both calcined and uncalcined ore. Pure colemanite's density is 2.42–2.43 g/ml, whereas calcite's is 2.71 g/ml, indicating that a gravity separation might be possible. However, in both laboratory and plant tests with 3- to 25-mm Turkish colemanite ore, the separation was poor. With +50-mm ore, a 15% recovery of 43% B_2O_3 was obtained. The liberation size was a problem, but because some calcite cocrystallized with colemanite, and the gangue was a clay–calcite mixture, the density difference was blurred (Ozkan and Lyday, 1995).

8.2.5.2 Flotation, Desliming

The flotation separation of colemanite from lower-grade and fine ores appears to be the most economical processing method. It can give reasonably high yields and selectivity, and its processing equipment and costs are much less than those for calcining. Sodium oleate ($C_{17}H_{33}COONa$) can be a good collector, but it is not selective in the presence of calcite. Sodium dodecyl benzene sulfonate [$CH_3(CH_2)_{11}C_6H_4SO_3Na$] is not quite as good a collector, but is reasonably selective when calcite is present. Floating a 48- to 200-mesh 50% mixture of calcite and colemanite with 10^{-4} mole/l of sulfonate gave a 90% colemanite product, but only a 40% yield. At 8×10^{-4} to 10^{-3}mole/l, the recovery was nearly complete, but there was no separation. Yarar (1985) and Yarar and Mager (1979a and b) reported concentrates of 41% B_2O_3 or higher with this reagent on Turkish colemanite ore. The flotation solutions had a pH of 9 ± 0.4, and contained 42.3 ppmB.

Barwise (1992), and Simon and Barwise (1993) suggested using ~1.5kg/ton of Na or NH_4 dialkyl (i.e., dinonyl, dilauryl or di-isodicyl) sulfosuccinate as the collector for both colemanite and ulexite Turkish ores. The branched chain nonyl salts were preferred. Mixing 50–80% of the collector with 2–30% water, 10–40% of a dihydric or >15C monohydric alcohol, and grinding to −60 mesh (desliming at −400 mesh) was recommended. Realgar and stibnite have been partially removed from −150-mesh ore by flotation with potassium amyl xanthate, mercaptobenzothiazole, or butyl xanthogen ethyl formate. In one test with 38% B_2O_3 Turkish ore, the As was reduced to 270 ppm, with an 82% B_2O_3 recovery by a sequential As–colemanite flotation. Sawyer and

Wilson (1973) found that above 75°C 80–98% of the arsenic could be floated with only a 2–8% borate loss. The arsenic either self-floated or could be assisted by C_2–C_5 xanthates or mineral oils. Other flotation articles are given in the References.

The *desliming* step on colemanite or ulexite ore usually results in a relatively high B_2O_3 loss. For instance, a 24.5% B_2O_3 Turkish colemanite ore had a 20% loss (and 35% B_2O_3) after one attrition stage when deslimed at -53μ (270 mesh). With higher-grade ore (28.4% B_2O_3), after four desliming stages at 270 mesh the grade was 40.5% B_2O_3, and the recovery 78.6% (Ozkan and Lyday, 1995). At Death Valley the loss is 15–25% at −65 mesh.

8.2.5.3 Magnetic, Electrostatic, Optical

Tests on the magnetic separation of dry Sijes (Argentina) colemanite ore indicated the presence of some magnetite, but most of the magnetic material (iron) was in the adhering clay and tuff. When 30- to 80-mesh colemanite was slowly passed under a strong electromagnet positioned above a right angle belt (with a 3-mm gap between the ore and the belt) the B_2O_3 yield was 92%, and the Fe_2O_3 content was reduced from 0.47 to 0.28%. The ore's B_2O_3 increased by 0.7–1.5% (Flores and Villagran, 1992).

In a series of laboratory and pilot plant tests the colemanite–ulexite slimes (10.7% B_2O_3) from leaching borax at U.S. Borax were air-dried and then crushed into three fractions: 4–10 mesh, 10–70 mesh, and −70 mesh. The first two were separately processed in three-stage magnetic separators with an average B_2O_3 recovery of 52%, and a 26% B_2O_3 product. The slimes' montmorillonite clay contained a small amount of pyrrhotite (FeS). In 1995, 14 million tons of slimes were in U.S. Borax's five solar ponds, and it was projected that a magnetic separation process could equal the mining cost (on a B_2O_3 basis) of borax. A 77 tons/hr plant was being considered (Downing, 1995).

Kaytaz, Onal, and Guney (1986) studied the *electrostatic* separation of Turkish colemanite, and felt that it could be moderately successful. Celik and Yasar (1995), and Yasar, Hancer, Kaytaz, and Celik (1994) achieved good electrostatic recoveries of 9- to 20-mesh pure-white colemanite in the laboratory at room temperature, and with beige colemanite at 50°C. With black colemanite the recovery at 80°C was only 60%. They noted that the conductivity of colemanite, borax, and ulexite largely depends on their impurities, particularly strontium and lithium, and that colemanite behaved as an insulator with both static and beam-type electrodes. *Optical sorting* tests on colemanite from Emet gave poor results, except for the removal of realgar (it is bright red). The minerals all looked the same to the machine because of the trapped (or surface-adhering) clay or tuff. Wetting the ore assisted in the color distinction, but greatly hindered its rejection (Ozkan and Lyday, 1995).

8.3 ULEXITE, PROBERTITE, AND OTHER BORATES

8.3.1 Argentina

The flow sheet of a typical boric acid plant processing ulexite in the Lerma Valley near Salta is shown in Fig. 8.13 (Pocovi, Latre, and Skaf, 1994). The ulexite is in the form of air-dried, hand-sorted nodules or slabs containing 22–25% B_2O_3, 4–15% Cl, 1–6% SO_4 and 9–34% acid insolubles. It is first

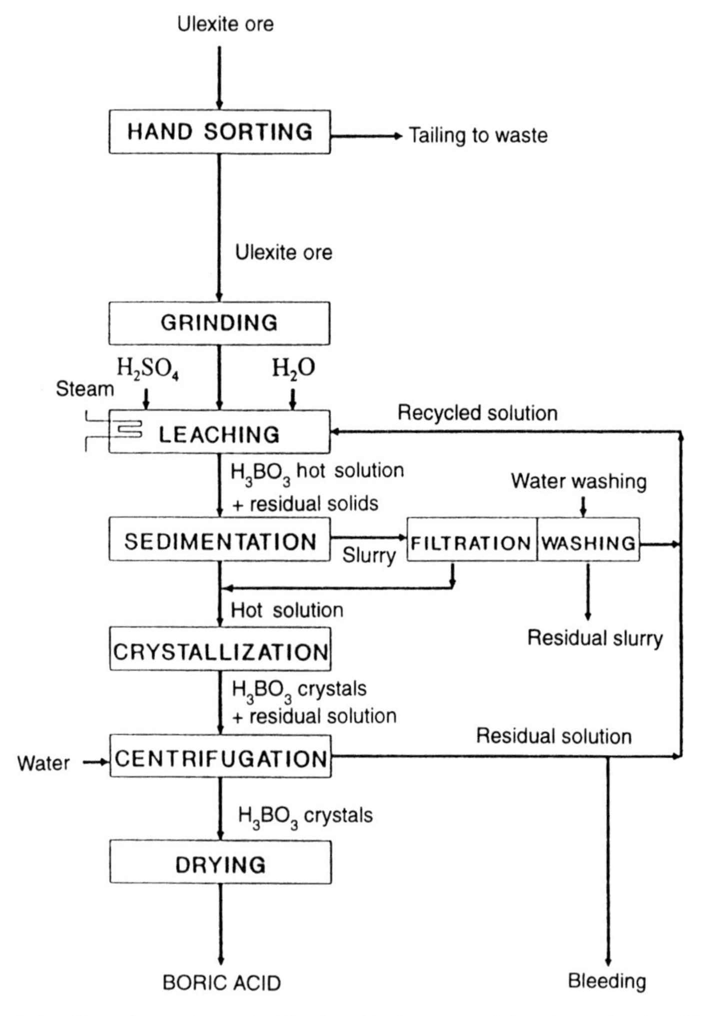

Figure 8.13 Flow sheet of a typical boric acid plant near Salta, Argentina in 1994. (From Pocovi, Latre, and Skaf, 1994; reprinted by permission of Chapman Hall.)

ground, and then batch leached with sulfuric acid at 60–80°C in baffled, agitated tanks with steam coils. When dissolved the residual slurry is sent (with a coagulant) to a steam-heated settler, and after it has settled the clear liquor is withdrawn. The insolubles are also removed, filtered, washed, re-pulped and sent to a tailings pond. The solution from the settler and the filtrate are pumped to agitated batch crystallizers where cooling water circulates through internal cooling coils. When cool, the residual brine is removed and partially recycled to the leach tank. The boric acid crystals are then repulped, pumped from the tank, centrifuged, washed, and dried in a parallel-flow hot-air rotary dryer. The product is kept below 95°C to prevent its dehydration to metaboric acid. The bleed steam from the end liquor removes the Na_2SO_4 formed during leaching, along with the NaCl and Na_2SO_4 that were in the ore. The disposal of this brine and the mud into tailings ponds has resulted in a serious pollution problem, and represents a major recovery loss.

These authors suggested that the operators crush, grind, and wash the ulexite at the salars before shipping it to Salta. Even with ulexite's low density of 1.7 g/cc, a hydrocyclone separation of a crushed ore slurry gave concentrates containing 36–40% B_2O_3, 0.2–0.9% Cl, 0.4–0.7% SO_4 and 4–16% acid insolubles, with a 78–87% B_2O_3 recovery. The slimes and water-soluble salts would be returned to the salar. As a second step, evaporators should be installed in the Salta plants to crystallize salt cake (Na_2SO_4) as a by-product rather than disposing of it (Pocovi *et al.*, 1994).

Among the individual operators, BoroSur S.A.'s boric acid plant in 1990 was 25 km from Buenos Aires, with a capacity of 550 metric tons/year, and plans for an increase to 300 metric tons/month (Norman, 1990). Ulexite (20–34% B_2O_3, 5–7% Cl, 3–4% SO_4, 15–20% insolubles, 5–10% water) from the Maggie mine was sent (1500 metric ton/month) to Salta in 1992 to produce 500 metric tons/month of boric acid. It was leached by a residual hydrochloric acid solution from a local sugar mill (Battaglia and Alonso, 1992).

8.3.2 Bolivia

In 1995 Corban S.A.'s Oruro plant received up to 5000 tons/month of 20% B_2O_3 ulexite from the Salar de Uyuni's Rio Grande area. It was first air-dried to 37–40% B_2O_3, and then ground to a fine size and reacted in a boiling slurry with crude trona mined from the nearby Salar de Collpa Laguna. When the reaction was complete the solution was settled, decanted, cooled, and crystallized. The borax was removed, centrifuged, washed, and sent to a gas-fired dryer, producing 90–100 tons/month of borax in 1995. The plant capacity was 200 tons/month of borax and 100 tons/month of pentahydrate. About 300 tons/month of 37% B_2O_3, and 200 tons/month of 40% B_20_3 ulexite were also shipped by trucks to the railhead at Santa Cruz, and from there to Corumba, Brazil or Iquique, Chile for distribution (Lyday, 1995). The plant and mine employed 40 full-time, and 200 part-time workers. A second company, Minera Tierra, employed 120 people, and in 1993 operated a 10,000-tons/year boric acid plant. Some of their ulexite was washed and roasted to a 42% B_2O_3 content (Lyday, 1994).

8.3.3 Chile

In the late 1800s all of the borax manufactured in Germany was from ulexite mined in Chile's Salars Ascotan, Maricunga and Pedernal, or Argentina's Rosario. Ascotan's ulexite was generally superior, with much less gypsum, and a B_2O_3 content of 34% compared to the 12–24% of the others. The ulexite nodules were first ground to a smaller size, and then to very fine particles in a "disintegrator." Next, 1.25 metric tons of the fine ore were placed in a vigorously agitated, boiling (by live steam injection) tank with 5–6.25 metric tons of water and a slight excess of soda ash. Trona or sodium bicarbonate could also be used, and were slightly favored, since if added in accidental great excess, metaborate would not be formed. A small excess of soda ash, however, did aid in speeding the borax crystallization. The leaching reaction tended to cause foaming, which was controlled by turning off the steam and spraying cold water on the boiling surface.

When the ore was dissolved, the steam and agitation were stopped and the mixture settled. The clear liquor at 30–35°Be (1.26–1.32 g/ml) was then pumped into 1000- to 1500-liter rectangular steel tanks and cooled for 3–4 days. (The density variability depended on the amount of sodium sulfate and other salts from the impurities and gypsum.) After crystallization was complete the brine was removed and recycled, except for a bleed stream, which was cooled in the winter to crystallize glauber salt ($Na_2SO_4 \cdot 10H_2O$), and then this end liquor, or all of the brine in the summer, evaporated to crystallize the non–borax salts. The residual liquor from the evaporators was sent back to the borax crystallizers. The mud from the leaching step was filtered, washed (with hot water), repulped, and again filtered. The wash water went to the leach tank, and the filtrate to the crystallizer.

The crude borax from this operation usually contained 40–50% borax, 40% sodium sulfate, and 10% salt. It was dissolved in boiling water to obtain a 30°Be (1.26 g/ml) solution, and the iron and organics oxidized and precipitated by sodium hypochlorite, and then settled. The clear liquor was decanted and sent to insulated 8- to 10-m^3 rectangular insulated steel vessels with wooden lids. This allowed very slow cooling to grow large, hard, colorless, and clear borax crystals. After 10–14 days the temperature was 33°C, and the residual brine was drained off. By this temperature control the other salts did not crystallize, and fairly pure borax was obtained. The residual liquor was either used for leaching or boiled to crystallize Na_2SO_4, and then cooled to recover borax. The 20-cm thick layer of borax was removed from the crystallizer walls, drained, washed, and then dried for 24 hr in a "drying chamber" at 30°C. The small crystals were screened and reprocessed, and the larger ones were "packed into 300- to 400-kilo barrels lined with blue paper" (to signify high purity) for shipment (Witting, 1888).

In more recent times, Quiborax trucks their ulexite to El Aquila, north of Arica, for processing. The plant had six diesel generators with a 2375-kWhr capacity, and two 6000 kg/hr boilers. A 200-m^3/day reverse osmosis unit generated freshwater from the plant's 1500-m^3/day brackish water wells (containing 1200–1500 ppmNaCl). The plant's capacity to wash, deslime in cyclones, filter, and dry the ore to make a 36.5% B_2O_3 product in 1995 was 30,000 tons/year. Some ore was also ground, screened, and reacted with sulfuric acid to produce boric acid in one of two agitated, steam-heated reaction vessels. The hot leach solution was settled, sent to 250-psi filter presses, and then cooled in agitated stainless steel, water-cooled tanks to crystallize boric acid. It was next thickened, centrifuged, washed, and dried in one of two inverted U-shaped flash dryers. A production capacity of 24,000–30,000 tons/year of 99–99.9% pure boric acid was claimed. The product was exported from Arica, Iquiqui or Antafagasta (Leiser, 1996; Lyday, 1995). Production of 18,000 tons/year of boric acid as a by-product from the large Salar de Atacama's Minsal potash plant was announced for 1998. It will probably be made by a salting-out process on the solar-evaporated potash end liquor as described on page 384.

8.3.4 China

The Chinese borate skarn deposits had the capacity for processing 120,000 metric tons/year of szaiblyite in 1995 (Norman, 1996), and the country's total capacity was 77,000–110,000 tons/year of borax and 10,000–18,000 tons/year of boric acid in 1994 (Lyday, 1994). A new plant with a capacity of 110,000 metric tons/year of boron–iron ore in Liaoning was announced in 1995, on the basis of their "recent success in separating boron from this type of ore." Su and Yu (1980) reported that in 1980 all the borate ore processing was being done by an alkaline digestion. This left a large residue ("muddy waste") of magnesia and other insolubles. There are many articles in the References on the processing of this type of ore.

8.3.5 Peru

The Inka Bor group at the Laguna Salinas in 1995 air-dried their ulexite at the salar, and then crushed, screened, and sent it to a rotary calciner. It was next cooled and passed through a strong magnetic field to remove some of the iron and clay. This was followed by screening, and about 5000 tons/year of a +10-mesh agricultural grade, and 15,000 tons/year of −10-mesh calcined ulexite could be shipped to customers directly from the salar. The company built medical facilities and housing for its workers at the Salar, equipped with television, communication, electricity, heating, and cooling. Some of the ulexite was shipped directly to a 12,000-tons/year boric acid plant in Lima where it was reacted with sulfuric acid in steam-heated, agitated reaction vessels. The hot slurry was next sent to a drum filter, and the filtrate polished in a filter press. The clear liquor was then cooled to crystallize boric acid, which was centrifuged, washed, dried, and packaged. Most of the products were shipped from the ports of Callao and Matarani (Leiser, 1996).

8.3.6 Russia

In 1939 three plants (Slavyansk, Buiskii, and Leningrad) produced borax and boric acid from Inder borate ore (Moshkareva *et al.,* 1971). The Slavyansk plant used a soda ash leach, whereas at Buiskii borax was produced by acid leaching and then reacting the boric acid with soda ash. The combined capacity of the three plants was about 25,000–30,000 metric tons/year of boron compounds (Gale, 1964). Some ascharite and hydroboracite ores were beneficiated from 15–25% B_2O_3 to 44% B_2O_3 by crushing, washing, desliming, and a short-contact leach to remove the NaCl. Lower-grade ores were upgraded to 18–20% B_2O_3 (occasionally 34% B_2O_3) by a density separation process.

Datolite ores from the skarn deposits with 3–15% B_2O_3 have been upgraded to >16% B_2O_3 by density separation or flotation. Ludwigite-type ores with 4% B_2O_3 have been upgraded to 10% B_2O_3 by a magnetic removal of the accompanying magnetite. The ore was then sintered with calcite, and the residue acid leached to produce 64,000 metric tons/year of B_2O_3 products in 1964, and 68,000 metric tons of B_2O_3 in 1968. An anhydrous borax plant was established in 1972 (Matterson, 1980). Kistler and Helvaci (1994) reported that low-grade borosilicates from the Bor deposit were first crushed and then processed in a complex plant employing magnetic separators, heavy media, and flotation. The concentrate was dried, acidified, and calcined before being converted to boric acid or borax.

8.3.7. United States

8.3.7.1 Death Valley Region: Boraxo, Billie

The ulexite–probertite ore from Tenneco's Boraxo mine averaged 28% B_2O_3 (70% equivalent ulexite), and was shipped directly by 22.7-metric-ton (25-ton) end-dump trucks the 61 km to their Dunn Siding, California plant. The ore size was −15 cm (6 in.), and it was first fed to a double rotor impactor in closed circuit with 2.5-cm (1-in.) screens. It was then conveyed to a Raymond mill to make either −70- or −200-mesh products, which were pneumatically conveyed to a 544-metric-ton (600-ton) or two 204 metric-ton (225-ton) silos. The impactor discharge could also be sent to a 10-mesh-screen and hammermill circuit. Each product analyzed 25–30% B_2O_3, with the finer ones loaded pneumatically, and the −10-mesh product by belt. They were shipped in 82-metric-ton (90-ton) rail cars (Evans *et al.*, 1976). The ulexite–probertite fraction of ore from the Billie mine was also shipped to the Dunn Siding plant (Norman and Johnson, 1980).

8.3.8 Processing

Many leaching processes have also been suggested for ulexite or other borate ores, as indicated in the References. An example of these is Demircioglu and Gulensoy's (1977) study of the alkaline (Na_2CO_3, with some $NaHCO_3$ to reduce metaborate formation) leaching of Turkish ulexite. They found that some forms of ulexite (i.e., "camel tooth," large masses of crystals) were much harder to dissolve than others (i.e., "compact" ore). When ground only to 20 mesh a preliminary roast at 390°C was required to obtain >90% yields for all of the ores with a 1-hour, 85–88°C leach. However, with fine grinding 98–99% yields of either ulexite could be obtained, but losses in the muds after filtration reduced the yields to 91–96%. The initial grinding step was best done in mills that cut as well as ground because of ulexite's fibrous nature. A typical process might require 100 parts of 33% B_2O_3 finely ground ulexite, 12 of Na_2CO_3, and 10 of $NaHCO_3$ to produce 117 parts of borax (Dupont, 1910).

When *priceite* was processed in the 1800s in France it was leached with soda ash (some boric acid or bicarbonate was also added) in heated autoclaves at 60 psi pressure overnight. The leached slurry was then settled and the mud filtered, washed, and discarded. The clarified leach solution was cooled to 56°C to crystallize pentahydrate, or to a lower temperature for borax (Dupont, 1910). The acid or alkaline leaching of ulexite has been extensively covered in the literature, as have magnesium borates and borosilicates (as from skarns), as indicated in the References.

8.3.9 Beneficiation

8.3.9.1 Flotation

Studies on the flotation separation of ulexite from colemanite have been made by Celik and Bulut (1996). They found that anionic surfactants [i.e., 6.7 10^{-5}M sodium dodecylsulfate (SDS)] easily floated colemanite at its natural pH of 9.3, but did not float ulexite. Cationic reagents [i.e., 10^{-4}M dodecylamine hydrochloride (DAH)] floated both minerals, but were adversely affected by the presence of negatively charged clay particles. In these tests perfect crystals of both ores were ground to 65–100 mesh, and conditioned 10 min. Hydroboracite was floated with isopropyl naphthalene sulfate, turpentine, kerosene, and starch. The ore was first deslimed with the aid of 500 g/ton of sodium silicate, and then conditioned with 100 g/ton of oleic acid, 400 g/ton of kerosene,

50 g/ton of pine oil, and the naphthalene sulfate. Yields of 81–92% were reported, of a 18–34% B_2O_3 product (Ozkan and Lyday, 1995). There have also been rather intensive studies on the flotation of borosilicate ores from the large skarn deposits (see the References).

8.3.9.2 Electrostatic, Magnetic, Others

Experimental tests by Fraas and Ralston (1942) have been made on the electrostatic separation of ulexite from synthetic bentonite mixtures. The "ore" was coated with benzoic acid at 150°C, and then passed through a cupric sulfide plate electrostatic separator. With 1.7% of 20- to 32-mesh ulexite, 61% was recovered in the first pass with a 95.5% purity; an additional 13.8% was recovered on the second pass with a 65.6% purity; and 4.9% more at an 18.8% purity on the third pass. Similar studies were made by Yasar *et al.* (1994). At Loma Blanca magnetic separation was used on ulexite, inyoite, and borax ores, first at the mine where the ore was dried, ground, and fed to magnetic separators to produce >26% B_2O_3 ore. Then it was trucked to Palpala (near Jujuy) where additional magnetic separation produced a >38% B_2O_3 product (Solis, 1996). At the Laguna Salinas, Peru, magnetic separation also removed some clay from 20,000t/yr of air-dried ulexite ore (Lyday, 1995).

Many studies (including pilot plant tests) on beneficiation procedures have been made for borosilicate and borosilicate–magnetite ores, including ascharite, danburite, datolite, garnierite, and ludwigite. The methods included density separators, heavy media, flotation, laser and x-ray luminescence, neutron absorption, and roasting. Rapid boron analytical procedures were also examined. Each of these subjects is considered in the References.

8.4 BRINE

8.4.1 Searles Lake

8.4.1.1 Evaporation (Main Plant Cycle)

In 1916 a plant was built at Searles Lake to process brine by plant evaporation, and large scale borax production was initiated in 1919. The lake's brine, a complex mixture containing about 35% dissolved solids (Table 8.2), is saturated with all of the solid phases present in each zone. Since the crystal and brine compositions vary somewhat, the wells have been drilled in the most favorable locations. Casings are cemented in the holes to within a short distance of the bottom, and pumps with 4.3-m (14-ft) suction capability are mounted in them to bring brine to the plants.

The plant evaporation process of the former American Potash & Chemical Corporation treated about 3 million gallons of upper structure brine per day in a continuous, cyclic process. Brine was first mixed with recycle liquor and

Table 8.2
Typical Brine Analyses for the Searles Lake Main Plaint Cycle (wt%)[a]

	Lake brine, 1960		Plant brine	
Constituent	Upper	Lower	ML-1	ML-2
KCl	4.90	3.50	11.67	11.82
Na_2CO_3	4.75	6.50	6.77	7.55
$NaHCO_3$	0.15	—	—	—
$Na_2B_4O_7$	1.58	1.55	9.77	7.16
$Na_2B_2O_4$	—	0.75	—	—
Na_2SO_4	6.75	6.00	1.93	1.99
Na_2S	0.12	0.30	1.00	1.01
Na_3AsO_4	0.05	0.05	—	—
Na_3PO_4	0.14	0.10	0.20	0.21
NaCl	16.10	15.50	6.77	7.38
H_2O	65.46	65.72	61.09	61.51
WO_3	0.008	0.005	—	—
Br	0.085	0.071	0.64	0.66
Li_2O	0.018	0.009	—	—
I	0.003	0.002	0.07	0.71
F	0.002	0.001	—	—

[a] Bixler and Sawyer, 1957; Garrett, 1960.

evaporated to the potash saturation point, and in so doing, crystallizing sodium chloride, licons ($NaLi_2PO_4$), and burkeite ($2Na_2CO_3 \cdot Na_2SO_4$). These salts were removed and hydraulically classified, allowing pure Li_2CO_3, Na_2CO_3, and Na_2SO_4 products to be obtained. The clear, concentrated liquor was next cooled to 38°C (100°F) in gentle, growth-controlled crystallizers, and the resulting potash removed. Since the borax also became saturated during this cooling step, it was only its high supersaturation tendency that kept it in solution.

The "mother liquor" (called ML-1) was then sent to special crystallizers where it contacted a thick seed bed to crystallize crude borax pentahydrate (see Figs. 8.14 and 8.15). The crystallizers had closed impellers at the bottom, and a large conical section on the top. The ML-1 was introduced in the middle, or the top of the lower cylindrical, high circulation, high-seed-density section. Pentahydrate crystals were removed from the bottom as a 36% slurry, and a relatively clear overflow brine came off the top of the upper conical (settling section). This provided considerable residence time (about 40 min) for both the liquor and the solids. The overflow liquor went to a thickener to remove

Figure 8.14 Searles Lake crude borax pentahydrate feed tank (left), crystallizer (center) and thickenter (bottom). (From Bixler and Sawyer, 1957; reprinted with permission from the American Chemical Society; copyright 1957.)

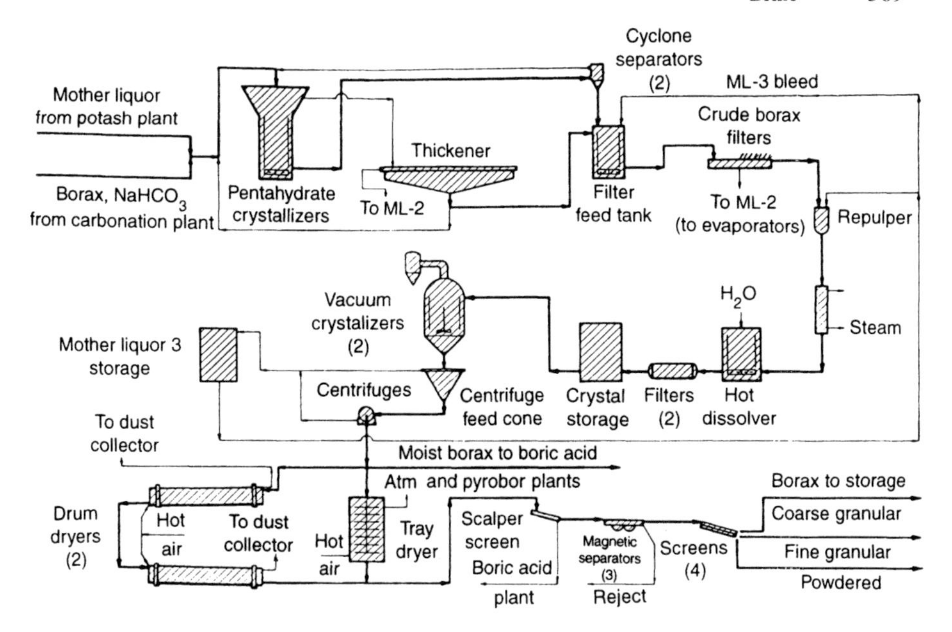

Figure 8.15 Flowsheet for production of borax from Searles Lake; evaporation process. (From Garrett, 1960; reproduced by permission of the Society for Mining, Metallurgy and Exploration, Inc.)

the last of its solids (occasionally 6–8%), and then (as ML-2) it was blended with fresh lake brine and returned to the evaporators. Borax from the separate carbonation plant (accompanied by some sodium bicarbonate to convert metaborate to tetraborate) joined the pentahydrate in the crystallizers. The lake brine contains some metaborate, and more was formed at the high temperature of the evaporators:

$$Na_2B_4O_7 + Na_2CO_3 \rightarrow 2Na_2B_2O_4 + CO_2 \quad (1)$$

To increase the borax yield per cycle (metaborate is very soluble), this reaction is reversed by bicarbonate:

$$2NaHCO_3 + 2Na_2B_2O_4 \rightarrow 2Na_2B_4O_7 + CO_2 \quad (2)$$

Foaming in the crystallizers or thickeners could be a problem in floating fine crystals out the top, but it was minimized by deaeration in the crystallizer feed tank, level control to keep the pumps under positive pressure, and the occasional use of a defoamant. The underflow stream from the crystallizers was further thickened in a liquid cyclone, and then sent with part of the thickener underflow to two 4-m (13-ft) flat bed, scroll-discharged filters. The remaining portion of the thickener underflow was returned to the crude pentahydrate crystallizer feed tank as seed crystals. The pentahydrate was washed on the filters with borax refinery end liquor (called ML-3) before it was discharged, and the filtrate returned to the thickeners. Analytical samples were taken on a 2- to 4-hr schedule to ensure the proper control of the process. Typical liquor concentrations are given in Table 8.2.

The crude pentahydrate was next repulped and redissolved in 88°C (190°F) refinery end liquor (ML-3), and further heated with steam sparging to 93°C (200°F) in the dissolver tanks. This brine was sent through polishing filters, and then to vacuum crystallizers where either high-purity borax [cooled to 49°C (120°F)] or pentahydrate [cooled to 60°C (140°F)] could be produced with heavy seed beds of either crystal. Caustic soda was added to increase the pH from 9.2–9.4 to 9.6–9.8 to form larger, chunkier crystals, and oleic acid to prevent crystal twinning and agglomeration. The product was originally dewatered in five continuous, automatic Sharples centrifuges to a moisture content of 3.5–6%, but later was separated in a screen bowl centrifuge. The crystals not going to the boric acid or anhydrous borax plants were dried in hot air [43–52°C (110–125°F)] rotary or Wyssmont shelf dryers, and then screened for shipment (Garrett, 1960).

8.4.1.2 Carbonation

The second process for recovering borax from Searles Lake brines was by carbonation, starting in 1926 at West End, 1946 at Trona, and 1976 at Argus. The three operations were slightly different, but they can still be described together. Primarily lower-structure brine, with its richer borax and sodium carbonate content, was introduced into the top of large carbonating towers. There it was contacted by compressed makeup CO_2 in flue gas, lime kiln gas, or CO_2 from an MEA unit, distributed into the base of the towers. One plant produced fine bubbles with agitators, and another used sieve plates to reduce the bubble size and provide a more even gas distribution. The absorbed CO_2 converts the Na_2CO_3 in the brine to $NaHCO_3$, which being only slightly soluble in the NaCl-saturated brine, crystallizes. The slurry from the primary towers is then sent to secondary towers where it absorbs CO_2 returned from the bicarbonate calciner, and the carbonation is complete (see Fig 8.16). The bicarbonate crystallized in the towers is centrifuged (or was filtered), and sent through equipment to convert it to high-purity dense soda ash. The remaining brine and centrate are blended with fresh lower-structure brine (with its metaborate content) to convert the pentaborate ($Na_2B_{10}O_{16}$) formed during carbonation back to tetraborate. The mixed brine is then sent to vacuum crystallizers where it is cooled in the presence of a heavy seed bed, and borax is crystallized. Slurry is withdrawn from the crystallizers, thickened, and centrifuged (or was filtered). The overflow is further cooled and sent to glauber salt crystallizers, which also produces a little more borax. In each plant the final brine is routed through heat exchangers to provide some cooling, and

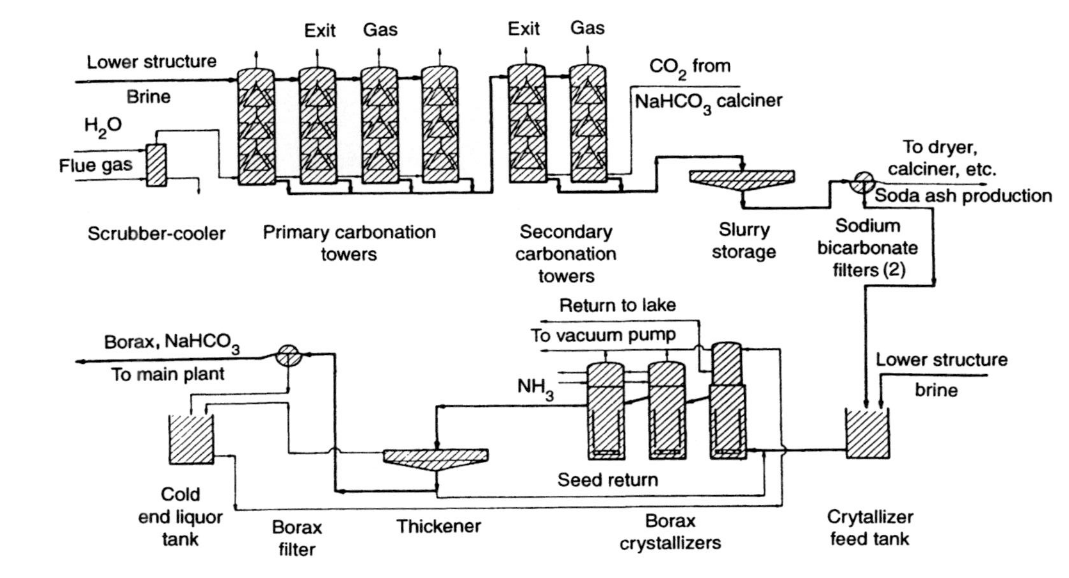

Figure 8.16 Flowsheet for production of borax from Searles Lake; carbonation process, old plant. (From Garrett, 1960; reproduced by permission of the Society for Mining, Metallurgy and Exploration, Inc.)

then returned to the lake. The crude borax is next redissolved to form pure products (Garrett, 1960, 1992). Lowry (1926) obtained one of the early patents on the process, and Helmers (1929) received a patent on the blending-of-brines concept.

8.4.1.3 Other Products

A full range of borate products had been made at Searles Lake, including borax, pentahydrate, anhydrous borax, boric acid, and B_2O_3. The boric acid process started with refined borax centrifuge cake and hot water being sent to a dissolving tank, along with heated recycle end liquor and part of the sulfuric acid. A pentaborate solution was formed, which was polish-filtered and further heated to 99°C (210°F). The remaining sulfuric acid was then added, converting the pentaborate to boric acid. This solution was sent to a vacuum crystallizer cooled to 30°C (86°F) by a barometric condenser operating with cooling-tower water. Boric acid was crystallized, and then withdrawn, centrifuged, and dried in a steam-tube dryer. Most of the process equipment was made of a special stainless steel to minimize corrosion. A USP grade of boric acid was also produced by dissolving technical boric acid, filtering the solution, and recrystallizing the product. A small excess of sulfuric acid was added to the refined product leach liquor in order to improve the crystal size and color.

The boric acid end liquor was sent through heat exchanger-condensers in the crystallizer's vapor head, and returned to the main plant cycle. There the by-product sodium sulfate was concentrated and recovered more economically than by installing a separate evaporator. (Its boric acid content also helped reduce metaborate in the ML-2.) The design of this single-pass system took advantage of two unique solubility features. First, for solutions saturated with both boric acid and sodium sulfate, there is maximum Na_2SO_4 solubility at 30°C (86°F; see Fig. 11.6). Therefore, a reaction solution cooled to this temperature has a maximum single-pass boric acid yield of 88%. Sodium sulfate saturation was maintained by recycling some of the end liquor. Secondly, the feed borax's solubility was greatly enhanced by adding only about 40% of the total sulfuric acid to form the very soluble sodium pentaborate ($Na_2B_{10}O_{16}$). If boric acid or sodium sulfate were allowed to crystallize immediately, it would be very hard to control their crystal size and purity. Consequently, enough end liquor was recycled to hold all of the boric acid in solution at 82°C (180°F) until the final sulfuric acid was added (Garrett, 1960; Garrett and Rosenbaum, 1958a, 1958b).

8.4.1.4 Dehydration

To save on shipping costs, and for some special purposes, considerable borax, and some boric acid is sold in the anhydrous form. They are produced by partially dehydrating borax or boric acid in rotary calciners, and then melting it in fusion furnaces. At the former American Potash & Chemical Company's plant, to make $Na_2B_4O_7$ three rotary-drum calciners first removed about 70% of the hydrate water from either wet centrifuge cake or dried granular borax. If borax fines or pentahydrate were used, the crystals puffed to the extent that some blew out of the calciners, and the calcine was not dense enough to make an adequate seal in the fusion furnaces. The borax first passed over a weigh-feeder and went to any of the three calciner storage bins. From there a variable-speed chain feeder withdrew and delivered it to the rotary calciners' feed screw. The three calciners were 2.4 m (8 ft) in diameter and 21.3 m (70 ft) long, and two were heated cocurrent, and one countercurrent by the 704–816°C (1300–1500°F) fusion furnace flue gas. Calciner dust was recovered first by dry and then by wet scrubbers, and the dust was sent to the borax/pentahydrate refinery. The flue gas left the calciners nearly saturated with water and at about 52°C (125°F), providing good thermal efficiency. The hot, partially dehydrated borax from the calciners then went to feed storage bins for the three fusion furnaces. It was next conveyed, along with a small amount of sodium nitrate (a bleaching agent for organics from the lake brine and oil burners), to the feed hoppers rotating around each furnace. The hoppers distributed calcined borax through a 20-cm (8-in.) gap between the furnace walls, allowing it to pile up and seal this opening. Plows and cutters then forced the calcined borax into the furnace.

Inside the furnaces a direct, oil-fired flame impinged on the borax sliding down the furnace walls, and the hot flue gas passed out the bottom (Fig. 8.17) on its way to the calciners. The borax almost completely dehydrated before melting and flowing to the seal lip at the base of the furnaces. Two furnaces had the melt flow into chain conveyor molds traveling at 6.7 m/min (22 ft/min) for the length of the building. (A very small amount of water was left in the melt to promote crystallization by minimizing its tendency to supercool). As the molds traveled they slowly cooled and crystallized. The molds were then dumped, and the crystals ground and sized quite easily (about 5% of the anhydrous borax still converted to the amorphous phase as it quickly cooled when poured into the molds). The melt from the third furnace was cooled and solidified on a chilled roll, producing the much more abrasive amorphous or glass form that was difficult to grind.

The furnace walls were lined with either cast or brick refractory material, and water jacketed to further protect the steel structure. However, molten borax attacks all commercial refractories, so a constant layer of calcine was required between the wall and the melt. The furnace temperatures were 1200–1430°C (2200–2600°F) at the fire box, and 980°C (1800°F) in the fusion zone [borax melts at 743°C (1370°F)]. Anhydrous borax was emptied from the molds onto a toothed roll that reduced it to −3.8-cm (1.5-in.) particles, which were sent to storage bins. The final grinding and screening was done

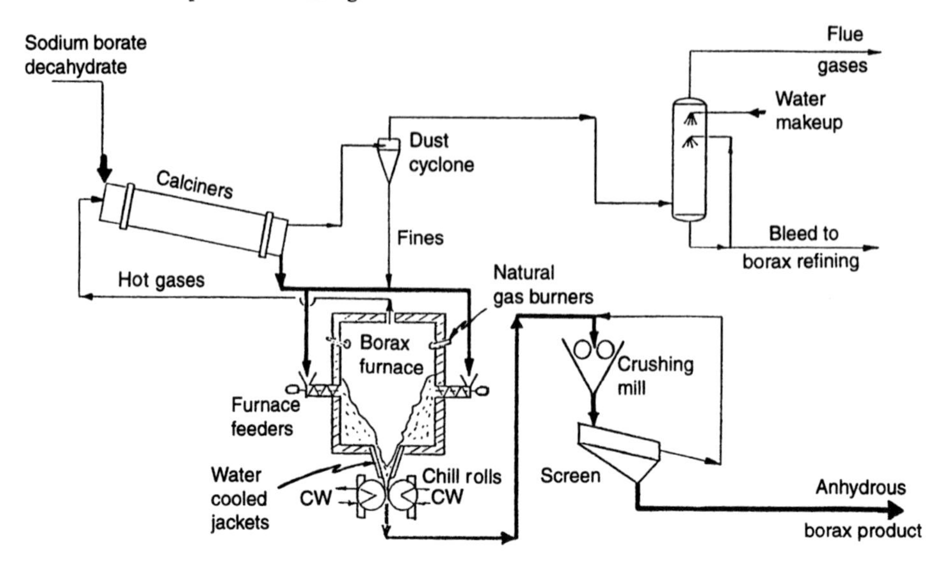

Figure 8.17 Searles Lake anhydrous borax plant flowsheet. (From Garrett, 1960; reproduced by permission of the Society for Mining, Metallurgy and Exploration, Inc.)

in a dry rod mill working with screens to produce three product sizes: +7, 7–16, and −16 mesh. Dust smaller than 80 mesh was recycled and remelted. The 1.6-mm- (1/16-in.-) thick sheet of anhydrous borax from the chilled roll was similarly crushed, screened, and stored (Bixler and Sawyer, 1957; Garrett, 1960). Anhydrous boric acid, or B_2O_3, could also be produced, but it was dehydrated and fused in a special calciner. The energy requirements to dehydrate boric acid to B_2O_3 have been estimated by Finlay (1952).

8.4.1.5 Late 1990s Operation

On April 1, 1996, the operator at that time, the North American Chemical Company, shut down the main plant cycle and produced boron products only from carbonation plant brine and solvent extraction. To partly make up for the lost borax capacity two changes were made: The brine flow rate was increased, and waste heat from a power plant's flue gas was used to preheat the "solution mining" injection brine. At the West End plant the 4000 gal/min carbonation/borax/glauber-salt plant's end liquor was heated to about 37°C (98°F), and the 3000 gal/min for the carbonation plant (it formerly was 2000 gal/min) and the 1000 gal/min (it was 667 gal/min) blending brine for the borax/glauber-salt plants at West End was withdrawn at about 27°C (80°F). This was a 6.7°C (12°F) increase, and raised the brine's $Na_2B_4O_7$ content from ~1.2% to 1.6–1.8%.

Brine from the carbonation plant was sent from Trona (the Argus plant) to West End, and blended with lake brine to a pH of 8.3–8.5. This mixture was first cooled with cooling-tower water, then with glauber salt end liquor at 16.4–16.7°C (61.5–62°F), and finally with ammonia coolers to 18.3°C (65°F) to crystallize borax. The crystallizers were arranged in two parallel lines, one with three units in series, and the other with two. About 85% of the borax crystallized in the first effects, 10 or 15% in the second, and 5% in the third. The final slurry was sent to three wet vibrating screens, with the oversized solids advanced to the centrifuges, the mid-sized fraction returned to the crystallizers to increase the slurry density, and the fines sent to a thickener. The underflow solids were then redissolved in the feed solution and recycled. The borax product was centrifuged and washed in three two-stage, 700-mm pusher-type centrifuges. The clarified brine from the thickener was sent to the glauber-salt crystallizers for further cooling. Six 1000- to 1500-ton ammonia compressors were used for the combined cooling load.

Part of the crude borax cake from the West End plant was transported 12 hrs during the evening on a private road in 41-metric-ton (45-ton) off-highway trucks to the Trona plant for conversion to purified borax and anhydrous borax. At West End the borax was converted to pentahydrate by dissolving the cake at 85–88°C (185–190°F) to form a 23–24% borax solution. It was then polish-filtered and recrystallized in bottom-agitated, draft-tube baffled crystallizers at 60°C (140°F), with cooling supplied by glauber-salt end liquor in barometric condensers. The pentahydrate crystals were withdrawn, dewatered, and washed in two-stage pusher centrifuges, dried in gas-fired rotary dryers, sent to storage, and loaded for shipment. The plant also converted brackish water to freshwater by reverse osmosis, and had a 19-megawatt gas turbine with a waste-heat boiler (and the injection brine heaters), large refrigeration units, and Evapco cooling towers containing 1.2-m (4-ft) ammonia condensing coils.

At the Trona plant, borax cake from West End was dissolved in water and borax end liquor (ML3), and sent through two (another was available as a spare) polishing filters (using a cellulosic-type filter aid). The clarified brine was next sent to vacuum crystallizers with 45-rpm internal "pachuca" impellers. Disodium hydrogen phosphate was added (instead of oleic acid) to increase the chunkiness of the crystals; a slight "excess alkalinity" was maintained with caustic soda; and 30–50 ppm of an antifoamant was often required. The crystallizers were cooled as far as possible with cooling-tower water in barometer condensers, which in the summer were occasionally 40–43°C (104–109°F). A slurry was withdrawn from the crystallizers, thickened, and then sent to a four-stage pusher centrifuge. The cake was washed with about 5 gal/min of wash water containing a small amount of boric acid to prevent caking and reduce the Na_2SO_4 to less than 10 ppm. The product was dried in a counterflow rotary dryer with steam-heated hot air and some direct gas firing, with the flue gas leaving at 43–49°C (110–120°F). The product was screened at +30 mesh, −30 mesh (98% +40 mesh), and −40 mesh. Cold end liquor from the crystallizer was used in the first effect's barometer condenser, and recycled.

In the anhydrous borax plant, only one of the three melting furnaces was usually operated, at a capacity of about 140 tons/day, and with the burner design changed to improve its fuel efficiency. Water jacketing of the furnace shell was also eliminated except at the lip where melted borax flowed onto a chilled roll. There was a 2–4 tons/day pilot plant to produce >99% B_2O_3, and a plant to make 1–1.5 tons/hr of a noncaking, very finely divided (20 lb/ft^3) boric acid powder for anticeptical use (De Nuz, 1996).

8.4.2 Owens Lake

During the early 1900s several companies operated small soda ash plants on Owens Lake, and a few such as the Pacific Alkali Company also recovered borax. They operated from 1926 to 1967 at Bartlett, California, employing the Kuhnert process. Brine from the lake was pumped 4 km (2.5 mi) to solar ponds where the borax content rose from about 2.6 to 4%. The brine was next carbonated with 14–15% CO_2 flue gas (for the makeup CO_2) in 16 carbonation towers, 24 m (80 ft) tall and 1.5 m (5 ft) in diameter. Sodium sesquicarbonate was formed; and after it was removed from the carbonated brine the end liquor was blended with solar pond brine, and cooled to obtain 1800 metric tons/year of borax. Columbia Southern (Pittsburgh Plate Glass) purchased the plant in 1944, expanded it 300% in 1958, and closed it in 1967 (Garrett, 1992).

8.4.3 Miscellaneous Processes

8.4.3.1 Solvent Extraction

Boron has a strong tendency to form ring structures with oxygen atoms, as indicated by the many borate compounds in nature. Similary, boron—oxygen—carbon bonding can be very strong, such as a hydrogen ion being liberated from an OH group (used for boron analyses) in mannitol in order to form B—O—C rings. This property also allows the selective extraction of boron from brines, or the removal of a boron contaminant. The first patent on this concept was by Garrett (1961), then Garrett, Weck, Marsh, and Foster (1963), and later many others (e.g., Weck, 1969). It was noted that some alcohols, polyols, acids, and ketones have borax extraction coefficients of up to 10 (at equilibrium the boron concentration in the solvent is 10 times that in the extracted brine). Specific organic polyols were then synthesized to achieve maximum boron extraction from basic solutions (it is much easier to extract from acidic solutions). They had high distribution coefficients (10–150), low solubility in water, good stability, high selectivity, and were easy to regenerate, stable, safe and inexpensive. The results were so successful that a large boric acid recovery plant was built in 1963 at Searles Lake to process plant end liquors or lake brines.

The specific solvent used in the plant has not been disclosed, but the saligenins were very effective, such as 4-tertiary butyl-6-methylol saligenin (with a coefficient of 122):

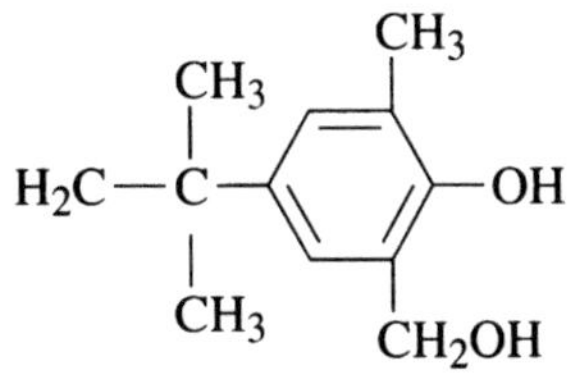

$$H_2C-C(CH_3)_2-C_6H_2(CH_3)(OH)(CH_2OH) \qquad (1)$$

The 1963 patent claimed to have favored 2-chloro-4 (1,1,3,3)tetramethylbutyl-6-methylol-phenol (with a coefficient of 100):

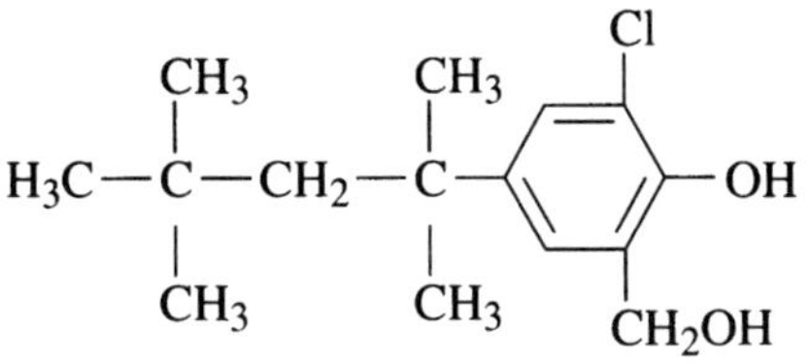

$$H_3C-C(CH_3)_2-CH_2-C(CH_3)_2-C_6H_2(Cl)(OH)(CH_2OH) \qquad (2)$$

As an example of the effectiveness of solvent (2), in one test at room temperature, 800 ml/min of solvent–carrier (one part solvent and three parts kerosene) contacted 500 ml/min of a 1.05% $Na_2B_4O_7$, pH 10 brine in a single stage. The loaded solvent was then stripped in four countercurrent stages with 800 ml/min of 1N sulfuric acid. The extract contained 6.1% H_3BO_3, 2.8% Na_2SO_4, 5.2% K_2SO_4, and no CO_3, Cl, or $S^=$. The extract was first evaporated at 90–95°C to crystallize the potassium–sodium salts, and then cooled to 35°C to crystallize boric acid. About 27 ppm of the organics remained in the stripped brine, and 44 ppm in the stripping solution. The (2) solvent had a better storage life in the boron complex form, loosing only 2% of its extraction capacity in 1 week at 60°C (below 25°C was recommended).

A general description of the process as it was first operated (Fig. 8.18) was given by Havighorst (1963) and Anon. (1963). About 11,000–15,000 l/min (3000–4000 gal/min) of end liquor was first contacted with the polyol–kerosene solvent in a single extraction stage, and the stripped brine (raffinate) was returned to the lake. The loaded solvent was sent to four stages of countercurrent stripping with dilute sulfuric acid, and the stripped solvent was returned to the brine extraction stage. The boric acid strip liquor was next passed through a semicontinuous carbon-adsorption column to remove the small amount of solvent and organics in the brine. The brine was then sent to a

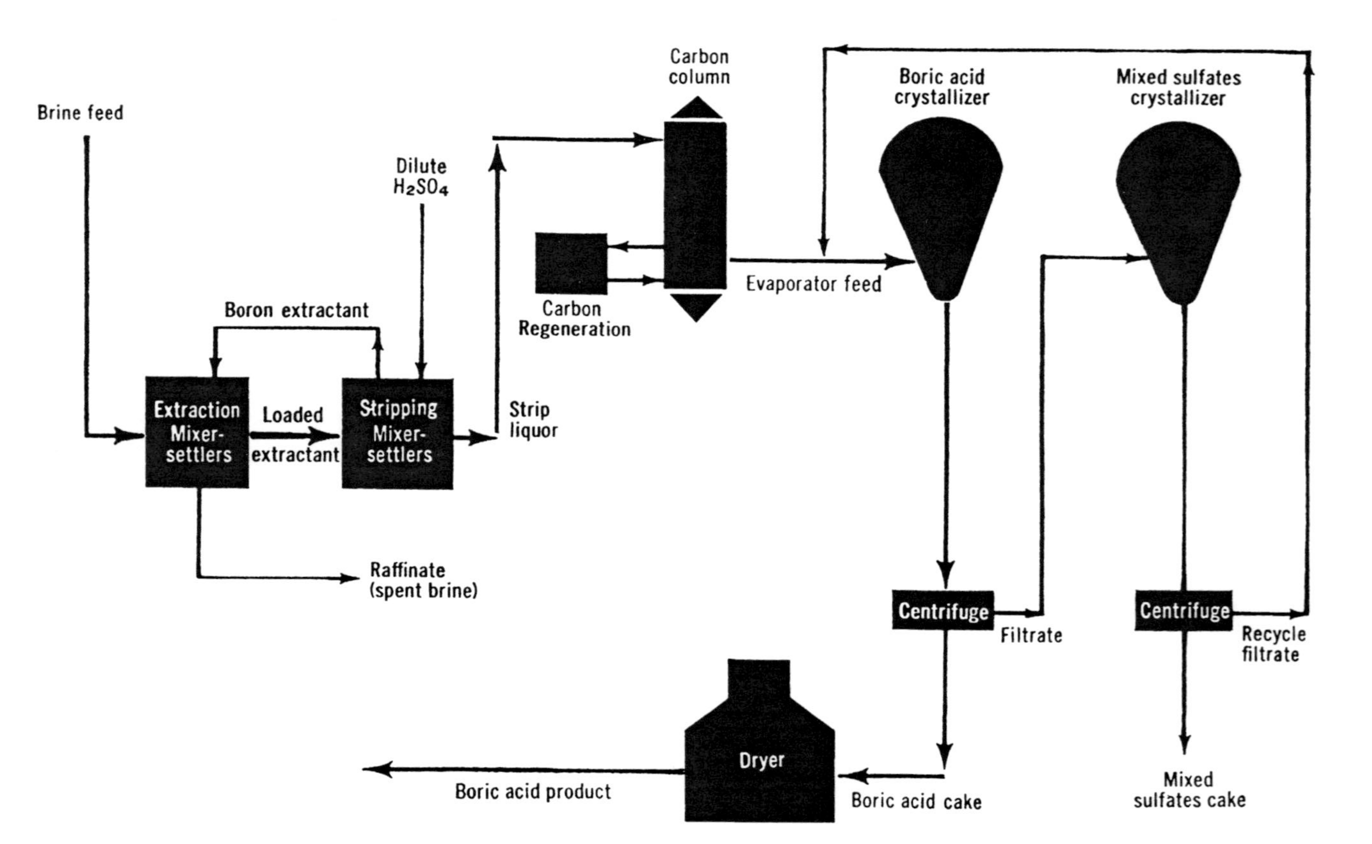

Figure 8.18 The Searles Lake borax solvent extraction process. (From Anon., 1963; reprinted with permission of the American Chemical Society; copyright 1963.)

Figure 8.19 Evaporator–crystallizers for the boric acid solvent extraction plant.

65°C growth-type glaserite–salt cake evaporator–crystallizer (Fig. 8.19), where it was joined by boric acid crystallizer overflow and centrate. The concentration was maintained just below boric acid's solubility, and a slurry was constantly withdrawn. The salts were centrifuged, washed, and sent to the main plant's glaserite reactor to form potassium sulfate, a valuable by-product. The mixed salt crystallizer's overflow brine and centrate was then sent to the boric acid crystallizer.

This unit was operated at the lowest temperature obtainable from cooling-tower water in a barometric condenser, nominally 35°C, and its brine was maintained at a concentration just below that at which the other salts would crystallize. A slurry of boric acid was constantly withdrawn, thickened, centrifuged, and washed. The product was then sent to a multiple hearth-type dryer for gentle drying (to prevent metaboric acid formation). A +99.9% product was obtained with only 0.05% SO_3 and 0.029% Na. The boric acid crystallizer's overflow stream and centrate were returned to the mixed sulfate evaporator to repeat the cycle. A bleed stream was occasionally sent to the extraction plant's acid stripper to control any H_2SO_4 buildup. The operator's skill, computer programs, and extensive instrumentation allowed the plant to have a rapid startup and essentially trouble-free operation. It very likely produces the most inexpensive boric acid in the world.

The solvent extraction plant by 1996 had changed slightly, and was operating well over its design capacity at 150 metric tons/day of H_3BO_3. The solvent loading was still done as a single stage, but originally the solvent–brine mixing was accomplished in four agitated tanks in series. It was found that the fourth vessel was unnecessary, and the third added very little borax and began to form an emulsion. Consequently, only two were used. The flow rates in the plant were about 1800–1950 gal/min of fresh brine at 34°C (94°F), and 3200 gal/min of a solvent–kerosine mixture. The brine from the loading tanks (at a pH of 8.9) was sent to three lines of settlers where the two phases were separated, and finally to an API settler. The stripped brine contained up to 100 ppm of organics, and the government had set limits of 17 ppm for it to be returned to the lake. Consequently, it was sent to three parallel lines (650 gal/min each) of Wemco Depurators (flotation-like cells) where the organics were removed as a froth or vapor. The former was collected and reused, whereas the vapor and fumes from every part of the plant were sent to the power plant's burners. The organics in the brine were reduced to less than 10 ppm.

The loaded solvent was next given five (the last one was a water wash) countercurrent mixing and settling stages (Fig. 8.20) at 38°C (100°F) with dilute sulfuric acid. The mixer–settler tanks were made of concrete with a fiberglass lining, and each of the other tanks were lined with fiberglass. Some emulsion constantly formed (partly from the solvent's degradation), so it was separated, treated to break the emulsion, and returned to the system. Since the solvent was acidic by nature, and always contained some entrained acid, it tended to precipitate a small amount of sodium bicarbonate from the brine. To minimize this, a few percent of water was added, but some still formed, along with the emulsion, so the plant had to be cleaned about once per year. This was difficult because of the tank covers, vents, and emergency quench system.

Figure 8.20 Solvent–acid mixers (to the right) and the totally covered settlers (to the left) of the Searles Lake boric acid solvent extraction plant.

The organic solvent was always maintained as the continuous phase, with the brine as droplets within it. If the phases were reversed, the flow would be disrupted, and there could be a major foam or emulsion problem. The plant consumed about 20,000 gal/24 hr of concentrated sulfuric acid, and added makeup solvent once per week. This required 4000 gal of the P_2O solvent and 6000 gal of kerosene. The solvent was manufactured adjacent to the extraction plant (Fig. 8.21) with nonyl phenol and chlorine as two of the reactants. The nonyl phenol provided greater solvent stability than the previously used P10 octyl phenol. The extraction plant was highly instrumented and partially computer controlled, requiring only one operator, who also handled the sulfuric acid receiving and unloading. A second operator was responsible for the P20 synthesis and its raw material handling (Mata, 1996).

The processing facilities for the boric acid strip solution are located across the Trona plant, far from the brine extraction unit. The brine enters the plant containing 10–15 ppm of organics, so it is first sent (downflow) through a 7.6-m (25-ft) adsorption column, 1.6 m (62 in.) in diameter, filled with 3-mm to 16-mesh activated carbon. Every few days a portion of the bed is flushed-out through six withdrawal pipes at the column's base, drained, washed with hot water (second effect condensate), and blown partially dry with air. It is next sent to a 1.2-m- (49-in.-) diameter, six-tray Hershoff furnace where the

Figure 8.21 The solvent manufacturing facilities for the Searles Lake boric acid solvent extraction plant.

organics are burned at 650–760°C (1200–1400°F). This regenerates the carbon, and it, along with 180 kg (400 lb) of makeup carbon, is sent to a holding tank to be slurried back into the column.

The brine is next sent to two hemispherically-domed evaporator–crystallizers (see Fig. 8.19) where it is joined by recycle liquor and circulated through titanium heat-exchanger tubes at 19,000 gal/min. Water is evaporated, and salt cake and glaserite are crystallized, which are removed, settled, and dewatered in a 1 × 1.5-m- (40 × 60-in.-) solid bowl centrifuge. In 1996 this cake was being repulped and returned to the lake. The overflow brine and centrate are then cooled in the second stage growth-type crystallizer circulating at 22,000 gal/min. Boric acid is crystallized and a slurry withdrawn, settled, and dewatered in a 900-mm, four-stage pusher centrifuge (Fig. 8.22). The product is washed in two of the stages, and the centrate and overflow brine are returned to the circuit. The cake is dried in a gas-fired rotary dryer and pneumatically transported to a storage silo. Caking is prevented by removing the transport air at the top and blowing cold air into the silo at its base. Normally, the product is 88% +60 mesh, but upsets can occur, perhaps caused by changes in the stripping sulfuric acid (i.e., the acid's copper content can randomly change from 30–400 ppm). The evaporator–crystallizer are made of 316 extra-low carbon (elc) stainless steel, and rubber-lined and acid-brick-

Figure 8.22 Boric acid centrifuge for the solvent extraction plant.

lined for the entire first stage, and for the lower 3 m (10 ft) of the second stage. Without the acid brick there was gradual rubber damage from the organics in the brine. Process lines are made of 316 elc stainless steel and must be replaced periodically (De-Nuz, 1996).

Several other commercial boron liquid extraction systems have been built, but they are for removing small amounts of impurities, as in the purification of magnesium chloride solutions (Norsk Hydro, 1974) for magnesium metal, or high-purity magnesium oxide. A similar process is used for the purification of lithium chloride solutions for high-purity lithium compounds or metal. In both cases boron can cause serious problems, so its nearly complete removal is important. Off-the-shelf diols or alcohols, with their 2–5 distribution coefficients, are employed because of the acidic solutions and much simpler removal demands. Usually, the boron is merely precipitated from the extractant.

8.4.3.2 Ion Exchange

The unique chemical properties that allow boron to be selectively removed by solvent extraction also apply to ion exchange resins. There is considerable literature on this subject (see the References), and experimental quantities of boron-specific resins have been available. The first of these was Amberlite XE-243R (Anon., 1965a; Kunin and Preuss, 1964; Lyman and Preuss, 1957), which contains pentahydroxyhexyl and tetrahydroxy pentyls, including sorbityl and mannityl (the preferred polyol was hydrogenated glucose in the presence of methylamine, or N-methyl–glucamine). The polyol was attached to a monovinyl resin base containing 0.5–50% $CH_2{=}C{=}$ groups. The resin weighed 40–45 lb/ft^3; had an exchange capacity of 2.4–2.5 meq/ml, 5.7 mg B/ml, or 0.36 lbB/ft^3 of resin; and contained 56–60% water. The capacity was flow-rate dependent, and the resin had to be regenerated with 10% H_2SO_4 (3 lb H_2SO_4/ft^3 of resin), followed by removal of the sulfate with 4% NaOH or NH_4OH (4 lb NaOH or 3 lb NH_4OH/ft^3). Suggested uses were removing boron from irrigation water or magnesium chloride brine. However, with most solutions there was no period of complete boron removal, the breakthrough curve was gradual, and the capacity dropped with each cycle. The elute concentration averaged about 0.2% H_3BO_3 in a large excess of H_2SO_4 (Chemtob, 1971). Very much improved boron-specific resins have been prepared by Garrett and Weck (1958). In one test on a Searles Lake end liquor (0.5% $Na_2B_4O_7$), the borax was completely removed, the resin capacity was 8 mgB/ml, and elution with pH 3 acid gave a 6% H_3BO_3 solution and completely regenerated the resin.

8.4.3.3. Salting Out

With a number of the boron-containing lakes (i.e., many of the playas in South America and China), a salting-out process for recovering boric acid would appear to be practical, especially if it were a by-product. Such a process has been developed by Garrett and Laborde (1983), starting with a solar-evaporated brine that had already produced the maximum K_2SO_4 KCl and/or lithium as the primary products of the operation. When sufficiently concentrated, boric acid can be crystallized (it is quite insolulble in a saturated NaCl brine) by adding sulfuric acid, either in a cyclic in-plant procedure, or by harvesting a boric acid mixture from solar ponds and separating the mixture by flotation.

Salting-out has been the method of boric acid production from the Chilean nitrate plants, and is usually combined with iodine recovery, where pH adjustment is also required. This production has been practiced only periodically and in rather small amounts. However, the large potash plant now operating on the Salar de Atacama has announced the production of 18,000 metric tons/year of boric acid beginning in 1998. In their pilot plant tests (Pavlovic, Parada, and Vergara, 1983), boric acid could be precipitated when its concentration was greater than 21 g/liter H_3BO_3. The optimum pH after acid addition was 2.0 for 25–50 g/liter of H_3BO_3 brine, requiring about 0.21 g of H_2SO_4/g H_3BO_3. A brief water wash gave high-purity boric acid. An identical process has been proposed by Gao, Yang, and Huang (1993) for a Chinese playa (Lake Da Chaidan). They suggested adding the acid to the brine when the MgB_3O_7 content was above 35 g/liter (48 g/liter of H_3BO_3). The acid requirement was 0.30 g/g H_3BO_3, and the optimum terminal pH was 3.0.

8.5 HEALTH AND SAFETY

8.5.1 Worker Studies: Toxicology

A number of reports have described the effects of borate dust on the health of borate processing plant and mine workers. Kistler and Helvaci (1994) noted separate U.S. OSHA and U.S. Borax studies at the Boron deposit on 800 employees who were exposed to borate dust. It was found that they had slightly healthier lung functions than the U.S. national average, indicating that there were no adverse lung or respiratory effects even from long-term exposure to borax, kernite, boric acid, or their dehydrated products. Kasparov (1971) also found no adverse health effects from boric acid dust on production workers, as was the case with the fibrogenic activity of borosilicate dusts (Kasparov and Yakubovskii, 1970). Studies on the sampling methods and exposure variability of dust contact by sodium borate workers have been made by Woskie, Shen, Eisen, Finkel, Smith, Smith, and Wegman.

More specific toxicologic information has been provided by Sprague (1972), who noted that borax and boric acid have a relatively low toxicity for animals (Table 8.3), with the official LD_{50} (the dosage at which half of the test animals die), as g/kg of body weight, being 6.05 for borax and 5.14 for boric acid. Chronic toxicity studies over a two-year period gave values of 5.00 and 3.00 g/kg LD_{50}, respectively, values only 2–3 times that of aspirin (salt is 3.75 g/kg). Substances with LD_{50}s greater than 2.0 are not classified as dangerous. The minimum lethal dose of borates for humans has not been established, but an intake of 4 g/day was reported without incident, and medical dosages of 20 g of boric acid for neutron capture therapy have caused no problems. Fish have survived 30 min in a saturated boric acid solution, and can survive longer in strong borax solutions.

Levinskas (1964) presented some slightly less optimistic LD_{50} data, noting a study in which 144 humans ingested 15–30 g of borax, or 2–5 g of boric acid, with a 50% mortality rate. However, in another study 15- to 30-g doses of boric acid were ingested by hundreds of patients with no serious toxic effects, indicating that toxicity varies with age and the individual. Ingestion symptoms for adults can be nausea, mild shock, vomiting, diarrhea, headaches, weakness, and excitement or depression. Children may have meningeal irritation, convulsions, coma, and delirium. An intense scarlatina form of rash (like "boiled lobster skin") may cover the entire body, including the mucous membranes, for a day or two.

Skin irritation with borate dusts or solutions is usually not a problem, with the borate cleaners and ointments being less irritating than pure soap. Occasionally, there is a mild allergic response, and with workers in borate plants or mines its alkalinity at high dust levels can cause irritation of the respiratory tract. In a study of 629 workers at a borax plant and mine, com-

Table 8.3

Toxicity of Borax, Boric Acid and Other Borates[a]

Sex & strain	Borax LD$_{50}$ (g/kg)	Borax Equiv. boron (gB/kg)	Borax Experimental limits (g/kg)	Boric acid LD$_{50}$ (g/kg)	Boric acid Equiv. boron (gB/kg)	Boric acid Experimental limits (g/kg)
A. LD$_{50}$ of borax and boric acids in animals, oral						
Long Evans male rats	6.1	0.69	3.5–10.4	3.2	0.55	—
Sprague Dawley female rats	5.0	0.57	4.3–5.8	4.1	0.71	3.6–4.6
Sprague Dawley male rats	4.5	0.51	4.1–5.0	3.5	0.60	3.0–4.0
Mice, guinea pigs[b]	>3.92 ± 1.42[c]	—	—	>3.27 ± 2.07[c,d]	—	—
B. As A, intravenous						
Mice[b]	1.32	—	—	—	—	—
Mice, rat, rabbit, dog[b]	—	—	—	1.6 ± 0.79	—	—
Rabbit, cat[b]	0.41 ± 0.34 (sodium perborate monohydrate)					
Rabbit[b]	>0.3[c] (sodium metaborate)					

C. As A, subcutaneous						
Mice, rat, dog, guinea pigs[b]	—	—	—	1.48 ± 0.53	—	—
D. As A, intraperitoneal						
Mice, rats[b]	—	—	—	>0.8[c]	—	—

E. Minimum lethal dose of borates for minnows

Substance	LD_{50}, (mg/l or 1000 ppm)	
	Distilled water	Hard water
Borax ($Na_2B_4O_7 \cdot 10H_2O$)	19,000–19,500	19,000–19,500
Anhydrous borax ($Na_2B_4O_7$)	3,000–3,300	7,000–7,500
Boric acid (H_3BO_3)	1,600–1,750	3,700–4,000
Sodium chloride	10,000	11,500–12,000

[a] Sprague, 1972.
[b] Levinskas, 1964.
[c] No toxic effect noted at this dosage.
[d] Not guinea pig, but including rat, rabbit, and dog.

plaints of symptoms of mouth, nose or throat dryness, dry cough, nose bleeds, sore throat, productive cough, shortness of breath, and chest tightness were common for workers with daily average exposures of 4.0 mg/m^3 of borate dust, and infrequent at exposures of 1.1 mg/m^3 (5.7 and 2.2, respectively, Woskie *et al.,* 1994). No damage was done, however, even with prolonged and/or high levels of exposure, except perhaps for some FEV_1 decline for cumulative high-level dust ($\geq$80 mg/m^3 years) exposure of smokers (Garabrant, Bernstein, Peters, Smith and Wright, 1985). The American Conference of Governmental Industrial Hygienists in 1980 recommended dust exposure limits of 1.0 mg/m^3 for anhydrous borax and pentahydrate, 5.0 mg/m^3 for borax, 2 mg/m^3 for boric acid, and 15 mg/m^3 for boric oxide.

There is little information on the mechanism of borate toxicity, other than a decreased oxygen uptake by brain tissue contacted by borates, but the effect is largely overcome by the addition of glucose or pyruvate. Borates may possibly interfere with protein formation and glutamine synthesis in brain tissue. In acute poisoning studies with animals, gross pathologic changes were minimal, but microscopic studies showed renal glomerular and tubular damage. There is no proven therapy for boric acid ingestion, but with laboratory animals large doses of Ringer's solution and plasma given intravenously have been very helpful, as has D-glucose. With infants, exchange transfusions and intravenous infusion of multiple electrolyte have been successfully used (Levinskas, 1964).

Normal boron blood levels in humans are on the order of 0.1–10 ppm, but 74.8 ppm was observed in an infant who drank a boric acid solution (with no visible sign of toxicity). Boron is rapidly excreted in unchanged form by the kidneys, and in one human test 500 mg of boric acid was injected intravenously. It was excreted rapidly in the urine, demonstrating a half-life of 21 hours and complete elimination in 96 hours. In chronic exposure tests, doses of 2500 and 5000 ppm of boric acid in the diet of mice over a 2-year period showed no evidence of carcinogenicity. Boron in the diet is normal at an ingestion rate of 10–20 mg/day, with foods varying from 0.002 mg (milk) to 0.279 mg (apple sauce) per helping. The boron is rapidly and completely absorbed, but there is no accumulation in the body (Raymond and Butterwick, 1992). When tested on a cellular level, borax and its ores were toxic only to mammalian cells (cytotoxicity; 50% plating efficiency). At a high concentration (800–3200 mg/liter of cells), they appeared to have no mutagenic effect, and were not significantly tumor forming (oncogenic; Landolph, 1985).

Borates, however, do have some toxicity to insects. High concentrations can control fly larvae in manure piles, dog runs, and the like, and powdered boric acid is effective in controlling cockroaches. Borates have been used as mild antiseptics in pharmaceutical applications for literally centuries, but are not strong enough for most modern uses. Sudden large amounts of borates can reduce the efficiency of sewage treatment organisms, but these organisms can adapt to more continuous high levels of borates. Boric acid impregnation of 430 ppm can protect wood against the house borer, and 1% impregnation protects against termites. Where the borate concentration is most critical, however, is in plants, since a very narrow margin exists between borates being an essential fertilizer and their being toxic (see Table 9.2; Sprague, 1972).

8.5.2 Environment

There have been a large number of studies on the effect of boron in the environment, such as summarized by Raymond and Butterwick (1992). With fish, the "lowest observable effect concentration (LOEC)" of boron was found with the early life stages of rainbow trout at 0.1 ppmB in reconstituted waters. However, in natural waters no effect was noted even at 17 ppm, and healthy rainbow trout have been found in California streams with 13 ppmB. Other fish species have shown a similar or higher boron tolerance. Low boron concentrations have been found beneficial for some fish growth, such as 0.4 ppmB raising carp production by 7.6%, and rohu benefiting from 1 mgB/fish/day. The range of several LD_{50} tests was 5–3000 ppmB for freshwater fish and 12–90 ppm for seawater fish. Amphibians have been found to respond to boron in a similar manner. A concentration of 30 ppmB was found to reduce the photosynthesis of half the phytoplankton tested, and 50 ppm decreased the growth rate of 5 of 19 species. In sewage tests 20 ppmB had no effect on activated sewage treatment, and 200 ppmB produced no inhibition of anaerobic sludge digestion.

Grazing animals have not been studied extensively, but no signs of toxicosis were found in heifers exposed to 120 ppmB in their drinking water, and 300 ppm was not acutely toxic. The safe tolerance level was estimated to be 40–150 ppmB in water. However, it was found that sheep developed enteritis when grazing on soils with 30–300 ppmB, drinking water with 1–20 ppmB, or eating 40 ppmB feed. In an extensive investigation of boron's effect on rivers and water courses in England, Waggot (1969) came to the conclusion that there was no toxicity problem, and that both natural boron removal (i.e., adsorption on clays, etc.) and sewage treatment appeared to be capable of maintaining a low boron concentration in the country's rivers and water courses (Raymond and Butterwick, 1992).

References

Abakumov, V. I., Bessonov, N. N., Kardashina, L. F., Shubin, A. S. (1988). Effect of several factors on the consumption of sulfuric acid in the decomposition of borate ores. *Zh. Prikl. Khim.* (*Leningrad*), **61**(3), 640–643.

Abbasov, A. D. (1979). *A Study of Borate Sorption on a Specially Synthesized Anion Exchanger,* pp 11–17, Mater. Konf. Molodykh. Aspir., 44 Publisher, Izd. "Elm," Baku, USSR.

Aksit, A. A. (1990). Phase transition anomalies in colemanite. *Ferroelectrics* **106,** 303–307.

Akdag, M., Batar, T., Ipekoglu, U., Polat, M. (1996). "Processing of Tincal and Colemanite Ores by Decrepitation, and Comparison with Wet Methods in Terms of Production Costs," Changing Scopes Minor, Process, 6th Proc. Int. Miner. Process Symp., Kemal, M. ed. Publ. Balkema, Rotterdam Neth., pp 389–393.

Albayrak, F. A., and Protopapas, T. E. (1985). Borate deposits of Turkey. In *Borates: Economic Geology and Production* (J. M. Barker and S. F. Lefond, eds.), pp. 71–85, Soc. Min. Eng., AIMMPE, New York.

Alekhin, A. M., Moiseeva, R. N., Ubizkaya, L. I., Lashkova, K. Y. (1985). Extraction of high-purity quartz from complex borosolicate ores. In *Pererab. Okislennykh Rud* (V. A. Chanturiya, and E. A. Trofimova, eds.), pp. 196–197, Nauka, Moscow.

Alkan, M., Kocakerim, M. M., and Colak, S. (1985). Dissolution kinetics of colemanite in water saturated by carbon dioxide. *J. Chem. Tech. Biotechnol.* **35A,** 382–386.

Alkan, M., Oktay, M., Kocakerim, M. M., Karagolge, Z. (1991). Dissolution kinetics of some borate minerals in CO_2-saturated water. *Hydrometallurgy* **26,** 255–262.

Allen, R. P., and Morgan, C. A. (1975, May 28). *Boric Acid.* German Patent 2,454,409.

Anac, S. (1988, March). Etibank's place in the production of industrial minerals in Turkey. *Industrial Minerals Suppl.* No. 246, pp. 25–29.

Andreeva, A. I., and Kuzin, C. A. (1937). The flotation of borax and boric acid in the processing of Inder borates. *J. Applied Chem., USSR* **10**(5), 845–852.

Anon. (1956). *U.S. Borax.* Company brochure.

Anon. (1958). U.S. borax has integrated a complex industrial plant. *Eng. Mining J.* **159,** 101–105 (April).

Anon. (1963, October 7). Chelating agent used to extract boric acid. *Chem. Eng. News,* pp. 44–45.

Anon. (1965a, December). *Amberlite XE-243.* Rohm and Haas Company, Philadelphia. [Brochure]

Anon. (1965b, August 23). U.S. borax & chemical will add a new plant to make boric oxide. *Chem. Eng. News,* p. 30.

Anon. (1966, April 11). It knocked boric acid costs for a loop. *Chem. Eng.,* p. 55.

Anon. (1976, June 7). Air support really cushions the cost of storage. *Chem. Eng.,* p. 23.

Anon. (1981, June). Boric acid plant goes onstream. *Chem. Eng.* **88**(11), 17.

Anon. (1995). China to process boron–iron ore. *Industrial Minerals,* No. 329, p. 61.

Anon. (1996a, July). The company behind the 20-mule team. *Compressed Air Mag.,* pp. 30–36.

Anon. (1996b). Borax production at Boron: Basic steps. *Borax, 20-Mule Team,* pp. 4–5, U.S. Borax Inc.

Anon. (1996c). The listening mine. *Borax Pioneer,* No. 6, pp 4–5, U.S. Borax Inc.

Ataman, G., and Baysal, O. (1973). Thermic reactions of some of the boron minerals and their effect on the atomic structure. In *Nomograph, Cong. Earth Sci., 15th Aniv. Turk. Republic* (S. Doyuran, ed.), pp. 541–569.

Aydin, A. O., and Gulensoy, H. (1984). Boric acid and borax from teruggite. *Doga, Ser. B.* **8**(1), 13–16.

Ayok, T., Ors, N., Ekinci, E., and Tolun, R. (1995). Reduction of chemical oxygen demand of Kestelek colemanite by chemically enhanced decrepitation. In *Flash Reaction Processes* (T. W. Davies, ed.), Vol. 282, pp. 359–371, Kluwer Academic Publishers, Netherlands, NATO ASI Ser. E.

Badeev, Y. S., Kozhevnikov, A. O., Nenarokomov, Y., Engel, R. I. (1973). Effectiveness of beneficiation in heavy suspensions of lean and low-grade ores. *Obogashch. Rud* **18**(5), 3–6.

Bailey, G. E. (1902). Borates-processes of manufacture. *The Saline Deposits of California, Calif. State Min. Bur., Bull. 24,* pp. 83–86.

Banateanu, G., Totoescu, D., Popovichi, T., Kissling, A. (1972). Boron extraction from colemanite. *Bul. Inst. Petrol, Gaze Geol., 1969.* (*Chim. Petrol.*) **18,** 71–78.

Barwise, C. H. (1992, June 16). *Froth Flotation of Calcium Borate Minerals.* U.S. Patent 5,122,290.

Battaglia, R. R., and Alonso, R. N. (1992). *Geologia y Mineria de Ulexita en el Grupo Minero Maggie, Salar Centenario, Salta,* pp. 241–251, 4th Cong. Nac. Geol. Econ., Cong. Latinamer. Geol. Econ., Cordoba, Argentina.

Beker, V. G., and Bulutcu, N. (1996, January). A new process to produce granular boric oxide by high temperature dehydration of boric acid in a fluidized bed. *Trans. Inst. Chem. Eng.* **74**(Pt. A), 133–136.

Bekturov, A. B., Naimushina, R. F., Konobritskir, E. G., Litvinenko, V. I. 1967. "Physiochemical Studies of the Processing of Natural Borates," Tr. Inst. Khim. Nauk, Akad. Nauk Kaz SSR, V. 16, pp. 137–165.

Birsoy, R. (1990). Stabilities of boron polyanions in Turkey's calcium, magnesiun and calcium–magnesium borates. *Turk Muhendislik Cevre Bilimleri Derg.* **14**(4), 618–627.

Bixler, G. H., and Sawyer, D. L. (1957, March). Boron chemicals from Searles Lake brines. *Ind. Eng. Chem.* **49**(3), 322–333.

Blumenberg, H. (1918, 1927). *SO_2 Leach Process for Colemanite.* U.S. Patents 1,259,718 (Mar. 19, 1918); 1,642,535 (Sept. 13, 1927).

Boratom. (1992, November 27). Borate mining and manufacturing expansion in Chile. *Boron Chemicals Processes, Chile S.A.,* Iquique, Chile.

Bozadzhiev, P. (1973a, 1973b). Decomposition of colemanite ore with phosphoric acid to monocalcium phosphate. *God. Vissh. Khim.-Tekhnol. Inst., Sofia* **21**(2), 67–77; Decomposition with monocalcium and double super phosphate, pp. 79–84.

Bozadzhiev, L., and Bozadzhiev, P. (1978). Structural and phase changes in colemanite during heating. *God. Vissh. Khim.-Tekhnol. Inst., Burgas, Bulg.* **13**(Pt. 1), 183–188.

Bozadzhiev, P., Slavov, L., and Ivanov, D.G. (1972). Crystallization of dicalcium phosphate during the decomposition of colemanite ore with phosphoric acid, II. *God. Vissh. Khim.-Tekhnol. Inst., Sofia* **15**(5), 213–225.

Budici, G. (1967, November 15). *Installation for Manufacturing Pure Boric Acid.* Romanian Patent 49,124.

Burger, A. (1914, August 25). *Process for the Production of Boric Acid.* U.S. Patent 1,108,129.

Burke, W. E., and deRopp, H. (1929, April 18). *Process for the Manufacture of Boric Acid.* British Patent 294,236.

Campbell, G. W., Wilkins, D. G., and Muench, J. T. (1977, August 9). *Fluid Bed Dehydration of Borax.* U.S. Patent 4,041,132.

Castor, S. B. (1993). Borates in the Muddy Mountains, Clark County, Nevada. *Nevada Bur. Mines, Geol., Bull. 107.*

Cebi, H., Yersel, E., Poslu, K., Behar, A., Nesner, R., and Langenbrick, H. (1994). Solid–liquid separation of Etibank borax plant effluents by centrifugal decanter. In *Prog. Miner. Process. Technol.* (Demiral, L. H. and Ersayin, S., eds.), pp. 513–516, Balkema, Rotterdam.

Celik, M. S., Atak, S., and Onai, G. (1993). Flotation of boron minerals. *Min. Metallurg. Proc.* **10**(3), 149–153.

Celik, M. S., and Bulut, R. (1996). Mechanism of selective flotation of sodium–calcium borates with anionic and cationic collectors. *Sep. Sci. Technol.* **31**(13), 1817–1829.

Celik, M. S., Uzunoglu, H. A., and Arslan, F. (1992a). Decrepitation properties of some boron minerals. *Powder Tech.* **79,** 167–172.

Celik, M. S., and Yasar, E. (1995). Effect of temperature and impurities on electrostatic separation of boron minerals. *Min. Eng.* **8**(7), 829–833

Celik, M. S., Saglam, H., Hancer, M. (1992b), Activation mechanisms of barium ions in colemanite flotation. *Proced. Fourth IMPS,* Antalya, Turkey.

Chebukov, M. F., and Ignateva, L. P. (1967). Wastes from boric acid production as additive during clinker grinding. *Tsement* **33**(1), 12–13.

Chebukov, M. F., and Ignateva, L. P. (1968). Building-grade gypsum from boric acid production wastes. *Stroit. Mater.* **14**(8), 27.

Chemtob, E. M. (1971, March 2). *Ion Exchange Process for Recovering Borates from Brine.* U.S. Patent 3,567,369.

Chemtob, E. M., and White, W. R. (1972, January 18). *Reagent Flotation of Borax from Salt Mixtures at Low Temperatures.* U.S. Patent 3,635,338.

Chumaevskii, V., Lebedev, V. Y., Barulin, E. P., Romanov, V. S., Fedosov, S. V., Kiselnikov, V. N. (1980). Introduction of a two-step cyclone dryer in borax production. *Khim. Prom-st. (Moscow)* **12,** 720–721.

Connell, G. A., and Rasor, J. P. (1939, April 25). *Process for Removing Shale from a Naturally Occurring Borate.* U.S. Patent, 2,155,784.

Corkill, F. W. (1937, October 26). *Process for Treating Hydrous Borate Minerals.* U.S. Patent 2,097,411.

Cramer, T. M., and Connell, G. A. (1933, September 19). *Process for the Production of Boric Anhydride from Boric Acid from Colemanite or the Like.* U.S. Patent 1,927,013.

Cui, C., Zhang, X., and Lui, S. (1994). Pig iron containing boron and boron-rich slag made from ludwigite in a blast furnace. *Kuangye (Beijing)* **3**(4), 29, 68–72.

Davies, T. W., Colak, S., and Hooper, R. M. (1991). Boric acid production by the calcination and leaching of powdered colemanite. *Powder Tech.* **65,** 433–440.

Dayton, S. H. (1957, June). $20,000,000 face lifting forges new era for Pacific Coast Borax. *Mining World* **19**(7), 36–45.

Demircioglu, A. (1978). Beneficiation of tincal and yield improvement at the Etibank Kirka plant, Turkey. *Kim. Sanayi* **26**(113–114), 26–32.

Demircioglu, A., and Gulensoy, H. (1977). The yield studies in the production of borax from the Turkish ulexite ores. *Chim. Acta Turc.* **5**(1), 83–91.

Demirhan, N., Avciata, U., and Afsar, H. (1995). Determination of the variables in boric acid production from borax by a carbon dioxide leach. *Chim. Acta Turc.* **23**(2), 99–102.

De-Nuz, A. (1996). Personal Communication, Process Engineer, North American Chemical Company, Trona, Calif.

Dickson, J. (1985). Etibank at Kirka: From ore to derivatives. *Industrial Minerals,* No. 210, pp. 65, 67.

Downing, T. L. (1995, March 7). *Beneficiation of Borate Ores.* SME Ann. Mtg., Denver, Colorado.

Dragila, V. (1969, March 7). *Boric Acid and Borax.* Romanian Patent 51,456.

Dragila, V. (1976, March 30). *Boric Acid, Borax, and Magnesium Sulfate from Boron Minerals.* Romanian Patent 60,097.

Dub, G. D. (1947). Owen's Lake; Source of sodium minerals. *AIME Trans.* **173,** 66–78.

Dublanc, E. A., Malca, D. A., and Leale, A. P. (1993). Industrial minerals of Argentina looking for investment. *Industrial Minerals (London)*, No. 312, pp. 25–36.

Dupont, F. M. (1910). The borax industry. *J. Ind. Eng. Chem.* **2,** 500–503 (Dec.).

Dwyer, T. E. (1963, September 10). *Recovery of Boron Compounds from Boron-Containing Ores.* U.S. Patent 3,103,412.

Echevarria, M. F., and Velazco, C. (1975). Preparation of borax from ulexite. *Bol. Soc. Quim. Peru* **41**(4), pp. 219–226.

Emrullahoglu, O. F., Kara, M., Tolun, R., and Celik, M. S. (1993). Beneficiation of calcined colemanite tailings. *Powder Tech.* **77**(2), 215–217.

Eric, R. H., and Topkaya, Y. A. (1987). *Laboratory-Scale Calcination Parameters of Colemanite Concentrate,* pp. 363–387, Pyrometallurgy '87 Symp., Inst. Min. Metall., London.

Ershov, E. I., Zaguraev, L. G., and Melnitskii, V. V. (1982). Preliminary concentration of datolite ore by a neutron absorption method. *Obogashch. Rud (Irkutsk)*, pp. 64–67.

Etibank. (1994, 1995). *Annual Reports,* 1995, pp. 21–32; 1994, pp. 21–24.

Etibank. (1996). *Product Specifications.*

Evans, J. R., Taylor, G. C., and Rapp, J. S. (1976). Borate deposits; colemanite and ulexite–probertite. *Mines and Mineral Deposits in Death Valley National Monument,* Calif. Div. Mines, Geol., Spec. Rpt. 125.

Ferro Corp. (1972, November 29). *Production of Sodium Pentaborate from Sodium Calcium Borate Ores.* British Patent 1,297,743.

Finlay, G. R. (1952, February). Calculated energy requirements of electric furnace products. *Chem. Can.* **4**(2), 25–28.

Flores, H. R., and Villagran, P. (1992). Borates in the Argentina Puna: A new application of magnetic separation. *Magn. Electr. Sep.* **3**(3), 155–169.

Fraas, F., and Ralston, O. C. (1942). Contact potential in electrostatic separation. *U.S. Dept. Interior, Bur. of Mines, Rept. of Investigations, R.I. 3667.*

Franke, E. (1934, March 6). *Process for the Manufacture of Boric Acid from Sodium Tetraborate.* U.S. Patent 1,950,106.

Gaft, M. L., Gorobets, B. S., and Malinko, S. V. (1979). Typochemical character of rare earth luminescence in calcium borosilicates. *Dokl. Akad. Nauk SSSR* **244**(5), 1211–1214.

Gaft, M. L., Ermolenko, V. I., Ershov, V. I., Litvintsev, E. G., Nazarov, V. V., Valshchikov, A. V. (1985). X-ray luminescence of borosilicate ores. *Obogashch. Rud (Leningrad)* **4,** 10–13.

Gale, H. S. (1912). The Lila C Borax Mine at Ryan, California. *U.S. Geol. Survey, Mineral Resources, Borax,* pp. 861–866.

Gale, W. A. (1964). History and technology of the borax industry. In *Boron, Metallo–Boron Compounds and Boranes* (R. M. Adams, ed.), pp. 1–27, Interscience Publishers, New York.

Gao, S., Yang, Z, and Huang, S. (1993). Recovery of Na_2SO_4, K_2SO_4, boric acid and lithium salt from Da Chaidan Salt Lake brine. *Seventh Symp. on Salt* **1,** 555–560.

Garabrant, D., Bernstein, L., Peters, J. M., Smith, T. J., Wright, W. E. (1985), Respiratory effects of borax dust. *Brit. J. Ind. Med.* **42,** 831–837.

Garkunova, N. V., Masalovich, N. S., and Trifonova, L. A. (1971). Extraction of boron from datolite by carbonic acid. *Tr. Ural. Nauch.-Issled. Khim. Inst.* **21,** 11–14.

Garrett, D. E. (1959). Plant tour. AIChE-AIME meeting.

Garrett, D. E. (1960). *Borax Process at Searles Lake," Industrial Minerals and Rocks,* 3rd ed., Ch. 7, Pt. 2, pp. 119–122. Am. Inst. Min, Met. Eng., New York.

Garrett, D. E. (1961, January 24). *Recovery of Boron Values.* U.S. Patent 2,969,275.

Garrett, D. E. (1978). Solution mining process using ion exchange. *U.S. Gov. Rep. Announc. Index* **78**(21).

Garrett, D. E. (1992). *Natural Soda Ash: Occurrences, Processing and Use.* Van Nostrand Reinhold, New York.

Garrett, D. E., Kallerud, M. J., and Chemtob, E. M. (1975, October 7). *Beneficiation of Salts Crystallized from Searles Lake Brine.* U.S. Patent 3,910,773.

Garrett, D. E., and Laborde, M. (1983). Salting-out process for lithium recovery. *Sixth Int. Symp. on Salt* **2,** 421–443.

Garrett, D. E., and Rosenbaum, G. P. (1958a). Crystallization of borax. *Ind. Eng. Chem.* **50**(11), 1680–1684.

Garrett, D. E., and Rosenbaum, G. P. (1958b, August 11). Crystallization. *Chem. Eng.* **65,** 125–140.

Garrett, D. E., and Weck, F. J. (1958). *Boron Ion Exchange Resins* and *Ion Exchange Process for the Recovery of Boron from Brines.* U.S. Patent Applications.

Garrett, D. E., Weck, F. J., Marsh, A. J., and Foster, H. R. (1963). *Boron Extractants.* U.S. Patent 3,111,383 (Nov. 19); British Patent 910,541 (Nov. 14, 1962).

Gay, T. E., and Hoffman, S. R. (1954). Borates. *Mines and Mineral Resources of Los Angeles County, Calif. J. Mines, Geol.* **50**(3–4), 506–508.

Genc, S., Aydin, A. O., and Sevinc, V. (1988). Solubility of the mineral teruggite in various solvents. *Marmara Univ. Fen Bilimleri Derg.* **5,** 115–123.

Golomzik, A. I., Komlev, A. M., Cherkashin, A. Y., Pekhova, L. P., Ropova, N. P., Suvorova, D. I. (1987). Pilot plant tests on complex beneficiation of borate–magnetite ore. *Kompleksn. Ispol'z. Miner. Syr'ya* **2,** 12–14.

Gorobets, B. S. (1988). X-ray luminescence spectra of minerals and ore-sorting criteria for ore beneficiation. *Mineral. Sb. (Lvov)* **42**(1), 74–80.

Griswold, W. J. (1967, March 14). *Method of Calcining and Classifying Borates.* U.S. Patent 3,309,170.

Gulensoy, H., and Kocakerim, M. M. (1978). Solubility of colemanite mineral in CO_2-containing water, and geological formation of this mineral. *Bull. Mineral Res. & Explor. Inst. Turkey, Foreign Ed.,* No. 90, pp. 1–19.

Gur, G., Turkay, S., and Bulutcu, A. N. (1994). The effects of process conditions on the flocculation of tincal slimes. In *Prog. Mineral Process. Technol.* (Demiral, L. H. and Ersayin, S., eds.), pp. 501–503, Balkema, Rotterdam.

Hamilton, F. (1919, 1921). Borax. *17th (and 15th)* Report of the State Mineralogist During *1920,* pp. 274–277; During 1915–1918, pp. 62–69, Calif. Div. Mines, Geol.

Hancer, M., and Celik, M. S. (1993). Flotation mechanism of boron minerals. *Separ. Sci. Tech.* **28**(9), 1703–1704.

Hancer, M., Celik, M. S., *et al.* (1993). Flotation of borax with anionic and cationic collectors in saturated solutions. *Proced. 13th Mining Cong. Turkey, Istanbul, Turkey,* pp. 519–528.

Hanks, H. G. (1883). *Report on the Borax Deposits of California and Nevada. Calif.* pp. 6–111, State Mining Bur., Pt. 2, Third Ann. Rept.

Hartman, G. J. (1994, 1996). *Fort Cody In Situ Borate Mining Project,* pp. 37–39, Proc. 12th Ind. Min. Int. Cong., Chicago, SME Ann. Mtg., Albuquerque, Preprint 94-125.

Havighorst, C. R. (1963, November 11). AP&CC's new process separates borates from ore by extraction. *Chem. Eng.,* pp. 228–232.

Hellmers, H. D. (1929, October 29). *Process for Recovering Borax from Brine.* U.S. Patent 1,733,537.

Hendel, F. J. (1949, September 14). *Improvements in Calcining Borax.* British Patent 629,171.

Huber, F. (1967, April 10). *Sodium Metaborate.* Austrian Patent 253,462.

Inoue, K., Nagabayashi, R., Hasegawa, M., Kinugasa, M. (1989). Effect of additives on the reduction behavior of garnierite. *Nisshin Seiki Giho* **61,** 10–19.

Ivanov, D., and Bozadzhiev, P. (1971). Kinetics of the decomposition of colemanite ore by phosphoric acid to produce boric acid and dicalcium phosphate. In *Miner. Torove, Nauch.-Tekh. Konf.* (D. Shishkov, ed.), pp. 131–136, Tekhnika, Sofia, Bulgaria.

Janik, W., Wardas, A., Chajduga, A., Kubiela, L., Marczewski, A., Reinelt, T. (1981, April 16). *Boric Acid from Ulexite Ore.* German Patent 3,029,349.

Jensen, R. C., and Schmitt, H. H. (1985). Recent changes and modifications at U.S. borax & chemical corporation. In *Borates: Economic Geology and Production* (J. M. Barker and S. J. Lefond, eds.), pp. 209–218, Soc. Min. Eng., AIMMPE, New York.

Ji, C., Liu, S., Cui, C. *et al.* (1988, March 9). *Manufacture of Borax.* Chinese Patent 87,101,578.

Johnson, F. R. (1936, December 22). *Magnetic Separation.* U.S. Patent 3,065,460.

Kalacheva, V. G., Karazhanov, N. A., Kim, G. E., Katz-David, G. G. (1980). Treatment of borate ores by oxalic acid. *Khim. Prom-st. (Moscow)* **9**(6), 355–356.

Karazhanov, N. A., Abakumov, V. I., and Kardashina, L. F. (1991). Kinetics of the sulfuric acid decomposition of complex borate ore. *Izv. Akad. Nauk Kaz. SSR, Ser. Khim.* **3,** 73–79.

Karazhanov, N. A., Kosenko, G. P., Sarsenov, S. K., Isabelkova, K. U., Savingkh, Y. G. (1981). Study of rates of dissolving and solubilities of potassium borate minerals in salt solutions. *Tr. Inst. Khim. Nauk, Akad. Nauk Kaz. SSR* **54,** 64–80.

Kasparov, A. A. (1971). Working conditions and health status of workers engaged in boric acid production. *Gig. Tr. Prof. Zabol.* **15**(8), 11–15.

Kasparov, A. A., and Yakubovskii, A. K. (1970). Fibrogenic activity of dusts of borosilicates of mixed composition. In *Orgaizm Sreda, Mater. Nauch. Konf. Gig. Kafedr* (S. N. Cherkinskii), Vol. 1, pp. 115–117, Pervyi Mosk. Med. Inst.

Kaverzin, E. K., Plyshevskii, Y. S., Nikolskii, B. A., Tkachev, K. V., Futoryanskii, A. Y., Strezhneva, I. I., Savinykh, Y. G., Kopylov, G. G., Alekhin, A. M., Kozhevnikov, A. D. (1978, November 25). *Borax.* Russian Patent 633,806.

Kayadeniz, I., Guelensoy, H., and Yusufoglu, I. (1981). Removal of arsenic from colemanite ores by vacuum calcination process. *Chimica Acta Turcica* **9**(1), 267–277.

Kaytaz, Y., Onal, G., and Guney, A. (1986). Beneficiation of bigadic colemanite tailings. *Proc. I, Int. Mineral Processing Symp.*, pp. 238–249, Izmir, Turkey.

Kelly, A. (1931, July 2). *Manufacture of Boric Acid.* British Patent 351,810.

Kemp, P. H. (1956). *The Chemistry of Borates.* Borax Consolidated Ltd., London.

Keys, C. R. (1910). Borax deposits of the United States. *Trans. AIME* **40,** 701–710.

Khoinov, Y. I., Lapshin, B. M., Semenovskii, S. V., Sokolskii, A. K. (1970). Measurement of pH in a reactor with borate concentrates and sulfuric acid. *Izv. Vyssh Ueheb. Zaved., Khim, Teknol.* **13**(12), 1819–1822.

Kistler, R. B., and Helvaci, C. (1994). Boron and borates. In *Ind. Min. and Rocks* (D. D. Carr, ed.), pp. 171–186, Soc. Min. Metal. Explor., Littleton, Colorado.

Knickerbocker, R. G., Fox, A. L., and Yerkes, L. A. (1940, June). Production of calcium borate from colemanite by carbonic acid leach. *U.S. Bur. Mines, R.I. 3525,* pp. 13–18.

Knickerbocker, R. G., and Shelton, F. K. (1940). Benification of boron minerals by flotation as boric acid. *U.S. Dept. of Interior, Bur. Mines, R.I. 3525,* pp. 3–13, U.S. Patent 2,317,413 (April 27, 1943).

Kocakerim, M. M., and Alkan, M. (1988). Dissolution kinetics of colemanite in SO_2-saturated water. *Hydrometallurgy* **19,** 385–392.

Koppe, S., Schulz, H., Grossmann, A., and Wulfert, H. (1979, April 18). *Recovery of Boron Compounds from Boron-Containing Ores.* East German Patent 135,185.

Kostenenko, L. P., Serebrova, N. N., Popova, N. P. *et al.* (1983). Composition and beneficiation of complex borate–magnetite ores of the Taezhnoe Deposit (southern Yakutia). *Mineral. Polezn. Iskop. Krasnoyarsh. Kraya,* pp. 28–32.

Kozhevnikov, A. O., Alekhin, A. M., Petrunina, S. I., Vlasova, V. N. (1971). Use of plant waste water for the beneficiation of datolite ores. *Sb. Tr. Leningrad Inzh.-Stroit. Inst.* **69,** 84–85.

Kozhevnikov, A. O., Alekhin, A. M., Petrunina, S. I., Ermolyuk, N. A., Rakhmanova, R. I. (1978). Implementation of a three-product drum separator for the enrichment of datolite ores in heavy suspensions. *Khim. Prom-st.* (*Moscow*) **7,** 542–543.

Krystek, A., Gorowski, J., Marczewski, A., Kandora, Z. (1994, February 28). *Manufacture of Boric Acid.* Polish Patent 163,138.

Kunin, R., and Preuss, A. F. (1964). Characterization of a boron-specific ion exchange resin. *I& EC Prod. R&D* **3**(4), 304–306.

Kuvatov, K. G., Elnova, T. V., Baimukhanova, S. K., Semashko, T. S., Bibik, O. A. (1989). Rapid procedure for atomic absorption and atomic emission determination of boron. *Izv. Akad. Nauk Kaz. SSR, Ser. Khim.* **2,** 7–10.

Lagov, B. S., Ershov, V. I., Alekhin, A. M., Khorkhordin, V. R., Moiseenko, A. F., Volchenko, G. V. (1989). Neutron-absorption separation as an effective method for processing datolite ores. *Gorn. Zh.* **14,** 34–37.

Landolph, J. R. (1985). Cytotoxicity and negligible genotoxicity of borax and borax ores to cultured mammalian cells. *Am. J. Ind. Med.* **7,** 31–43.

Lei, Z. (1994, March 23). *Manufacture of Calcium Borate from Boron Ore.* Chinese Patent 1,084,137.

Leiser, T. (1996). *An Overview of Selected South American Borate Producers,* pp. 31–35, 12th Int. Min. Internat. Cong.

Lekhanov, O. N., and Ryabets, E. N. (1985). Automatic control of a boron ore flotation. *Gorn. Zh.* **9,** 55–56.

Lekhanov, O. N., Ryabets, E. N., Aleksandrova, Z. V., Zhilkina, S. A. (1990, April 23). *Control of Datolite Ore Flotation.* Russian Patent 1,558,487.

Lesino, G., and Saravia, L. (1992), Solar ponds in industrial boric acid production. In *Sol. Energy World Cong. Proc. Bienn. Cong. Int. Sol. Engery. Soc.* (M. E. Arden, et al., eds.), pp. 4043–4046, Pergamon, Oxford, U.K.

Levinskas, G. J. (1964). Toxicology of boron compounds. In *Boron, Metallo-Boron Compounds and Boranes* (R. M. Adams, ed.), pp. 693–765, Interscience Publishers, New York.

Liberman, V. I., and Eigeles, M. A. (1967). Flotation properties of slicate minerals from boron-containing skarns. *Sovrem. Sost. Zadachi Selek Flotatsii Rud,* pp. 240–249.

Liu, S., Xi, Z., Ji, C., Cui, C., Zhang, X. (1990). New process for preparing borax from ludwigite. *Dongbei Gongxueyuan Xuebao* **65,** 139–143.

Lowry, M. V. (1926, February 16). *Process for Producing Borax and Sodium Bicarbonate from Lake Brines.* U. S. Patent 1, 573, 259, 2p.

Lyday, P. A. (1992, December), History of boron production and processing. *Ind. Minerals (London),* pp. 19, 21, 23–25, 27–28, 31, 33–34, 37.

Lyday, P. A. (1994, 1995). *Boron.* Industrial Mineral Survey, U.S. Bureau of Mines, Washington, D.C.

Lyman, W. R., and Preuss, A. F. (1957, November 19). *Boron-Adsorbing Resin and Process for Removing Boron Compounds from Fluids.* U.S. Patent 2,813,838.

Lynn, L. (1974, August 13). *Process for Treating Borocalcic Ores.* U.S. Patent 3,829,553.

Martynyuk, Y. L. (1979). Ammonium carbonte method for processing borosilicate raw material. *Khim. Prom-st.* **5,** 280–287.

Mata, M. G. (1996). Personal communication. Chief Engineer, Trona Facility, North American Chemical Company, Trona, California.

Mathis, P. (1972, April 4). *Process for the Solubilization of Calcium Borates in Boron Minerals.* U.S. Patent 3,653,818.

Matsaberidze, T. G., Voitsekhovskaya, N. F., Lominadze, D. L., Dolendzhishvili, T. G. (1968). "Decomposition of Tourmaline without Flux Additives," Tr. Tbilis, Gos. Univ., V. 126, pp. 149–157.

Matterson, K. J. (1980). Borate ore discovery, mining and beneficiation. *Inorg. Theoret. Chem.* **5**(Sect. A3), 153–169.

May, F. H., and Levasheff, V. (1958, 1962). *Recovery of Borate Values from Calcium Borate Ores.* U.S. Patents 2,855,276 (October, 7, 1958); 3,018,163 (January 23, 1962).

Mehltretter, C. L., Weakley, F. B., and Wilham, C. A. (1967). Boron-selective ion exchange resins containing d-glucitylamino radicals. *I&EC Prod. R&D* **4**(3), 145–147.

Mehta, S. K., Khajuria, H. A., Sayanam, R. A., Upadhaya, J. M., Krishnaswami, S. P. (1976). Borax extraction by utilizing geothermal energy. *Trans. Indian Ceram. Soc.* **35**(2), 5N–9N.

Meisner, L. B. (1994). Methods using laser beams for enrichment and phase analysis of mineral raw materials. *Razved. Okhr. Nedr.* **4,** 12–14.

Meisner, L. B., Maltseva, N. I., Korolev, V. A., Lagov, B. S., Churbakov, V. F. (1990). Laser method to determine the concentration of mineral phases. *Obogashch. Rud (Leningrad)* **5,** 31–33.

Miles, D. E. (1973, January 23). *Rotary Apparatus for Treating Colemanite Ore.* U.S. Patent 3,712,598.

Mokrousov, V. A., Lileev, V. A., Lagov, B. S. (1973). Use of neutron ore radiation for ore beneficiation. *Gorn. Zh.* **149**(6), 60–62.

Mori, M. (1980). Study of wet-process phosphoric acid production: 4. Effects of boron on the hydration of hemihydrate. *Gypsum Lime* **166,** 90–94.

Morley, H. B., Skrzec, A. E., and Shiloff, J. C. (1969, July 8). *Process for Producing Boric Acid from Alkali Metal Borates.* U.S. Patent 3,434,359.

Moshkareva, G. A., Masalovich, V. M., Masalovich, N. S. (1971). Titrimetric determination of sulfates in solution obtained during production of borax from hydroboracite concentrate. *Tr. Ural. Nauch.-Issled. Khim. Inst.,* No. 21, pp. 94–99.

Nazarov, V. V., Ershowv, V. I., and Zaitseva, N. I. (1991). Mathematic modeling of the beneficiation of bulk datolite. *Obogashch. Rud (S. Peterburg)* **5,** 27–31.

Nezhad, Z., Manteghian, M., and Tavar, N. S. (1996). On the confluence of dissolution, reaction and precipitation: The case of boric acid production. *Chem. Eng. Sci.* **51**(11), 2547–2552.

Nies, N. P. (1980). Alkali metal borates: Physical and chemical properties. *Inorg. Theoret. Chem. (Longman, New York)* **5**(Pt. A), 343–501.

Nikol'skii, B. A. (1980). Pilot plant testing of autoclave decomposition of datolite ore. *Tr. Ural'ski. N.-i Khim. In-ta* **51,** 66–69.

Nippon Denko, (1985, January 8). *Feroboron Production in an Electric Furnace.* Japanese Patent 60,002,649.

Norman, J. C. (1990, 1991, 1992, 1993, 1994, 1995, 1996). Boron: A review of the year's activities. *Mining Eng.* **43**(7), 740–741 (1991); **44**(7), 699–700 (1992); **45**(7), 718–720 (July, 1993), **46**(7), 661 (1994); **47**(7), 660 (July, 1995); pp. 41–42 (July, 1996).

Norman, J. C., and Johnson, F. C. (1980). The Billie borate ore body, Death Valley, California. In *Geology and Mineral Wealth of the California Desert* (Fife and Brown, eds.), pp. 268–275, South Coast Geol. Soc., Santa Ana, California.

Norsk Hydro (1974, May 30). *Purification of Aqueous Magnesium Chloride.* British Patent 1,354,944.

Nozhko, E. S., Kononova, G. N., Avdeeva, E. S., Ksenzenko, V. I. (1978). Macrokinetic characteristics of the sulfuric acid decomposition of ascharite. *Izv. Vyssh. Uchenbn. Zaved., Khim. Teckhnol.* **21**(9), 1303–1306.

O'Brien, P. J., and Chettle, R. V. (1951, March 20). *Process for the Manufacture of Boric Acid from Sodium Borate.* U.S. Patent 2,545,746.

Ozakan, S. S., and Lyday, P. A. (1995, March 6–9). *Physical and Chemical Treatment of Boron Ores.* SME Annual Mtg., Denver, Colo., Preprint No. 95-186.

Ozkan, S. G., and Veasey, T. J. (1994). The effect of slime coatings on colemanite flotation. In *Prog. Mineral Proc. Tech.* (Demirel, L. H. and Ersayin, S.,eds.), pp. 205–210, Balkema, Rotterdam.

Ozkan, S. G., *et al.* (1993). Flotation studies of colemanite ores from the Emet deposits of Turkey. *Proc. 13th Mining Cong. Turkey, Istanbul,* pp. 451–458.

Pak, H. J., and Kim, H. Y. (1976). Review of borax production by statistical analysis: 1. Multifactorial analysis in borax extraction processes. *Choson Minj. Immin Kong. Kwahag. Tongbo* **24**(1), 21–27.

Papke, K. G. (1976). Evaporites and brines in Nevada playas. *Nevada Bur. Mines, Geol., Bull 87.*

Patel, K. P., Oza, M. R., Rao, K. M., and Seshadri, K. (1967). Borax and by-products from Puga Valley saline deposits. *Salt Res. Ind.* **4**(3), 81–82.

Pavlovic, P. Z., Parada, N. F., and Vergara, L. E. (1983). Recovery of potassium chloride, potassium sulfate and boric acid from the Salar de Atacama brines. *Sixth Int. Symp. on Salt* **2,** 377–394.

Peterson, W. D. (1974, December 24). *Removal of Boron from Water.* U.S. Patent 3,856,670.

Petrov, B. A., and Erokhima, Z. V. (1969, August 4). *Boric Acid Production.* Russian Patent 247,266.

Peyo Garcia, R., Ochoa Bendicho, V., and Guillen, P. M. (1987, July 1). *Continuous Manufacture of Sodium Perborate Tetrahydrate from Borax and Sodium Hydroxide.* Spanish Patent 554,624.

Plyshevskii, Y. S., Tkachev., K. V., and Nikol'skii, B. A. (1976). *Complex Processing of Borosilicate Raw Material.* Deposited Doc., VINITI 4088-76.

Pocovi, R. E., Latre, A. A., and Skaf, O. A. (1994). Improved process for the concentration of ulexite, and boric acid production. *Hydrometall. '94, Pap. Int. Symp.*, pp. 1025–1034, Chapman and Hall, London.

Polendo-Loredo, J. (1988, July 12). *Beneficiation of Colemanite and/or Howlite Ores.* U.S. Patent 4,756,745.

Polonskii, S. B., and Beloborodov, V. I. (1980). Improvement of the flotation benefication of datolite ore. *Obogashch. Rud (Irkutsk)*, pp. 98–102.

Popa, I., Andreia, O., Procopiu, D., Constantin, G. (1970, July 9). *Borax by Alkaline Treatment of Colemanite, Pandermite and Other Borocalcite Minerals.* Romanian Patent 52,241.

Proks, M., Hrazdira, M., Bures, B., Ruzicka, O., Gelnar, S. (1985, October 15). *Apparatus for Borax Dehydration.* Chech. Patent 220,155 B.

Rao, K. M., Patel, K. P., Oza, M. R., Seshadri, K., Datal, D. S. (1966). Process for recovery of borax from tincal. *Chem. Age India* **17**(12), 1014–1016.

Rasor, C. M. (1924, March 25). *Process for Separating Colemanite from its Gangue.* U.S. Patent 1,487,806.

Ratobylskaya *et al.,* (1978). *Preparation of a Sodium Tripolyphosphate Solution for the Flotation of Boron Containing Ores.* Russian Patent 712,129.

Ratobylskaya, L. D., Alekhin, A. M., Moiseeva, R. N., Ubizkaya, L. I., Lashkova, K. Y., Petrumina, S. I. (1981a). Scientific basis for the design of a closed water recycle system in the complex beneficiation of datolite ores. *Intensif. Prots. Obag. Miner. Syrya,* pp. 219–225.

Ratobylskaya, L. D., Alekhin, A. M., Moiseeva, R. N., Ubizkaya, L. I., Lashkova, K. Y., Petrumina, S. I. (1981b). *Preparation of High Quality Datolite Concentrate from High Iron Silicate–Carbonate Ores.* SPSL 738 Khp.-D 81.

Raymond, K., and Butterwick, L. (1992). Ecotoxicology of boron. In *The Handbook of Environmental Chemistry* (O. Hutzinger, ed.), Vol. 3, Pt. F, pp. 294–318, Springer-Verlag.

Ren, X. (1992). Chemical processing of ludwigite-type boron ores. *Kuangchan Zonghe Liyong* **4,** 48–51.

Ridgway, R. R. (1933, January 3). *Method and Apparatus for Electrically Fusing Nonconducting Materials.* U.S. Patent 1,893,106.

Rize, D. F., *et al.* (1971, January). The flotation of boric acid from acidified Inder ores. *The Soviet Chemical Industry,* No. 1.

Saiko, I. G., Kononova, G. N., and Zakgeim, A. Y. (1985). Study of the dehydration kinetics of natural borates. *Zh. Prikl. Khim. (Leningrad)* **58**(3), 654–656.

Sawyer, D. L. (1969, September 23). *Preparation of Boric Acid.* U.S. Patent 3,468,626.

Sawyer, D. L., and Wilson, M. (1973, October 30). *Flotation of Arsenic Minerals from Borate Ores.* U.S. Patent 3,768,738.

Schmutzler, G., and Kircheisen, J. (1967). Datolite, a raw material for boric acid production. *Chem. Tech. (Leipzig)* **19**(8), 488–491.

Schuelke, D., *et al.* (1987). KLM's boric acid reclamation system (BARS). *Waste Manage. '87* **3,** 123–126.

Shevchenko, V. I., and Mikhailov, M. A. (1969). Dependence of the conditions of pyrohydrolysis of ascharite ores on their chemical composition. *Izv. Sib. Otd. Akad. Nauk SSSR, Ser. Khim. Nauk* **2,** 129–32.

Shiloff, J. C. (1972, March 21). *Boric Acid Production.* U.S. Patent 3,650,690.

Shishko, I. I., and Shabalin, K. N. (1967). Performance test for a three-section furnace with fluidized beds and down-flow grids used for the roasting of datolite ores. *Khim. Prom-st. (Moscow)* **43**(9), 700–702.

Simon, J. M., and Barwise, C. H. (1993, August 24). *Separation of Calcium Borate Minerals from Other Calcium-Containing Minerals.* U.S. Patent 5,238,119.

Skrylev, L. D., *et al.* (1978). Flotation separation of borate ions. *Zh. Prikel. Khim.* **79**(52-3), 708–710.

Smith, P. R., and Walters, R. A. (1980, February). Production of colemanite at American Borate Corp.'s plant near Lathrop Wells, Nevada. *Mining Eng.* **32**(2), 199–204.

Solis, A. R. (1996). *Loma Blanca in Argentina.* Universidad Nac. De Juju, Inst. Invest. Tech. Miner. Indust., Juju.

Solvey et Cie. (1971, October 1). *Boric Acid Production by Stripping Aqueous Solutions of Ammonium Borate Obtained by Carbo-amoniacal Attack of Borate Ores.* Belg. Patent 766,912.

Spivakova, O. M. (1982). Multiple uses of water in datolite ore beneficiation. *Nov. Met. Soor. Vodo. Och. Stroc. Vod,* pp. 5– 9.

Sprague, R. W. (1972). *The Ecological Significance of Boron.* U.S. Borax Research Corp.

Stern, D. R., and Uchiyama, A. A. (1959, July 7). *Continuous Electrothermic Production of Boron Oxide.* U.S. Patent 2,893,838.

Stupachenko, P. P., and Karya, A. L. (1980). Use of wastes from beneficiation of boron-containing ores as raw material for silicate building materials. *Ispolz. Otk. Pop. Prod. Pro. Stro. Mater. Izd. Okhr. Okr. Sredy* **2,** 6–7.

Su, Y. F., Yu, D. Y., and Chen, S. D. (1980). Process development of boron recovery from ascharite. *Int. Solv. Extr. Conf., Assoc. Ing. Univ. Luge. Belg.,* Vol. 2, Paper 80-57.

Taylor, D. S. (1953, 1956). *Production of Boric Acid and Anhydrous Sodium Sulfate.* U.S. Patents 2,637,626 (May, 1953); 2,746,841 (May 22, 1956).

Taylor, D. S., and Connell, G. A. (1953, 1956, 1957). *Stabilization of Borax Supersaturation in Solutions.* U.S. Patents 2,662,810 (Dec. 15, 1953); 2,774,070 (Dec. 11, 1956); 2,785,952 (March 19, 1957).

Tkachev, K. V., Kaverzin, E. K., and Guseva, T. A. (1980). Production of boric acid from Turkey colemanite ore. *Tr. Ural'sk. N.-i. Khim. In-ta* **51,** 74–78.

Tkachev, K. V., Plyshevskii, Y. G., and Kozhenvikov, A. O. (1973). Industrial production of borax. *Khim. Prom. (Moscow),* **49**(8), 600–601.

Tolun, R., Emir, B. D., Kalafatoglu, I. E., Kocakusak, S., Yalaz, N. (1984, April 24). *Sodium Hydroxide and Boric Acid Production by Electrolysis of Sodium Borate Solutions.* U.S. Patent 4,444,633.

Travis, N. J., and Cocks, E. J. (1984). *The Tincal Trail.* Harrap, London.

U.S. Borax & Chemical Corporation. (1962, January 24). *Apparatus for Producing Boric Oxide.* British Patent 887,640.

U.S. Borax & Chemical Corporation. (1977, February 23). *Flotation of Boric Acid from Sodium Sulfate Obtained During the Processing of Borate Ores.* British Patent 1,465,299.

Utine, M. T. (1973). Beneficiation of colemanite ores by decrepitation. *Madencilik* **12**(2), 37–38.

Validz, K., Sengil, I., and Aydin, A. O. (1991). The use of the waste matter from the colemanite concentrator at Kutahya-Emet as a glaze raw material by adding calcium oxide and silica. *Kim. Sanayi* **33**(165–168), 63–75.

Vasconi, I, and Mazzinghi, P. (1988). *Boric Oxide Preparation by the Controlled Dehydration of Boric Acid.* British Patent 2,192,625.

VerPlanck, W. E. (1956). History of borax production in the United States. *Calif. J. Mines Geol.* **52**(3), 273–291.

Vinogradov, E. E., Azarova, L. A. 1995. "Effect of a Solution's Salt Composition on the Conversion of Natural Borates," Zh. Neorg. Khim. V. 40, No. 3, pp. 418–422.

Vyas, M. H., Sanghavi, J. R., and Seshadri, K. (1976). Boric acid from borax. *Salt Res. Inst.* **2**(1), 15–22.

Waclawska, I., Stock, L., Paulik, J., Paulik, F. 1988. "Thermal Decomposition of Colemanite," Thermochimica Acta, V. 126, pp 307–318.

Waggott, A. (1969). An investigation of the potential problem of increasing boron concentration in rivers and water courses. *Water Res.* **3,** 749–765.

Walters, R. A. (1975). *Tenneco Oil's Colemanite Milling Operations Near Lathrop Wells, Nevada,* pp. 33–35, Nevada Bur. Mines and Geology, Rept. 26.

Wang, J. (1993). Distribution of boron mineral resources and their processing technologies. *Kuangchan Zonghe Lijong* **3,** 16–24.

Weck, F. J. (1969, November 18). *Boron Extractant Compositions.* U.S. Patent 3,479,294.

Wilson, D., and Burwell, B. T. (1975, February 11). *Processing Colemanite Ore.* U.S. Patent 3,865,541.

Winkler, J. (1907). Factors in boric acid manufacture. *Proc. Am. Chem. Soc. for 1907* **29,** 1366–1371.

Witting, F. (1888). On the manufacture of borax from boronatrocalcite. *Soc. Chem. Ind. J.* **7,** 748–749 (Nov. 30).

Woskie, S. R., Shen, P., Eisen, E. A., Finkel, M. H., Smith, T. J., Smith, R., Wegman, D. H. (1994). The real-time dust exposure of sodium borate workers: Examination of exposure variability. *Am. Ind. Hyg. Assoc. J.* **55**(3), 207–217; Nasal dose. *Am. Occup. Hyg.* **38**(Suppl. 1), 533–540.

Wu, S., and Xi, S. (1991). Reverse flotation of boron ores. *Feijinshukuang* **6,** 17-18, 21.

Xu, J. (1988, January 13). *Decolorization in Boric Acid Production.* Chinese Patent 87,103,625 A.

Xu, E. (1990, April 25). *Production of Borax in a Wet Mining Grinder.* Chinese Patent 1,041,577.

Yale, C. G. (1905). Borax. *U.S. Geol. Survey, Mineral Resources for the U.S. in 1904,* pp. 1023–1028.

Yang, C., Sun, H., and Wu, R. (1982). Combined alkali method for borax manufacture. *Huaxue Shijie* **23**(5), 132–133.

Yarar, B. (1973). Upgrading of low grade colemanite by flotation. *Kim. Muhendisligi* **6**(62), 33–42.

Yarar, B. (1985). The surface chemical mechanism of colemanite–calcite separation by flotation. In *Borates: Economic Geology and Production* (J. M. Barker and S. J. Lefond, eds.), pp. 219–233, Soc. Min. Eng., AIMMPE, New York.

Yarar, B., and Mager, J. (1979a). Dressing of boron ores and flotation of colemanite. *Przem. Chem.* **58**(2), 98– 101.

Yarar, B., and Mager, J. (1979b). Enrichment of boron ores, and a flotation process for colemanite. *Przem. Chem.* **79**(58-2), 91–101.

Yasar, E., Hancer, M., Kaytaz, Y., and Celik, M. S. (1994). Mechanism of electrostatic separation of boron minerals. In *Progress in Mineral Processing Technology* (H. Demirel and S. Ersayin, eds.), pp. 89–93, Balkema, Rotterdam.

Yong, C., Wu, Z., and Chen, F. (1985). New technology of boric acid production. *Huaxue Shijie (Shanghai)* **26**(9), 322–324.

Young, G. J. (1924). Mining borax in the Muddy Range. *Eng. Mining J.-Press* **117**(7), 276.

Yu, S. (1983). Roasting test of ascharite. *Huaxue Shijie* **24**(6), 162–164.

Zhang, K., Xie, J., Liu, J., *et al.* (1988, August 3). *Process for the Simultaneous Production of Synthetic Ammonia and Borax.* Chinese Patent 87,101,172.

Zhang, P., Guo, Z., Lin, H., Sui, Z. 1995. "Crystallization of the Boron Component Bearing $MgO\text{-}B_2O_3\text{-}SiO_2$ Slag," Trans. Nonferrous Met, Soc. China, V. 5, No. 4, pp. 45–48, 78.

Zhantasov, K. T., Grant, E. B., Ospanov, E. S., Starichenko, V. G., Asanbaev, N. D., Labaev, D. P., Sitnikov, V. M. (1990). Strengthening effects of boric acid production wastes on phosphorite pellets. *Aktual. Vopr. Poluch. Fosfora Soedin. Ego Osn.,* pp. 21–23.

Zhao, Q. (1990). Separating iron from boric iron ore by selective reduction. *Dong. Gong. Xuebao* **65,** 122–126.

Chapter 9 Uses of Borates

9.1 GLASS

Borates and their derivatives find an unusually large range of uses, as indicated in Tables 9.6, 10.7 and Fig. 10.3, and briefly discussed in the following section. The production of borosilicate glass, however, is by far the largest single use, amounting to about 55% of the total U.S. consumption in 1990. Borates enter into glass production of fiberglass for insulation, fabrics and reinforcement, and also as specialty glass products for household, laboratory, optical, heat resistance and many other uses. Boron's effect on glass was initially established in the late 1800s, and by the late 1930s its usefulness in fiberglass was also well recognized. By the 1940s fiberglass, high-durability and performance glass, and specialty glasses were being made in large continuous furnaces. Examples of various glass compositions using borates are given in Table 9.1.

In general, boron considerably reduces the thermal expansion of glass, provides good resistance to vibration, high temperatures and thermal shock, and improves its toughness, strength, chemical resistance and durability. It also greatly reduces the viscosity of the glass melt as it is being made. These features, and others, allow it to form superior glass for many industrial and specialty applications. E-glass was the initial textile fiberglass, and other compositions such as C- and D-glass were later used. Iron, arsenic, and sulfate are usually undesirable constituents in the boron source. The high-boron glasses greatly improved the fiberglass for both fabrics and insulation. Modern fiberglass is both very strong and resistant to destructive conditions such as elevated temperatures, burning, chemicals other than alkalies, bacterial and fungal attack, and high-energy radiation. However, it will not elongate under stress, and is comparatively brittle. Much of it is employed in batting form as nonflammable thermal insulation, and it is the load-bearing fiber imbedded in heat and/or catalyst-curable resin for many structural and other applications. It is woven into fabrics for higher-temperature industrial uses, or for decorative purposes when nonflammability and resistance to sunlight are important, such as with draperies and curtains.

Molten glass is sufficiently viscous for it to be drawn into uniform fibers, which are then cooled and solidified (Figs. 9.1 and 9.2). To reduce brittleness, the filaments are made smaller (filament fineness is known as *denier*) than organic fiber, and special protective coatings are added so they will not cut and abrade each other when woven into fabrics. The extremely high surface-

Table 9.1

Typical Glass Compositions Containing Boron (wt%)[a]

	D-Glass[b]	[c]	[d]	Pyrex[b]	[e]	[f]	[g]	[h]	E-Glass[b]		[i]	C-Glass[b]
B_2O_3	<23	22.5	15.0	13.5	13.0	12.0	11.5	9.6	7.0	6–8	7.0	5
SiO_2	72–75	65.0	74.0	80	81.0	3.0	69.5	74.7	55.0	53–55	61.0	60–65
CaO	—	—	—	—	—	—	—	0.9	21.8	17–22	9.0	14
Al_2O_3	—	5.2	1.0	2	2.0	11.0	—	5.6	15.0	14–15	4.0	3.5–6
MgO	<3	—	—	—	—	—	—	—	0.5	1.0	3.5	3
Na_2O	—	7.2	4.0	4.5	4.0	—	9.0	—	0.5	0.8	14.5	10
Fe_2O_3	—	—	—	—	—	—	—	—	—	0.3	—	0.5
TiO_2	—	—	—	—	—	—	—	—	—	0.5	—	—
PbO	—	—	6.0	—	—	74.0	—	—	—	—	—	—
BaO	—	—	—	—	—	—	3.0	2.2	—	—	—	—
K_2O	—	—	—	—	—	—	7.0	0.5	—	—	1.0	—

[a] Gagin, 1985.
[b] Russell, 1991.
[c,d] Electronic glass.
[e] Corning 7740 heat and chemical-resistant glass.
[f] Solder glass for sealing electrical components.
[g] Optical lenses.
[h] Older chemical-resistant glass.
[i] Soft glass for making insulation by rotary fiberization; C-, D-, and E-Glass are for fiberglass.

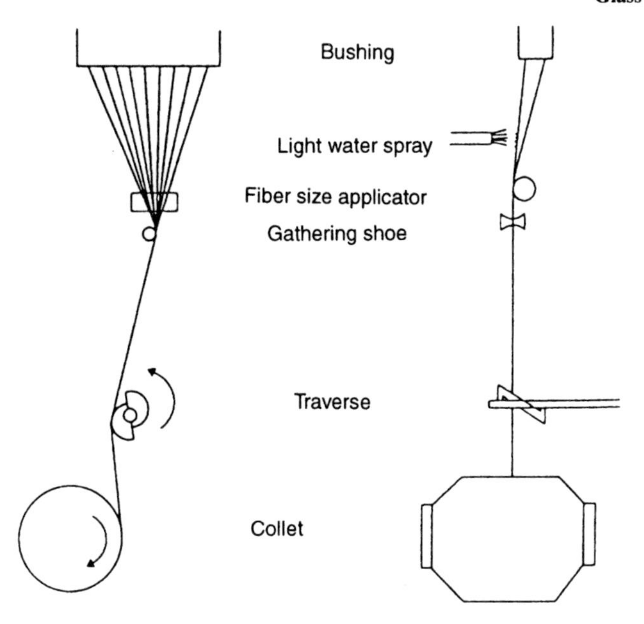

Figure 9.1 Extruding molten borate glass fibers from nozzles fed by a glass furnace. (From Russell, 1991; reprinted by permission of Industrial Minerals.)

to-volume ratio that results from use of such small filament sizes also requires that the surfaces be protected against the action of alkalies, even those as mild as soaps and detergents (Mark and Atlas, 1966).

E-glass is an alkali-free calcium–magnesium borosilicate, and when used for textile fiber glass its raw material requirements are more demanding than in the production of insulation. The silica is first ground to a very fine powder

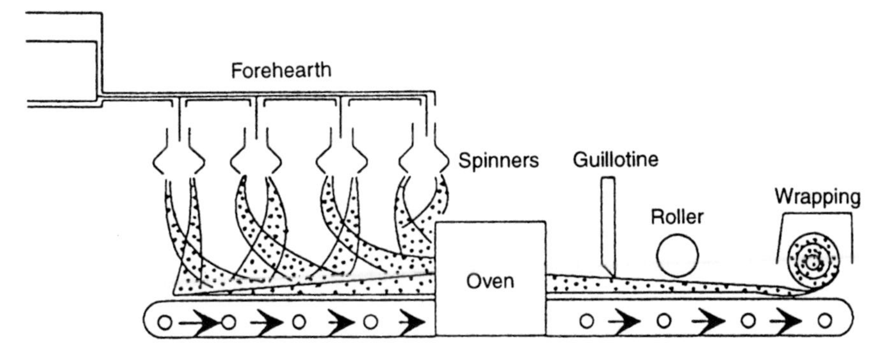

Figure 9.2 Glass-melting furnace, fiber-extruding nozzles, and finishing equipment to form fiberglass insulation mats. (From Russell, 1991; reprinted by permission of Industrial Minerals.)

to speed its melting, while the limestone, fluorspar, colemanite, and boric acid can be used with larger particle sizes. The raw materials are first weighed and blended before being fed into the furnace. The temperature in the upper portion of the furnace is about 1600°C, and the glass is withdrawn from its base at about 1300°C (the furnace is nominally at a temperature of 1400–1450°C). Because boric acid has significant volatility, the dust from the flue gas, which can contain up to 30% B_2O_3 and 20% Na_2O, is collected and recycled.

When melted to a uniform viscosity (partly controlled by the amount of boron), the glass flows through a submerged throat into a canal and a forehearth that feeds the various production lines. The melt is then forced through as many as 200 electrically heated platinum alloy bushings (nozzles) per furnace. There are usually 500 or more individual filaments drawn from tiny orifices in each nozzle, usually 6–9 μ in size, although they can be as small as 1.5 μ. The filaments are then rapidly cooled and coated with a binder or sizing compound from sprays, or by being rolled over a sizing applicator. The filaments are next combined on a high-speed winder to form a single strand as a textile yarn, which can be formed into mats, chopped strands, or other products (Russell, 1991; Roskill, 1993).

In another process the glass from the furnace flows into a rapidly rotating dish (spinner) with several thousand small holes in its perimeter. As it rotates the molten glass is forced through the holes and formed into filaments ("fiberized"). The fibers are cooled and sprayed with a resinous binder to make a "wool," which is collected on a conveyor. The speed of the spinner determines the product's grammage (weight per unit area). The wool is next sent through an oven at 280°C for curing, and then trimmed, chopped, and either stacked or rolled before final packing. If the wool is for pipe insulation, the uncured glass is separated into pelts on the forming conveyor and converted into pipe sections by a winding machine, which wraps the pelt around a heated mandrel with the wall thickness of the pipe determined by a set of counter-rollers. The formed sections are then passed through a curing oven and split; faced with cloth, paper or aluminum foil, and packaged (Russell, 1991).

The two basic boron products used in the 1940s for glass were boric acid and borax, but as the demand increased, an effort was made to reduce the delivered costs of the borates. Anhydrous borax became widely used, but by the 1970s the standard purified product for E-glass in textiles became five-mole borax (the pentahydrate) because it is much easier and cheaper to produce. For insulation fiberglass, colemanite and ulexite began to be used, usually containing 34–48% B_2O_3. Colemanite melted better in existing furnaces, so the processing and blending problems to produce a uniform, adequately pure product were then given more attention. Washing the ore for partial clay removal (and/or flotation) helped to remove impurities, increase the B_2O_3 content and improve the ore's consistency.

The choice of boron raw materials was also influenced by the size and type of glass-melting facilities. E-glass is produced in large gas-fired furnaces (such as 68 metric tons/day) with heat exchangers for preheating the combustion air. Using colemanite in place of limestone and boric acid allowed faster batch melting, sometimes increased furnace output, and yielded better fuel efficiency. Ulexite is usually a replacement for borax, but its effect on batch melting was not as pronounced as that of colemanite in the production of E-glass, since it replaces some of the soda ash, a very active flux. When purity or color are critical, the refined boric acid or sodium borates are used because both colemanite and ulexite usually contain much more iron, arsenic, sulfate or other harmful impurities.

All commercial fiberglass has some vulnerability to water, which tends to destroy the bond with the organic binder. Partly for this reason, and for added strength and ease of use, textile fiberglass is often mixed with plastics. This reinforces and protects the fibers from brittle failure, and allows a wide variety of applications to such things as sporting equipment, roofing, shingles, storage tanks, reinforcing mats, electric appliances, components of automobiles and aircraft, and many other products. They benefit from the light weight, high tensile strength, high modulus of elasticity and chemical stability of the fiberglass (Gagin, 1985).

As the boron content increases in many glass compositions a phase separation occurs during heating, which is the basis for the high-silica Vycor products. After phase separation, regular melting and forming techniques are used, but later heat treatment forms a highly disseminated, interconnected borate phase that can be leached with hot acids. The porous structure is then sintered above 899°C (1650°F) to a solid, transparent 96% silica glass with a good resistance to acids and thermal shock. The presintered porous glass may also be used as a desiccant or as a catalyst support.

In another application, sodium borates can be used in producing high-silica glass for improved acid-resistance by replacing some of the soda ash. This glass, such as Nos. e and h in Table 9.1 has a lower alkali content, which is good for most chemical uses. The borax in glass retains a trigonal planar shape and forms a network structure with the silica. This reduces the thermal expansion of vitreous silica more than network-modifying ions, making it more resistant to thermal shock. Borates also reduce the surface leaching of water or alkalies by inhibiting the removal of alkali ions, which can destroy the silica network. This causes frosting of the glass surface, which is detrimental in many products.

There are a large number of commercial applications for the borate-containing glasses. For instance, the borosilicate glass used for industrial glass piping, reaction columns and gauge glass is designed to withstand thermal shock and retain its dimensional stability to 400°C (750°F). The piping is easily assembled with flanged fittings, and is available in a wide range of sizes. It is very chemical resistant and has smooth, pore-free surfaces, which improve the fluid flow and resists scaling (Shelley, 1994). Other specialty glasses have been developed with a wide range of B_2O_3 contents. Sodium vapor lamps may contain up to 36% B_2O_3, and low x-ray absorption glass up to 83% B_2O_3. Most, however, are in the 1–34% B_2O_3 range, and include glass for electron tubes, optical fibers and filters, pharmaceutical applications, laboratory and kitchen ware (such as Pyrex), vacuum flasks, sealed-beam headlights, high-performance lights, electrical equipment with seals to metals, high-voltage insulators, and many others. Borax also enhances the quality of art and optical glasses (Anon., 1995b).

9.2 ABRASIVES AND REFRACTORIES

The boride compounds are very hard (about 9 on Mohs' scale; diamonds are 10), and are used as abrasives and refractories. They have a basic chemical formula with a limited range of compositions, and a high thermal and electrical conductivity. These compounds can be produced by reacting mixtures of powdered metal with boron at 1800–2000°C, and then compressing and sintering the residue into the desired shapes. Very pure borides may be prepared by sintering in a vacuum or an inert atmosphere near the melting point. They also may be produced by reducing B_2O_3 or KBF_4 with aluminum, magnesium or potassium, but the borides are impure, and further purification is difficult. High-purity borides also may be deposited on a filament in an atmosphere of boron tribromide, hydrogen and methane. Boride coatings are prepared with arc plasma or high-powered carbon dioxide lasers. Technical-grade borides with 67–76.5% boron are produced by reacting B_2O_3, finely divided coke, and the desired metal in an electric furnace.

Boron carbide (pure B_4C contains 78.3% boron) may be produced in the same manner at 1400–2300°C, but without the metal, or at lower temperatures (1400–1800°C) with Mg in a H_2 atmosphere, and then leached with HCl and boiling HF. The latter product is very hard and has good electrical conductivity as well as a high compressive strength and melting point (2450°C). It is chemically inert, but remains stable in an oxidizing environment only to about 800–1000°C. It is used as a polishing agent, for sandblast nozzles and in nuclear shielding. A composite of boron carbide and fiberglass has been developed that can stop a 30-caliber bullet at point-blank range, and was used for seats in the AH-10 Cobra attack helicopter. A silicoboron carbonitride ($Si_3BC_{4.3}N_2$) has extraordinarily high thermal stability (i.e., up to 2000°C; Anon., 1996d).

Boron nitride (BN; "white graphite") is produced by the thermal decomposition of boron–nitrogen compounds, such as $B(NH_3)_3$ and $BF_3 \cdot NH_3$, has a hexagonal graphite-like platelet structure, and sublimes above 2980°C. It is similar to graphite in directional properties, has good machinability, high thermal conductivity, excellent resistance to thermal shock, a low density, and is an electrical insulator. It resists air oxidation up to 1400°C and is not wetted by many molten metals, slags, or glasses, but it can hydrolyze. It is available as a solid, powder, or aerosol, and its uses include crucibles, molten metal nozzles and lubricants (Luehrsen and Ott, 1990). A cubic boron nitride, Borazon, is produced by applying high temperatures (1400–1700°C) and pressure (1 million psi) to the hexagonal BN. It is used for cutting, honing, lapping, and polishing ferrous metals and superalloys. It has a higher cutting rate and continuous heavy-duty ability than tungsten carbide, and does not require coolants.

Borides of chromium, hafnium, titanium, zirconium, and many other metals are high-melting-point materials with very good strength, hardness, wear resistance, electrical conductivity, and resistance to chemical attack. Titanium boride (TiB_2) has been used as anvils, bearings, bearing liners, jet nozzles and crucibles. Aluminum boride (AlB_{12}) is a substitute for diamond dust in grinding and polishing, while zirconium diboride (ZrB_2) is used in spray nozzles for metal atomization (Lyday, 1985; Roskill, 1993).

9.3 AGRICULTURE

9.3.1 Function of Boron in Plants

Boron is essential to plant growth, being one of the 16 basic plant nutrients. However, high concentrations of boron are also toxic, resulting in a relatively narrow range of concentration between too much and too little (Table 9.2). The exact function of boron in plants is still unclear, but it has been related to several processes, such as the translocation, or the control of the amount of various organic compounds within the plant. It appears that the complexing ability of boron with poloyls facilitates the movement of sugars, and a significant increase in the oxygen uptake by root tissue was observed with the addition of 5 ppmB to a sucrose foliar spray. Also, there was a greater distribution of sucrose in the plant when a 10 ppmB solution was applied to the leaves of bean and tomato plants. Boron is known to enhance the effects of sugars on the hormone action in plants, the amount of photosynthesis, the rate of absorption of CO_2 from the air, and the growth of plant roots.

Another important function of boron is with cell growth and structure. A deficiency alters the cell walls, making most of them thinner, except for the phloem parenchyma and ground parenchyma cell walls, which are made thicker. This appears to result from an alteration in the condensation of carbohydrates into wall material. A boron deficiency in sunflower tissue resulted in less lignification and a reduction in RNA, causing changes in protein and nucleic acid syntheses. Without adequate supplies of boron the growth of cell walls also may be altered. With bean plants a boron deficiency reduced

Table 9.2

Boron in the Soil Solution and Tissue Analysis for Best Plant Growth[a]

	Boron concentration in soil solution for best growth (ppm)	Lowest soil solution concentration for injury (ppmB)	Best tissue analyses (ppmB)	Typical application rates[b] (lbsB/acre)
A. Sensitive Plants				
Bermuda grass	—	—	7–20[b]	1–2[b]
Blackberry (*Rubus* sp.), Raspberry	Trace, <0.5[c]	1, 0.3–0.5[d]	30–60[b]	—
Cherry (*Prunus arium L.*)	1, 0.5–0.75[c]	5, 0.3–0.5[d]	25–60,[b] 30[e]	1–2[b]
Citrus, Lemon (*Citrus limonia osbeck*)	Trace, <0.5[c] 0.5–0.75[b]	1, 0.3–0.5[d]	30–100[b]	1–2[b]
Cowpea (*Vigna sinensis, Torner, Savl.*)	Trace, 0.5–0.75[c]	5, 0.3–0.8[d]	—	—
Elm (*Cimusamericana L.*)	1, <0.5[c]	1	—	—
Fig (*Ficus carica L.*)	1, 0.5–0.75[c]	5, 0.5–0.8[d]	—	—
Grape (*Vitisrinifera L.*)	1, 0.5–0.75[c]	5, 0.5–0.8[d]	40–60,[b] 50[e]	1.5–3[b]
Jerusalem artichoke (*Helianthus tuberosus L.*)	1, 0.75–1.0[c]	1	—	—
Kidney bean (*Phaseolus vulgaris L.*)	1, 0.75–1.0[c]	1, 0.8–1.0[d]	—	—
Larkspur (*Delphinium* sp.)	1, 0.5–1.0[c]	5	—	—
Lupine (*Lupinus hurtwegi Lindl.*)	1, 0.75–1.0[c]	5	—	—
Pansy (*Viola tricolor L.*)	Trace, 0.5–1.0[c]	5	—	—
Peach (*Prunus persica L. Batsch*)	1, 0.5–0.75[c]	5, 0.3–0.5[d]	—	—
Pears	0.75[b]	—	30–60,[b] 50[e]	1–2[b]

Persimmon (*Diospyros kaki L.f.*)	1, 0.5–0.75[c]	1, 0.5–0.8[d]	—	—
Plum	0.75[b]	0.3–0.5[d]	30–60[b]	2–4[b]
Strawberry (*Fragaria* sp.)	1, 0.75–1.0[c]	5, 0.8–1.0[d]	25–50[b]	1–2[b]
Violet (*Viola odorata L.*)	Trace, 0.5–1.0[c]	5	—	—
Walnuts	1[b]	0.5–0.8[d]	40–100,[b] 400[e]	2–4[b]
Zinnia (*Zinnia elegans Jacq.*)	Trace, 0.5–1.0[c]	1	—	—
Bean, mung; cucumber; garlic; peanut; sesame; sugar cane; sunflower; wheat	0.75–1.0[c]	0.5–0.8[d]	6–20,[b] 5,[e] 8–20[g]	0.5–1[b]
Almonds; apricot; avocado; figs, Kadota; pecan	0.5–0.75[c]	0.3–0.8[d]	20–45,[b] 20–70[g]	—
B. Semi-tolerant plants				
Alfalfa (*Medicago sativa L.*)	10, 4–6[c]	15, 4–6[d]	30–70,[b] 80[e]	1–4[b]
Apples	1[b]	—	30–50,[b] 25[e]	1–3[b]
Barley (*Hordeum vulgare L.*)	Trace, 2–4[c]	5, 1–2[d]	6–20[b]	—
Broccoli; brussel sprouts; cucumber	1–2[c]	—	25–50,[b] 30–60[e]	1–3,[b] 3–4,[f] 2–5[c]
Cabbage (*Brassica oleracea var. capitata L.*)	1, 2–4[c]	10, 1–2[d]	25–50,[b] 40[e]	1–3,[f] 2–3,[b] 3–4[f]
Calendula (*Calendula officinalis L.*)	Trace	5	—	—
California Poppy (*Eschscholtzia californica Cham.*)	5, 2–4[c]	5	—	—
Carrot (*Daucus carota L.*)	Trace, 1–2[c]	10, 1–2[d]	25–70,[b] 30–200[e]	1–2,[b] 2–3[f]

continues

Table 9.2 (*continued*)

	Boron concentration in soil solution for best growth (ppm)	Lowest soil solution concentration for injury (ppmB)	Best tissue analyses (ppmB)	Typical application rates[b] (lbsB/acre)
Cauliflower	2–4[c]	—	30–60,[b] 25–50[e]	2–3,[b] 2–5,[e] 3–4[e]
Celery (*Apium graveolens L.*)	15, 2–4[c]	15, 1–2[d]	25–50[b]	2–3,[b] 1–3,[e] 2–3[f]
Collards	—	—	30–60[b]	1–3[b]
Corn (*Zea mays L.*)	1, 2–4[c]	5, 2–4[d]	10–20,[b] 25[e]	1–2,[b] 0.5–1[f]
Eggplant	—	—	40–50[b]	1–2[b]
Kentucky bluegrass (*Poa pratensis L.*)	5, 2–4[c]	1	—	—
Lettuce (*Lactucca sativa L.*)	5, 2–4[c]	1, 1–2[d]	25–50,[b] 27–43[e]	1–2[b,f]
Lima bean (*Phaseolus tunatus L.*)	Trace, 0.75–1.0[c]	1, 0.8–1.0[d]	30–40[b]	1–2[b]
Milo (*Sorghum vulgare Pers.*)	Trace	5	—	—
Mustard (*Brassica sp.*)	1, 2–4[c]	10	—	—
Oats (*Avena sativa L.*)	5, 2–4[c]	5, 4–6[d]	8–20[g]	—
Onion (*Allium cepa L.*)	Trace, 0.5–0.75[c]	1, 0.5–0.8[d]	25–50[b]	1–2,[b] 0.5–1[f]
Parsley (*Petroselinum crispum, Mill., Nym.*)	5, 4–6[c]	15, 4–6[d]	—	—
Pea (*Pisum sativum L.*)	1, 1–2[c]	5, 0.8–1[d]	25–50[b]	1[b]
Peanut	0.75[b]	—	25–50[b]	0.5–1,[b] 0.5–0.75[g]
Potato (*Solanum tuberosum L.*), pumpkin	1, 1–2[c]	1, 1–2[d]	20–40[b]	1–2,[b] 0.5–1[f]

Radish (*Raphanus sativus L.*)	1, 1–2[c]	10, 1–2[d]	20–50,[b] 45[e]	1–2[b,f]
Red pepper (*Capsicum frutescens L.*)	Trace, 1–2[c]	5, 0.8–1	40–100[b]	1[b]
Rhubarb, spinach	—	—	40–60[b]	1–2[b,f]
Rutabaga	—	—	25–60[b]	1–2[b]
Soybean	0.5[b]	—	20–60[b]	0.5–1[b,g]
Squash	—	2–4[d]	30–40[b]	1–2,[b] 0.5–1[f]
Sweet potato (*Ipomoca batatas, L., Lam.*)	Trace, 0.75–1.0[c]	5, 0.5–0.8[d]	20–40,[b]	1[b] 0.5–1[f]
Tobacco (*Nicotiana tomentosa Ruiz and Pav.*)	15, 2–4[c]	10, 2–4[d]	20–50,[b] 100[e]	0.5–1[b,g]
Tomato (*Lycopersicon esculentium Mill.*)	10, 4–6[c]	5, 4–6[d]	30–80,[b] 100[e]	1–2[b,f]
Vetch (*Vicia atropurpurea Desf.*)	5, 4–6[f]	5	—	1[b]
Watermelon	—	—	—	1–2,[b] 0.5–1[f]
C. Tolerant plants				
Artichoke (*Cynara scolymus L.*)	5, 2–4[c]	5, 2–4[7]	—	—
Asparagus (*Asparagus officinalis L.*)	5, 10–15[c]	25, 10–15[d]	40–65[b]	1–2[f]
Common beet (*Beta vulgaris L.*)	5, 4–6[c]	15	30–70,[b] 80[e]	1–2,[b] 3–4[f]
Cotton (*Gossypium hirsutum L.*)	10, 6–10[c]	10, 6–10[d]	30–50,[b] 30–80[e]	0.5–2,[b] 1–2[e]
Leaf beet (*Beta vulgaris var. cicla L.*)	5	25	—	—
Muskmelon (*Cucumis melo L.*)	5	15	25–50[b]	1–2[b]
Oxalis (*Oxalis bowiei Herb.*)	10	—	—	—
Parsnips	—	—	25–70[b]	1–2,[b] 2–3[f]

continues

Table 9.2 *(continued)*

	Boron concentration in soil solution for best growth (ppm)	Lowest soil solution concentration for injury (ppmB)	Best tissue analyses (ppmB)	Typical application rates[b] (lbsB/acre)
Sorghum	6–10[c]	4–6[d]	1–2[b]	8–20[b]
Sugar beet (*Beta vulgaris var. crassa Alef.*)	5, 4–6[c]	15, 4–6[d]	30–70,[b] 50[e]	1–3[b]
Sunflower	—	—	1–2[b]	40–60[b]
Sweet clover (*Melilotus indica L., All.*)	5, 2–4[c]	10, 4–6[d]	20[h], 25–50[b] 45[e]	0.5–2,[b] 1–2[h]
Sweet pea (*Lathyrus odoratus L.*)	10, 2–4[c]	15	—	—
Turnip (*Brassica rapa L.*)	5, 2–4[f]	25, 1–2[d]	25–60,[b] 50[e]	1–3,[b] 2–4,[e] 3–4[f]

[a] Sprague, 1972.
[b] Segars, 1987.
[c] Maas, 1986.
[d] Raymond and Butterwick, 1992 (quoting 1929 and 1944 references).
[e] Anon., 1994.
[f] Mack, 1986.
[g] Garratte, 1983.
[h] Flannery, 1985.

the cell division and decreased the movement of glucose into pectin in the root sections. There was an increased movement into cellulose, indicating the possibility that boron is involved in cell wall bonding, or the stretching phase of cell elongation.

Finally, boron is involved in enzymatic reactions such as the conversion of glucose-1-phosphate into starch by starch phosphorylase. It may also have an effect on the synthesis of plant hormones regulating plant growth. There appears to be less nucleic acid in boron-deficient tissue, and it can be increased by adding nucleic acid to the plants' soil solution. Without adequate boron in peanut plants, the total nitrogen and certain amino acids were higher than normal, but the protein and DNA were reduced.

9.3.2 Boron's Quantitative Effects on Growth

Table 9.2 contains guidelines on the tolerance of plants to boron, the optimum boron concentration in the soil solution and plant tissue for best growth, and the lowest concentration for injury. It is evident that different experts suggest widely different boron requirements and tolerances, probably on the basis of local conditions and the availability of other nutrients. With soybeans, an increase in potassium induced a boron deficiency, and more boron was required if phosphorus was low or the nitrate was high. The presence of high calcium also can induce a boron deficiency, and increased nitrogen application on cotton can sometimes be ineffective without increased boron. The effect of boron in the soil solution on the growth of a few plants is illustrated in Fig. 9.3. In general, soil solution and plant tissue analyses correlate well (Fig. 9.4).

The concentration of boron in plant leaves may vary from 2 to 3875 ppm, and mild leaf injury may occur at or below the boron concentration that results in the greatest growth. Boron may accumulate in the older leaves, causing yellowing or burning, even though the boron supply to actively enlarge meristematic tissues is insufficient for their most rapid growth. In general, tissue levels less than 15 or 20 ppmB are associated with deficiency, while levels in excess of 200 ppm are required for symptoms of boron excess.

It appears necessary to move the boron in the soil solution primarily by liquid flow to the root area, rather than by diffusion (the mechanism of most other nutrients). Once boron reaches the plant root, transport across selective cytoplasmic membranes into the root cells occurs, even though the boron concentration in the soil solution is generally much lower than in the root tissue. The rate of uptake for boron is also much slower than that of other nutrients, possibly because of the need for boron to react with the root, or be adsorbed by it, and then to be desorbed into the plant's fluids. It then must be transported to utilization and immobilization sites in other portions of the plant. The uptake of boron can be reduced by a factor of 4 as the soil solution

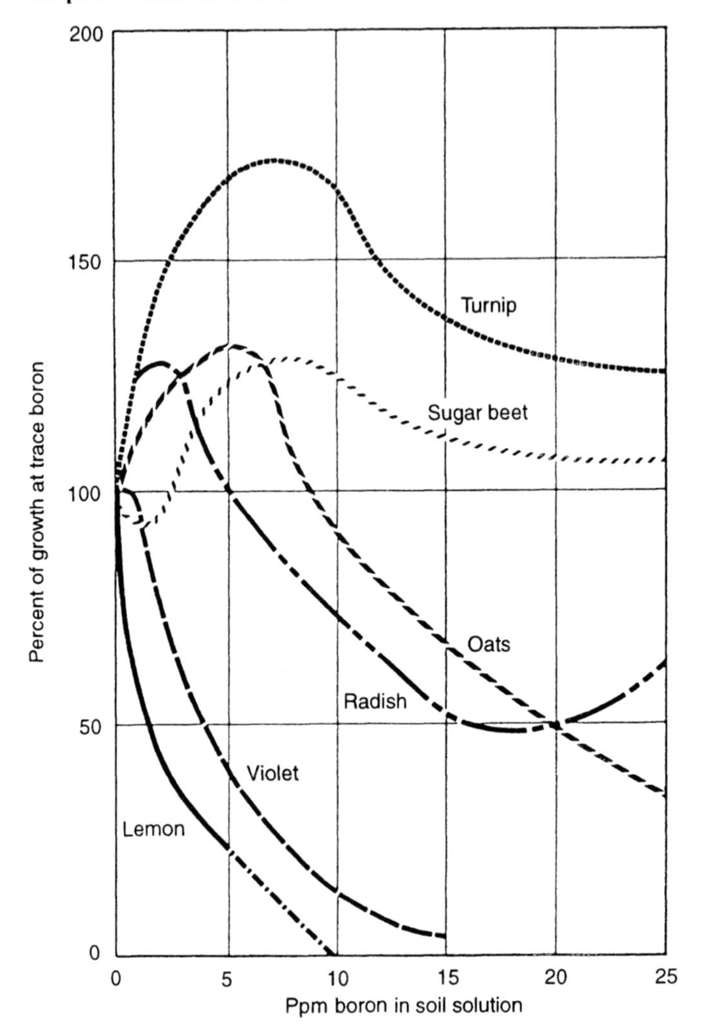

Figure 9.3 Effect of boron concentration in the soil solution on plant growth. (From Sprague, 1972; drawing courtesy of U.S. Borax Inc.)

changes from pH 4 to pH 8, and increases over the range of 10–30°C (followed by a sharp reduction above 35°C). Increase in light intensity also increases the rate of boron uptake.

Irrigation water with 5 ppmB in California has damaged walnut and citrus orchards, resulting in rules or regulations for the water's boron concentration. No more than 0.3 to 1 ppmB is suggested for water used on sensitive plants, and less than 1–2 ppmB is recommended for semitolerant plants. These recom-

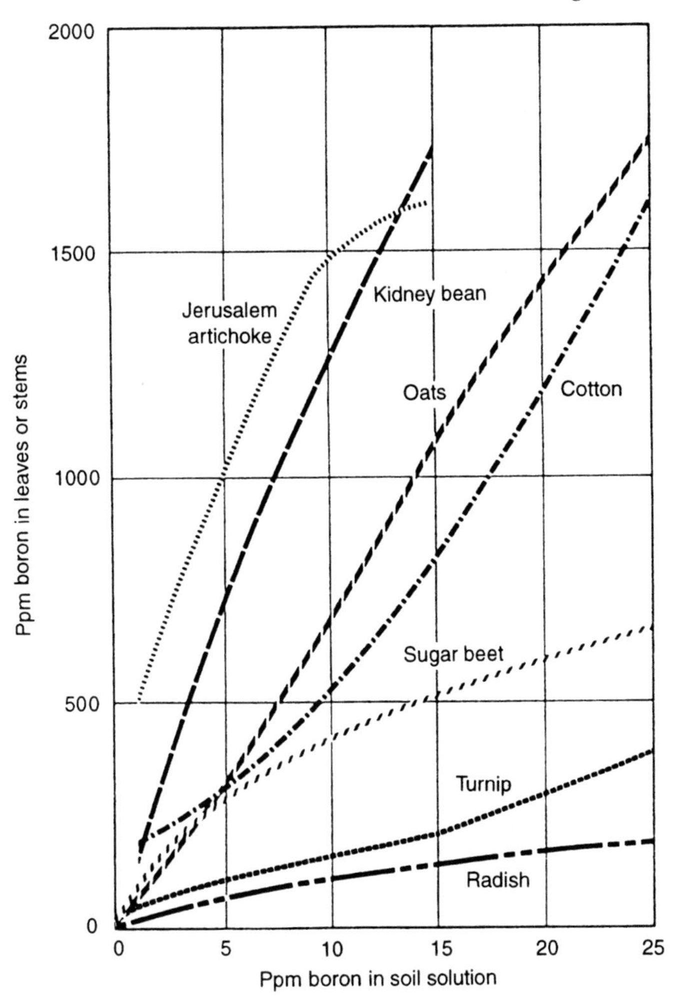

Figure 9.4 Boron concentration in plant tissue related to the content in the soil solution. (From Sprague, 1972; drawing courtesy of U.S. Borax Inc.)

mendations do not allow for differences in soils, climate, use conditions, or other growth factors, and thus are usually the limits for the most adverse conditions (Sprague, 1972).

9.3.3 Boron Deficiency Symptoms and Fertilization Rates

9.3.3.1 Natural Boron in Soils

Some of the common symptoms of boron deficiencies are listed in Table 9.3. The average boron content in all plants, on a dry matter basis, is 20 ppmB

Table 9.3

Boron Deficiency Symptoms[a]

	Visible symptoms
Field crops	
Alfalfa	Death of terminal bud, rosetting, yellow top, little flowering, poor pod set and yellowing with short internodes in new growth.
Clover	Poor stands, growth and color. Reduced flowering and seed set.
Corn	Short, bent cobs, barren ears, bland stalks and poor kernel development.
Cotton	Shedding of squares and young bolls, ringed or banded leaf petioles with dieback of terminal buds, ruptures at base of squares, dark fluid exuding from ruptures, internal discoloration at base of boll, small half-opened bolls and green leaves until frost. Hard to defoliate.
Peanut	Dark hallow area in center of the nut called "hollow heart."
Soybean	Yellow leaves, chlorotic between veins, downward curling of leaf tips, crinkling of leaves, dieback of tips, no flowering and roots stunted.
Sugar beet	Yellowing or drying of leaves, cracking of the leaf midrib, brown discoloration of internal tissue ("black heart" or "hart rot"), and rotting of the crown.
Sweet corn	Elongated, watery or transparent stripes later becoming white on newly formed leaves, and dead growing points.
Tobacco	Leaf puckering and deformed buds.
Fruit crops	
Apple	Pitting, skin discolored, cracking and corking core.
Apricot	Twigs die back and fruit fails to set.
Citrus	Thickened rind, gum pockets near axis, discolored patches, die back and rosetting.
Grape	"Hen and chick" symptom, and dead main shoots.
Pear	Blossom blast, pitting, internal corking and bark cankers.
Strawberry	Pale chlorotic skin of fruit, cracking and die back.
Walnut	Die back from shoot tips, and leaf fall.
Vegetable crops	
Beet (red)	External spotting, cracking and canker.
Broccoli	Hollow stems, internal discoloration and brown curds.
Cabbage	Hollow stem, watery areas, heads hollow and stunted plants.
Carrot	Reddening of leaves and root splitting.
Cauliflower	Leaves curled, hollow stems, curds dwarfed and brown.
Celery	Stem crooked, cracked, striped brown and the heart blackened.
Lettuce	Stunted growth, discoloration and brittleness of the leaves.
Potatoes	Black spots in the interior.
Tomato	Thickened leaves, brittle leaves and the fruit fails to set.
Turnip	Hollow center or brown head, and watery areas.
Radish	Pale roots, brittle stems, watery flesh and flecked coloration.

[a] Flannery, 1985; Anon., 1994.

(Rosenfelder, 1978). In animal tissues the boron content has been found to be about 1 ppm. Boron is not known to have an essential biochemical function in humans or animals, although one of its functions may be a role in the body's ability to use calcium (Kistler and Helvaci, 1994). The natural boron content in soils (the solid phase, not in solution) varies from approximately 10 ppm to more than 300 ppmB, with the U.S. average at about 30 ppm. Much of the boron in the top layers of soil comes from decayed plant tissue, and the total boron is usually not "available" to the plants. Most of the boron occurs as very insoluble minerals, and boron is also complexed by, or adsorbed on, organic material and soil particles, with much of the adsorption being on fine clay fractions, particularly illite or micaceous high-iron and aluminum clays. Adsorption is influenced by the pH, with the maximum at pH 8.5–9. Thus, with reduced adsorption and increased dissolving power, the greatest boron "availability" occurs at pH values of 5.5–7. The correlation of total boron content in the soil with plant growth is very erratic.

There are two laboratory procedures for extracting "available" boron from soils. A 1 to 2 ratio of soil to water may be boiled for 5 minutes under reflux, and the solution separated and analyzed, or a soil can be saturated with water, allowed to equilibrate for 24 hours, and then filtered. The boron concentration in both tests is proportional to be that of the soil solution, and may be correlated with optimum growth conditions for different plants. Examples of soils that may have a boron deficiency are those formed from igneous rocks (light-textured sandy soils low in organic matter), acidic soils containing peat or a high-organic content, alkaline soils, and heavily irrigated soils. An excess of boron may occur in soils derived from parent rock that is rich in boron, some marine sediments, and some arid soils. Soils in both classes occur frequently in the western United States (Anon., 1980, 1995a).

9.3.3.2 *Application Methods*

Broadcast application of boron fertilizers before planting, during dormant periods, or after any cutting is generally preferred. Split applications may be best for heavy clay soils, irrigated crops, or for maximum economic efficiency. Sideband rates of 0.25 kgB/hectare (0.2 lbB/acre) should be placed 5–8 cm (2–3 in.) to the side and below the planting level, or 0.25–0.5 kgB/hectare may be applied through irrigation water, alone, or combined with other nutrients. Finally, boron may be foliar applied, with or without a number of insecticides, fungicides, or other nutrients during the growing season, but rates should not exceed 0.25–0.5 kgB/hectare (Segars, 1987).

9.3.3.3 *Herbicidal Applications*

Because of borax's relative low toxicity to man and animals, and because it has phytotoxicity comparable to that of sodium chlorate ($NaClO_3$) or arsenic trioxide (As_2O_3), higher concentrations are used as nonselective herbicides. Combinations with sodium chlorate are even superior to either ingredient alone. The maximum effectiveness, and where the fire or explosion hazard is removed, is at a borate-to-chlorate ratio greater than four. When used alone, the borates are usually applied at rates ranging from 0.24 to 0.49 kgB/m^2 (5 to 10 lbB/100 ft^2) for annuals, and the effects last for one growing season. Borates can also be combined with one or more organic herbicide, supplying a biostatic effect to reduce microbial breakdown (Sprague, 1972).

9.4 CLEANING COMPOUNDS AND BLEACHES

Borax's mild alkalinity allows it to emulsify oil and greases, and to reduce the surface tension of water, which aids in loosening dirt particles. Borax also reacts with some organics to form esters, and has a mild bactericidal action. This combination gives borax a strong but gentle cleansing action for personal use, and on many types of fabrics, surfaces, and contaminants. Borax may be combined with surfactants or abrasives for scouring powders, soap bars, or many other cleaning applications, such as in automotive cooling systems (Oberhofer, Benko and Drozd 1976). Sodium perborates are very popular as a laundry cleanser in countries that employ comparatively hot washing machine water. Either the monohydrate or tetrahydrate is employed in various formulations (Table 9.4A) at the beginning of the wash cycle. It functions both as a mild alkali and a controlled oxidizing agent, loosening dirt and gently removing (oxidizing) stains and chemically reacted contaminants from the cloth. It is more powerful than chlorine-type bleaches at temperatures above 55°C, and is less likely to harm some fabrics. About 790,000 tons/year were used in Europe and Japan in 1985 (Lyday, 1985; Raymond and Butterwick, 1992). Activators have been added, such as tetracetyl-ethylenediamine, alkyloxyben-

Table 9.4

Typical Composition of Several Borate Products[a]

A. Heavy-duty household detergent (wt.%)			
Sodium perborate	20–35	Optical brightner	0.1–0.5
Sodium tripoly phosphate	35–45	Amount used	7–8 g/l
Surfactant	10–15	Liquor ratio	1–5 kg/l
Sodium silicate	3–5		
B. Nickel-plating electrolyte (g/l)			
	Nickel sulfate	240	
	Nickel chloride	45	
	Boric acid	30	

[a] Rosenfelder, 1978.

zene sulfonate, or pentaacetylglucose, to increase its effectiveness at lower temperatures (i.e., 40°C), resulting in the U.S. market growing by about 4%/ year in the late 1990s. Washing temperatures are also slowly being reduced elsewhere from 60 to 40°C.

Sodium perborate is produced by reacting borax (usually the pentahydrate) with sodium hydroxide to form sodium metaborate, heating, and then adding hydrogen peroxide. Stabilizers such as magnesium sulfate and silicates are usually present, and the perborate is next crystallized, centrifuged, and dried (Rosenfelder, 1978). The crystals usually are washed at about O°C as they are centrifuged or filtered, making the active oxygen recovery almost quantitative. The tetrahydrate ($NaBO_3 \cdot 4H_2O$) has 10.5% active oxygen, while the monohydrate ($NaBO_3 \cdot H_2O$), usually produced by drying the tetrahydrate, has 15.5%. Other methods of producing sodium perborate are by reacting sodium hyperoxide ($NaHO_2$) with boric acid (Lyday, 1985), or by the electrolysis of a borax solution containing sodium carbonate.

Because the monohydrate has a higher active oxygen content, it is increasingly replacing the tetrahydrate, particularly in "compact" washing powders. Both hydrates are white crystalline or powdery materials with a good storage stability. They lose only a few percent of active oxygen content in a year's normal storage. The perborates are also used for pulp-mill and textile bleaching, as well as in dye oxidation, tooth powders, laundry bleaches, dishwashing powder, and household surface, denture, and other special-purpose cleaners. The tetrahydrate solubility (as g/100gH_2O) is about 2(23 g/liter) at 20°C, 3(37 g/liter) at 30°C, and 30 at 60°C; the monohydrate is 15 g/liter at 20°C and 24 g/liter at 30°C. A trihydrate can form that is less soluble, but it is rarely obtained. The pH of the perborate solutions is about 10.1–10.4, and is only slightly affected by changes in concentration. The monohydrate rapidly hydrates with moisture to the tetrahydrate, and the tetrahydrate melts at about 63°C with decomposition. The monohydrate is stable at this temperature. Above 100°C the active oxygen is lost, and intermediate compounds such as the peroxyborates are formed. Perborates are shipped in barrels with an ICC yellow label (oxidizing material), and although not regarded as hazardous, they should not be mixed or heated with combustibles (Edwards and Curci, 1967; Raymond and Butterwick, 1992). Other borates can also stabilize enzymes in liquid laundry detergents (Anon., 1996c).

9.5 FIBERS AND COMPOSITES OF BORON AND BORIDES

Composites of boron or boride fibers (as mats, whiskers, chopped or continuous) in a matrix of plastics, ceramics or metals have great strength and a high modulus of elasticity. The initial applications for these advanced composites were for the military, primarily in air or spacecraft. Boron fibers in plastics offer a stiffness-to-density ratio six times that of aluminum and titanium, and they have been used in jet-engine compressor blades, wing flaps, rudders, floor beams and sheathing. They combine high temperature resistance, flexibility, and light weight with strength and ease of fabrication (Kistler and Helvaci, 1994). A boron–epoxy skin on the horizontal stabilizers of the F-14 Tomcat, F15 Eagle, and B1 bomber decreased the weight of the planes by 91 kg (200 lb) compared with equivalent parts of titanium. About 15.9 metric tons (35,000 lb) of boron filament were used, but later replaced by the less expensive carbon composites. In the space shuttle a weight saving of 137 kg (300 lb) was obtained by using boron–epoxy reinforcement. Other applications include sports equipment such as golf club shafts, tennis rackets, fishing rods and bicycle frames, for which the higher cost can be justified by superior performance.

Boron fibers are made by chemical vapor deposition on a 1-mil tungsten or carbon filament, to become 4–8 mils (100–200 μ) thick. Boron trichloride is deposited on the wire at 1300°C, and reduced by hydrogen: $2BCl_3 + 3H_2 \rightarrow 2B + 6HCl$. The filament, with a density of 2.3 (on carbon) to 2.6 (on tungsten), is processed into a tape held in position by semicured resin. Wide tapes 1.2 m (48 in.) across use a scrim cloth for support. The fibers are compatible with the resin (but not metal matrices), with the amount controlled to 50 vol%. Boron filaments may also be interwoven into fabrics with polyester, glass, graphite, Kevlar, and other reinforcements to provide specific mechanical and physical properties. They can also be bonded at 5000 psi and 500°C to aluminum alloy foils. Strength in several directions is supplied by cross-plies of the filaments. Boron epoxy composites have tensile strengths of up to 230,000 psi, and compressive strengths of 360,000 psi, both superior to those of other materials such as fiber glass, silicon carbide, aluminum oxide, and Aramid fibers. High-strength graphite does have the same tensile strength, but less than 50% of boron's compressive strength (Lyday, 1985).

9.6 FLAME RETARDANTS

Boric acid, borax and pentahydrate have been used to make inexpensive cellulose insulation material. Shredded newspaper can be used as blown-in insulation for attics and walls, but has a lower insulation value than fiberglass, and in time compresses to lose much of this effect. However, if the cellulose is treated with a borate solution (primarily boric acid) and then well dried before it is installed, this problem is considerably reduced, and it becomes reasonably fire resistant, toxic to bacteria (anticomposting), and unpalatable for rats, mice, and insects. The borates react with the hydroxyl groups of the cellulose to form a very thin but uniform borate film that is quite stable. With most insulations of this type, however, the boric acid or pentahydrate is dry-mixed with the newspapers (i.e., 40-wt% paper, 7% borax pentahydrate, and 3% boric acid) as they are shredded in a hammermill (Rosenfelder, 1978). Its insulating effectiveness is thus somewhat reduced, but it retains its low flammability properties, since the borates promote the formation of water, which absorbs heat and lowers the temperature during heating. It later forms a glassy substrate that inhibits burning.

Wood particle chips also may be mixed with boric acid before the urea–formaldehyde resin is added to form flame-resistant particleboard. The particle- or pressboard will give off little smoke or flame when heated, and "afterglow" is prevented. Similarly, boric acid can be milled with cotton to flameproof cotton mattresses. In many fireproofing applications zinc borate ($2ZnO \cdot 3B_2O_3 \cdot 15H_2O$) has a greater flame retardancy than borates used alone. To produce it, zinc oxide is reacted with just enough boric acid solution to form a paste, and then dewatered by centrifuging, dried in a gas furnace, and ground. The zinc borate does not affect the color of plastics, and is only slowly affected by high temperatures. When heated, zinc borate promotes the formation of char and inhibits the release of combustible material (Lyday, 1985). The addition of aluminum trihydrate to the zinc borate forms a synergistic mixture more effective than either material alone, particularly in reducing the fire's smoke.

9.7 FUELS

Boron and its hydrides have the highest heat of combustion per unit weight of all the elements, 25,120 compared with 18,500 Btu/lb for aviation gasoline, causing it to be considered as a fuel for space or aircraft. However, their cost and harmful end products have prevented commercialization. Diborane (B_2H_6, a gas), pentaborane (B_5H_9, a liquid), and decaborane ($B_{10}H_{14}$, a solid) were all considered. As a totally different application, 10-hydroxy-9,10-boroarophenathrene is a highly effective antioxidant additive for hydrocarbon fuels, and sodium borohydride has been used for the same purpose in jet bomber fuel by the U.S. Navy. As little as 1 gal/5000 gal of certain organoboron compounds is used in sterilizing hydrocarbon fuel storage systems to reduce corrosion and prevent microorganisms from growing and clogging filters (Anon., 1996a). Other fuel-related uses are boron trichloride or fluoride used as catalysts in petroleum refining, and boron–nickel catalysts (resistant to sulfur poisoning) used in converting carbon monoxide to fuels. Finally, coal averages about 75 ppmB, and 10,500 metric tons/year of boron ends up in fly ash or enters the atmosphere (Lyday, 1985).

9.8 GLAZES, FRITS, AND ENAMELS

One of the earliest uses of borates, and one that is still important even though it does not consume much borax, is the production of glazes and frits to impart color and texture, as well as heat, chemical, or wear resistance to appliances, ceramics, and tile. Borax assists in the production of smooth, hard, resistant, blemishless, and craze-free ceramic surfaces. Early in the eighteenth century ceramic enamels consisting of a low-melting borax glass to which pigments were added began to be used, and ceramic painting techniques became quite sophisticated. With pigmented enamels, variations in tint could be achieved by laying on the color more or less thickly (impasto technique), and the style of overglaze enamels came to resemble oil painting. New colors were developed, and new flux compositions for enamels that contained bismuth oxide and borax additions to the basic lead or lead alkali silicates evolved, causing the decorating process to become increasingly complex. One of the earliest pigments was "Egyptian blue," a copper–calcium silicate manufactured by grinding together quartz, malachite, lime and borax (as a "flux"). This technique is still in use today, with the ingredients intimately mixed and then heated to form a liquid. After the melt is cooled the excess borate is dissolved in acid, and the frit is milled to obtain the fine particles required for effective light scattering (Kingery and Vandiver, 1986).

Most boron compounds are soluble to some extent, so the 3–15% B_2O_3 desired in most glazes (Table 9.5) normally is used in this insoluble glass, or fritted form. The frits melt easily, and the boron intensifies the coloring effect.

Table 9.5

Examples of Borate-Containing Ceramic Frits (wt%)

	Typical 1000–1020°C M.P.[a]	Leadless tile frit[b]	Lead borosilicate[b]	Corrosion-resistent glaze[c]
B_2O_3	10.8	13	21	3.7–7.3
SiO_2	39.5	63	12	1–5 (spodumene)
Al_2O_3	19.8	8	—	—
CaO	17.9	8.3	—	5–15
Na_2O	10.5	6.6	—	11.7–17.6
K_2O	1.5	—	—	5–15 (Mo ore)
ZnO	—	1.1	—	0.5–2 (Co_2O_3)
Pb	—	—	67	0.5–2 (NiO)

[a] Fournier, 1977.

[b] Rosenfelder, 1978.

[c] Wang and Shao, 1990.

Very soft (low melting point) borax frits (850°C) can be used with the addition of 3% bentonite for raku ceramics. Borax also can be added to salt to assist in salt-glazing, although it will slightly change the glaze characteristics. A property of boric oxide that makes it very useful in earthenware glazes is that in amounts less than 10% it has a negative coefficient of expansion, which helps with craze resistance (Fournier, 1977). Small amounts also help in making glazes heal over and melt more smoothly, and in forming a tougher surface. Commercial tableware formerly was glazed with lead oxide–boric oxide frits because of their smoothness, freedom from pits or other blemishes, long firing range, and good wearing properties (Rhodes, 1975). However, they are now being replaced by high-borate, no-lead mixtures because of lead's toxicity problems.

Vitreous (porcelain) enamel is made by a single or two-coat dipping, spraying, or electrostatic process with a borate glaze to give appliances and other products a glass-like surface with excellent thermal, corrosion, and abrasion resistance. Wall and floor tile also may use a highly reactive borosilicate glaze that allows a wide range of decorations (Rosenfelder, 1978). Molten borax-containing glazes ("glass") are employed in coating metal vessels and pipes for corrosion protection (Wang and Shao, 1992).

9.9 MEDICINE

Boric acid and sodium borate are mild antiseptics that inhibit gram-negative bacteria, and boric acid has long been used as an eye wash. Some heterocyclic boron compounds inhibit tumor growth, while borax and some of its ores are highly effective in killing parasites such as oncomelania (Xu, 1990). Organic compounds such as the closo icosahedral anion $[^{10}B_{12}H_{11}SH]^{2-}$ (that can attach to tumorous tissue, and with a high ^{10}B content) are injected into cancer patients and then exposed to a high-neutron source. This generates intense energy in a very localized area, destroying the tumor (Greenwood, 1991). In one study 5 years after the boron treatment 33% of the patients were alive, which is an improvement over other therapy (Lyday, 1985). The short-range alpha particles from the ^{10}B and low-energy neutrons have also been used for microsurgery in previously inoperable areas of the brain (Barth, Soloway, Fairchild and Brugger, 1992).

In other medical applications, some boron compounds reduce serum cholesterol and other harmful proteins. Potassium borohydride (KBH_4) has been used to manufacture hydrocortisone, prednisone, and prednisoliene (corticosteroids used in arthritis therapy), and in the synthesis of vitamin A. Sodium borohydride ($NaBH_4$) assists in the manufacture of hormones. Sodium and potassium borates are frequently used in face creams, lotions, dusting powder, ointments, hair preparations, mouth washes, and emulsifiers in medical or cosmetic formulations (Kistler and Helvaci, 1994).

9.10 METALLURGY

The addition of 0.001% to 0.003% boron to steel reduces the amount of nickel, chromium, or molybdenum required in many alloys. The boron is added in the form of 10–17% B ferroboron pellets produced by the aluminothermic reduction of a borate compound (Zambrano, 1985). Because boron is rapidly oxidized it is the last material added to the molten steel, and usually aluminum is also added to react with the steel's oxygen instead of the boron. Boron strengthens high-performance, low-alloy steels by precipitating certain carbides, and by its effect on crystal structure. Being small, boron tends to locate in the interstitial space between the metal atoms of the alloy, causing it to be harder and stronger than the parent metal. It is most effective in low-carbon and fine-grained steels. About 1,000,000 tons/year of boron-containing carbon steels and 300,000 tons/year of boron alloy steels were produced in United States in 1980. The largest tonnage was in Cr and Cr-Mo types (90,000 tons/year each) and carbon steel plate (200,000 tpy). A typical application was with cold-forged parts that subsequently were to be heat treated, since they do not require all of the heat-treating steps needed by other alloys, and develop better machining characteristics. Steels containing less than 4.75% B can be forged, and steels with up to 6% B can be cast. In some cases, however, boron causes poorer surface characteristics, less tolerance to heat treating, loss of hardening ability during carburizing, and in excess, brittleness (Porter, 1980).

Boron master alloys (0.1–2.5% B) are also used in aluminum and titanium smelting as grain refiners (crystal nucleating agents to form a fine, uniform structure; Sussman and Evans, 1985). An iron–boron–silicon alloy sprayed on a cold object, or a rapidly cooled melt produces an amorphous glass–metal (Metglas) coating superior to sprayed metals in strength (up to 600,000 psi), magnetism (low-magnetic hysteresis), and corrosion resistance. It is also 85% more efficient in energy transformers than stainless steel. An iron–boron–rare earth (neodymium and praseodymium) alloy is an excellent magnet for automobile starter motors, stereo speakers, computer disk drives, and telecommunication printers. Boron can also be sprayed or diffused onto metals (boronizing) to produce a tough corrosion- and abrasion-resistant surface coating. Boron alloy journal pins used in oil-drilling bits have a 15% increased weight-bearing capacity, and can remain in the drill hole up to three times longer, thereby reducing round-trip time for bit replacement. Many electroplating-bath compositions include boron chemicals as buffers, cleansers, and agents to reduce the deposit's pitting and porosity (Table 9.4B). There are a wide variety of other metallurgical applications for boron and borates, as seen in the References.

9.11 NUCLEAR APPLICATIONS

In nuclear reactors the fission of radioactive material produces heat and a variety of alpha and beta particles, gamma rays and neutrons. The most effective materials for shielding the neutrons are boron (especially ^{10}B), hydrogen, lithium, polyethylene and water. Most of the shielding materials produce secondary gamma rays, which then require heat removal and further shielding. Boron is unique in its ability to absorb thermal neutrons and produce only a soft gamma ray and an easily absorbed alpha particle. The gamma rays are effectively absorbed by dense materials such as lead, steel or concrete, while the alpha and beta particles are stopped by thin sheets of metal. The ^{10}B isotope has most of the desired neutron-capture capability, and may be separated by ion exchange or the fractional distillation of boron trifluoride (BF_3) or its dimethyl ether complex $[(CH_3)_2O \cdot BF_3]$. The distillation column in one plant was 45 m (148 ft) tall, metal-packed, and operated under a partial vacuum. Boron trifluoride–dimethyl ether enriched in ^{11}B was obtained at the bottom of the column, and >90% ^{10}B collected at the top. The ^{10}B was then precipitated as potassium fluoborate and converted to B_2O_3, boric acid, ferroboron or elemental boron.

Boron carbide also is a widely used neutron-absorbing material available as pellets encapsulated in stainless steel tubes for use in control rods, or as bricks or a core between cast aluminum for shielding. Its neutron-shielding ability and high melting point makes the handling, transportation, and storage of spent fuel elements possible. Other reactor control rods are made from steel containing a minimum of 2% B, and 1% B polyethylene (as bricks and slabs that can be machined) is used as a shielding material. Borates (such as colemanite) can also be added to concrete or structural ceramics to increase their ability to absorb neutrons (Tarasevich, Isaeva, Kuznetsov and Zhenzhivist, 1990; Yarar and Bayuelken, 1994).

9.12 MISCELLANEOUS

There is a very wide range of other applications for borate products. Many boron compounds are useful in organic synthesis. Boric acid is used in nylon to control the oxidation of cyclohexane to cyclohexanol and cyclohexanone, and is a buffer during the dyeing of nylon carpet. Boron tribromide is used in the production of photopolymer-covered silicon chips for the manufacture of printed circuits, and boron trifluoride is used in the production of butyl

Table 9.6
A Partial List of Borate Uses[a]

Abrasives	Glass
Adhesives	Glazes
Alloys	Goldsmithing
Antiseptics	Hair creams
Bactericide	Herbicides
Bleaches	Hydraulic fluids
Boron filaments	Insecticides
Buffering	Leather tanning
Catalysts	Lubricating oil additives
Cement	Magnets
Ceramics	Medical applications
Cleaning compounds	Metallurgical applications
Corrosion inhibitor	Metal hardening
Cosmetics	Nuclear applications
Detergents	Nylon
Disinfectants	Organic synthesis
Dystuffs	Paints and pigments
Electrical insulation	Pharmaceuticals
Electrolytic refining	Photography
Electronic components	Plastics
Electroplating	Plating solutions
Enamels	Polymer stablisers
Enzyme stabilization	Pulp bleaching
Eye wash	Purifying speciality chemicals
Fertilizer	Pyrotechnics
Fiber optics	Refractories
Fiber glass	Shampoos
Textiles	Soil sterilant
Insulation	Swimming pool sanitizer
Composites	Taxidermy
Structures	Textile finishing
Fire, flame retardants	Textile dyes
Fluxes	Transformers
Frits	Waste treatment
Fuel additives	Wax emulsifier
Fuel (high energy)	Wire drawing
Fungicides	Wood preservative

[a] Anon., 1996c; numerous others.

rubber from isobutene. Boron tribromide and trichloride are used as catalysts in literally dozens of other organic reactions. Boron tetrahydrofuran and borane-methyl sulfide can reduce amides to amines, and carboxylic acids, aldehydes, and ketones to alcohols. Boric acid is used in the production of quinizarin by the reaction of phthalic anhydride and chlorophenol, and in the separation of terpene alcohols. Boric acid reacted with phenol, and the resulting ester treated with paraformaldehyde is used to manufacture the perfumery chemical salicylaldehyde. Sodium borohydride is a powerful reducing agent, such as with carboxylic acids, in organic synthesis (Gribble, 1996).

Zinc borate and disodium octaborate tetrahydrate are widely used as wood preservatives because of their antimicrobial and insecticidal properties (Anon., 1996b). The treatment is by a brief immersion of the timber in hot concentrated aqueous borate solutions. The borates diffuse into the wood during the dip and subsequent seasoning, leaving the wood fully preserved. The wood maintains its original color and odor, and there is no toxicity hazard to man or animals. The two borate compounds may also be mixed with wood chips before the addition of the resin and wax for hot pressing to form wood composites (Laks, 1995). An unusual application of this treatment is in the periodic spraying of an octaborate solution on the water-logged timbers of sunken ships being raised and reconstructed to prevent the wood's destruction by the wharf-borer beetle (Greenwood, 1991).

Borax is highly soluble in ethylene glycol and is useful in car antifreeze formulations, brake fluids, and hydraulic systems as a corrosion inhibitor for ferrous metals. Aqueous solutions have replaced chromates in railroad and other diesel coolants. Boron-doped silicon is used in diodes, semiconductors, transistors, and microcircuitry. A boron coating of yttrium hexaboride and erbium dodecaboride is used to trap solar radiation. Teflon cooking utensils use boron as a bond coat. Borates may also be used to improve the performance of chlorine as a swimming pool sanitizer (Anon., 1995b). There are many other uses for borates, some of which are indicated in Table 9.6 and the References.

References

Anon. (1980, May–June). Boron for crop production. *Fertilizer Progress,* pp. 24–26.

Anon. (1994). *Boron in Agriculture.* Potash and Phosphate Institute, Atlanta, Georgia.

Anon. (1995a, summer). Effective boron management. *Fluid J.* **3**(3, Issue 10), 20–23.

Anon. (1995b). Pooling chlorine with boron. *Borax Pioneer,* No. 5, pp. 10–11, *Borosilicate Glass,* pp. 12–13.

Anon. (1996a, July). The company behind the 20-mule team. *Compressed Air Mag.* pp. 30–36.

Anon. (1996b). Discovery rediscovered: Borates to rescue of historic ship. *Borax Pioneer,* No. 6, pp. 18–19.

Anon. (1996c). *Borax; 20 Mule Team.* U.S. Borax Inc., Valencia, Calif.

Anon. (1996d, September 2). Boron-containing ceramic has enhanced thermal stability. *Chem. and Eng. News* **74**(36), 22.

Anon. (1996e, October). Germany will build a plant to vitrify radioactive wastes. *Chem. Eng.,* p. 19.

Barth, R. F., Soloway, A. H., Fairchild, R. G., Brugger, R. M. (1992). Cancer, V. 70, pp. 2995.

Cheek, L., Wilcock, A., and Olsen, N. (1985). Stability of a boron-based smolder retardant finish for cotton upholstery fabrics. *Text. Res. J.* **55**(5), 271–277.

Cui, C., Zhang, X., and Liu, S. (1994). Pig iron containing boron and boron-rich slag made from ludwigite in blast furnaces. *Kuangye* (*Beijing*) **3**(4), 29, 68–72.

Edwards, J. O. and Curci, R. (1967). *Peroxides and Peroxy Compounds," Kirk-Othmer Encyclopedia of Chemical Technology,* Vol. 14, pp. 758–760, Interscience Publishers, New York.

Flannery, R. L. (1985, May–June). Understanding boron needs in crop production. *Fertilizer Progress,* pp. 41–45.

Fournier, R. F. (1977). *Illustrated Dictionary of Practical Pottery,* pp. 26–27, 52, Van Nostrand Reinhold, New York.

Gagin, L. V. (1985). Borates in glass. In *Borates: Economic Geology and Production* (J. M. Barker and S. J. Lefond, eds.), Ch. 18, pp. 267–268, Min. Eng., AIMMPE, New York.

Garrette, G. B. (1983, January–February). Guidelines for using boron soil and tissue analysis. *Fertilizer Progress,* p. 21.

Greenwood, N. N. (1991). "She burns green, Rosie—We're rich." *Royal Inst. Proc.* **63,** 153–174.

Grey, J. (1977, February 15). *Fire Retardant Composition.* Canadian Patent 1,005,202.

Gribble, G. W. (1996, December). A reduction powerhouse. *Chemtech* **26**(2), 26–31.

Inoue, K., Nagabayashi, R., Yamauchi, T., Hasegawa, M., Kinugasa, M. (1990). *Fluxed Nickel Oxide Ore Mixtures for Firing; Manufacture of Nickel Alloys.* Japanese Patents JP03253520, 5p., Nagabayaski, R., Yamauchi, T. 1990 JP04063241, 4p.

Kingery, W. D., and Vandiver, P. B. (1986). *Ceramic Masterpieces,* pp. 41, 216, 170, The Free Press, New York.

Kistler, R. B., and Helvaci, C. (1994). Boron and borates. In *Ind. Min. & Rocks* (D. D. Carr, ed.), 6th ed., pp. 171–186, Soc. Min Met. Explor., Littleton, Colorado.

Laks, P. (1995). Protecting wood composites. *Borax Pioneer,* No. 5, pp. 7–9.

Lu, S. (1990). *Solid Combustion Aids for Coal.* Japanese Patent CN1092459.

Luehrsen, E., and Ott, A. (1990, February 14). *Manufacture of Boron Nitride Nozzles For Molten Metals.* European Patent 354,304.

Lyday, P. A. (1985). End uses of boron other than glass. In *Borates: Economic Geology and Production* (J. M. Barker and S. J. Lefond, eds.), Ch 7, pp. 257–268, Soc. Min. Eng., AIMMPE, New York.

Maas, E. V. (1986). Salt tolerance of plants. *Applied Agricultural Res.* **1**(1) 12–26.

Mack, H. J. (1986, April). Boron. *Agrichemical Age,* pp. 12A, 281.

Mark, H. F., and Atlas, S. M. (1966). Man-made fibers. *Kirk-Othmer Encyclopedia of Chemical Technology,* Vol. 9, p. 159, Interscience Publishers, New York.

Nepin, Y. N., and Sapunova, N. A. (1991). Composition of sulfate liquors obtained in autocausticization with natural boron-containing ores. *Bum. Prom-st.* **2,** 14–15.

Oberhofer, A. W., Benko, J. J., Drozd, J. C. (1976, May 25). *Cleaner for Automotive Engine Cooling Systems.* U.S. Patent 3,959,166.

Porter, L. R. (1980). The present status and future of boron steels. *Boron in Steels.* Metall. Soc. AIME, N.Y., pp. 199–211.

Qui, Z. (1994). Application of molten salts in the smelting and casting of nonferrous metals in ancient China. *Youse Jinshu* **44**(1), 68–70.

Raymond, K., and Butterwick L. (1992). Perborate. *Handbook of Environmental Chemistry* (O. Hutzinger, ed.), Vol. 3, Pt. F, pp. 305–310, Springer-Verlag, New York.

Rhodes, D. (1975). *Clay and Glazes for the Potter,* pp. 94, 110, Chilton Book Co., Radnor, Pennsylvania.

Rosenfelder, W. J. (1978, June 17). The industrial uses of boron chemicals. *Chemistry and Industry,* pp. 413–416.

Roskill, (1993). *The Economics of Boron.* Roskill Information Service Ltd., London.

Russell, A. (1991, November). Minerals in fiber glass. *Industrial Minerals,* pp. 27–41.

Segars, W. I. (1987, March). Don't pass by boron. *Farm Chemicals,* pp. 60–64.

Shelley, S. (1994, October). Borosilicate ductwork prevents fire propagation. *Chem. Eng.,* p. 179.

Sprague, R. W. (1972). *The Ecological Significance of Boron.* U.S. Borax Research Corp., Valencia, Calif., Ward Ritchie Press, New York.

Stoughton, W. J. (1990). *Smelting of Fluxed Ores for Aluminum Recovery.* U.S. Patent 5,332,421.

Sussman, R. C., and Evans, L. G. (1985, October 2). *Boron Alloy.* European Patent 156,459.

Tarasevich, B. P., Isaeva, L. B., Kuznetsov, E. V., Zhenzhurist, I. A. (1990). Boron-containing structural ceramics protecting against neutron radiation. *Steklo Keram.* **5,** 17–19.

Wang, L., and Shao, W. (1990). *Coasting Process for Metallic Pipes with Molten Glass-Based Glazes.* Chinese Patent CN1061363.

Wells, F. L., Schattner, W. C., and Ekwell, L. E. (1971). Manufacture of dissolving pulps by extraction in sodium hydroxide-borax solutions. *Tappi* **54**(4), 525–529.

Xu, G. (1990). *Composition for Killing Oncomelania.* Chinese Patent CN1077589.

Yarar, Y., and Bayuelken, A. (1994). Investigation of neutron shielding efficiency and radioactivity of concrete shields containing colemanite. *J. Nucl. Mater.,* V. 212–215 (Pt. B), pp. 1720–1723.

Zambrano, A. R. (1985, April 9). *Ferroboron.* U.S. Patent 4,509,976.

Zhuruli, M. A., Mazmishvili, S. M., Tsinadze, P. S., Sumongulov, Z. A., Mchedlidze, T. A., Martynov, S. V. (1992). Dust-containing briquets for smelting manganese alloys. *Izobreteniya* **31,** 96.

Chapter 10 Borate Industry Statistics

10.1 WORLD BORATE PRODUCTION

10.1.1 Argentina, Bolivia, Chile, and Peru

Argentina is the world's third largest borate producer, having buried deposits at Tincalayu, Sijes and Loma Blanca, and as with the other Puna Region countries, numerous ulexite playas. The playas were worked intermittently from perhaps the 1700s until the early 1900s when the borate prices were comparatively high and could justify operations in this remote, cold region with high transportation costs. Then after several decades with very little production, the playas again began to be operated when more modern truck transportation greatly reduced the haulage costs for product and supplies, and demand in the fiberglass industry gave ulexite a renewed value. Labor costs have remained low for the area's indian workers (there are few other jobs in the area), and inexpensive by-product sulfuric acid from the nearby copper smelters has encouraged some conversion of ulexite to boric acid. When open pit mining was initiated on the buried deposits it also led to much larger, more consistent, and less expensive production from these deposits (Table 10.1).

10.1.2 Turkey

Turkey has by far the world's largest borate reserves, but until the 1990s they had less purified borate capacity than the United States. Their near-shipping-port priceite (pandermite) deposits began to be operated in the late 1800s, and continued as the country's only borate producers until 1950–1970 when the large colemanite, ulexite, and borax deposits were discovered and slowly commercialized. Initially only crude priceite, colemanite or ulexite were shipped, but in ever increasing tonnage. Then washing facilities were installed to upgrade the products, and finally large scale purified borax, borates, and boric acid facilities were built. The total borate production is listed in Table 10.2A, and the individual operations in Table 10.2B.

10.1.3 United States

Borate production and product value figures for the United States are listed in Table 10.3, and shown graphically in Fig. 10.1. In the initial period (1864–

Table 10.1

Borate Production in South America (the Puna Region; 1000 metric tons)[a]

Year	Production	Year	Production	Year	Production	Year	Production
1. Argentina							
1995	245	1986	192?	1977	83	1969	29.2
1994	215	1985	158	1976	81	1968	21
1993	146[b]	1984	143	1975	77	1967	16
1992	125	1983	114	1974	73	1964	12
1991	116	1982	124	1973	63	1961	7.6
1990	144	1981	125	1972	55	1957[c]	36
1989	261?	1980	156	1971	35	1956[d]	45
1988	270?	1979	134	1970	32	1955[e]	26
1987	185?	1978	127				
2. Chile							
1995	90	1985	5.0	1965	4.60	1910	30
1994	86	1984	3.6	1964	3.30	1903	15.73
1993	117	1983	1.0	1963	2.98	1902	14.33
1992	203?	1982	—	1962	3.81	1901	11.55
1991	97	1981	3.0	1961	0.16	1900	13.18
1990	132	1980	3.28	1960	2.92	1899	14.95
1989	131	1979	3.05	1959	5.76	1898	7.03
1988	32	1978	26.4	1958	8.4	1897	3.17

1987	13	1977	4.5	1957	5.8	1896	7.49
1986	6	1966[f]	0.40	1956	9.2	1895	4.53
				1913	50	1894	6.70
3. Peru							
1995	30	1988	15	1981	16	1901	4.16
1994	30	1987	23	1980	20.9	1900	7.08
1993	37	1986	23	1979	11.8	1899	7.64
1992	27	1985	10	1978	6.4	1898	7.18
1991	26	1984	10	1977	5.5	1897	11.85
1990	20	1983	10	1913	38.0	1896	1.18
1989	25	1982	14	1903	2.58	1895	4.00
				1902	5.06	1894	0.80
4. Bolivia							
1995	7	1992	23	1989	10	1902	0.59
1994	10	1991	14	1988	1	1901	3.07
1993	12	1990	3	1903	1.21		

[a] U.S. Bureau of Mines; Matterson, 1980.
[b] As ulexite. Some was converted to boric acid; plants' capacity 30,000 metric/year (Pocovi, Latre and Skaf, 1994).
[c] Borax only.
[d] 20,000 metric ulexite, <25,000 metric borax.
[e] 13,000 metric ulexite, 23,000 metric borax.
[f] From this date and earlier the ulexite was sold as 33% B_2O_3.

Table 10.2A

Borate Production in Turkey, 1000 Metric Tons[a,b]

Year	Production		Year	Production	Year	Production	Year	Production
	Concentrates[c]	Refined[c]						
1995	1130	234.0	1974	728	1953	6.44	1929	13.5
1994	1213	255.0	1973	636	1952	13.7	1928	14.9
1993	1217	242.5	1972	618	1951	12.1	1927	17.9
1992	1038	216.1	1971	571	1950	9.76	1926	15.6
1991	1210	—	1970	304	1949	7.08	1925	16.1
1990	1250	—	1969	323	1948	5.31	1924	11.0
1989	1174	—	1968	266	1947	3.61	1898–1901	Σ39.9
1988	1231	—	1967	229	1946	2.21	1897	11.4
1987	980	—	1966	225	1945	5.03	1896	12.6
1986	1010	89.3	1965	171	1941	5.04	1895	9.08
1985	924	59.0	1964	128	1940	5.02	1894	9.10
1984	907	75.1	1963	88.1	1939	15.2	1893	9.00
1983	618	67.2	1962	114	1938	4.06	1892	9.00
1982	655	47.4	1961	88.6	1937	4.66		
1981	758	65.2	1960	71.0	1936	6.48		
1980	643	63.1	1959	43.3	1935	5.08		
1979	628	51.4	1958	69.4	1934	7.52		
1978	472	47.8	1957	27.4	1933	7.55		
1977	344	64.4	1956	29.3	1932	4.99		
1976	330	56.7	1955	42.2	1931	6.50		
1975	349	37.4	1954	13.3	1930	5.46		

[a] Production includes ore and refined products.
[b] U.S. Bureau of Mines.
[c] 1992–1995, and refined are from Etibank, 1995, 1994; 1975–1986 refined are from Anac, 1988.

Table 10.2B

Recent Turkish Plant and Mine Capacity and Production (metric tons/year)[a]

1. Banderma Borax and acid works				
Capacity: Borax deca and pentahydrate	55,000			
Boric acid	135,000			
Sodium perborate	20,000			
Production:	1995	1994	1993	1992
Borax decahydrate	27,099	30,090	20,000	18,121
Borax pentahydrate	11,167	9,273	7,500	5,901
Boric acid	47,000	46,100	35,000	23,203
Perborate	11,167	15,067	20,000	13,923
2. Bigadic colemanite works				
Capacity: Colemanite concentrate	200,000			
Uxexite concentrate	200,000			
Production:	1995	1994	1993	1992
Colemanite mined	181,950	159,600	250,000	214,800
Colemanite concentrate	140,560	93,900	163,900	143,100
Ulexite mined	267,400	328,150	250,000	200,700
Ulexite concentrate	173,000	198,200	171,000	135,000
3. Emet colemanite works				
Capacity: Colemanite concentrate	500,000	(Hisarcik 450,000; Espey 60,000)		
Production:	1995	1994	1993	1992
Colemanite mined	655,000	700,000	630,000	470,000
Colemanite concentrate	351,425	305,250	261,600	169,425
4. Kestlek colemanite works (reported with Bigadic from 1995 on; only 1994 listed here)				
Production: Colemanite mined		116,895		
Colemanite concentrate		37,956	(capacity 100,000)	
Lumpy colemanite		19,281		
5. Kirka borax works				
Capacity: Borax concentrate	600,000	Anhydrous 60,000		
Borax pentahydrate	160,000			
Production:	1995	1994	1993	1992
Borax (tincal) mined	658,800	733,000	915,000	842,000
Borax concentrate	465,000	558,000	620,000	590,000
Borax pentahydrate	137,617	154,449	160,000	155,000
6. Totals	1995	1994	1993	1992
Concentrates	1,129,985	1,212,587	1,216,500	1,037,525
Refined products	234,050	254,979	242,500	216,148

[a] Etibank, 1996.

Table 10.3

Borate Production in the United States (1000 metric tons)[a]

Year	Borate production		Value, thousand$	Year	Borate production		Value, thousand$
	Total	B_2O_3			Total	B_2O_3	
1996	1150	581	519	1936	—	85.5	—
1995	1190	728	560	1935	—	74.5	—
1994	1110	550	443	1934	—	66.4	—
1993	1060	574	373	1933	—	50.9	—
1992	1010	554	338.7	1932	—	49.6	—
1991	1240	626	442.5	1931	—	48.7	—
1990	1090	608	436.2	1930	—	48.4	—
1989	1114	562	429.8	1929	—	(57)	—
1988	1149	578	429.7	1928	—	(48)[b]	—
1987	1256	625	475.1	1927	—	(43)	—
1986	1135	571	426.1	1926	—	(44)	—
1985	1151	577	404.8	1925	—	45.5	—
1984	1243	606	456.7	1924	—	(45)	—
1983	1185	579	439.2	1923	—	(48)	—
1982	1122	552	384.6	1922	—	(38)	—
1981	1346	673	435.4	1921	—	(41)	—
1980	1405	712	366.8	1920	—	(47)	—
1979	1445	726	310.2	1919	—	(41)	—
1978	1413	707	279.9	1918	—	(43)	—
1977	1335	668	236.2	1917	—	(48)	—
1976	1133	573	184.9	1916	—	(45)	—
1975	1065	548	158.8	1915	—	(41)	—
1974	1077	563	128.3	1914	—	(38)	(1,000$)
1973	1114	604	113.6	1913	52.77	(24)	1,491.5[c]
1972	1019.1	551.8	95.88	1912	38.47	(25)	1,127.8[c]
1971	951.8	516.4	89.86	1911	48.48	(24)	1,569.2[c]
1970	946.4	510.9	86.83	1910	38.51	(27)	1,201.8[c]

1969	927.3	500.9	81.26	1909	37.67	(25)	1,534.4[c]
1968	875.5	471.8	76.54	1908	22.73	(18)	975.0[c]
1967	810.9	430.0	69.82	1907	48.05	(29)	1,121.5[c]
1966	787.4	420.0	68.21	1906	52.88	(31)	1,182.4[c]
1965	733.6	386.4	64.18	1905	42.12	(25)	1,119.2[c]
1964	705.5	368.2	60.87	1904	41.50	(24)	698.8[c]
1963	636.4	335.5	54.98	1903	31.24	(19)	661.4[c]
1962	588.2	308.2	49.34	1902	49.73	18.19[c]	2,538.6[c]
1961	548.2	284.5	46.94	1901	30.77	21.12[c]	1,021.1[c]
1960	582.4	294.5	47.55	1900	26.38	23.49[c]	1,013.3[c]
1959	563.6	285.7	46.15	1899	21.83	18.51[c]	1,139.9[c]
1958	480.2	241.5	38.31	1898	13.91	8.30[c]	1,153[c]
1957	491.9	244.8	38.04	1897	17.6	7.27[c]	1,080[c]
1956	495.2	243.5	32.81	1896	12.31	6.14[c]	675.4[c]
1955	840.5	266.5	24.36	1895	6.13	5.42[c]	595.9[c]
1954	718.6	209.5	26.71	1894	5.95	5.23[c]	807.8[c]
1953	650.2	193.9	17.67	1893	(5)	3.60[c]	593.3[c]
1952	530.8	153.7	14.11	1892	(6)	5.02[c]	838.8[c]
1951	784.4	219.1	20.03	1891	(6)	3.88[c]	640.0[c]
1950	588.9	173.6	15.89	1890	4.32	2.91[c]	480.2[c]
1949	425.1	126.5	11.51	1889	4.00	0.877[c]	145.5[c]
1948	409.9	122.5	11.15	1888	3.56	1.277[c]	196.6[c]
1947	456.3	132.5	11.84	1887	5.00	0.923[c]	116.7[c]
1946	391.5	118.0	9.576	1886	4.44	1.168[c]	173.5[c]
1945	296.3	95.1	7.635	1885	3.64	0.856[c]	155.4[c]
1944	252.4	83.4	6.580	1884	3.18	0.926[c]	198.7[c]
1943	233.3	79.6	6.402	1883	2.95	0.818[c]	265.5[c]
1942	206.1	70.5	5.734	1882	1.93	0.665[c]	201.3[c]
1941	273.9	86.5	6.786	1881	1.84	0.627[c]	189.8[c]
1940	221.2	73.5	5.643	1880	1.75	0.554	149.2
1939	223.0	74.4	5.690	1879	0.72	0.331	65.4
1938	196.1	59.0	4.739	1878	1.27	0.334	66.3
1937	226.3	98.2	7.233	1877	1.69	0.903	193.7

continues

Table 10.3 *(continued)*

Year	Borate production		Value, thousand$	Year	Borate production		Value, thousand$
	Total	B_2O_3			Total	B_2O_3	
1876	2.35	1.300	312.5	1868	0.029	0.011	22.4
1875	2.47	1.062	289.1	1867	0.200	0.073	156.1
1874	1.82	0.832	259.4	1866	0.183	0.067	132.5
1873	0.909	0.468	255.4	1865	0.114	0.042	94.1
1872	—	0.127	89.6	1864	0.011	0.004	9.48
1869 –1871	0	0	0				

[a] Various authors.
[b] ( ) means rough estimate.
[c] In California only; total borate, not B_2O_3.

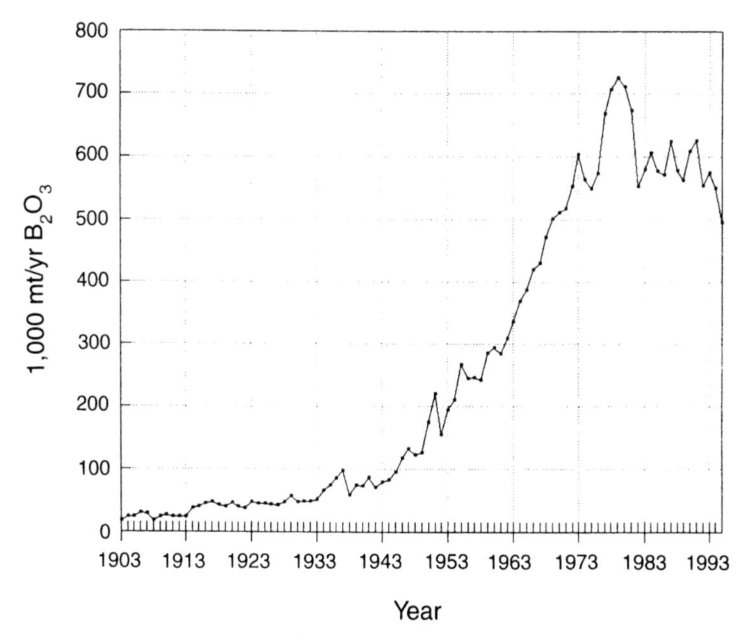

Figure 10.1 United States borate production. (From Lyday, 1995.)

1868) the production was entirely from Borax Lake, California, with the 1864 price being \$858/ton. The price fell steadily to \$110/ton in 1895, and to a low of \$17/ton in 1904. Next there was some production from Little Borax Lake (1872–1873), followed by the "playa period" in Nevada and California (1872–1909), and then by operation of the colemanite mines of California and Nevada (1889–1927). After the production of refined products from Searles Lake brine in about 1919, and discovery of the Boron deposit in 1925, the sale of refined borates rapidly increased.

10.1.4 Other Countries; World Production

Estimates of the total world borate production, and that of various smaller producers are shown in Table 10.4. The lists are somewhat uncertain, however, because of combining crude and refined products, and difficult guesses for various of the countries' production. It indicates a strong peaking of sales in 1988, and plateauing since then (to 1996), or even showing a small decline. However, the apparent drop-off is primarily a function of more concentrates or higher-purity products being sold, and the influence of some declining markets (e.g., cellulosic fire proofing) and normal business cycles. It has been predicted that a general 0–2%/year increase in production should resume in the late 1990s.

Table 10.4

World Borate Production, and that of Miscellaneous Countries (1000 metric tons/yr)[a]

Year	Production, 1000 mt	Year	Production, 1000 mt	Year	Production, 1000 mt	
					Total	Est. B
China[b]		**India**				
1996	180	1975	0.20[c]	1976	182	
1995	140	1901	0.16	1973	182	
1994	188	1900	0.22	1972	182	
1993	155	1899	0.25	1971	144	
1992	127	1898	0.18	1970	141	
1991	93	1897	0.28	1969	141	
1990	75	1896	0.34	1968	140	
1989	74	1895	0.40	1967	138	
1988	95	1894	0.37	1966	67[b]	
1987	27			1965	65[b]	
1986	27			1964	64[b]	
1985	27	**Iran**		1958	480	
1984	27	1993–1996	1/yr	1957	492	
1983	27	1992	<0.5	1956	859	
1982	27	1991	2	1955	840	
1981	27	1990	1			
1980	27	1989	<0.5			
1979	27	1988	2			
1978	27	1987	1			
1977	27			**World Total**		Est. B
1976	23			1996	2980	—
1972	32	**Italy**		1995	2890	—
1971	32	1990–1996	2/yr	1994	2860	—
1970	31	1977	5.9	1993	2690	—
1969	30	1957	3.6	1992	2670	—
1968	27	1956	3.7	1991	2960	402
1967	27	1955	3.7	1990	2910	411
		1903	2.58	1989	2968	390
		1902	2.76	1988	2994	391
		1901	2.55	1987	2685	366
Finland		1900	2.49	1986	2511	341
1980	2.47	1899	2.67	1985	2505	341
1979	3.17	1898	2.65	1984	2524	340
1978	4.62	1897	2.70	1983	2240	304
		1896	2.62	1982	2275	307
		1895	2.63	1981	2564	352
Germany		1894	2.75	1980	2615	362
1990–1996	2/yr			1979	2763	352
1976	4.2			1978	3123	365
1975	4.2	**Russia (Kazakstan)**		1977	2338	333
1974	4.2	1993–1996	80/yr	1976	2345	277
1903	0.16	1992	130	1975	1997	282
1902	0.20	1991	160	1974	2231	312
1901	0.18	1990	180	1973	2108	295
1900	0.23	1989	200	1972	1908	277

Table 10.4 (*continued*)

Year	Production, 1000 mt	Year	Production, 1000 mt	Year	Production, 1000 mt	
					Total	Est. B
Germany		**Russia (Kazakstan)**		**World Total**		
1899	0.18	1988	200	1971	1736	254
1898	0.23	1987	200	1970	1455	218
1897	0.20	1986	200			
1896	0.18	1985	200			
1895	0.15	1984	200			
1894	0.18	1983	200			
		1982	200			
		1981	200			
		1980	200			
		1979	200			
		1978	200			
		1977	182			

[a] Various authors.
[b] B_2O_3.
[c] Announced new capacity.

10.2 BORATE RESERVES

Very uncertain estimates of the world's borate reserves are shown in Table 10.5. The values have been taken from recent articles on the deposits, annual reports, or U.S. Geological Survey estimates, and in some cases probably represent optimistic guesses of proven and inferred resources, regardless of their economic recovery potential or ore grade.

10.3 BORATE PLANT CAPACITY

Estimates for the rated plant capacity for refined products from the world's borate producers are shown in Tables 10.2B and 10.6. The numbers are based on recent estimates or projections, and some probably overstate the actual capabilities of the plants.

10.4 UNITED STATES CONSUMPTION OF BORATES

The consumption of borates in the United States, as tabulated by end use, is listed in Table 10.7, and shown graphically in Fig. 10.2. The averages of end uses in 1982–1992 are shown in Fig. 10.3. There is a rather surprising variability

Table 10.5

Estimated World Borate Reserves (million metric tons)

Country/deposit	Ore	B_2O_3
Argentina		
Loma Blanca	20	2.7
Sijes	20	4
Tincalayu	30	4.8
Salars, ulexite	30	6
Brine (H. Muerto)	—	3
Subtotal	100	20.5
Bolivia		
Salars, ulexite	20	5
Brine (Uyuni)	—	10
Subtotal	20	15
Chile		
Nitrate deposits	—	56
Salars, Surire	34.5	5
Others, ulexite	25	5
Brine (Atacama)	—	10
Subtotal	59.5	76
China		
Liaoning	400	40
Other skarns	50	5
Playas, borates	30	10
Brine	—	10
Subtotal	480	65
Mexico		
Mesa del Amo	10	1
Vitro-U.S. Borax	100	10
Tubutama	30	2.5
Subtotal	140	13.5
Peru		
Salar de Salinas	10	3
Other salars	10	2
Subtotal	20	5
Russia		
Dalnagorsk	600	54
Inder	34	1
Satimola	43	4
Others	100	10
Subtotal	777	69
Turkey		
Bigadic	935	330
Emet	545	200
Kestelek	7	3
Kirka	520	140
Subtotal	2007	673
United States		
Boron,		
Na-borates	113	26
Ca-borates	198	20
Subtotal	311	46
Searles Lake	—	43
Death Valley	38	8
Hector	120	7
Owens Lake	—	3
Salton Sea	—	5
Four Corners	94	10
Muddy Mountains	3	1
Subtotal	566	123
Yugoslavia		
Jarandol Basin	30	6
Raska	10	2
Subtotal	40	8
Others	100	10
World total	4310	1078

Table 10.6

Estimated World Borate Refined Product Capacities (1000 metric tons)

Country/location	Capacity	Product	Country/ location	Capacity	Product
Argentina			**Turkey**		
Salta, others (combined)	30	Boric acid	Banderma	55(95)[a]	Borax or pentahydrate
Campo Quyano	38	Borax or pentahydrate		135(150)[a]	Boric acid
Bolivia	10	Boric acid		20	Sodium perborate
	4	Borax or pentahydrate	Bigadic	200	Colemanite concentrate
				200	Ulexite concentrate
Chile					
Salar operations	30	Boric acid	Kestelik	100	Colemanite concentrate
	10	Mixed borates	Emet		
Atacama	(18)[1]	Boric acid	Hisarcik	450	Colemanite concentrate
China			Espey	60	Colemanite concentrate
Liaoning	110	Borax	Kirka	600	Borax concentrate
	18	Boric acid		160(320)[a]	Pentahydrate or borax
Other (from playas)	40	Mixed borates		60	Anhydrous
Peru			**United States**		
Salar operations	12	Boric acid	Boron	160(200)[a]	Boric acid
	15	Mixed borates		520	Mixed borates
Russia			Searles Lake	30	Pentahydrate or borax
Dalnagorsk	250	Boric acid		33	Anhydrous
	50	Mixed borates	West End	27	Pentahydrate
Other	50	Mixed borates	Death Valley	45	Colemanite concentrate
			Hector	(30)	Synthetic cole.

[a] Total after expansion announced in 1996.

Table 10.7

United States Consumption of Borates by End Use[a] (metric tons B_2O_3)[b]

End Use	1994	1993	1992	1991	1990	1989	1988	1987	1986	1985	1984	1983	1982
1. Agriculture	21,200	14,800	9,186	5,712	5,787	14,583	14,362	13,474	13,474	13,639	13,639	12,895	9,800
2. Borosilicate glasses	27,400	64,500	29,288	29,504	27,905	30,600	34,136	28,016	27,965	31,481	29,471	31,488	27,100
3. Enamels, frits, glazes	15,400	10,300	8,230	5,878	7,882	8,149	9,923	11,241	10,686	11,177	10.156	10,160	10,400
Fire retardants:													
4. Cellulosic insulation	15,800	9,670	9,451	11,662	11,741	11,920	11,167	11,792	17,197	24,111	26,500	27,476	28,300
5. Other	1,360	286	1,394	1,769	1,401	509	641	968	379	271	1,593	1,194	1,700
6. Insulation fiberglass	97,000	106,000	124,950	82,395	95,059	91,736	103,355	111,968	107,420	94,082	106,774	83,121	52,500
7. Metallurgy	1,950	1,870	2,624	3,518	2,554	3,040	5,175	3,839	2,808	3,006	3,756	3,427	3,100
8. Miscellaneous	23,800	32,600	23,130	16,693	14,586	14,729	28,824	18,850	25,092	20,474	18,741	21,989	19,900
9. Nuclear	395	8,870	455	546	546	611	679	536	981	989	1,006	955	600
10. Soaps & detergents	14,000	13,800	38,460	22,848	24,230	25,145	26,446	22,046	22,271	22,316	26,095	27,646	24,500
11. Distributors[c]	51,300	37,200	37,575	36,264	71,670	73,460	75,352	77,334	33,880	40,134	39,296	35,892	34,500
12. Textile fiberglass	26,000	21,300	60,173	45,067	55,698	40,499	45,420	34,950	45,120	65,259	63,518	53,443	28,700
13. Total	296,000	321,000	344,916	261,856	319,059	314,981	355,480	335,015	307,273	326,939	340,601	309,687	241,900

[a] Includes imports.
[b] U.S. Bureau of Mines.
[c] Sold to distributors, with the end use unknown.

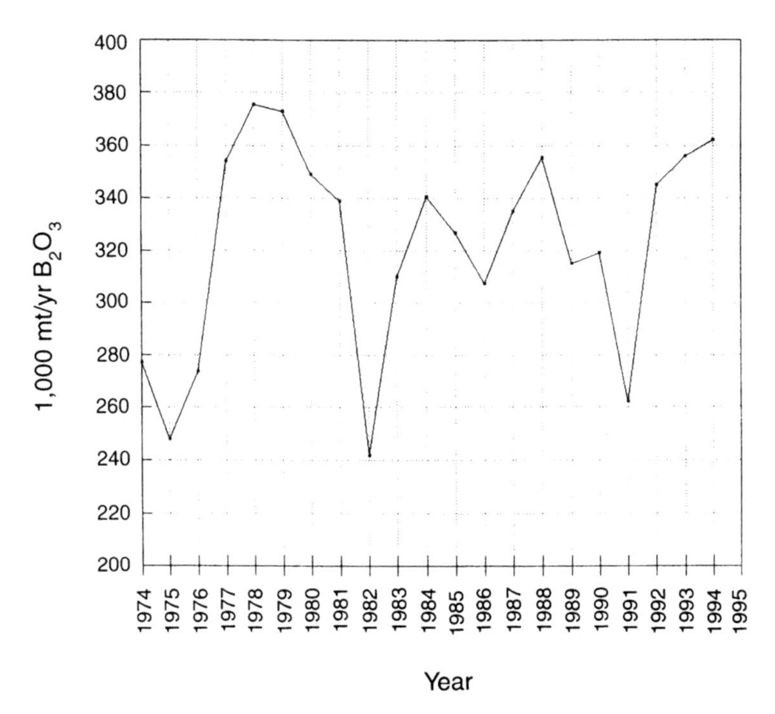

Figure 10.2 United States borate consumption. (From U.S. Bureau of Mines.)

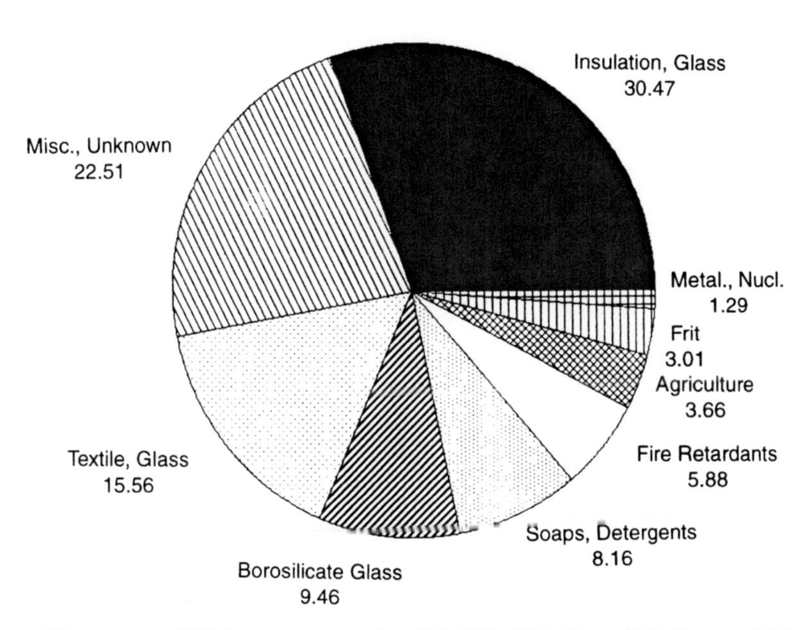

Figure 10.3 Average U.S. borate consumption (%) 1982–1992. (From U.S. Bureau of Mines.)

in the end use with time, and the total U.S. borate consumption appears to have a 7- to 9-year cycle, with no net growth (even some decline) from 1974 to 1994. The use of borates in fiberglass for insulation and fabrics is by far the largest sales area (46% of the total), but miscellaneous small uses at 22–25% of the total consumption appears to be the highest growth area. It is also predicted that U.S. borate consumption should have an average growth of 2%/year.

10.5 UNITED STATES BORATE PRICES

The U.S. year-end list prices (actually, the average yearly price for the early period) for borax pentahydrate are listed in Table 10.8, and shown graphically

Table 10.8

Time–Price Relationships for Borax Pentahydrate[a,b]

Year	December 31 U.S. price, $/mt	Year	December 31 U.S. price, $/mt
1958	62.7	1977	130
1959	63.8	1978	141
1960	63.8	1979	184
1961	63.8	1980	186
1962	63.8	1981	205
1963	63.8	1982	222
1964	63.8	1983	222
1965	63.8	1984	229
1966	68.2	1985	236
1967	68.2	1986	243
1968	73.7	1987	249
1969	73.7	1988	259
1970	82.5	1989	249
1971	82.5	1990	249
1972	83	1991	247
1973	88	1992	272
1974	108	1993	304
1975	116	1994	324
1976	121	1995[c]	324
		1996	375

[a] Borax pentahydrate, technical, granular, 99.5%, bulk, carload, FOB works.

[b] U.S. Bureau of Mines.

[c] Yearly average price increase 1958–1995 = 4.5%.

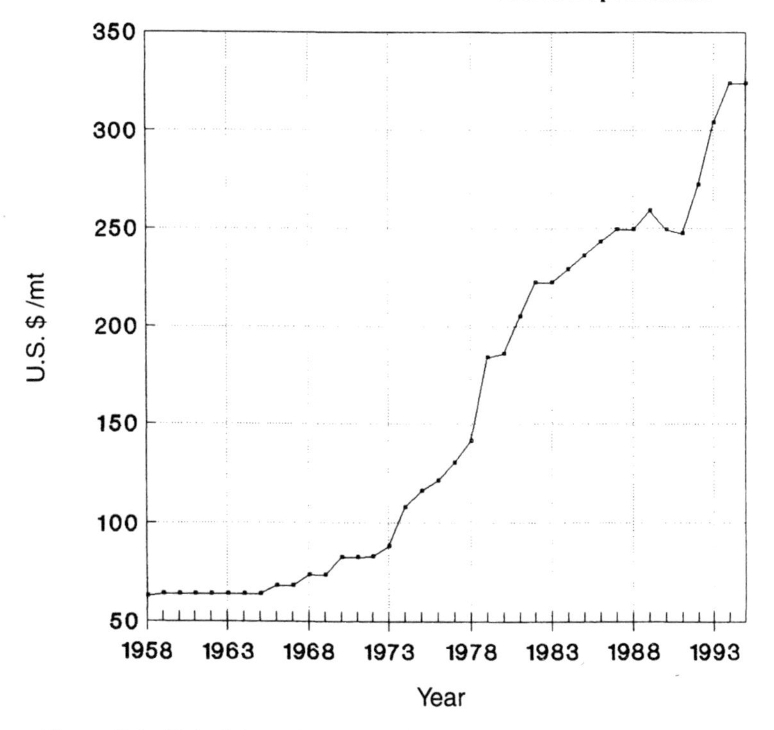

Figure 10.4 United States borax pentahydrate prices. (From Lyday, 1995.)

in Fig. 10.4. This product has had an average compounded price increase of 4.5%/year, and for 1967–1994, an increase of 5.8%/year. This rate of price increase is fairly typical of that for the other borate products.

10.6 UNITED STATES EXPORTS AND IMPORTS

Figure 10.5 is a graph of the U.S. borate exports, indicating that about $\frac{2}{3}$ of the U.S. production is exported. The amount of U.S. imports is comparatively small, with the specific amounts being (as 1000 metric tons/year): borax 9, boric acid 16, colemanite 45, and ulexite 153 in 1995, with a total net value of about $59 million.

10.7 PRODUCT SPECIFICATIONS

Typical borate product specifications are shown in Tables 10.9, 10.10, and 10.11. Even though differences do appear among suppliers, basically each of

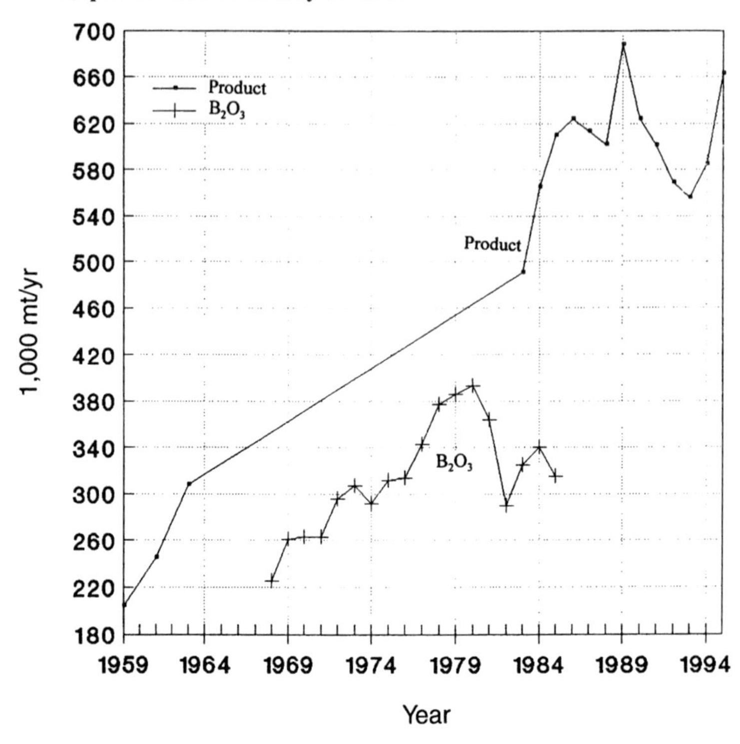

Figure 10.5 United States borate exports. (From Lyday, 1995.)

the companies listed produces very highly refined borate products. One notable difference shown in the tables is in the bulk densities of the products, but this almost certainly reflects merely a difference in the testing method. The products are shipped in bulk, using 1-ton bulk bags, or 50- and 100-lb (25- and 50-kg) plastic-lined paper bags.

Table 10.9

Typical Grades and Analyses of Turkish Borate Products[a]

1. Colemanite	Bigadic			Kestelek			
Particle size (mm)	25–125	3–25	0.2–3	300	25–100	−25	−3
B_2O_3 (%)	42	36	27	42	42	38	36
CaO (%)	26–29	27–29	29–30	25–29	26.5–28	25–26	24–25
SiO_2 (%)	2–4	3–5	7–10	2.5–4	2.5–4.8	6–10	15
SO_3 (%)	0.1–0.6	0.2–0.8	0.4–1.0	0.10–0.30	0.15–0.20	0.50	0.50–0.65
As (ppm)	13–23	13–23	13–23	20–40	25–45	35–55	35–55
	Espey			Hisarcik			
Particle size (mm)	25–100	3–25	0.2–3	25–100	3–25	0.2–3	
B_2O_3 (%)	42	38	36	42	38	36	
CaO (%)	24–26	25–28	23–26	24–27	23–27	23–26	
SiO_2 (%)	3–6	4–7	5–10	3–6	4–7	5–10	
As (ppm)	600–1000	600–1200	600–1200	1800–3200	2200–3500	2200–3800	
	2. Ulexite		3. Borax Products				
	Bigadic		Tincal	Borax		Pentahydrate	
Particle size (mm)	3–125	0.2–3	3.5% + 6 18% 4–6 50% 1–4 28% .2–1	Crystalline 0.5% max +1;4% max −0.075	Powder 0% + 1; 15% max −0.075	10% max +1; 0.5% max −0.075	Etibor −48 3% max +1.2;4% max−0.075
B_2O_3 (%)	38	25	32	36.47 min		47.75 min	47.80 min
Na_2O (%)	4–6	3–5	15	16.48		21.41	21.60
CaO (%)	16–18	17–26	—	Purity 99.9% min		99.9% min	100%
SiO_2 (%)	1–3	5–13	—	—		—	—
	4. Anhydrous borax	5. Boric acid Crystalline	Powder	6. Sodium perborate			
Particle size (mm)	0.07–1.40	1% max +1; 4% max −0.075	1% max + 0.425; 20% min −0.075	0.1–1.0 pH 10.30 ±0.05			
B_2O_3, %	68.9	56.25		22.7 min			
Na_2O, %	30.7	—		20.1 min			
SO_4 (ppm)	—	Normal, 500 max; Low S, 99.9 max		Active oxygen 10.0% min			
Purity (%)	—	Normal 99.8	Low S, 99.9				

[a] Etibank, 1996.

Table 10.10

Typical Chemical and Screen Analyses of Searles Lake Borate Products[a]

A. Chemical analyses (wt %)

			Anhydrous		Boric acid	
	Borax	Pentahydrate	Glass	Pyrobor	Granular	Powdered
Purity	100.0[b]–105	101.5[b]–103.5[c]	99.0[b]–99.8	99.0[b]–99.8	99.8[b]–101.5	99.8[b]–101.5
B_2O_3	36.5–38.3	48.5–49.5	68.5–69.1	68.5–69.1	56.2–57.1	56.2–57.1
$Na_2B_4O_7$	52.8–55.4	70.1–71.5	99.0–99.8	99.0–99.8	—	—
Na_2O	16.3–17.1	21.6–22.0	30.5–30.7	30.5–30.7	—	—
H_2O (cryst.)[d]	44.6–47.2	28.5–29.9	—	—	—	—
Na_2SO_4	—	0.002–0.05[c]	0.002–0.07[c]	0.002–0.10[c]	0.02–0.15[c]	0.03–0.50[c]
Cl	0.01–0.05[c]	0.02–0.08[c]	0.03–0.10[c]	—	35–90[c] ppm	40–150[c] ppm
Fe (ppm)	—	2–15[c]	10–50[c]	10–50[c]	—	—
Magnetic Fe (ppm)	—	—	0.2–2.5[c]	0.2–2.5[c]	—	—
Excess Na_2O	—	—	—	0–0.5	—	—

B. Screen analyses (cumulative wt %)

	Borax				Anhydrous		Boric acid
Tyler mesh	Standard	40/200 mesh	Extra coarse	Pentahydrate	Glass, Pyro	Pyro 30/100	granular
10			0–1[c]	12 mesh 0–2[c]	12 mesh 0–2[c]		
20			20–60	1–40	10–50	0–2	0–1[c]
30	0–5[c]		60–85	10–65	20–75	0.1–10[c]	
40	10–40	0–5[c]	90–99				2–50
50				35–95			
60	50–80	30–55					15–90
100	75–95	65–85[c]		65–98	65[b]–95	70[b]–90	65–98 Powder 15–35[c]
200		90–97			85–99	80–98	90–99

[a] North American Chemical Company, 1994.
[b] Minimum standard specification.
[c] Maximum standard specification.
[d] Water of crystallization.

Table 10.11

Physical Properties of Borate Products[a,b]

	Borax	Pentahydrate	Anhydrous	Boric acid	Sodium perborate
Molecular weight	381.372	291.296	201.219	61.833	153.860
PerCent B_2O_3	36.510	47.800	69.198	56.297	22.624
Density (g/cc)	1.73	1.815	2.36	1.435	1.7
Turkey: Bulk Density (g/cc)	Crystalline 0.813 Powdered 0.750	0.958 Etibor48 0.98	1.25	Crystalline 0.800 Powdered 0.700	0.65–0.75
NACC: Bulk Density (g/cc)	Standard 0.90[c]–1.04 Powdered 0.83[c]–0.99 Crystalline 0.96[c]–1.04	0.913–1.041	1.089–1.153	0.849–0.929	—
NACC: Angle of Repose (°)	33	33	31	34	—

[a] Molecular weights taken from 1989 NBS values.
[b] Etibank, 1996; North American Chemical Company, 1994.
[c] Minimum standard specification.

References

Absalom, S. T. (1979). Boron. *Mineral Commodity Summaries,* U.S. Bureau of Mines.

Albayrak, F. A., and Protopapas, T. E. (1985). Borate deposits of Turkey. In *Borates: Economic Geology and Production* (J. M. Barker and S. J. Lefond, eds.), Soc. Min. Eng., AIMMPE, New York, pp. 71–87.

Anac, S. (1988, March). Etibank's place in the production of industrial minerals in Turkey. *Industrial Minerals Suppl.,* No. 246, pp. 25–29.

Anon. (1994, July 25). Borates. *Chem. Marketing Reporter* **246**(4), 37.

Anon. (1996). *Etibank Product Specifications.*

Bailey, G. E. (1902). Borates. *The Saline Deposits of California, Calif. State Mining Bureau, Bull. 24,* pp. 33–90.

Dickson, J. (1985). Etibank at Kirka–from ore to derivatives. *Ind. Minerals,* No. 210, pp. 65, 67.

Etibank. (1994, 1995). Annual reports. *Sihhiye Cihan Sok.,* No. 2, 06443, Ankara, Turkey.

Etibank. (1996). Etibank products. *Cihan Sokak,* No. 2, Sihhiye 06443, Ankara, Turkey, 42p.

Harben, P. W., and Dickson, E. M. (1984, October). An overview of the economics and market outlook for borates. *Ind. Min.,* No. 205, pp. 19, 21–23, 25.

Huang, J. (1996). Personal communication, Sr. Vice Pres., Tech. and Develop., North American Chemical Co., Trona, California.

Lyday, P. A. (1987). Boron. *Minerals Yearbook,* U.S. Bur. Mines.

Lyday, P. A. (1994, 1995, 1996). *Boron, Annual Review.* Mineral Commodity Summaries; Mineral Industry Survey, U.S. Bureau of Mines (since 1996 the U.S. Geol. Survey), (August).

Norman, J. C. (1993–1995). Boron: A review of year's activities. *Mining Eng.* **43**(7), 661.

North American Chemical Company. (1994, November). *Product Technical Information,* Overland Park, Kansas, Bulletins 2200, 2600, 2601, 5100, 5101 and 5500.

Papke, K. G. (1976). Evaporites and brines in Nevada playas. *Nevada Bureau of Mines and Geology, Bulletin 87.*

Pocovi, R. E., Latre, A. A., Skaf, O. A. (1994). "Improved Process for Concentration of Ulexite, and Boric Acid Production," Hydrometall., '94 Pap. Int. Symp., Publ. Chapman and Hall, London, pp. 1025–1034.

Roskill. (1993). *The Economics of Boron.* Roskill Information Services Ltd., London.

U.S. Bureau of Mines. Boron. *Yearly issues of the Mineral Commodity Summaries,* pp. 6–17 (This is the same reference as Lyday 1987–1996, during those years).

Wendel, C. (1978). *Special Report on Borate Resources.* Mining, Minerals Div., U.S. Nat. Park Svc., Washington, D.C.

Yale, C. G. (1905). Borax. *Mineral Resources of the U.S. for 1904,* pp. 1017–1022, U.S. Geol. Soc.

Chapter 11 Phase Data and Physical Properties of Borates

11.1 SOLUBILITY DATA

11.1.1 Sodium Borates

At least 27 currently known sodium borates can be crystallized in the laboratory or occur naturally as minerals, not including the anhydrous forms of the nine families of salts listed in Table 11.1. There are similar numbers of borates containing other cations. This makes possible an exceedingly large number of potential phase diagrams and lists of physical properties. Consequently, because of space limitations, only a limited amount of the most important data is presented in this chapter. Fortunately, only eight of the sodium borates are well known, and the solubility data for six of them, in addition to that for boric acid, is shown in Fig. 11.1. Their corresponding phase data, as reported by several investigators, is listed in Table 11.2. The data for borax on a larger scale is shown in Fig. 11.2. More detailed phase data and physical properties of the borates may be found in the References that follow.

Table 11.1

Some of the Known Sodium Borates[a]

$Na_2O:B_2O_3:H_2O$	Formula	Name, mineral
2:1:1, 5	$Na_4B_2O_5 \cdot XH_2O$	—
1:1:0.5, 2, 4, 6, 8, 12	$NaBO_2 \cdot XH_2O$	Sodium metaborates
1:2:1, 2, 4, 5, 10	$Na_2B_4O_7 \cdot XH_2O$	Sodium tetraborates; 10 borax, 5 tincalonite, 4 kernite
2:5.1:3.5	$Na_4B_{10.2}O_{17.5} \cdot 7H_2O$	Ezcurrite; Suhr's borate
2:5:1, 3, 4, 5, 7	$Na_4B_{10}O_{17} \cdot XH_2O$	5 Nasinite, 3 biringuccite; Auger's borate
3:8:10	$Na_6B_{16}O_{27} \cdot 10H_2O$	—
1:3:2	$NaB_3O_5 \cdot 2H_2O$	Sodium triborate, ameghinite
2:9:11	$Na_4B_{18}O_{29} \cdot 11H_2O$	—
1:5:0.5, 1, 2, 4, 10	$NaB_5O_8 \cdot XH_2O$	Sodium pentaborates; 10 sborgite

[a] Nies, 1980.

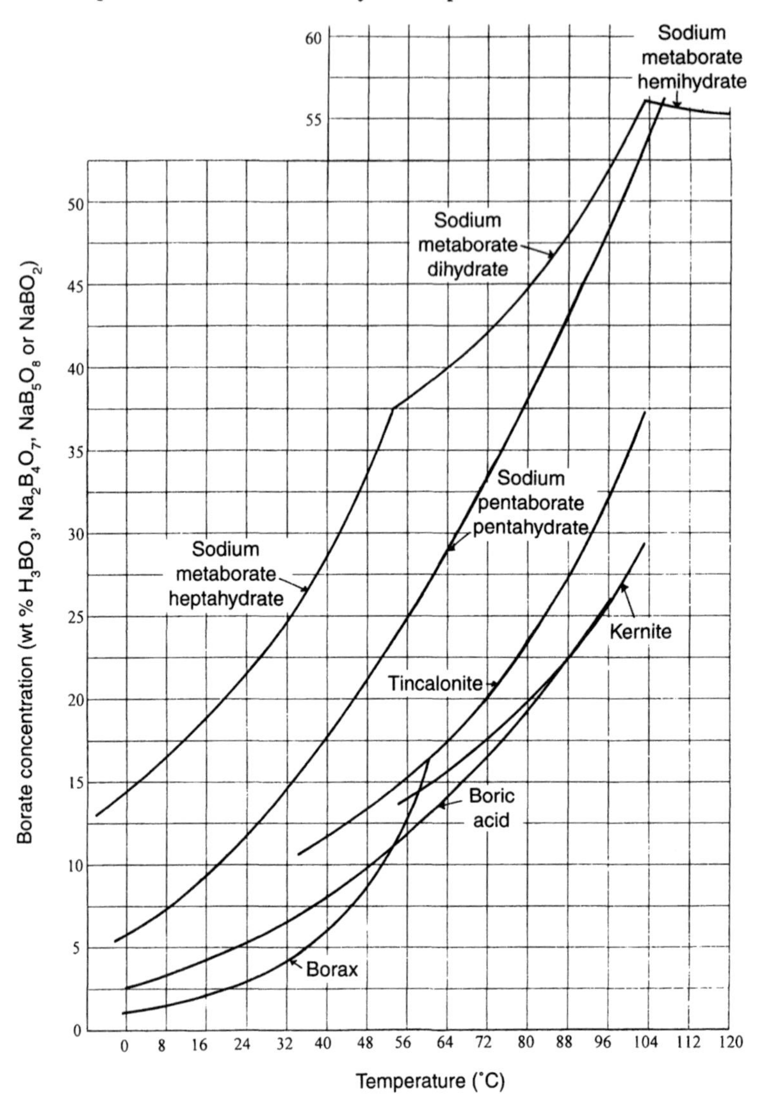

Figure 11.1 Solubility of boric acid and several sodium borates.

Table 11.2

Solubility Data for Boric Acid and Several Sodium Borates (wt %)[a]

Temperature, °C	Boric acid, H_3BO_3[b]	Concentration as $Na_2B_4O_7$[c] Borax	Tincalonite	Kernite	Sodium pentaborate as $Na_2B_5O_8$	Sodium metaborate as $NaBO_2$
−5.77	—	—	—	—	—	13.1[d,e]
−1.7	—	—	—	—	5.5[d,f]	—
−0.76	2.466[d]	—	—	—	—	—
−0.42	—	1.03[d]	—	—	—	—
0	2.52	1.07	—	—	5.75	14.45
5	2.95	1.33	—	—	6.65	15.70
10	3.50	1.64	—	—	7.80	17.05
15	4.17	2.02	—	—	9.15	18.5
20	4.72	2.54	—	—	10.60	20.2
25	5.43	3.13	—	—	12.20	21.9
30	6.23	3.85	—	—	13.85	23.9
35	7.19	4.76	10.97	—	15.65	26.0
40	8.17	6.00	11.90	—	17.50	28.75
45	9.12	7.55	12.85	—	19.60	31.4
50	10.27	9.54	13.93	—	21.80	34.6
53.3	—	—	—	—	—	37.2[g,h]
55	11.55	12.25	15.06	13.68	24.30	37.7
58.0[g]	—	14.28	—	14.28	—	—
60	12.97	16.00	16.38	14.79	26.85	38.75
60.6[g]	—	16.55	16.55	—	—	—
65	14.42	—	17.85	15.88	29.60	40.0
70	15.75	—	19.49	17.12	32.35	41.4
75	17.41	—	21.20	18.41	35.00	42.9
80	19.19	—	23.38	19.85	37.84	44.7
85	21.01	—	25.60	21.28	40.93	46.7
90	23.27	—	28.06	23.05	44.2	48.8
95	25.32	—	31.10	25.10	47.6	51.2
100	27.53	—	34.63	27.60	51.0	54.0
102.8[i]	—	—	36.73[i]	—	—	—
103	—	—	—	—	—	56.1[g,j]
103.3	29.27[i]	—	—	—	—	—
105	—	—	—	—	54.7	56.0
107	—	—	—	—	56.3[i]	—
110	—	—	—	—	—	55.7
115	—	—	—	—	—	55.5
120	—	—	—	—	—	55.4
120.2	—	—	—	—	—	55.35[i]

[a] Data is graphically smoothed from Blasdale and Slansky, 1939; Bowser, 1964; Menzel and Schultz, 1940; Nies, 1980; Nies and Campbell, 1946; Nies and Hulbert, 1967; and Suhr, 1936.

[b] Smoothed date of Blasdale and Slansky, 1937; Liley and Gambill, 1973; and Sprague, 1980.

[c] $Na_2B_4O_7$ converts to borax by multiplying by 1.8953; pentahydrate × 1.44765; the 4hydrate × 1.35812.

[d] Cryohydric point (ice-borate point).

[e] $NaBO_2 \cdot 4H_2O$.

[f] $NaB_5O_8 \cdot 5H_2O$ throughout.

[g] Phase transition (two-phase) point.

[h] $NaBO_2 \cdot 2H_2O$.

[i] Boiling point.

[j] $NaBO_2 \cdot 5H_2O$.

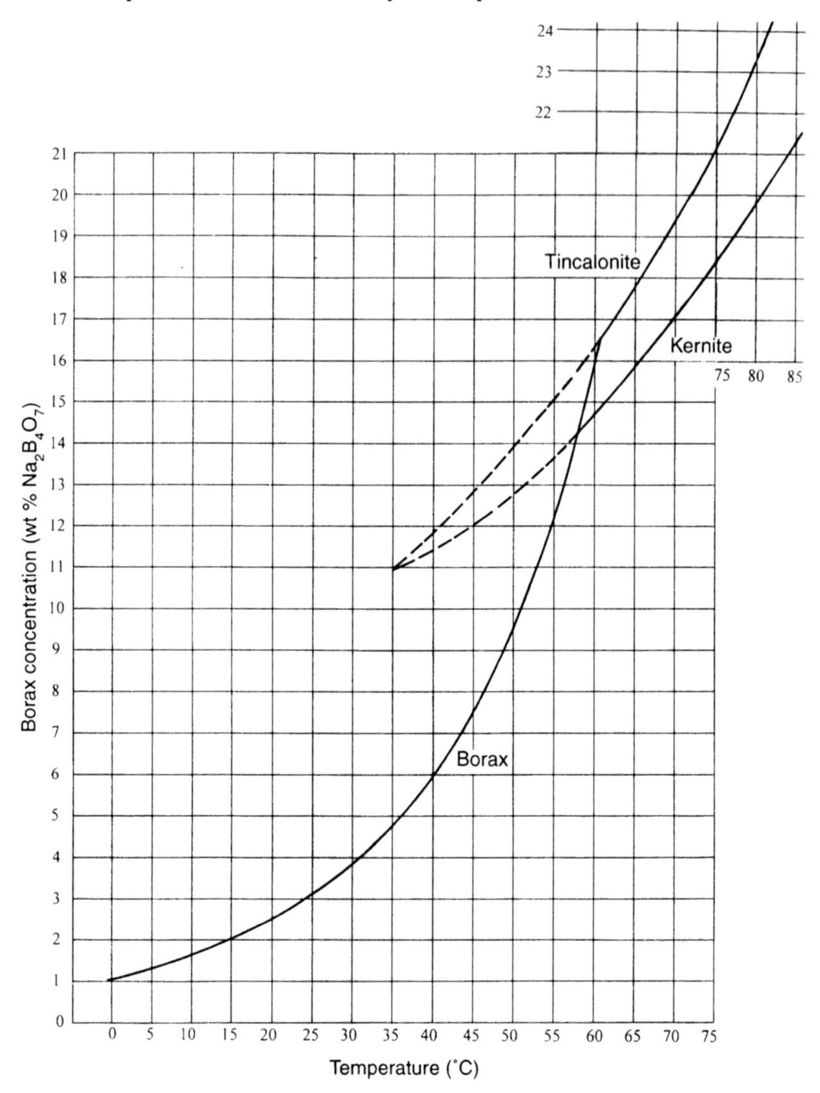

Figure 11.2 Borax solutility on a larger scale.

Table 11.3

Solubility of Borax and Boric Acid in Saturated NaCl[a] Solutions[b] (wt % $Na_2B_4O_7$ or H_3BO_3)

Temperature (°C)	Sodium borates			
	Borax	Kernite	Tincalonite	Boric acid
0	0.26			1.23
5	0.32			—
10	0.40			—
15	0.51			—
20	0.64			2.10
25	0.82			—
30	1.10[d]			—
35	1.43			3.15
37.5[c]	1.66	1.66	1.88	—
38	—	—	1.89	—
39[d]	1.87	1.67	1.93	—
39.8[c]	1.97	1.89	1.97	—
40	2.02	1.73	1.98	—
41	2.14	—	—	—
42		—	2.06	—
45		1.88	2.22[d]	—
45.8		—	2.26[d]	—
50		2.05	2.52	4.36
55		2.27	2.88	—
60		2.46	3.30	—
65		2.75	3.84	—
70		3.05		—
70.8		3.17		—
71		—		75° 7.79
80		6.98[e]		—

[a] The saturated NaCl concentration with the borates varies from 25.22–26.18; avg 25.68%.
[b] Suhr, 1938.
[c] Transition temperatures: Borax-kernite 37.5°C; borax-tincalonite 39.8°C.
[d] Bowser, 1964.
[e] Gale, 1964.

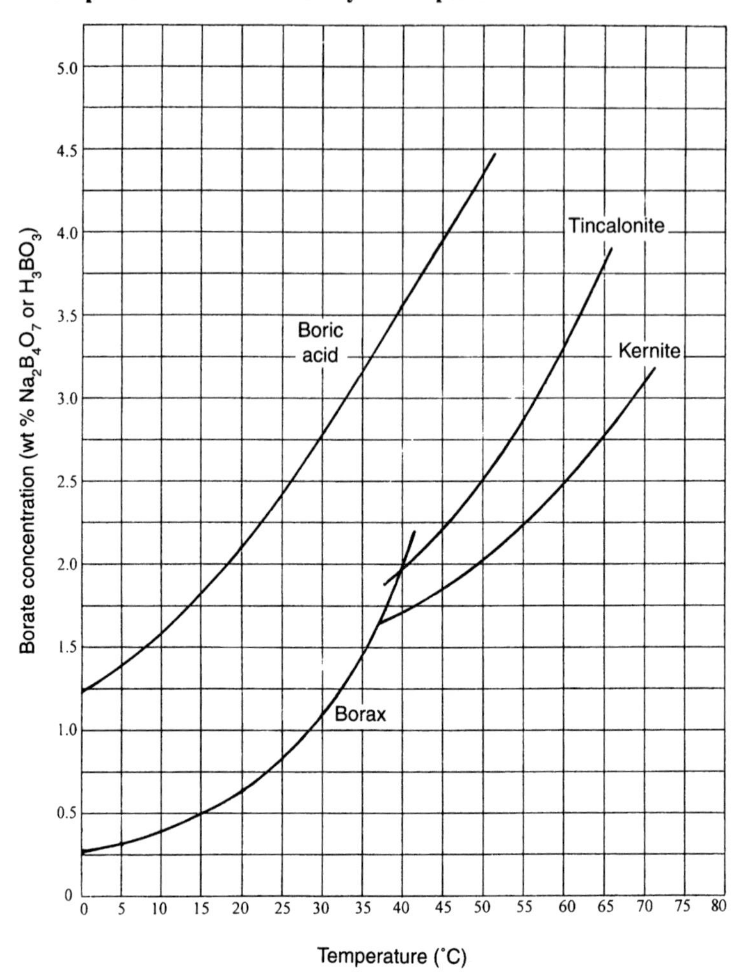

Figure 11.3 Sodium borate and boric acid solubility when the solutions are saturated with sodium chloride (wt %). (From Suhr, 1933.)

Table 11.4

The Solubility of Borax Solutions as Effected by the Presence of Sodium Chloride (wt %)[a]

Temperature (°C)	$Na_2B_4O_7$	NaCl	Solid phase
30	0	26.6	NaCl
30	1.10	26.1	Borax + NaCl
30	1.10	15.3	Borax
30	1.10	6.19	Borax
30	1.27	4.62	Borax
30	1.85	1.89	Borax
30	3.20	0.72	Borax
30	3.80 (3.85)	0	Borax
45	0	26.9	NaCl
45	2.60	21.51	Borax
45	2.54	19.24	Borax
45	2.53	17.43	Borax
45	3.50	7.77	Borax
45	4.57	4.24	Borax
45	6.36	1.33	Borax
45	7.59 (7.55)	0	Borax
45	1.83	26.18	Kernite +NaCl
45	1.95	24.40	Kernite
45	2.11	22.76	Kernite
45	2.35	20.91	Kernite
45	2.24	25.80	Tincalonite +NaCl
45	12.85	0	Tincalonite

[a] Bowser, 1964.

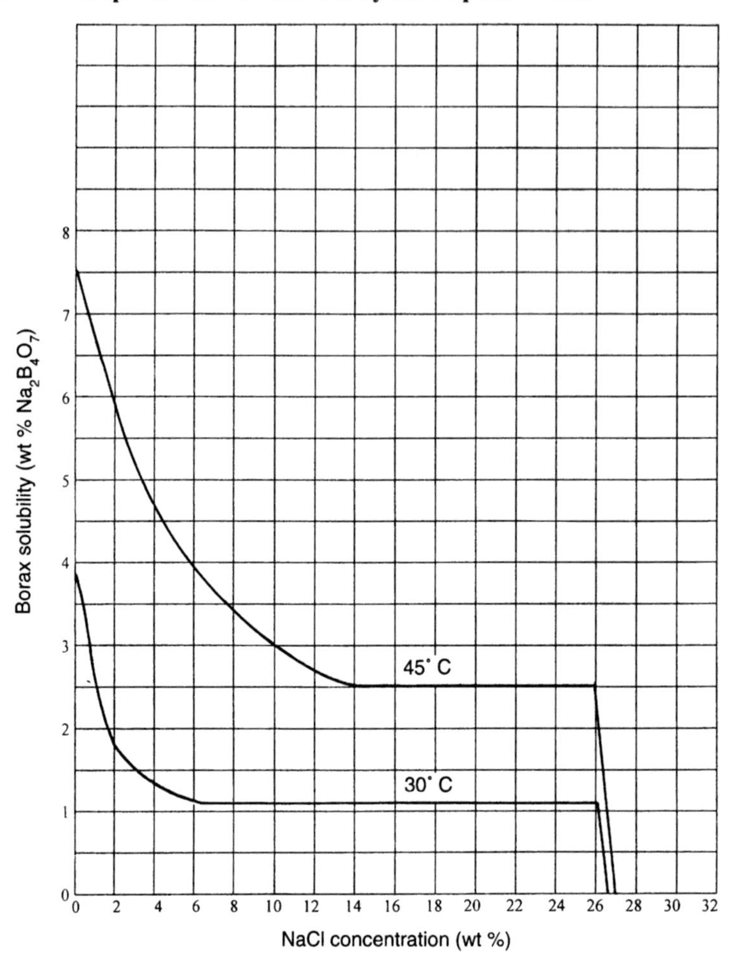

Figure 11.4 Solubility of borax as effected by sodium chloride.

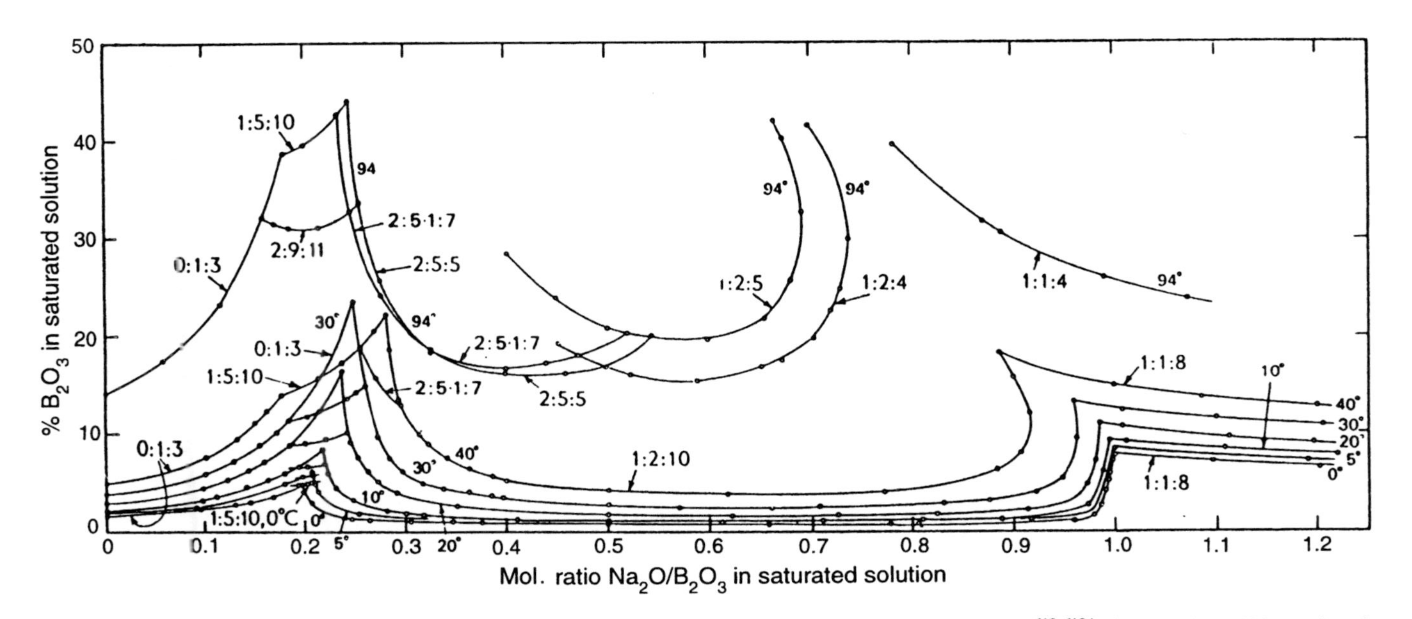

Figure 11.5 Solubility isotherms in the system Na_2O-B_2O_3-H_2O at 0°, 5°, 10°, 20°, 30°, 40°, 50° and 94°C[619, 619A]. (From Nies, 1980; reprinted by permission of Addison Wesley Longman Ltd.)

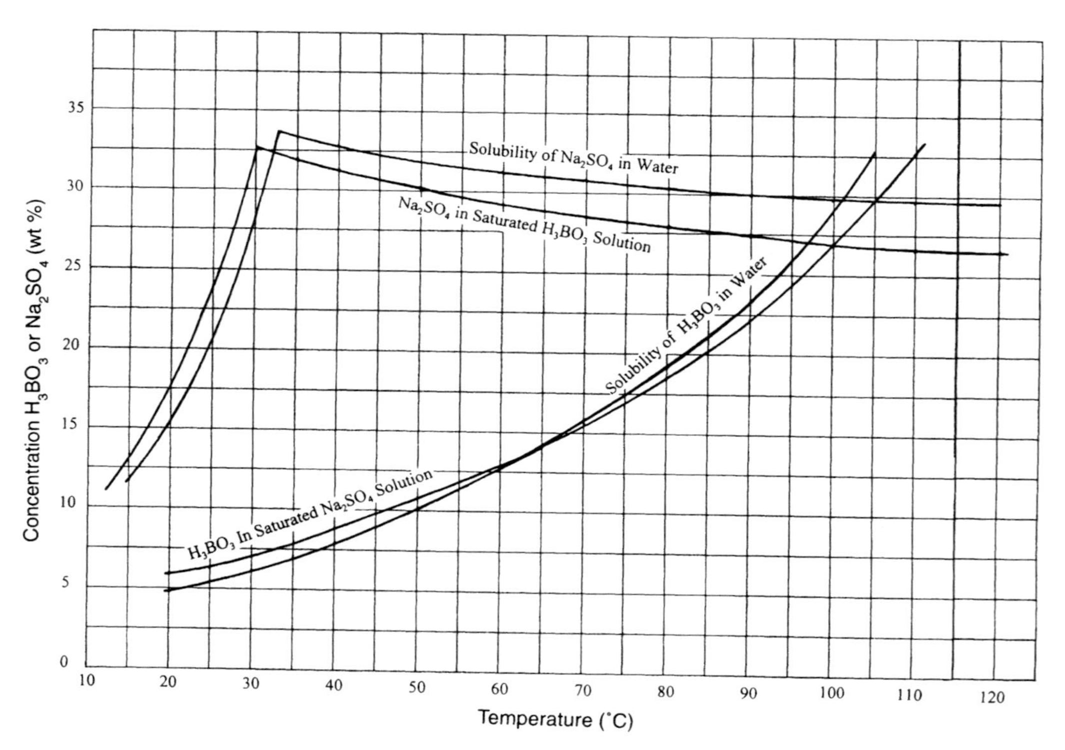

Figure 11.6 Mutual solubility of boric acid and sodium sulfate. (From Suhr, 1938.)

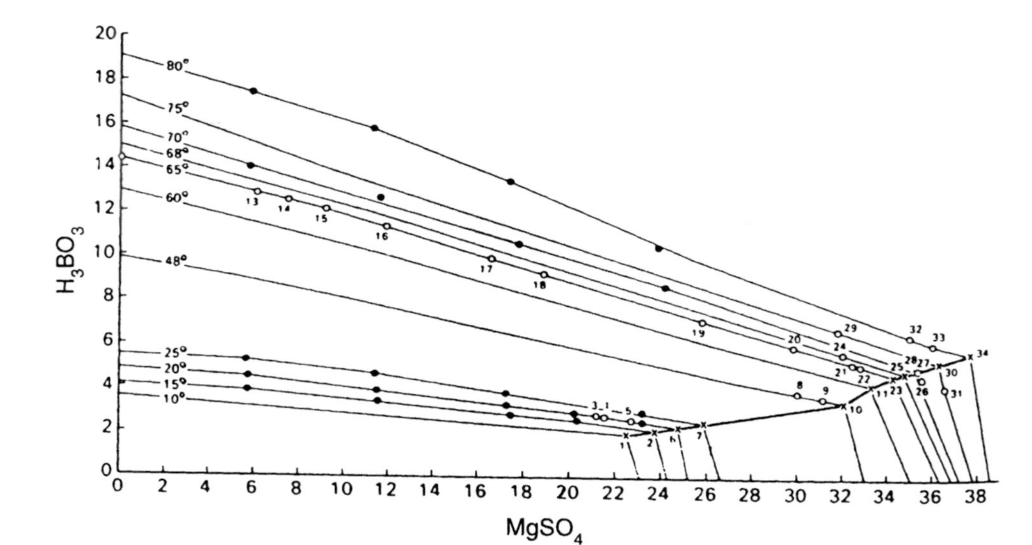

Figure 11.7 Phase data for the system H_3BO_3-$MgSO_4$-H_2O (wt%). (From Sprague 1980; reprinted by permission of Addison Wesley Longman Ltd.)

Table 11.5

pH of Diluted Borax Solutions[a]

Temperature (°C)	Molality (wt%)						
	0.01 (0.38)	0.025 (0.95)	0.05 (1.91)	0.075 (2.86)	0.1 (3.81)	0.125 (4.77)	0.150 (5.72)
0	9.46	9.46	9.50				
5	9.39	9.39	9.43				
10	9.33	9.33	9.35				
15	9.27	9.27	9.29				
20	9.23	9.23	9.24	9.27	9.29	9.32	
25	9.18	9.17	9.18				
30	9.14	1.13	9.14	9.17	9.19	9.20	9.21
35	9.10	9.10	9.11				
40	9.07	9.07	9.06				
45	9.04	9.03	9.04				
50	9.01	9.00	8.99				
55	8.99	8.98	8.98				
60	8.97	8.96	8.94				
70	8.93						
80	8.89						
90	8.85						
95	8.83						
20[b]	9.24	9.24	9.25	9.28	9.31		

[a] Smoothed average of: Manov *et al.*, 1946, Nies, 1980.
[b] Wt% values of pentahydrate.

Table 11.6

Solubility of Borates in Various Organic Solvents (wt% at 25°C)

	Borax	Pentahydrate	Boric acid
Glycerol, C.P.	50.26	43.58	22.2
99%	51.6		18.2
98.5%	52.6		
Ethylene glycol	43.02	36.58	
Propylene glycol	28.8		
Diethylene glycol	18.65	14.91	
Methyl alcohol	5.4[a]		20.2
Ethyl alcohol			
95%			11.2
50%	0.38	0.29	
100%			0.24
Propyl alcohol			7.18
Isobutyl alcohol			5.26
Isoamyl alcohol			4.31
Acetone	0.64		0.6
Ethyl acetate	0.14		

[a]36°C.

Table 11.7

The pH of Dilute Boric Acid Solutions at 25°C[a] (wt % concentration)

H_3BO_3 molality	pH	H_3BO_3 molality	pH	H_3BO_3 molality	pH
0.02 (0.124)	5.38	0.25 (1.55)	4.60	0.60 (3.71)	3.90
0.04 (0.247)	5.28	0.30 (1.85)	4.48	0.65 (4.02)	3.83
0.06 (0.371)	5.20	0.35 (2.16)	4.36	0.70 (4.33)	3.77
0.08 (0.495)	5.12	0.40 (2.47)	4.26	0.75 (4.64)	3.72
0.10 (0.618)	5.04	0.45 (2.78)	4.16	0.80 (4.95)	3.68
0.15 (0.927)	4.88	0.50 (3.09)	4.07		
0.20 (1.24)	4.74	0.55 (3.40)	3.98		

[a] Smoothed average of Kemp, 1956; Sprague, 1980 and others.

References

Anovitz, L. M., and Hemingway, B. S. (1996). Thermodynamics of boron minerals: Structural, volumetric and thermochemical data. In *Boron Mineralogy, Petrology and Geochemistry* (Grew, E. S. and Anovitz, L. M., eds.), Vol. 33, Ch. 5, pp. 181–262, Mineral. Soc. Amer.

Barner, H. E., and Scheuerman, R. V. (1978). *Handbook of Thermochemical Data for Compounds and Aqueous Species,* pp. 32, 50. Wiley-Interscience, New York.

Blasdale, W. C., and Slansky, C. M. (1939). The solubility curves of boric acid and the borate of sodium. *J. Am. Chem. Soc.* **61,** 917–920.

Bowser, C. J. (1964). The system Na_2O-B_2O_3-H_2O. *Geochemistry and Petrology of the Sodium Borates in the Non-Marine Evaporite Environment.* Doctoral Dissertation, Univ. Calif. Los Angeles, pp. 18–109.

Chase, M. W., Curnutt, J. L., Hu, A. T., Prophet, H., Syverud, A. N., Walker, L. C. (1974). JANAF thermochemical tables, 1974 supplement. *J. Phys. Chem. Ref. Data* **3**(2), 351–371.

Christ, C. L., Truesdell, A. H., and Erd, R. C. (1967). Borate mineral assemblages in the system Na_2O-CaO-MgO-B_2O_3-H_2O. *Geochim. et Cosmochim. Acta* **31,** 313–337.

Dulat, J. (1980). Boric oxide. *Inorganic and Theoretical Chemistry,* Vol. 5, Pt. A, Sect. A4, pp. 170–213, Boron–Oxygen Compounds, Longman, New York.

Erd, R. C. (1980). The minerals of boron. *Inorganic and Theoretical Chemistry,* Vol. 5, Pt. A, Sect. A1, pp. 7–71, Boron-Oxygen Compounds, Longman, New York.

Gale, W. A. (1964). Heterogenous equilibria in aqueous systems of inorganic borates. In *Boron Metallo-Boron Compounds and Boranes* (R. M. Adams, ed.), Ch. 2, pp. 29–52, Interscience Publishers, New York.

Kracek, F. C., Morey, G. W., and Merwin, H. E. (1938). The system water–boron oxide. *Am. J. Sci.* **35A,** 143–171.

Kurnakova, A. G., and Nikolaev, A. V. (1948). The solubility isotherm of the system Na_2O-CaO-B_2O_3-H_2O at 25°C. *Akad. Nank. U.S.S.R. Izv. Otd. Khim. Nauk.* **1,** 377–382.

Manov, G. G., DeLollis, N. J., and Acree, S. F. (1944, October). Ionization constant of boric acid and the pH of certain borax–chloride buffer solutions from 0 to 60°C. Research Paper RP1609. *J. Res. Natl. Bur. Stds.* **33**(4), 287–308.

Manov, G. G., DeLollis, N. J., and Acree, S. F. (1946, June). Effect of sodium chloride on the apparent ionization constant of boric acid and the pH values of borate solutions. *J. Res. Natl. Bur. Stds. Res. Paper RP1721* **36**(6), 543–558; Also No. 33, (1944), RP1609, p. 287; No. 34 (1945), RP1632, p. 115.

Menzel, H., Schulz, H. (1940). Zur Kenntnis der Borsauren und Alkalisalze X. Der Kernit (Rasorit) $Na_2B_4O_7 \cdot 4H_2O$. *Zeit. Anorg. Allg. Chem.* **245,** 157–220.

Newkirk, A. E. (1961). Preparation and chemistry of elementary boron. *Borax to Boranes,* pp. 27–41, Advances in Chemistry Series, No. 32, American Chemical Society, Washington, D. C.

Newkirk, A. E. (1964). Elemental Boron. In *Boron, Metallo-Boron Compounds and Boranes* (R. M. Adams, ed.), Ch. 4, pp. 233–299, Interscience Publishers, New York.

Nies, N. P. (1980). Alkali metal borates: Physical and chemical properties. *Inorganic and Theoretical Chemistry,* Vol. 5, Pt. A, Sect. A9, pp. 343–501, Boron-Oxygen Compounds, Longman, New York.

Nies, N. P., and Campbell, G. W. (1964). Inorganic boron–oxygen chemistry. In *Boron Metallo-Boron Compounds and Boranes* (R. M. Adams, ed.), Ch. 3, pp. 53–231, Interscience Publishers, New York.

Nies, N. P., and Hulbert, R. W. (1967, July). Solubility isotherms in the system sodium oxide-boric oxide-water. *J. Chem. Eng. Data* **12**(3), 303–313.

Nikolaev, A. V., and Chelishcheva, A. G. (1940). The 25°C isotherm of the systems: CaO-B_2O_3-H_2O and MgO-B_2O_3-H_2O. *Comp. Rend. (Doklady) Acad. Sci. USSR* **28**(2), 127–130.

Post, B. (1964). Refractory binary borides. In *Boron Metallo-Boron Compounds and Boranes* (R. M. Adams, ed.), Ch. 5, pp. 301–371, Interscience Publishers, New York.

Sprague, R. W. (1980). Properties and reactions of boric acid. *Inorganic and Theoretical Chemistry,* Vol. 5, Pt. A, Sect. A6, pp. 224–320, Boron-Oxygen Compounds, Longman, New York.

Stull, D. R., and Prophet, H. (1972). *JANAF Thermochemical Tables,* Natl. Bur. Stds., No. 37, 2nd ed., pp. B-$C_{10}O_{17}Pb_2$, Washington, D.C.

Suhr, H. B. (1933–1938). *Borates Phase Systems.* American Potash & Chemical Co., Trona, California.

Index

CPSIA information can be obtained at www.ICGtesting.com
Printed in the USA
BVOW030158280911

272276BV00003B/18/A